Statics and Dynamics of Components and Systems

Statics and Dynamics of Components and Systems

Robert E. Keller, Ph.D.
Los Altos, California

John Wiley & Sons, Inc
New York London Sydney Toronto

Library of Congress Catalogue Card Number: 77-121907

ISBN 0-471-46760-X

Printed in the United States of America.

10 9 8 7 6 5 4 3 2 1

To my parents
in appreciation of
my education

Preface

The writing of this book, which has been extended over a period of about 10 years, has been influenced by many people and events. I believe the reader will find a brief history of the project to be useful in understanding the nature and emphasis of the final product. At the time when Professor John Arnold was establishing the Design Division of the Mechanical Engineering Department at Stanford, there was considerable academic interest in the design aspect of engineering education. Many people felt that design was being neglected and that students were being given a completely distorted view of real-life engineering activity. My own feelings were very strong on this subject. My education to that point had included far too much rigid, traditional technology and there had been far too little emphasis on the imaginative aspect of engineering. Consequently, Professor Arnold's MIT course, *Creative Engineering*, had been a very welcome and invigorating part of my Master's degree program.

Professor James Adams (presently head of the Stanford design program) and I were to be the first Stanford Ph.D. candidates in the design area. We had the worrisome problem of conceiving programs which would truly be design centered. Being young and zealous, I was determined to break with all the traditions and conventions which might compromise my pursuit of the essense of design. I was firmly of the opinion that the various areas of advanced calculus had little to do with design. Similarly, I felt that in-depth studies of physics were not appropriate. I did have the vague notion that logic, topology and other such esoteric pastimes of mathematicians had some pertinence. It seemed that a relatively broad knowledge of engineering practice, beyond that of mechanical engineering, would be fruitful. Therefore, I set out on a crazy-quilt program of coursework that included electronics, noneuclidean geometry, switching theory, and probability. John Arnold warned me that such a program would be controversial and would likely get me into troublesome circumstances in the future. He was right. However, I am glad that I pursued my own course and I hope that the present book will give at least some indication of the value of an unconventional academic program.

The book began as a series of notes on those subjects which I was taking at the university and which seemed potentially useful to our design students. The electronics courses made me particularly aware of the field of device characterization, which appeared to be unknown in mechanical engineering. The circuit-by-circuit static analysis techniques used by electrical engineers made me wonder if a more general approach might not be possible, and perhaps useful in other branches of engineering. The switching theory had obvious application in hydraulics and mechanical design and this theory was brought into our design courses. Both analog and digital computer simulation techniques included potential for computer optimization of designs and was also incorporated into our courses.

The book started as a bound set of notes entitled *Concepts in System Engineering* which was submitted for publication. Reviewers pointed out that the book did not have a logical organization, which was certainly true. I decided to add several subject areas which would solidify the presentation. The present coverage of techniques for handling differential equations was developed during this time. However, it was still clear that the collection of topics were not effectively combined.

John Arnold died tragically during his sabbatical in Europe and several years later my own teaching career came to an end. I joined the Electronics Research Laboratory of Hewlett-Packard in Palo Alto. There I had the opportunity to design a number of electronic circuits, and became further convinced that there were no essential differences between design in mechanical engineering and electrical engineering. My interest in the subject of analogy (which had originally been developed by Professor Henry Paynter of MIT) became even stronger.

When I once again turned my attention to the book, I realized that I was attempting to write about four kinds of physical things: components, systems, controllers, and computers. Similarly, there were four conceptual areas which differentiated the chapters. These were: qualitative considerations, static analysis, linearized dynamic analysis, and nonlinear dynamic analysis. A 4×4 matrix display of these items produced 16 well defined and logically related subjects for the book chapters. Most of these subjects were already completed, and I immediately set out to fill in the missing ones. At the same time, stimulated by the new plan, I rewrote the material on characterization and static analysis. The results looked fine, but had unfortunately expanded into a book of some 900 pages. The publisher's suggestion that we print only the first half at this time was difficult to accept, but obviously sensible, and necessary, for any realistic distribution of the book. Thus, we reserved all the material on controllers and computers for a future volume. Here and there, the reader will find references to computation and control which seem to transcend the plan of the present book. It is because of the partition of the original book that these references occurred, and they remain in the book today.

Another problem arose when we realized that there were nearly 400 illustrations to be included in the book. Formal rendering of these figures would have delayed publication and increased cost out of proportion to the increase in its clarity and functional utility. For this reason, most of the illustrations are reproduced directly from my original sketches. It is hoped that the unusual number of illustrations made available to the student more than compensates for the informality of their appearance. Frankly, I feel that the sketches convey my ideas and attitudes more effectively than professionally inked drawings could have done. The statement *From the Engineer's Sketchbook* has been inserted on the illustrations as a reminder of the source of the drawings.

The book represents my current feelings toward the subject of system design engineering. I believe that the application of analogy is one of the most useful techniques of design. Furthermore, I believe that there is negligible difference, in the area of system design, between mechanical, electrical, and hydraulic power and they should be taught together rather than separately. Finally, I am becoming aware that the type of preparation which is done for a computer simulation of a system or machine may be of more value to the designer than the result produced by the computer. I hope this book supports these opinions.

Los Altos
March 1971

ROBERT E. KELLER

Special Symbols, Notation, and Conventions

The following summary of special symbols, notation, and conventions used in this book is presented here for the reader's convenience.

1. Variable Symbols

Unless otherwise specified in the text material, the upper case symbols F, V, M, Ω, I, V, W, and P are used throughout the book as follows:

(a) F to represent force and V to represent velocity in translational mechanical systems.

(b) M to represent torque and Ω to represent rotational speed in rotary mechanical systems.

(c) I to represent current and V to represent voltage in electrical systems.

(d) W to represent flow rate and P to represent pressure in hydraulic systems.

The redundant use of V for velocity and voltage is unfortunate, but it seems a better choice than any alternative, considering conventional practice. The use is always clear in context.

2. Sense of Variables

Positive voltage, current, and pressure are defined as in normal engineering practice. Positive rotational speed is defined by a right hand rule and the symbol $-\!\ominus\!\!\rightarrow$ oriented parallel to the shaft or shafts under consideration. A shaft which rotates clockwise when viewed in the direction indicated by the arrow is taken as having a positive rotational speed.

Positive torque is defined as being the kind of torque existing in the shaft of a screwdriver when a typical woodscrew is being driven. Thus, "tightening" torque is taken as positive, and "loosening" torque is taken as negative. This is an important convention for the development of many ideas in the book.

The torque and speed sense conventions above avoid the use of clumsy and often misleading $\overset{\curvearrowright}{+}$ symbols on rotary mechanical system schematics.

For translational mechanical systems, the sense of a force is defined in terms of a transmitting bar, tape, or cable being in "compression" or "tension." This is parallel to the usage for rotary mechanical systems and avoids the need for free-body diagrams in the analysis of systems of machines. Positive velocities are defined by a simple arrowhead symbol.

3. Unspecified Power

In many of the developments of the book it is convenient to discuss components or systems without reference to the specific form of power involved; that is, whether we are dealing with mechanical, electrical, or fluid power. In this case, the symbols Γ and Δ are used in place of the symbols in Section 1 (Variable Symbols) as follows:

Γ takes the place of M, F, I, and W.
Δ takes the place of Ω, V (velocity), V (voltage), and P.

Note that the product $\Gamma\Delta$ is proportional to transmitted power, as are the products $M\Omega$, FV, IV, and WP. Also, note that those variables in the Γ group are involved in conservation type laws (momentum, charge, and mass) while those in the Δ group are always measured on a relative basis. For further discussion see the introductory portion of Chapter 3.

In the schematic diagrams for cases involving the unspecified power, the system connections are indicated as dashed lines with the component symbols having irregular boundaries, rather like the outline of a potato. (See Figures 2.1, 2.15(a), and 3.4.) This is in contrast to the specific connector symbols and component outlines utilized for real systems. (See Figure 1.32.)

4. Instantaneous, Average, Excursion, and Normalized Variables

Generally, no distinction has been made between "total instantaneous" and "average" (or dc) values of variables, as is often done in the field of electronics. Considering the variable symbol X, we take X to mean the static or average value of the variable in Chapters 3 and 4—where the subjects are static characterization and static analysis. In the remaining chapters, X stands for the total instantaneous value except when the discussion specifies it to be the static or average value, often by subscripting a Q (for quiescent) or 0 (for time zero).

Excursion variables, indicated by lower case symbols, are used extensively and are defined as follows:

$$x = X - X_0$$

where

x is the excursion variable,
X is the total variable,
X_0 is a reference value of X.

Again, no distinction is made between static and instantaneous excursions; the distinction becomes obvious in each discussion.

Normalized variables are indicated by a super-star (asterisk) and are defined as follows:

$$X^* = \frac{X}{X_{\text{REF}}}$$

where

X^* is the normalized variable,
X is the total variable,
X_{REF} is a reference value for X.

Relative measurements of variables are indicated by double subscripts and defined as follows:

$$X_{21} = X_2 - X_1$$

where

X_{21} is the relative variable;
X_2 and X_1 are the (Δ type) variables, at measurement points 2 and 1, respectively, measured with respect to an external reference using the sign conventions given earlier.

5. Representation of General Algebraic Functions

In general, the following form is used to represent an algebraic function:

$$\mathscr{F}(X, Y, Z).$$

This is understood to be some function of three independent variables X, Y, and Z. In many cases, the $\mathscr{F}$ is replaced by another convenient symbol such as $\mathscr{E}$, $\mathscr{P}$, $\mathscr{Z}$, etc. In Chapter 3, and elsewhere, when the functions giving the static characteristics of a component are represented, the symbol for the dependent variable also identifies the function yielding the variable. For example, in

$$M_A = M_A(\Omega_A, M_B)$$
$$\Omega_B = \Omega_B(\Omega_A, M_B)$$

the first line reads: "The torque M_A is a function M_A of the speed Ω_A and the torque M_B." This form provides some economy in linearization procedures; the dual usage of the symbol (such as M_A above) has represented no problem for students in the author's classes.

6. Schematic and Computing Diagrams

The majority of system diagrams in the book are of the schematic diagram variety wherein the connections between the representations of the components represent routes for the transfer of power. However, in some cases, systems are represented by a completely different kind of diagram—derived from the mathematical relationships enforced by a system—in which the connections represent signal, not power paths. These are equivalent to the so-called "block diagrams" of control system engineering and are called "computing diagrams" here. It is always important to understand which is which, to know that there are no "inputs" or "outputs" on a schematic diagram, and to realize that there is no power balance involving the variables of a computing diagram. (See Figures 9.1(a) and (b).)

Contents

1 System components

In this chapter, the objective is to develop an overview of system components. No attempt is made to give an encyclopedic coverage of components; instead, it is shown that a relatively small number of functions are carried out by components and that the system designer can effectively think in terms of function in place of some particular physical realization of a component. In presenting the argument, both relatively detailed descriptions of the construction of various devices and an indication of their static operating characteristics are given. It is hoped that, in this way, the reader will be aided in seeing the importance of functional analogy while not losing his appreciation for the practical significance of construction details. At the end of the chapter, the material is summarized in the form of a chart in which schematic symbols for mechanical, hydraulic, and electrical components are displayed according to function. This chart has been prepared in such a way that mechanical networks can be derived from electrical or hydraulic networks, or vice versa, by a straightforward, direct substitution process. This leads to the approach taken in the following chapter, where the essential operating similarities of a variety of simple networks are discussed.

A by-product of the chapter is a simple definition for the word "system" as used in this text. By *system*, this author means things which are constructed by combining the kinds of components discussed in this chapter.

1.1 Functional Categorization of Components

The functional analogy between components and systems in different physical media has been recognized and utilized for various purposes for many years. Especially prior to the widespread use of computers, simply constructed electrical networks were used to simulate the dynamic behavior of difficult-to-construct and expensive-to-instrument mechanical systems. The analogy between electrical and thermal conduction has been used to solve geometrically complex problems in heat transfer. But most importantly, analogy is one of the basic "stocks in trade" of the professional inventor; the ability to recognize and apply analogy appears to be one of the strongest attributes of those persons whom we recognize as creative.

The organization of this chapter is based on the inventive application of analogy. Specifically, we attempt to categorize system components in such a way that, if a kind of component exists in some but not in all fields, the vacancy will be exposed. This, then, represents an opportunity for invention; several modest examples are given. Also, the categorization is intended to expose the analogy of various systems, thereby providing another avenue for invention. In order to accomplish this goal, some apparently peculiar considerations must be made. For example, the role of voltage in electrical components and systems must be considered equivalent to that of speed in mechanical things, not torque as might be expected. Also, the way in which inertia influences mechanical devices and the significance of the "terminal" at which machines are bolted down tend to be surprising.

In order to accomplish this categorization, the author made a list of more than one hundred system components. This list was then studied, manipulated, and worried about over a period of a year. The approach presented here is the result of that work, and must be considered as one individual's interpretation of a matter which is not without subjective aspects. In particular, the choice of counting physical terminals instead of conceptual "energy ports," the sign conventions, and the symbols for mechanical devices are arbitrary. The naming of functions and the relative importance given various functions reflects the author's interests and background. The particular examples selected expose some further personal bias; those manufacturers' descriptions of components—included in an effort to bring a feeling of reality to the material—exemplify this since they were chosen as much for convenience as for product quality. In summary, this is a non-mathematical discussion; there are no proofs, criticism is expected and welcome, and the usefulness of the discussion depends on the reader's attitude and objectives.

The functions which can be carried out by any two-terminal device are relatively easy to determine. By "a two-terminal device" we mean some component which has two shafts, two pipes, or two wires connected to it. We require that there be no other place where force, torque, fluid, or current, respectively, can be delivered to the device and that the two shafts be parallel in mechanical devices or the two pipes be co-linear in fluid devices. Two extreme possibilities for the function of such a device are that it act like a perfect conductor (shaft, pipe, or wire) or that it act like a perfect non-conductor (broken shaft, clogged pipe, or cut wire). If it is somewhere in between these extremes, it is a dissipator of energy (friction damper, orifice, resistor), and the power which is delivered to it is converted to heat and is transmitted across the surface of the component or conducted away by the fluid in a hydraulic component. If it is like a conductor sometimes and like a non-conductor at others, it may be one of three general kinds of devices. One kind changes its state of conduction in the region where zero power is conducted and serves to direct the flow of power; the devices which do this are over-running clutches, check valves, and diodes, respectively. Another kind changes at some present level and is used to prevent excessive power transfer to or from some system component. Some of these are self-destructing, such as fuses and shear-pins; others can operate repeatedly, such as pop safety valves which open at a preset pressure and close when the pressure has been reduced a given amount. The third kind of device of this nature has a "memory." It would appear to be a non-conductor until too much speed difference, pressure, or voltage appears between its terminals; it then changes to a conductor and remains so until the torque, fluid flow, or current level takes a value which leads to a change back to the non-conducting state. (Devices with memory are not discussed in this chapter; they are considered in detail in Chapter 2.)

An important group of two-terminal components act as energy storage devices. Some of these are very good conductors when the force, torque, fluid flow, or current is steady. An example is an inductor, which may be built to have a very small dc resistance. Others are almost perfect non-conductors under steady conditions. For example, a capacitor will not transmit direct current. If a device is purely an energy storage element, one of the above conditions must be met. Otherwise, it can be shown that an infinite amount of energy could be stored in the element. Therefore, this distinction is useful and convenient for the categorization of energy-storage elements.

The last basic group of two-terminal elements is the opposite of the dissipator group. These convert heat into mechanical, fluid, or electrical energy and are of considerable interest to researchers today, the direct conversion of thermal to electrical energy being of very great significance to the development of nuclear power. The thermocouple and sun-cell are examples of this class of device.

Generally, various kinds of two-terminal devices can be connected together, in series and parallel arrangements, to construct new two-terminal devices with functions that may, or may not, fit into the above categories. To the best of the author's knowledge, the foregoing breakdown includes all the components in which a single physical mechanism is in play; in this sense the list is complete and satisfactory for our purposes here.

An important group of components is represented by the "adjustable" elements. These have four terminals, two of which are mechanical and provide a means by which the geometry of the component can be changed. There are always two mechanical terminals so that the adjusting force or torque will have a suitable reaction. (This was overlooked in some of the early designs for tools for spacetravelers; finally, someone pointed out that the astronaut instead of the bolt would be rotated by a certain power wrench and the matter was reconsidered.) Typical adjustable components are the variable dissipators such as globe valves and rheostats. There are also mechanically controllable connectors, such as clutches and toggle switches, as well as some mechanically variable storage elements, such as tuning capacitors for radios. (The second group is not very common and is not given much attention here.)

Another important group is the mechanical actuators which convert electrical or fluid power into a translational or rotational stroke—as opposed to motors which give continuous rotation. Also in this group are the devices, such as flyball actuators, which simply transform the nature of mechanical power. Again, this is a four-terminal group in which at least two of the terminals are mechanical. These, of course, provide the stroke force or torque and the associated reaction.

On the author's long list of components to be categorized, quite a few of the entries were simply mechanically variable dissipators or connectors driven by one or another of the actuators. Once this is recognized, the list of possibilities becomes quite lengthy, and it is obvious that such constructions are not basic components but simply kinds of combinations not usefully categorized.

In the actuator group, it was recognized that an important distinction existed according to whether or not the primary terminals (as opposed to the actuator terminals) represent a conductive or a non-conductive element. It can be shown that, for any of the actuators, if all the inefficiency can be removed, one or the other case must prevail. This is because, when the actuator is not in motion but is supplying a torque or force, there is no output power. The torque or force can only be the result of some voltage, current, pressure, rotational speed, or what-have-you; in order for there to be no input power, one of the primary terminal variables must go to zero, implying a perfect conductor or perfect non-conductor in the primary. This leads to a realization that there are two physically possible actuators in each media. Examples of these, not all of which are well known, are given.

Perhaps the most important group of components discussed is the three-terminal, variable dissipators. This includes the basic control components, such as clutches, pilot-operated hydraulic valves, and transistors, as well as a variety of others. Since these components have three terminals, the laws of conservation of momentum, matter, or charge require that all be of the same media and be colinear if mechanical or hydraulic. A number of examples are discussed in detail, and particular care is taken to explain the design of a mechanical analog of the transistor, this being a most useful device which could be put to good use if well understood.

If we attempt to determine all the possible functions which might be carried out by a three-terminal device, this being worth exploring for invention possibilities as well as in search of a particularly effective viewpoint, we quickly recognize that two of the terminals can provide one of the previously listed two-terminal functions while the third, acting with one or the other of the other two, will act as another two-terminal device. Except in trivial, uninteresting cases, the power conduction or storage in the pairs will be interdependent so that one will have control over the other. Also, it is obvious that there is always something special about one of the three terminals since it must be a member of both paris. The terminal variables at this terminal are often selected as references in system design; this is pointed out to be quite sensible because it leads to the simplest representation of device behavior.

The group of devices including the gearbox, transformer, and hydraulic intensifier is shown to be a four-terminal group which carries out the "exchange" function. In these, if the undesired losses due to friction, resistivity, etc., are neglected, there is no dissipated power. Also, there is a constant ratio between one pair of variables and the inverse between the other. For example, in a gearbox with a 2:1 speed reduction, the torque is increased by a factor of two if the inefficiencies of the unit are neglected. The power input equals the power output, and the device acts simply as an intermediary in a system, allowing otherwise incompatible components to operate together. The function is similar to that of currency exchange at the borders of countries; little is lost in the exchange, nothing is gained, but the system would be inoperative without it.

A somewhat related group of four-terminal devices converts one form of power to another. In these, there are two terminals of one kind and two of another. Ideally, the power supplied by one pair is equal to that taken by the other. In some of these converters, there is a definite ratio between some of the variables. For example, in a positive displacement pump, there is a certain ratio between flow rate and shaft speed. If the inefficiencies are neglected, an inverse ratio exists between the pressure difference and the shaft torque. Hence, the "positive converters" provide an exchange function as well as the conversion. Other converters, such as turbines, do not display this exchange function, and are categorized in a different group. Certain insight into this situation is again available from a consideration of the conductivity of the primary of the converter; this forms a part of the discussion.

Although components with more than four terminals exist (the adjustable converter is one type, the separately excited dc motor is one example), their consideration develops no new ideas and is not carried out here. The primemovers, or fuel convertors, are in this group. These are discussed to some extent in example problems and in later chapters.

This summary of the following discussion has been included to give the reader an indication of the approach to be taken. It should be borne in mind that the examples which are discussed are to illustrate this line of thought, not to show ideal designs for the components or to thoroughly explain the physics of their operation. In most of the examples, the drawings are the author's own sketches and are not intended to represent manufacturable equipment. The hydraulic devices are shown greatly oversimplified, flow forces and multi-function provision being neglected in all cases. In the discussion of the electronic devices, especially the semiconductor components, either the reader's good faith or some outside study will be necessary to develop a satisfactory understanding of the physics. For those interested, detailed information on the components discussed may be obtained by writing those manufacturers listed in Thomas' Register, or similar publications, under the appropriate heading.

1.2 Dissipators

Several functionally similar mechanical couplings are indicated in Figure 1.1. The first, at (a), is an eddy current coupling in which permanent magnets provide the basic coupling mechanism. The magnets are arranged radially with alternating orientations. A steel drum encloses the magnet assembly and a flux path, as indicated by the dashed arrows, is developed. If the drum rotates with respect to the magnets, the copper liner moves through the flux fields, individual elements of the liner experiencing a constant reversal of the flux direction and hence the induced voltage. The net effect is for currents to be established in the liner, these taking more or less circular paths in the material. The currents can be large, because of the high conductivity of the liner, and considerable heat may be generated. This power comes from the mechanical power transmitted by the shaft and is proportional to the slip speed (the difference in speed between the two shafts) and the transmitted torque. A momentum balance shows that, for this and all similar couplings, in the absence of angular acceleration, the shaft torques must be identical. Generally speaking, the torque and slip speed increase together; obviously, when one is zero so is the other.

A basic form of hydraulic coupling is shown at Figure 1.1(b). This coupling has a vortex-like fluid path which includes a large number of radial separators (not shown). If the left-hand part of the coupling is rotated, the fluid is forced to move in the direction shown due to centrifugal force. This fluid, which has a tangential velocity component, impinges against the stationary separators of

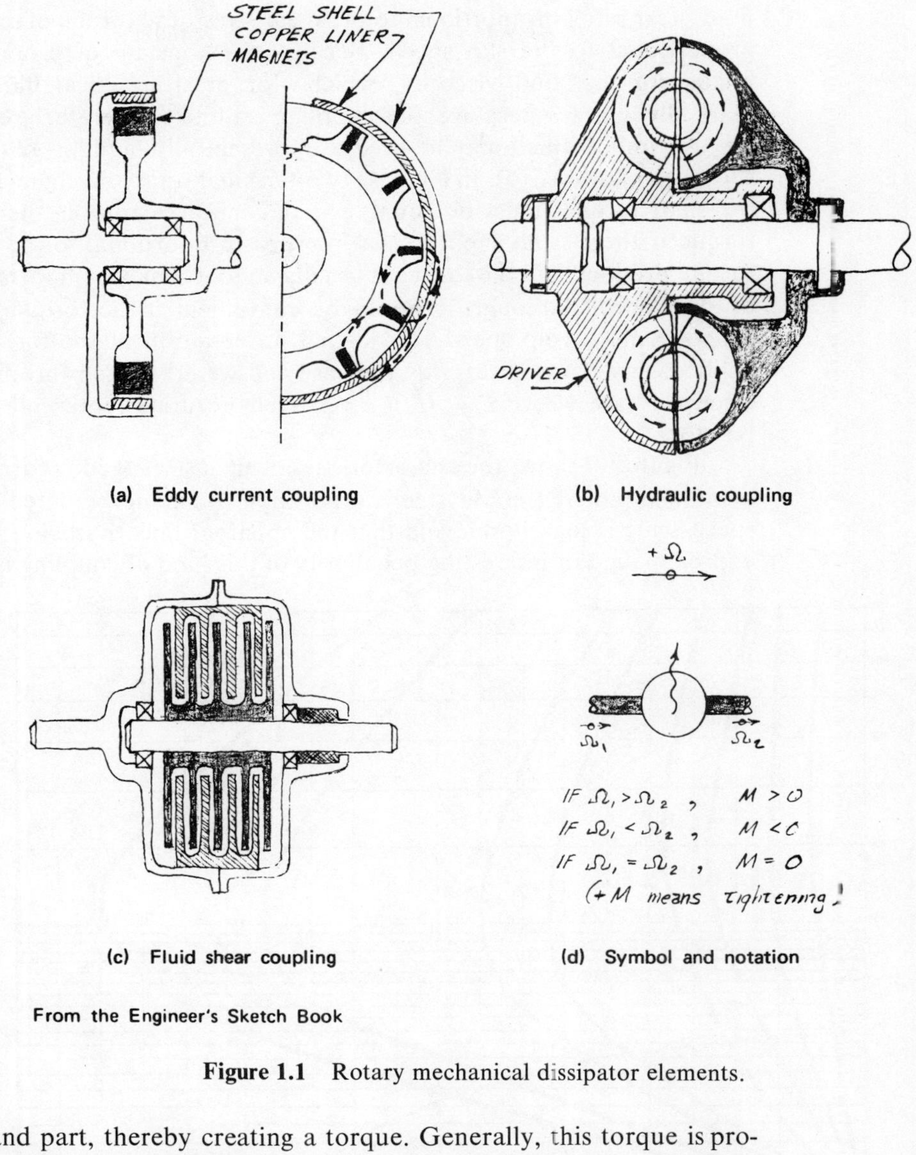

(a) Eddy current coupling

(b) Hydraulic coupling

(c) Fluid shear coupling

(d) Symbol and notation

From the Engineer's Sketch Book

Figure 1.1 Rotary mechanical dissipator elements.

the right hand part, thereby creating a torque. Generally, this torque is proportional to the slip speed and must be zero when the slip speed is zero. Heating occurs here because of the fluid friction and the high turbulence; the heating power is of course proportional to the product of slip speed and torque.

A coupling based on the shear of fluid is shown at Figure 1.1(c). A number of closely spaced discs are fastened alternately to one shaft and the other. Since fluid shear rate is proportional to the shear stress, the torque of the coupling is proportional to the slip speed. Because the constant of proportionality is related to the fluid viscosity, which—for most practical fluids—changes markedly with temperature, this coupling is quite temperature sensitive.

A symbol which may be used to represent all couplings of this nature is shown in Figure 1.1(d). In all these two-terminal rotary mechanical couplings, the shaft torques must be equal if there is no acceleration, there can be no torque if there is no slip, the heat power is proportional to the slip-speed—torque product, and the torque is usually a monotonically increasing function of slip speed. Although it would be convenient if the torque were simply proportional to slip speed $\Delta\Omega$, so that an equation of the form $M = b\,\Delta\Omega$ applied, this is not necessarily the case and we are not so fortunate as in the electrical case where $E = IR$ is a relatively accurate representation of most resistors.

Figure 1.2 shows the characteristics of eight sizes of eddy current coupling manufactured by the Maysteel Corporation. It should be noted that the slip speed scale is logarithmic and that the apparent linearity below 200 RPM is approximate. An interesting peculiarity of this kind of coupling is that torque

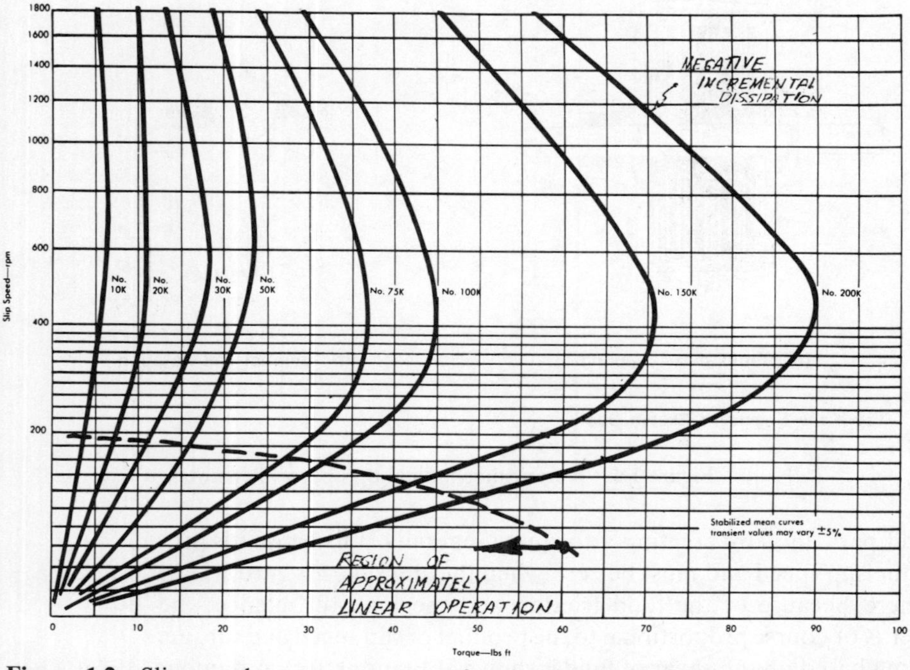

Figure 1.2 Slip speed vs. torque curves for the Gleason "Tormag" eddy current coupling. Courtesy of Tormag Products Div., Maysteel Corp.

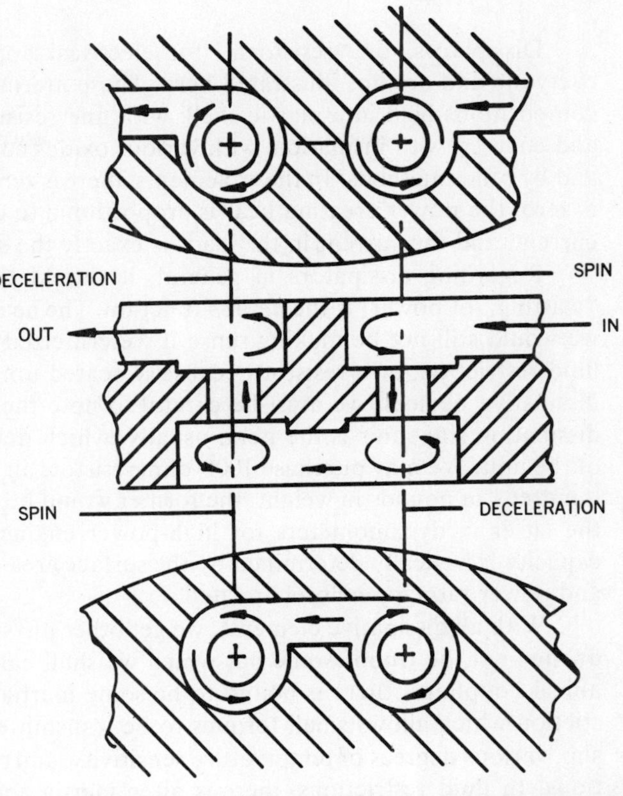

A. Above are three cross sections of a portion of a Visco disc. They show how the flow repeatedly passes through the same disc. The flow path begins on one side at the center and ends on the opposite side at the center of the disc.

B. The exit slot from each deceleration chamber is in a direction which is opposite to the direction of spin. This forces the liquid to come to rest before it makes its exit from the deceleration chamber.

C. On the smaller sizes each Visco disc is fabricated from the three photo etched plates. On the larger sizes this disc is of one piece construction.

D. The Visco disc is covered both top and bottom with a flat lapped disc. (not shown)

E. The Visco discs and their covers are rigidly and permanently clamped in the cartridge.

Figure 1.3 Manufacturer's description of a fluid restriction based on vortex generation. Courtesy of The Lee Co.

decreases with slip speed above a certain level; this can lead to system instability in some troublesome cases, or it can be used to advantage in others.

Dissipators for hydraulic systems take many forms. A simple orifice, a capillary tube, a long pipe, a sharp bend, or tube fitting—all dissipate hydraulic power. From a practical point of view, it is desirable to make such dissipators small in size when they are intentionally used in a system; one way of accomplishing this is indicated by Figure 1.3, which is self-explanatory.

Dissipators, or "resistors," for electrical applications are familiar to everyone and are not illustrated here. These are made with carbon and clay compositions held in a plastic shell, with fine resistive wire wound on a core and covered with insulation, with various oxides coated onto a glass cylinder, and by other methods. In these elements, there is zero current when the voltage is zero, the power creating heat is proportional to the product of voltage and current, and the current in the leads is exactly the same.

Regarding dissipators in general, it should be noted that heating (the "wasting" of power) is the *desired* function. The heat may be troublesome, but we would still not be making sense if we criticized a mechanical coupling, a fluid restriction, or a resistor because it heated up. In the specification of all dissipative devices, we must be careful to note the required amount of heat dissipation since this is the item, usually, which determines the physical size of the unit. We may purchase 1000-ohm resistors all the way from one ounce to hundreds of pounds in weight; the former would be used in pocket radios and the latter in dynamometers for high-power engines. Since the heat transfer capacity is basically determined by the surface area of the component, the size and power rating are closely related.

With all dissipative elements, we get other physical properties which may, or may not, be troublesome but which we shall call imperfections. In mechanical couplings, there is bound to be some inertia, or flywheel effect, some stiction which allows small torques to be transmitted even though there is no slip, various degrees of temperature sensitivity, and other non-dissipative functions. In fluid restrictions, there is often inertia and temperature sensitivity. In wire-wound resistors, there is quite an appreciable amount of inductance with regard to radio frequency applications, and in other kinds of resistors temperature effects may be significant.

Not all dissipative components have a steadily rising characteristic relationship between the variables; the eddy-current coupling is one example of this. A particularly useful group of dissipators is that in which one of the variables remains approximately constant. In Figure 1.4 is shown such a device for hydraulic applications; in this component, known as a "relief valve," there is no flow until the pressure attains a preset value. At that point, the valve opens and will transmit quite a range of flows with no significant change in the pressure. For the valve shown in the figure, the pressure is applied at the *top*. When the force exceeds that created by the preloaded spring, the conically shaped plug moves *downward* and opens the flow path. These devices are used both to prevent overpressures in system and to provide a reference pressure where constant operation is desired.

The symbol and operating characteristics for an analogous electronic

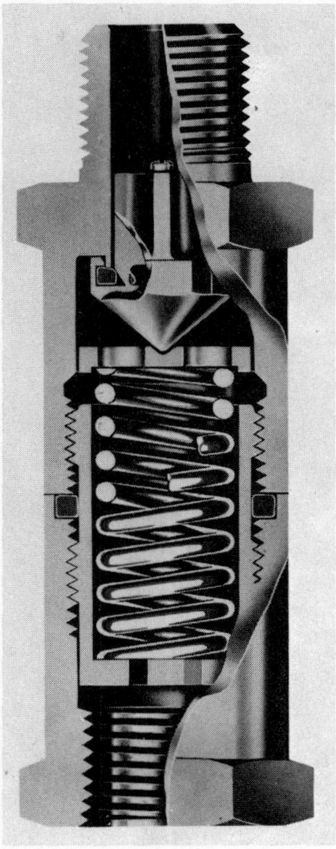

(a) One form of relief valve for hydraulic system service

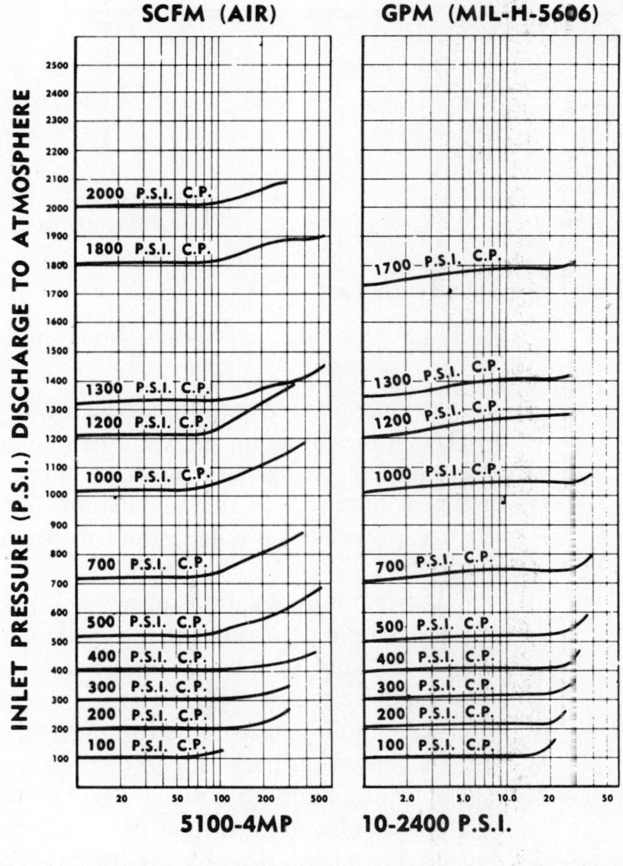

(b) Characteristics of the above valve ("C.P." is cracking pressure)

Figure 1.4 A representative hydraulic relief valve. Courtesy of Circle Seal Products Co. Inc.

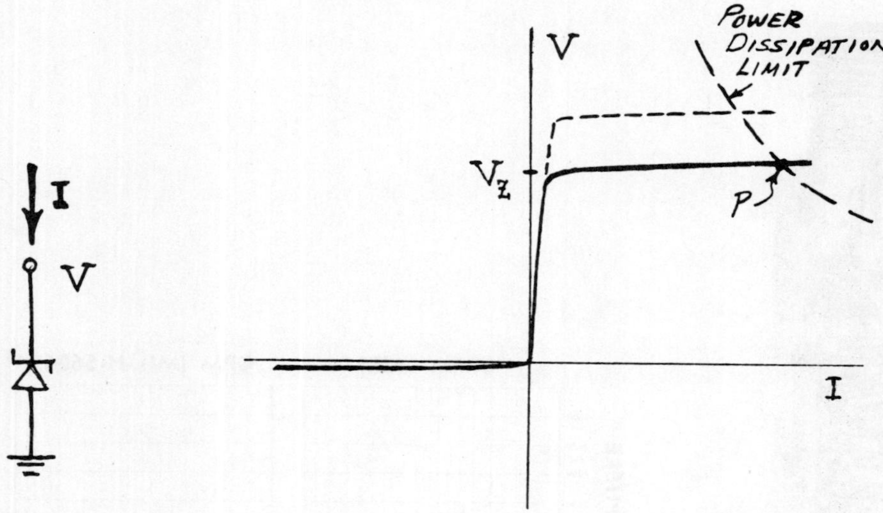

(a) Symbol

(b) Characteristics

From the Engineer's Sketch Book

Figure 1.5 A constant voltage dissipator: the Zener diode.

device, known as a Zener diode, are shown in Figure 1.5. If a current I is forced to flow in the direction shown, a certain voltage V_z appears across the device and does not change to a marked extent as the magnitude of the current is changed. Similarly to the applications of the relief valve, the Zener diode is used to prevent overvoltages and to provide a voltage reference for control purposes.

The major applications of normal dissipative elements are in the control of the dynamic performance of a system through interaction of the dissipator with some kind of energy-storage component. For example, a rotary mechanical drive which includes both a dissipative coupling and a flywheel can give very smooth start-up performance of a motor-driven system. If the coupling is chosen properly, it can operate with relatively little slip in the steady power transmission condition, thereby yielding good system efficiency. Dissipators are also used to limit the power that may be taken from a source or delivered to a load. If we place a 10-ohm resistor in series with a 12-volt battery, it is impossible to draw more than 1.2 amps from the combination. Similarly, a fluid coupling can prevent excessive torques from being developed in a mechanical system of high inertia; if something jams, the coupling will slip until motors and other units slow down or stop. Another important application of dissipators is in conjuction with variable dissipators in amplifier systems; this will be shown in the next chapter.

1.3 Directors

A class of devices which has considerable engineering usefullness is the "director" group. These serve, generally, to "route" the transmission of power in a system; the mechanical device transmitting only one kind of torque (tightening or loosening), the hydraulic and electrical devices transmitting flow and current, respectively, in only one direction. A very common application of such a mechanical device is on the starter of an automobile; the starter can drive the motor if the motor is turning slowly, but once the motor starts, the device disengages so that the starter does not continue to rotate. Functionally, the device steers the power so that it may be transmitted from the starter to the motor, but not vice versa.

Several kinds of mechanical directors, which are known as "over running clutches," are shown in Figure 1.6. At (a), shoes lined with a friction material such as on automobile brakes are held against a drum by links. The angle β between these links and a radial line is important; if it is less than arctan (f), where f is the coefficient of friction, the device is self-locking as required. If it is not, the device is simply a dissipator. In operation, imagining the shaft on which the spider is located to be coming out of the paper, when tightening torque is applied the shoes lock onto the drum, thereby transmitting the torque to the shaft on which the drum is mounted. When loosening torque is applied, the shoes drag along the surface of the drum and there is no tendency for them to lock. Hence, only a small amount of friction torque is transmitted and the speed difference between the spider and the drum can be large. Springs, not shown in the figure, would probably be used to maintain a light amount of shoe pressure at all times, in order to insure locking under tightening conditions.

Another form of over running clutch is shown at Figure 1.6(b). This form uses hard steel rollers which wedge into the tapered cavity between the drum and spider when tightening torque is applied. Small springs behind each roller insure quick response upon torque direction changes. Due to its simplicity, this is a practical form for the clutch and is widely used. Another form is shown at (c). In this, which is suitable only for light loads and slow operation, the spider is made of molded plastic material. The action depends on the friction angle as in the shoe type unit.

A symbol which may be used for all kinds of overrunning clutches is shown at (d). The diagonal hachures may be thought of as acting like the links in the shoe-type clutch; they can be put in compression by friction but will

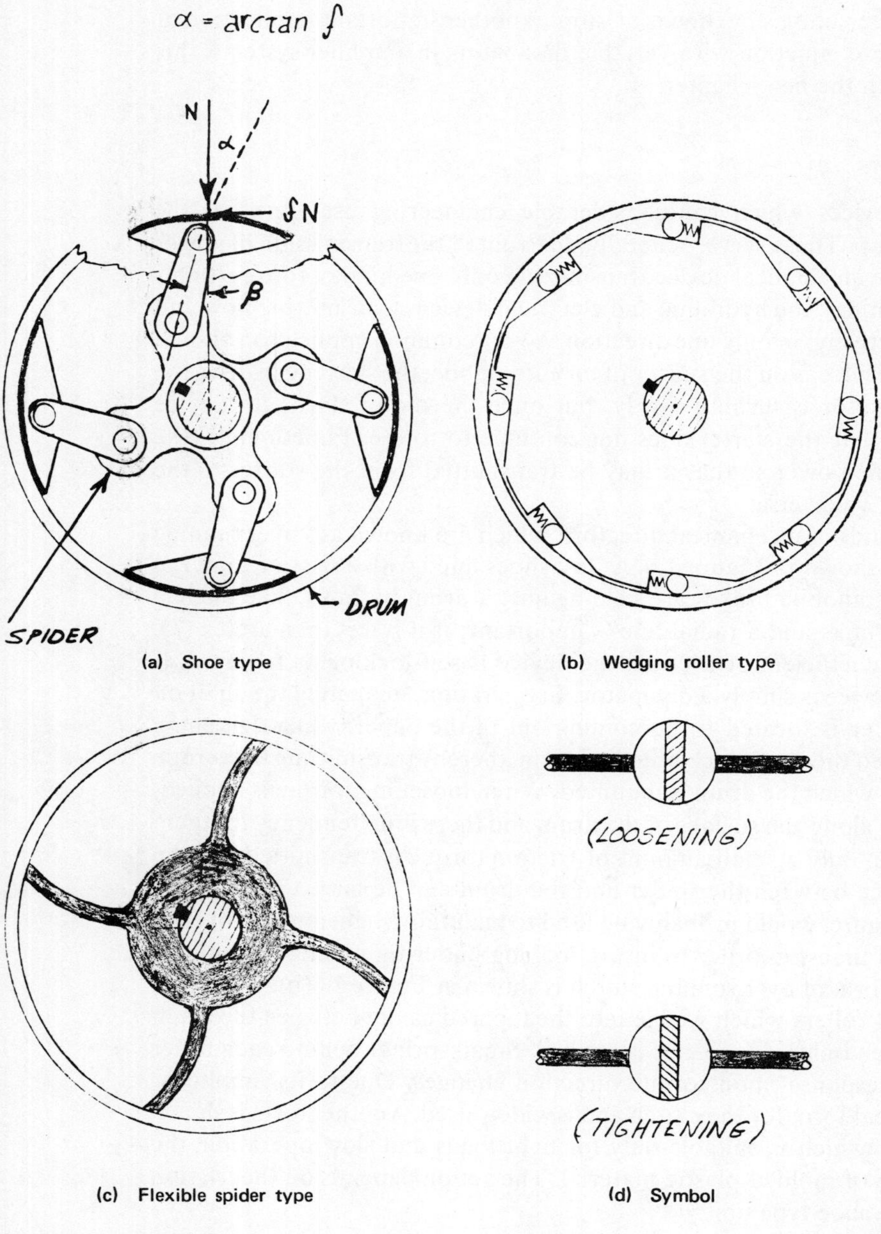

(a) Shoe type

(b) Wedging roller type

(c) Flexible spider type

(d) Symbol

From the Engineer's Sketch Book

Figure 1.6 Rotary mechanical director elements: "Over-running clutches."

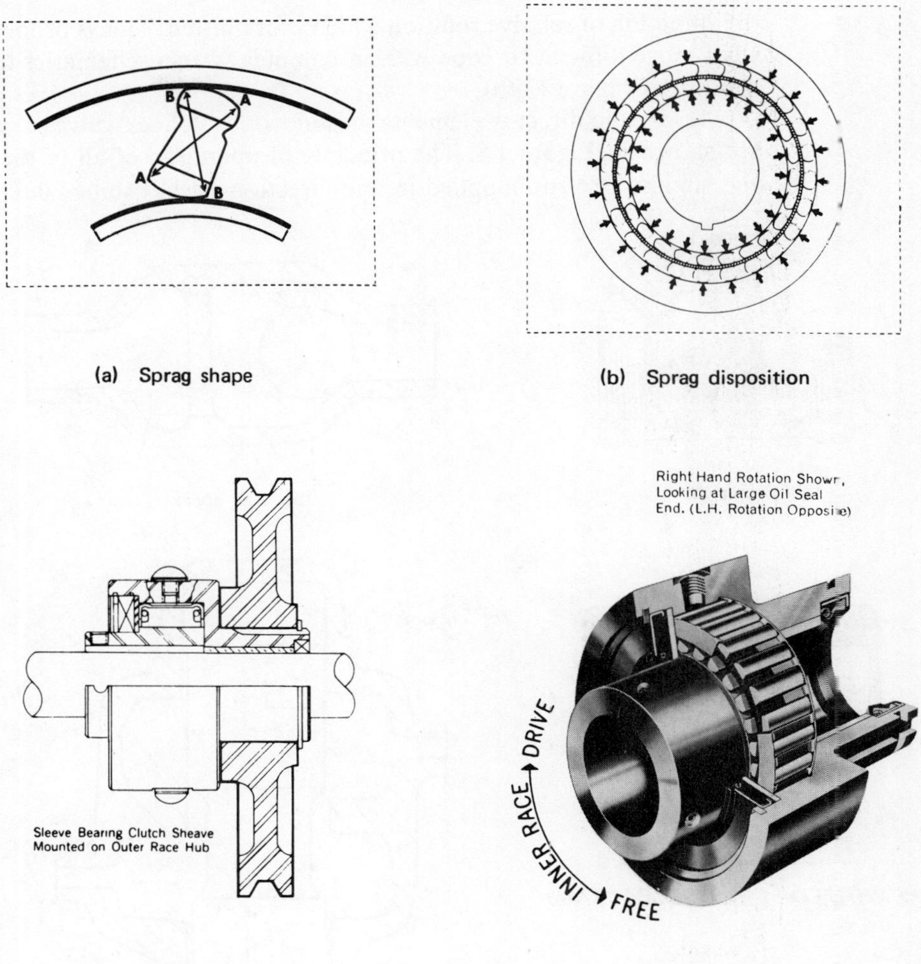

(a) Sprag shape (b) Sprag disposition

(c) Cross-section (d) Pictorial section

Figure 1.7 A commercially available sprag-type over-running clutch. Courtesy of Formsprag.

disengage whenever there is a tendency for tension to develop. Of course, over-running clutches can be constructed to transmit either tightening or loosening torque, and two forms of the symbol are required as shown.

The construction details of one commercially available overrunning clutch are shown in Figure 1.7. In this, a variation on the wedging roller arrangement, sprags are used to perform the locking function. It may be noticed that this manufacturer prefers to specify the nature of the clutch according to the direction of relative rotation which is premitted. This is proper, but somewhat more difficult to cope with in complex system schematics than the convention of Figure 1.6(d).

Hydraulic director elements are known as "check valves;" several types are shown in Figure 1.8. The principle of operation of all of these is simple and similar. Pressure applied in one direction pushes some kind of plug onto

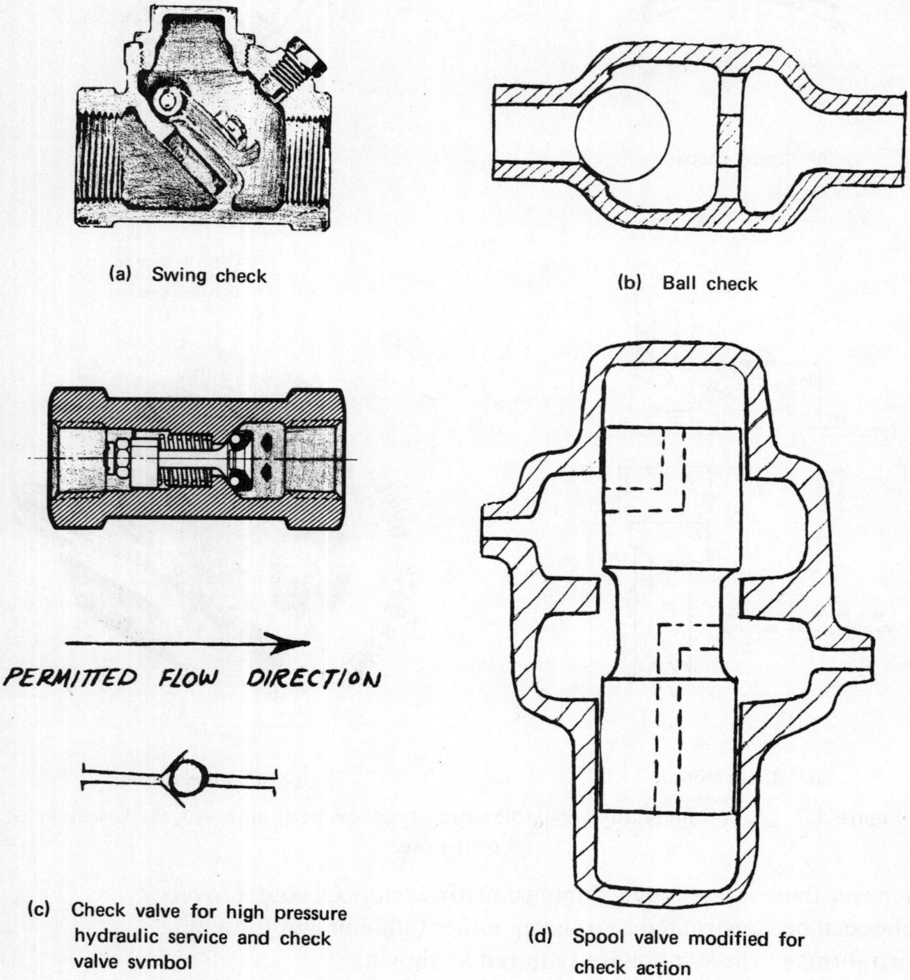

From the Engineer's Sketch Book

Figure 1.8 Some check valve constructions. Courtesy of (a) The Lunkenheimer Co. and (c) Racine Hydraulics.

a seat, closing the valve. Flow in the opposite direction tends to move the plug away from the seat, thereby opening the valve and permitting the flow at a small pressure difference. The symbol for the valve, shown as part of the (c) portion of the figure, relates to the ball-type construction in an obvious way. The spool-type construction, shown at (d), utilizes passageways inside the valve spool to allow the differential pressure to act. This is not a common way of constructing the valve, but is included here to show the versatility of the construction, more use of which will be made in later sections.

Electrical directors can be constructed in a variety of ways. Until a few years ago, gas-filled tubes with a heated cathode or stacks of selenium rectifier cells were widely used. Semiconductor technology has made these units obsolete since smaller packages, lower cost, and better performance can be achieved with germanium or silicon devices.

A typical silicon unit is shown and described in Figure 1.9. The fundamental basis for the operation of the device is the peculiar behavior of a junction between two kinds of "imperfect" silicon. One kind has about one atom in ten-million replaced by an atom with five valence electrons (silicon has four). Arsenic and antimony are frequently used. The "extra" electron does not take part in the bonds of the crystal structure and is free to act as a conductor. Since this kind of material has a free negative aspect, it is called "*n*-type" silicon. The other kind of silicon has about one atom in ten-million replaced by an atom with only three valence electrons. Aluminum, gallium, and indium are frequently used. This results in there being a vacancy in the electron arrangement in the crystal. This vacancy is relatively mobile and, being the absence of a negative aspect, appears to be a positive aspect in the conductive portion of the crystal behavior. Hence, this is called *p*-type material.

At a junction between the two, the mobile electrons in the *n*-type material move across the boundary to fill the vacancies in the *p*-type material. This leaves the *n*-type material with a net positive charge and the *p*-type with a net negative charge, since both were uncharged prior to the migration of the electrons. If we now attempt to force electrons into the *p*-type material and out of the *n*-type material by the use of an external current source, the *p*-type will take on more and more of a negative charge as the new electrons continue the original vacancy-filling process. Similarly, the *n*-type will continue to become more positive as we draw off even more of its mobile electrons. This continues until the voltage difference between the two materials equals the externally applied voltage, at which time no significant current flows. In this condition, the diode is much like a charged capacitor.

If we now attempt to force electrons into the *n*-type and out of the *p*-type materials by using an external current "pump," the net positive charge of the *n*-type is lowered and the net negative charge of the *p*-type is also lowered.

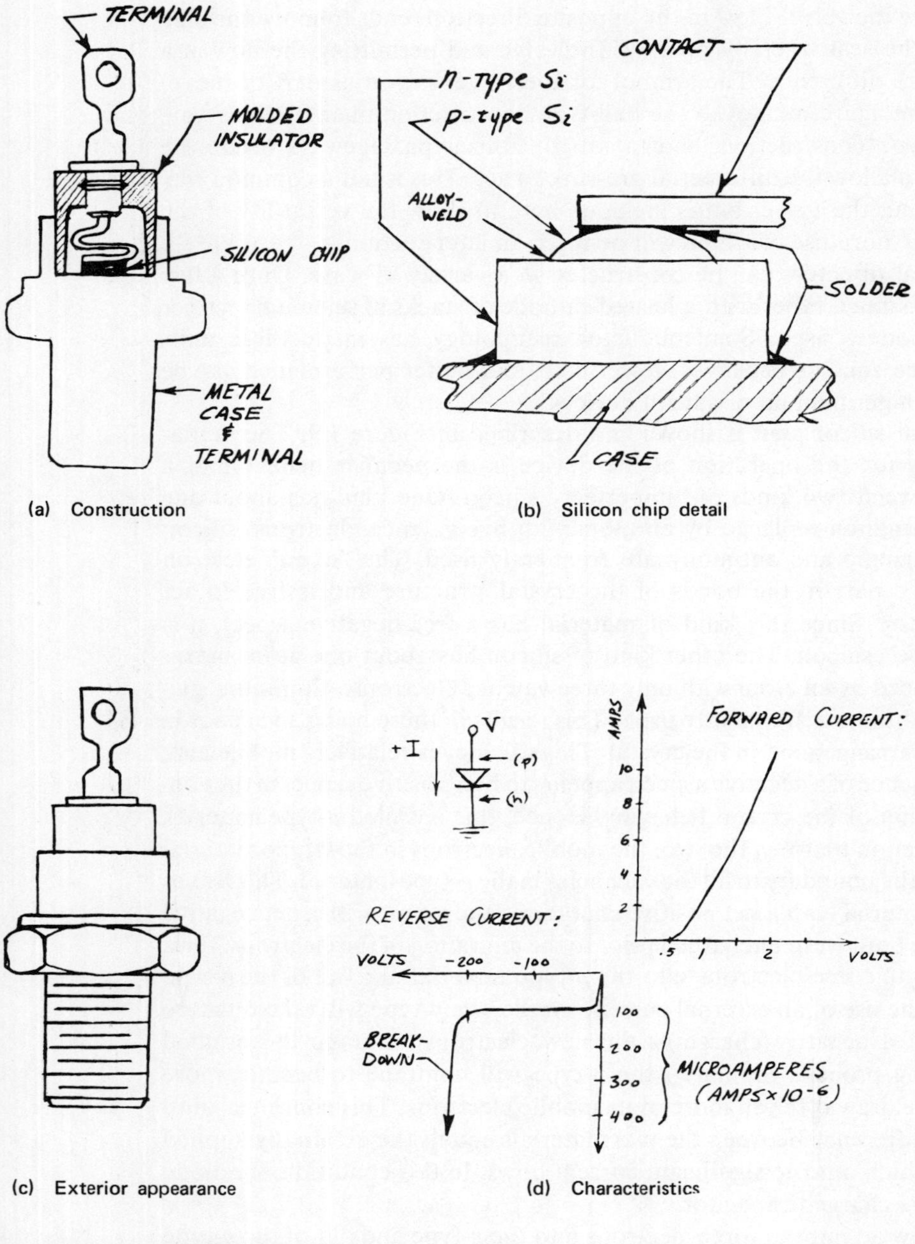

(a) Construction

(b) Silicon chip detail

(c) Exterior appearance

(d) Characteristics

From the Engineer's Sketch Book

Figure 1.9 A semiconductor diode with 10-ampere, 200-volt rating.

These eventually reach zero, all the vacancies in the *p*-type are vacant again and all the excess electrons in the *n*-type are again free. These act to attract electrons from the *n*-type to the *p*-type, and large currents can flow.

The actual piece of silicon in the diode of Figure 1.9 may be only 50 to 100 thousandths of an inch square. It is soldered to a metal case, which helps to carry away the heat generated due to the inefficiencies of the device. Another contact is made, on the opposite side of the chip, by an S-shaped wire which can bend as thermal changes distort the parts of the diode. Externally, provisions are made for connecting the unit into a circuit. The schematic symbol and performance of the diode are shown at (d). The direction of free current flow is from the *p* to the *n* side of the chip; the symbol indicates this direction through its arrow-like nature. When the diode is conducting, there is a small voltage across it, usually from 0.6 to 2.0 volts in magnitude. Ideally, this would be zero; its existence is the reason that there is some heating of the device. In the reverse condition, a very small leakage current exists, usually on the order of 100 millionths of an ampere (100 microamps). When the reverse voltage is made too large, the diode may break down and be destroyed. Although the leakage current also gives rise to heating of the diode (in proportion to the product of it and the reverse voltage), the effect is small compared to the heating in forward conduction and is normally not of concern.

The specification of director elements is essentially the same in each media. First, there must be a rating of the amount they can conduct (torque, fluid flow, current). Then there must be a rating of their reverse blocking capability (maximum reverse speed difference, pressure, voltage). Usually, the system designer will need to know the degree of imperfection in their conduction (slipspeed at rated torque, pressure drop at rated flow, voltage drop at rated current) and the degree of imperfection in their non-conduction (torque at maximum reverse speed, fluid leakage at maximum reverse pressure, current leakage at maximum reverse voltage). It will be assumed that these ratings are consistent with the ability of the device to transfer away the heat due to the imperfections without reaching destructive temperatures. Of course, the thermal environment in which the device is applied has a significant influence on this; the manufacturer will usually specify the conditions under which his ratings were established. In high-performance systems, the peripheral imperfections, such as clutch inertia and diode capacitance, may also require specification.

Perhaps the most common application of directors is in the conversion of alternating power to continuous power, as in an automobile battery charger or the drive of a socket wrench. (Examples are given in section 1 of Chapter 2.) In a broader sense, directors are fequently applied when a "logic" type of condition must be imposed upon a system. This is the essence of digital computation, where many diodes are used.

1.4 Protectors

A group of devices used for the protection of expensive system components is shown in Figure 1.10. These devices are chosen so that they are destroyed before the components to which, or from which, they deliver power are damaged. They are readily constructed, inexpensive, and easy to replace. The shear pin coupling shown at (a) fails when the transmitted shaft torque becomes too great; it is frequently used in motor-boat drive shafts to protect the expensive propeller. The (hypothetical) brittle band coupling at (b) will cease to transmit power if the shaft speed exceeds the amount at which the centrifugal force on the cylindrical pins is sufficient to break the band. It should be noted that this is actually a three-terminal device since it is sensitive to speed relative to an inertially fixed reference frame; it is not perfectly analogous to the other protectors shown in the figure. The well-known electrical fuse is shown at (c); this device ceases to conduct if the current becomes so large that the small resistance of the fuse causes overheating and melting of the material. The hydraulic device shown at (d) conducts until the fluid velocity becomes so large that the momentum change caused by the ball is sufficient to roll it up the incline into the pipe inlet. Once it starts to clog the pipe, the pressure jams it in further, forming a tight seal as the lead is forced to conform to the pipe contour. This kind of device is often applied to the water-level glass of a steam boiler. Normally, the flow into and out of the glass is small and the ball is undisturbed. However, if the glass is accidentally broken, allowing steam and water to rush out of the two connections to the glass, the valves close. This prevents the loss of further steam, reduces the possibility of injury to persons in the vicinity, and allows someone to get close enough to close the shut-off valves between the glass and the boiler. Repair can then be undertaken. In this application, the ball is usually hard and accurately ground. The pipe inlets are also carefully constructed, perhaps from brass, and the valve is not destroyed by its action.

Components may also be protected by re-routing the power. This is accomplished by devices such as those shown in Figure 1.11. At (a) is shown a hypothetical arrangement which could serve to clamp a shaft if its speed became excessive. The principle of the device is that a rubbing shoe is held by a spring against an inner drum. The force on this shoe would be constant so the power dissipated by friction would have to be proportional to the speed difference between the shafts. Therefore, the heating and the expansion of the inner drum would be proportional to this difference and the gap, which is shown much exaggerated, would close if it became too great. Once the inner drum starts to rub on the outer one, there would be even more heating and expansion, causing the device to lock up quite rapidly. Although we assume the action is so severe that it causes destruction of the unit, this might not be necessary.

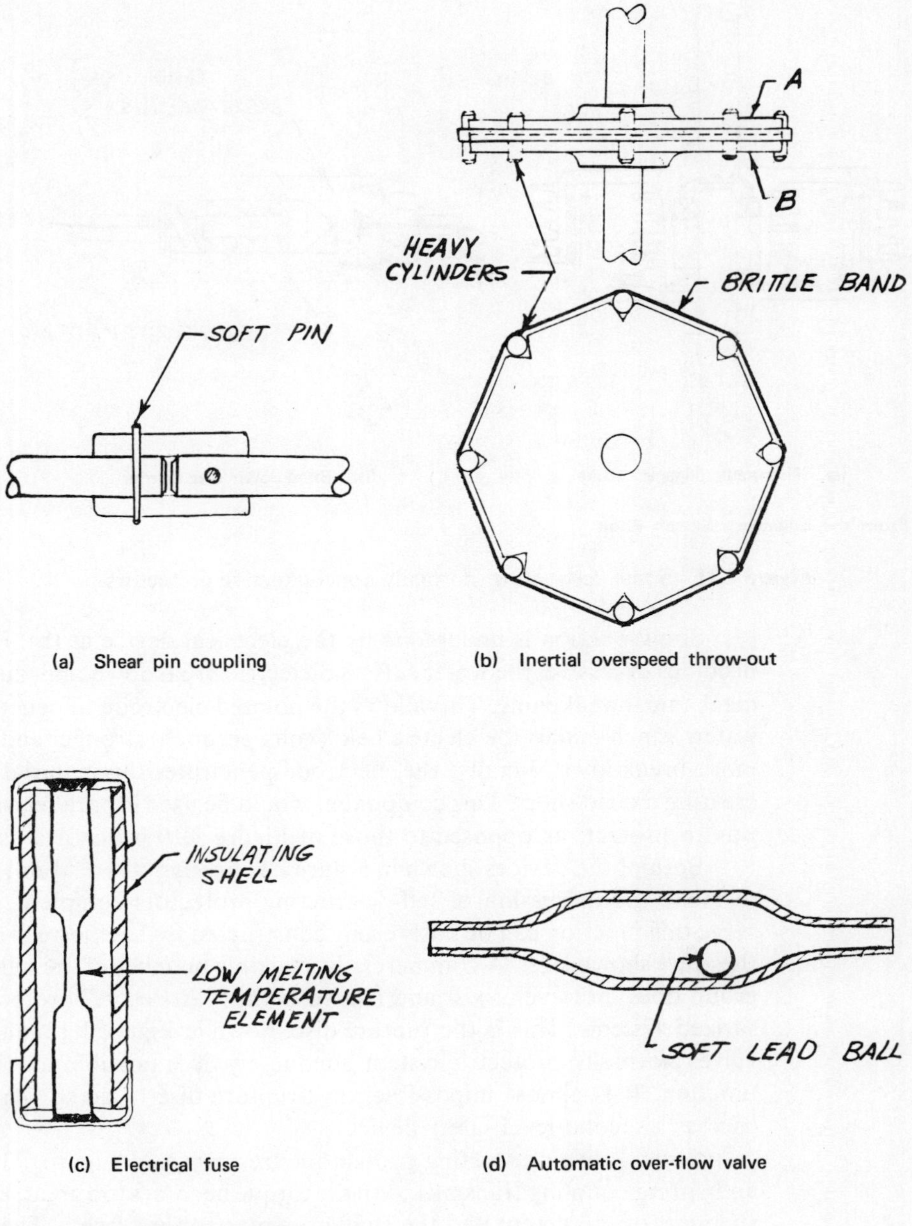

(a) Shear pin coupling

(b) Inertial overspeed throw-out

(c) Electrical fuse

(d) Automatic over-flow valve

From the Engineer's Sketch Book

Figure 1.10 Destructing normally conductive protectors.

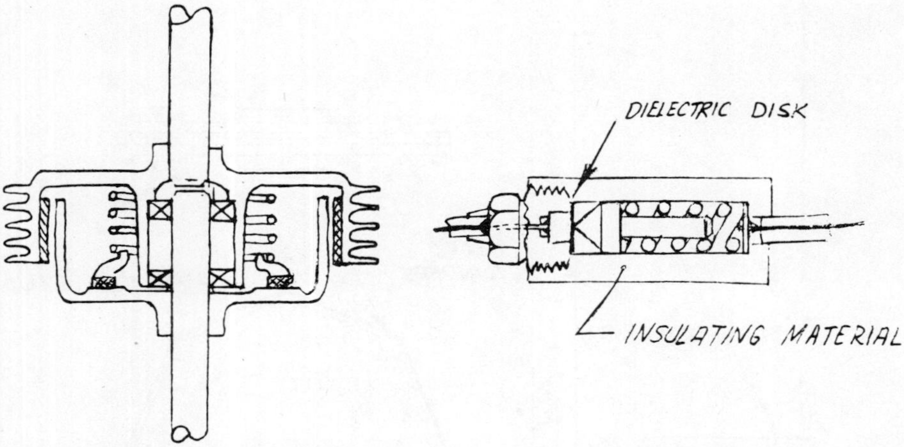

(a) Thermally engaged brake

(b) Breakdown disc barrier

From the Engineer's Sketch Book

Figure 1.11 Some destructing, normally nonconductive protectors.

Similar action is undergone by the electrical device at (b). If the voltage becomes excessive, the wafer suffers dielectric breakdown and, due to heating, mechanical weakening. This allows the pointed electrode to penetrate into the wafer, which makes the electric field (volts per inch) stronger and causes even more breakdown. Finally, the electrode penetrates the material completely, creating a dead short. This component would be used in parallel with the unit it was to protect, as opposed to those of Figure 1.10 which are used in series.

Both of the devices shown in Figure 1.11 are hypothetical and are included to make the discussion of self-destructing protectors complete. Units which serve this function can be fairly easily constructed so that they will not destruct; these are shown later. A commercially available member of the self-destructing group does, however, exist and is used as the "last-resort" protection of pressurized systems. This is the rupture disc shown in Figure 1.12; although relief valves normally protect a system adequately, it is possible for them to malfunction. It is almost impossible for a rupture disc to do so. Therefore, it is used as a second-level safety device.

Several non-destructing protectors are shown in Figure 1.13. The ball-and-spring coupling transmits until the torque becomes too great; the balls then snap out of the detent and the shafts run more or less freely. The unit has the desirable feature that it is quite noisy when it has opened up; the operator is quickly informed of the nature and location of trouble in his system. The neon tube shown at (b) conducts current quite well after the voltage has exceeded

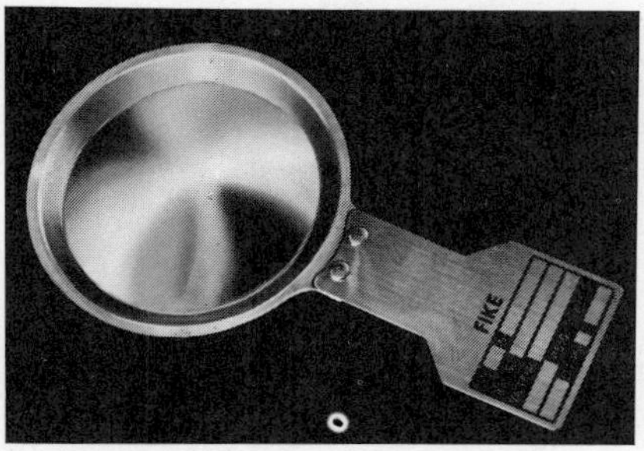

(a) Prebulged rupture disc.

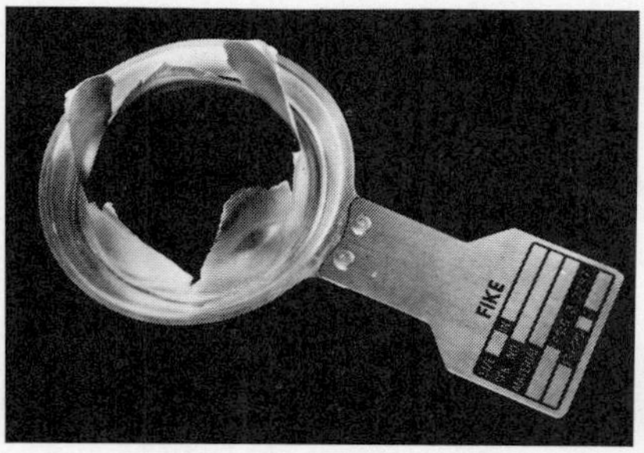

(b) Disc after rupture

(c) Rupture disc assembly

Figure 1.12 A commercially available, destructive, over-pressure protector. Courtesy of Fike Metal Products Corp.

the value at which the gas ionizes; the nature of its characteristic is indicated at (c). Incidentally, the characteristic is symmetric about the origin. A snap-action hydraulic valve of this class is shown at (d); when the pressure P becomes too great, the spool snaps out of the ball-and-groove detent, and the valve

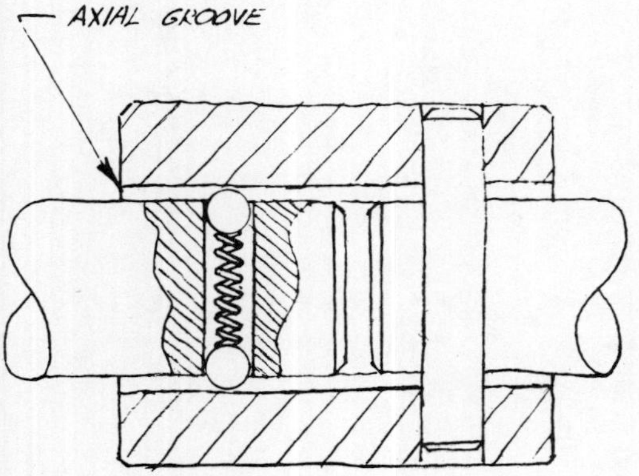

(a) Ball-and-spring coupling

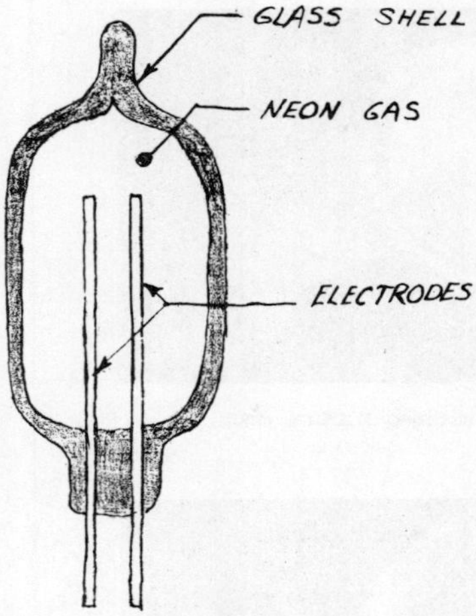

(b) Neon tube

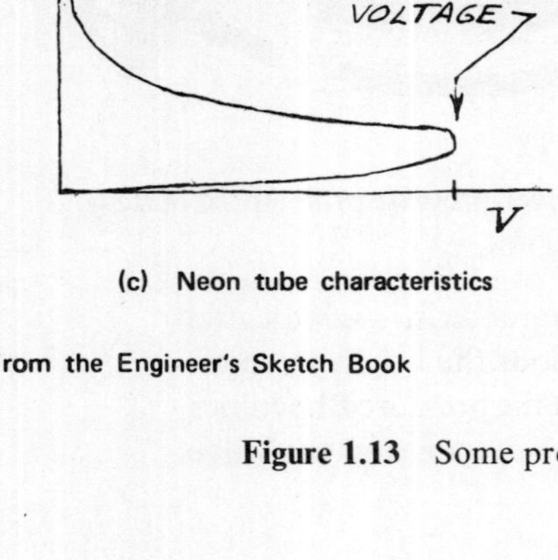

(c) Neon tube characteristics

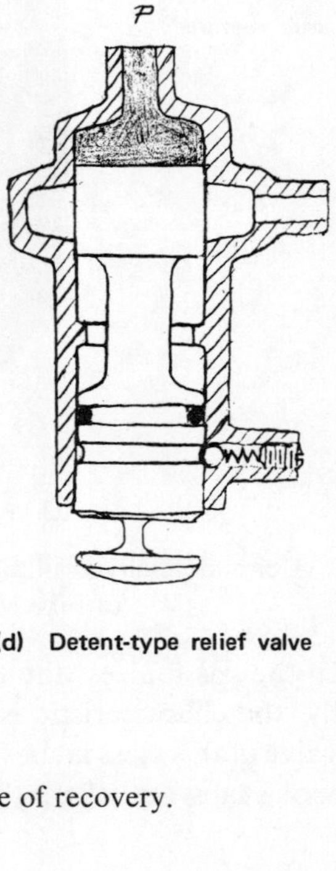

(d) Detent-type relief valve

From the Engineer's Sketch Book

Figure 1.13 Some protectors capable of recovery.

remains wide open until manually reset. This kind of "toggle" action is the basis of a variety of other protector elements.

In the specification of these units, the degree of conduction (torque, flow rate, current) and the amount of non-conduction (speed difference, pressure, voltage) must be given as well as the level at which the state of conduction is to change. For example, we must make sure that the flow rate through a broken rupture disc will be enough to reduce the system pressure in the worst case. This could be a run-away pump, an uncontrollable furnace fire in a boiler, or even an explosion. As a case in point, one wall of an explosives plant is normally made of flimsy construction; nothing less is adequate to protect the structure as a whole. The author is acquainted with one such installation where the wall in question is on a river bank; this simplifies the clean up problem before a replacement wall is constructed.

1.5 Energy Storage Elements

Although energy storage elements are responsible for the extremely important and challenging dynamic considerations of system engineering, they will be covered quickly here; five chapters of the text are devoted to their consideration. It may be pointed out, however, that just as these elements are responsible for troublesome dynamic aspects in system performance, they represent the designer's way of establishing the dynamic performance he desires. They are a mixed blessing.

Practically every engineering component includes some energy storage. To consider the extreme case, we may note that a shaft has inertia and torsional springiness; a pipe carrying fluid includes the inertia of the moving fluid as well as the elastic effects of the pipe and the compressibility of the fluid itself; a wire forms a weak capacitor with respect to ground, and the current it carries creates a magnetic field about the wire so that some inductance appears. In many cases, these influences are so weak that they do not affect system performance; in others, notably where the connector is "long," the effect may be very marked.

Energy-storage devices for intentional engineering application are built so as to localize, as much as possible, the kinds of energy storage found in connectors. Generally, we try to store as much energy as possible in a given volume, we try to make the energy-storage process as efficient as possible by keeping the spurious dissipation to an absolute minimum, and we try to minimize the non-desired kind of energy storage. For example, in a spring, we design for high levels of strain, we select materials with little hysteresis, and we try to keep the inertia to a minimum. This allows the device to be easy to use from a practical point of view and also makes the engineering prediction of its performance in a system as straightforward as possible. With an electrical capacitor, the size is kept to a minimum by using the highest possible voltage per inch of thickness in the dielectric material, the dissipation is minimized by choosing a dielectric with extremely low leakage currents, and the inductance is kept small (without very much trouble) by avoiding current paths, for the charging and discharging conditions, which lead to strong magnetic fields. Similarly, in an inductor as much flux as possible is developed in the iron and the winding resistance is kept as small as possible.

The symbols and equations for four ideal energy-storage elements are shown in Figure 1.14. The analogy of their performance is obvious. (Torque, force, current, and flow rate are considered analogous; velocity, voltage, and pressure are also.) Since there is no common engineering device which provides this kind of action in hydraulic systems in a direct way, the element shown at (d) represents a mechanical inertia which is reflected into the hydraulic line by some sort of positive-acting converter. This converter must make the speed of the inertia proportional to the fluid flow rate; many devices do just this. The proportionality constant in this case depends on the speed-to-flow-rate conversion factor, and is not specified here. All these devices act as ideal conductors in the case that the transmitted quantity (torque, force, current, or flow rate) is not changing with respect to time; they have no influence on system performance in this case and can be replaced by pure conductors for purposes of static analysis.

In real devices of this class, the mathematical relationship between the variables is not the simple proportionality shown in the figure, but is a more general function such as

$$\mathbf{\Gamma} = \mathscr{F}\left\{\int_0^t \mathbf{\Delta}\, d\tau\right\} + \mathbf{\Gamma}_0$$

where $\Gamma = M, F, I$ or W (in Fig. 1.14),
$\Delta = \Omega, V, V,$ or P (in Fig. 1.14).

This provides for the non-constant stress/strain behavior of the spring material, the magnetic saturation of the inductor iron, or imperfections in the hydraulic connection to the mechanical inertia. Generally, however, it is possible to build devices which have a fairly wide range of approximately linear behavior and, even beyond this, do not behave so irregularly as to be intractable. (Unless they destroy themselves, which does happen.)

In specifying these components, we must give the required transmission level (torque, force, current, or flow rate), the value of the proportionality constant, and then see to the matters of imperfection. This might involve the

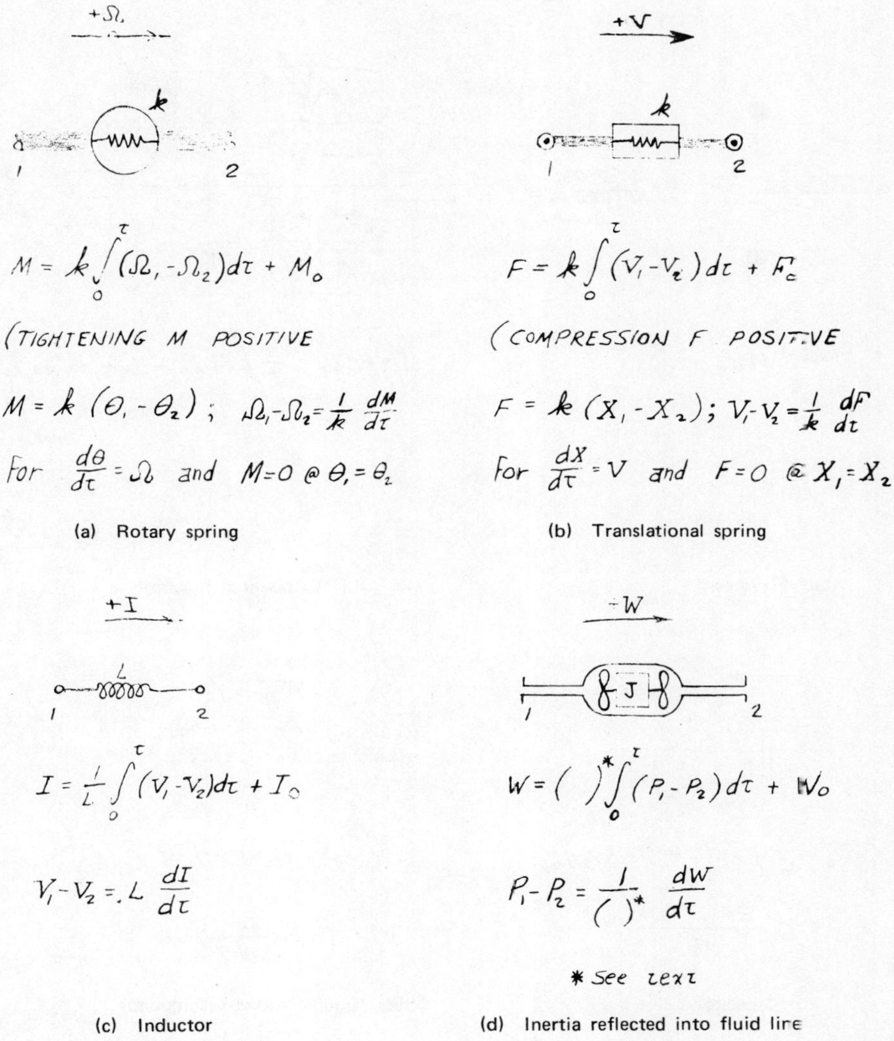

(a) Rotary spring (b) Translational spring

(c) Inductor (d) Inertia reflected into fluid line

From the Engineer's Sketch Book

Figure 1.14 Energy storing devices which are static conductors.

accuracy of the proportionality constant, the extent of linear performance, and the expected dissipation.

A group of energy-storage elements which are non-conductors under steady conditions are indicated in Figure 1.15. Again, the analogy is obvious.

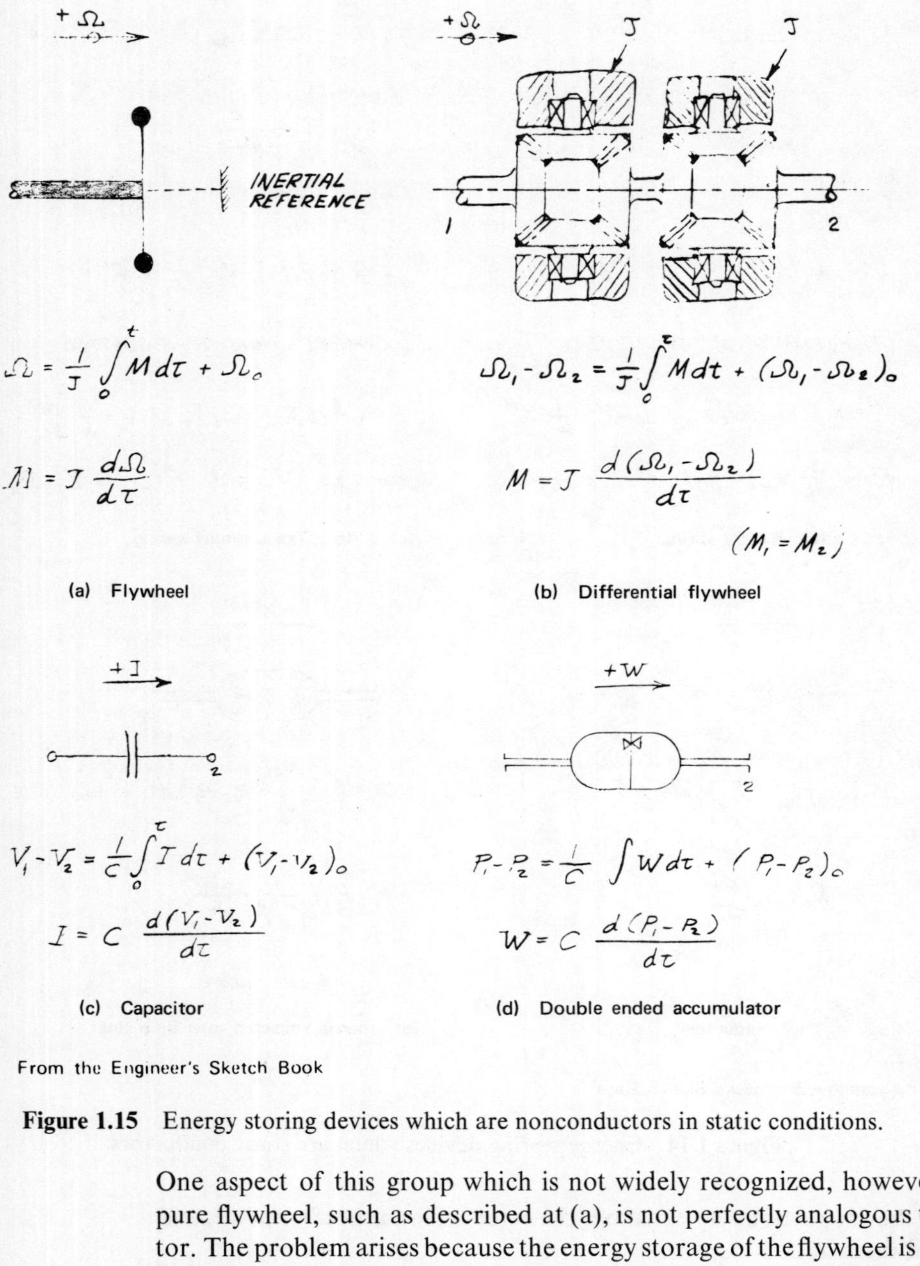

From the Engineer's Sketch Book

Figure 1.15 Energy storing devices which are nonconductors in static conditions.

One aspect of this group which is not widely recognized, however, is that a pure flywheel, such as described at (a), is not perfectly analogous to a capacitor. The problem arises because the energy storage of the flywheel is determined by its speed relative to an earth reference, while the capacitor storage depends on the voltage difference between the plates. In constructing or analyzing analogous circuits, which is one of our objectives here, we cannot replace one by the other. A mechanical device which does have perfect analogy to a capacitor is shown at (b) in the figure; this has two flywheels which spin about the shaft axis, but which are connected to the shaft through a pair of differential bevel-gear arrangements. It is easy to show that the torque at shaft 1 is exactly equal to that at 2 (if friction is neglected), and that this torque is proportional to the rate of change of the speed difference. The application of this apparently odd device is shown in several examples in the next chapter.

The hydraulic accumulator shown in Figure 1.15(d) is also somewhat unconventional. Normal accumulators have only a single connection and must be considered to have a virtual connection to the atmosphere. (They are often built with a bladder containing pressurized gas; once they have been pressurized, the terminal is sealed off. However, the pressurizing conditions must be known for the performance to be described mathematically.) In order to retain the generality of this discussion, the two-sided unit, which can be constructed from a double acting piston that works against springs, has been shown. All these troubles can be avoided if we always ground one side of a capacitor, are satisfied with accumulators which appear as if they had one terminal returned to the reservoir, and need the influence of inertia only with respect to an earth reference. However, this represents an artificial and unnecessary constraint on the designer, as will be shown by several later examples.

The imperfections and specifications of these units follow such a similar pattern to those for the elements of Figure 1.14 that their consideration is left to the reader.

1.6 Two-Terminal Generators

In some applications, weather monitoring satellites being a notable example, it is necessary to convert heat in some form into mechanical, fluid, or electrical power by a direct method. This implies a two-terminal device, and some items of this nature do exist. The bimetallic strip may be used to power a queer kind of engine, temperature gradients may be used to pump fluids, and thermocouples can provide electrical power directly from the existence of a temperature difference.

A particularly useful device of this class is shown in Figure 1.16; this is the solar cell used in the "wings" of a spacecraft in order to generate electrical power from sunlight. The characteristics of a typical cell are shown at (b) in the figure. Along either of the axes, there is zero power. In the generation region, the maximum power point is where the product of voltage and current is the highest; this would be used as the operating point in most applications.

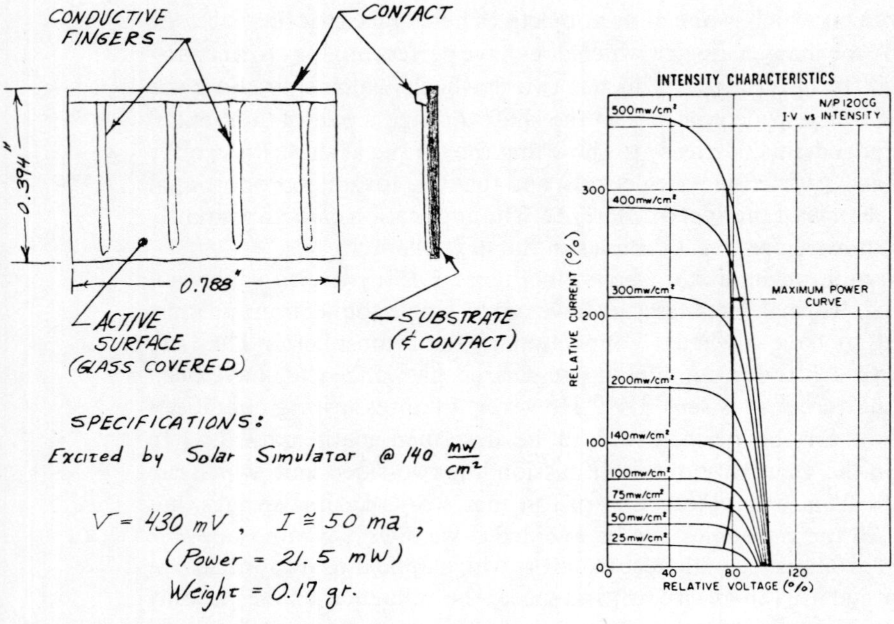

(a) Outline and specifications

(b) Characteristics

From the Engineer's Sketch Book

Figure 1.16 Solar cell intended for spacecraft applications. Courtesy of Semiconductor Div., Hoffman Electronics Corp.

We shall see some very similar conditions when we examine the mechanical characteristics of an electrical motor or, in general, the output characteristics of the group of four-terminal converters discussed in sections 1.12 and 1.13.

1.7 Mechanically Variable Dissipators

In many situations, it is necessary to be able to adjust the amount of conduction or degree of conductivity which exists between two terminals. This implies a four-terminal device, two of which are used to supply the power necessary to perform the adjustment. If we want either total or zero conductivity, the device has the nature of a relay; these are discussed in the following section. In this, we consider devices which are intended to serve as dissipators and which may, or may not, provide the extreme ranges of conductivity.

A variety of such components is shown at Figure 1.17. The mechanical device at (a) is a simple friction drive wherein the normal force is developed

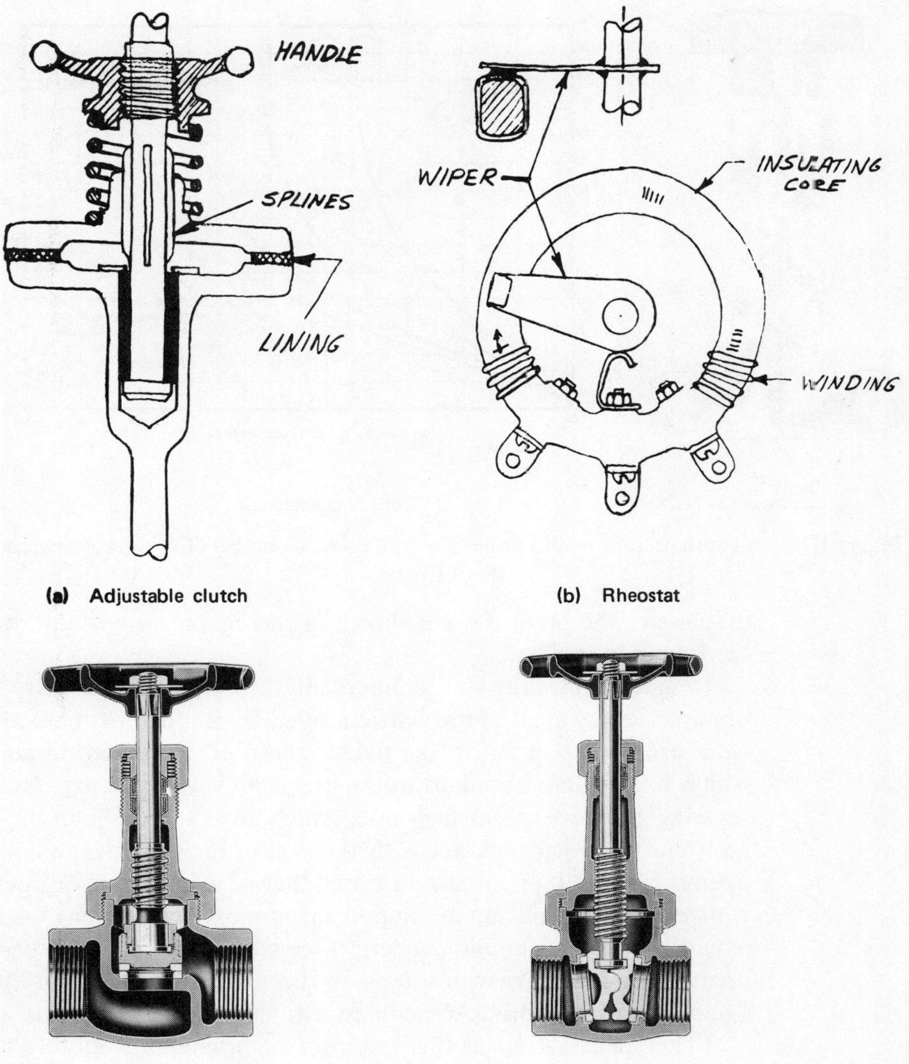

(a) Adjustable clutch

(b) Rheostat

(c) Globe valve

(d) Gate valve

Figure 1.17 Some mechanically variable dissipators. Courtesy of Lunkenheimer Co. for (c), (d).

by a spring through a screw adjustment. The construction has been somewhat simplified by using the upper shaft for two purposes—the drive as well as for providing the reaction against the adjustment torque; this makes the unit a three-terminal device and a degenerate form of the general class being

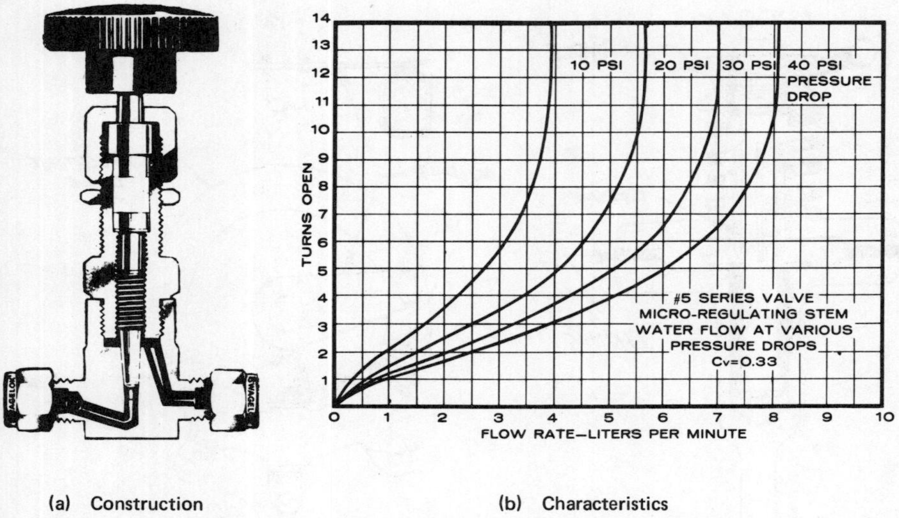

Figure 1.18 A commercially available needle-type valve. Courtesy of Whitey Research Tool Co.

discussed. The other devices shown in the figure are well known and will not be described in detail.

Figure 1.18 shows a commercially available needle valve used for the precision adjustment of the restriction in a hydraulic line. The valve characteristics are shown at (b) in the figure; these are presented in an unusual way which has become standard in the precision valve industry. Here, instead of showing pressure versus flow rate, which gives the designer an indication of how the valve will interact with the rest of his system, the number of turns open is plotted. It might also be noted that no information on operating torque is given; this would not be important in most applications of the valve, but would be of considerable concern if we were to design a motor-driven valve-operating device. We will attend to this general subject more thoroughly in Chapter 3 which discusses component characterization in a general way.

The specification of this group of components is closely related to the specification of fixed dissipators. The worst-case power dissipation must be determined, the maximum transmitted quantity given (torque, current, fluid flow rate), and the desired characteristic relationship described. If the imperfections, such as slip in the clutch or leakage in the valves when they are supposed to be non-conducting, are important, their maximum allowable values must be determined. Usually, the stroke required for full operation is given and it is assumed that the required torque is appropriate to the application.

These dissipators form the basis of many remotely controlled valve-like system components. In these units, some kind of actuator replaces the manual control of the basic unit. It is in the design of these components that the actuating force or torque information becomes important. In some cases, especially with hydraulic valves, there is a tendency for the device to actuate itself in the direction of higher conduction; this can lead to instability and generally poor performance.

1.8 Mechanically Controlled Connectors

If variable dissipation is not required for system control, simple devices which either connect or do not connect the system can be applied. A group of these is shown in Figure 1.19; each has no practically useful condition in which it is

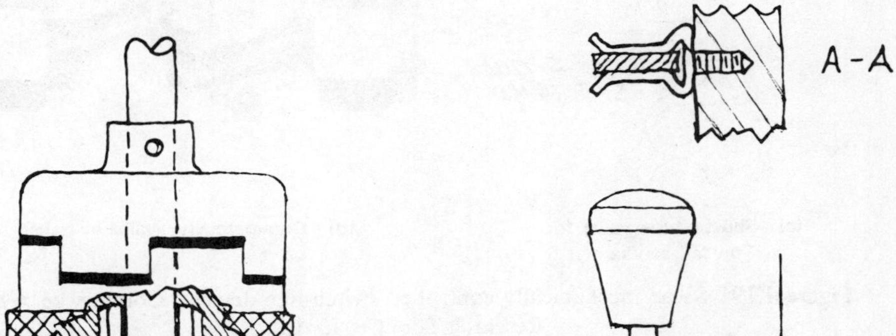

Figure 1.19 *(continued overleaf)*

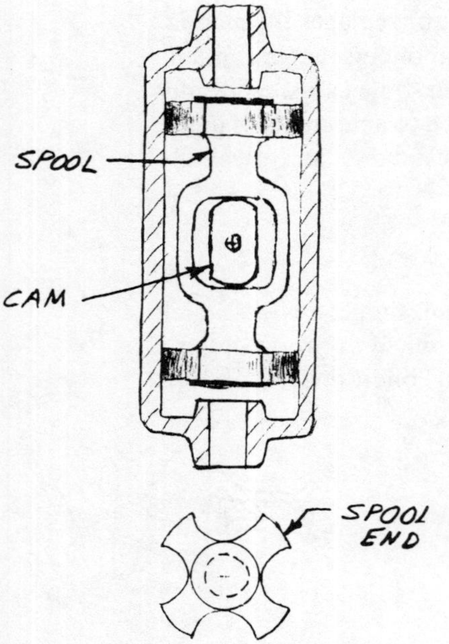

(c) Shuttle-type valve for "on-of" service

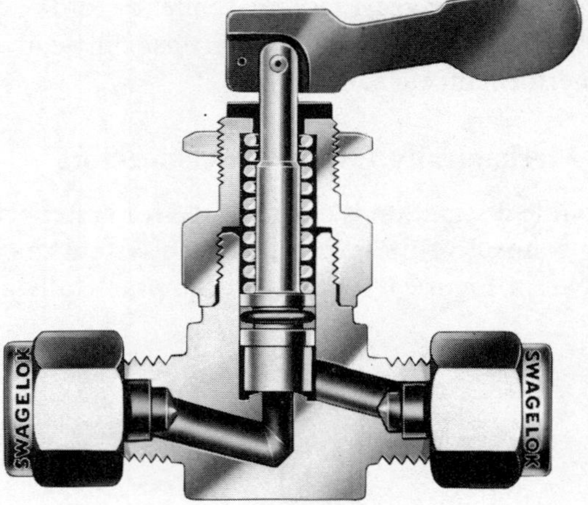

(d) Commercially available toggle valve

Figure 1.19 Some mechanically controlled switch-like devices. Courtesy of Whitey Research Tool Co. for (d).

neither a good conductor or a good non-conductor. The valve shown at (c) is operated by a cam which holds the plug centered when aligned with the flow. If the cam is rotated 90°, the plug is free to act as a double-sided check valve and no flow can be transmitted. At an intermediate position, the forces on the plug tend to rotate the cam to the open position so that no useful intermediate condition exists. The "toggle valve" shown at (d) is operated by an over-center spring arrangement, much as in a light switch. This device is designed to be mounted on a control panel, and allows hydraulic networks to be switched as readily as electrical circuits.

1.9 Mechanical Actuators

In many systems, the final objective is to control the movement of some object. In others, the operation of the system depends on the automatic actuation of a mechanically variable dissipator or a mechanically controlled connector. The devices for accomplishing these functions may be classified as "actuators." They are known as linear actuators if the stroke is translational, and as rotary actuators if angular; in both cases a stroke of limited magnitude is implied—if continuous motion is possible, the device is a motor.

Several such actuators are shown in Figure 1.20. In these, if the output connection is made immovable with some sort of locking apparatus, the input side connections appear to be joined by a very good conductor. This differentiates the group from another to be discussed. The device shown at (a) derives its force from the torque in a shaft. This torque causes a clockspring to be wound and the relative motion of its ends to be applied to a screw. The resulting translational motion is transmitted, through a pair of bearings, to members A and B which are actuator terminals. This is one way in which a torque gage may be constructed. The device shown at (b) in the figure utilizes the momentum change in a fluid stream to create a force; with careful design and fabrication such a device can have very small pressure drop between the inlet and outlet. The well-known electromagnet is shown at (c) and a representative set of characteristics at (d).

The group of actuators shown in Figure 1.21 appear to have a good nonconductor between the input terminals when the output motion is blocked.

The flyball actuator shown at (a)—which has a virtual connection to an inertially fixed reference frame—spins quite freely in the absence of acceleration.

The electrostatic actuator shown at (b) derives its force from the attraction between oppositely charged plates. Although the force levels are low, the device is useful because of its simplicity, the low mass with which it may be constructed, and the fact that it draws no current in the static state. The principle is applied in some high-frequency speakers for audio systems.

A closely related device, which depends on the mechanical changes in certain crystals (piezo-electric) under electrical field excitation, is shown at (c). This device has very small strokes, but can produce large forces; a variation on it is constructed like a beam—along the lines of a bimetallic strip—and produces larger deflections at the expense of force. This has been applied, in a case familiar to the author, to the construction of an electronically driven Braille character generator. The system allows the blind to read books which have been recorded, letter by letter, on magnetic tape.

The unit shown at (d) is a hydraulic actuator with a long history and rather obvious operating principles. It appears in many pressure regulators where it operates a mechanically variable valve.

In actuators of both types, the force is a function of the stroke. For the electromagnet [Figure 1.20(c)], the effect is quite pronounced because the reluctance of the magnetic path changes very strongly as the air-gap magnitude changes (X is the length of the air-gap). The flyball actuator has obvious sensitivity to stroke since the radius at which the balls spin changes as a

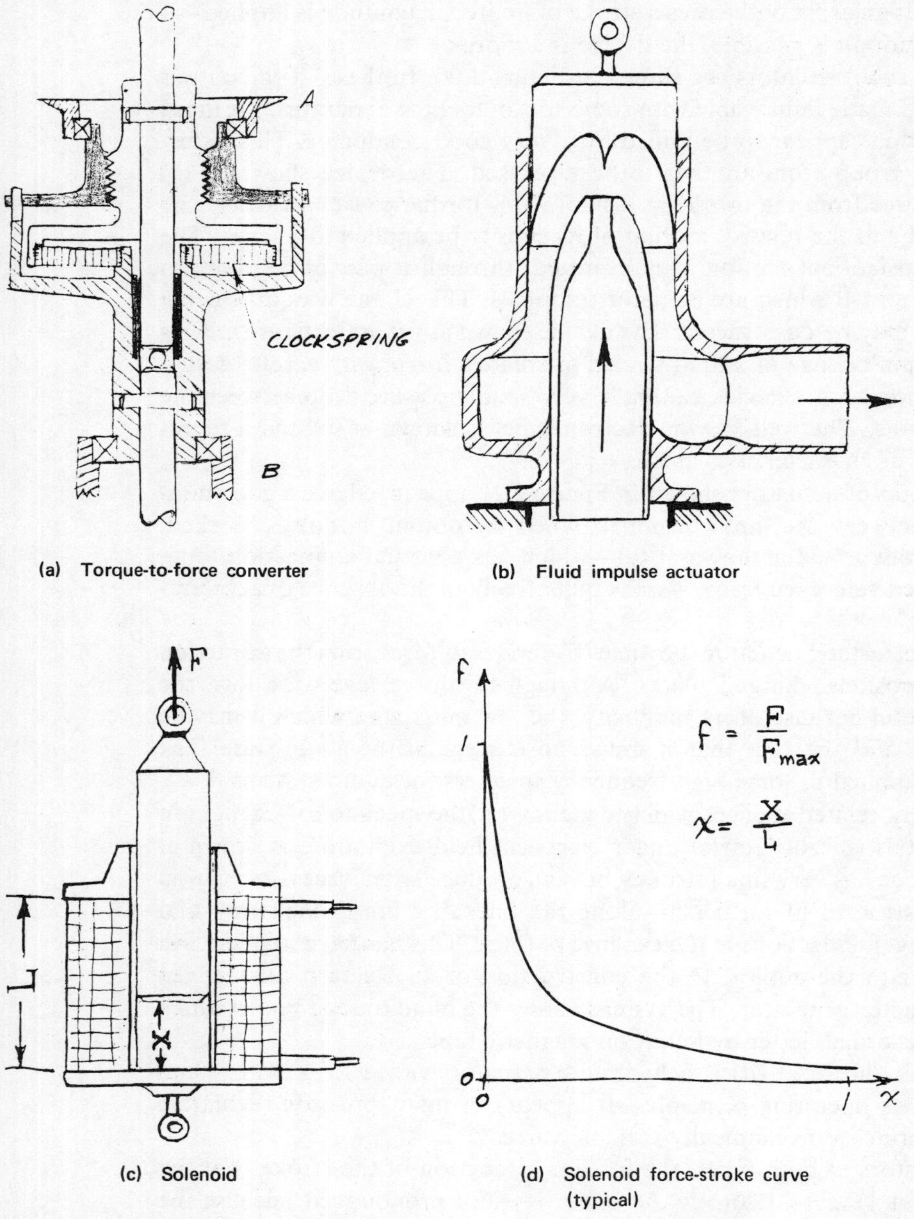

(a) Torque-to-force converter

(b) Fluid impulse actuator

(c) Solenoid

(d) Solenoid force-stroke curve (typical)

From the Engineer's Sketch Book

Figure 1.20 Some mechanical actuators with conductive primaries.

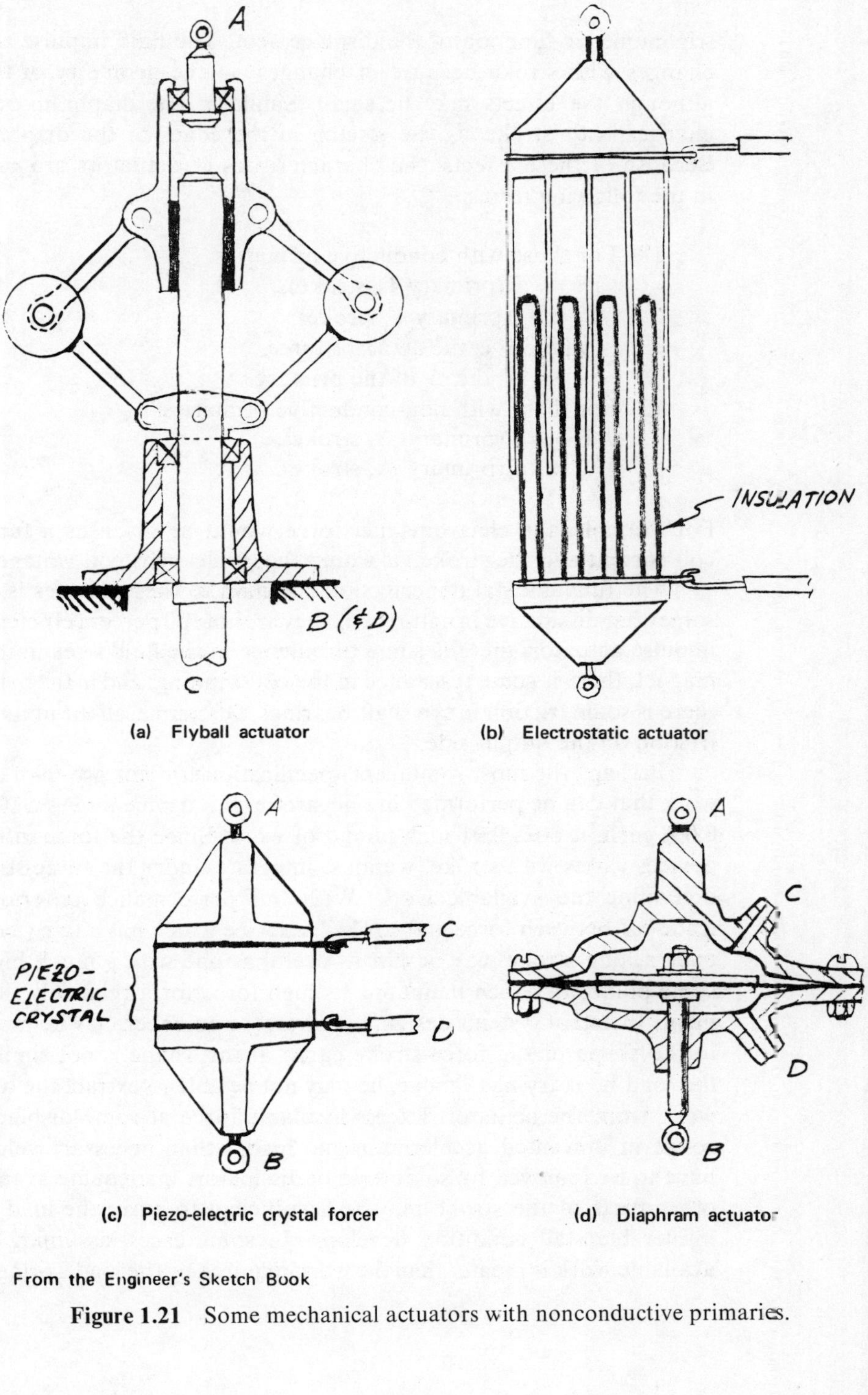

Figure 1.21 Some mechanical actuators with nonconductive primaries.

trigonometric function of the displacement. The fluid impulse actuator force changes with stroke because of changes in the geometry of the fluid flow, although the effects may be small. Similarly, the diaphram actuator force changes with stroke as the stretch in the edges of the diaphram changes. Because of these effects, the characteristics of actuators are normally given in the following forms:

(1) For those with conductive primaries:
$F = \mathcal{F}_1(\text{primary } \boldsymbol{\Gamma}, \text{stroke}),$
$\boldsymbol{\Delta} = \mathcal{F}_2(\text{primary } \boldsymbol{\Gamma}, \text{stroke}),$
where F is the actuator force,
$\boldsymbol{\Delta}$ is the $\boldsymbol{\Delta}$ of the primary.

(2) For those with non-conductive primaries:
$F = \mathcal{F}_3(\text{primary } \boldsymbol{\Delta}, \text{stroke}),$
$\boldsymbol{\Gamma} = \mathcal{F}_4(\text{primary } \boldsymbol{\Delta}, \text{stroke}).$

For example, the electromagnet force would be given as a function of the coil current and the stroke, as would the (undesired) coil voltage drop.

The fundamental imperfection common to these devices is that they are somewhat dissipative in nature; i.e., they are not 100 percent efficient. In the fluid impulse actuator, there is some turbulence in the fluid stream; in the electromagnet, there is some resistance in the coil winding; and in the flyball actuator, there is some friction in the shaft bearings. Of course, all the units exhibit some friction on the output side.

Perhaps the most significant specification for this group of devices is the work that can be performed in one stroke. If a device moves a 10-pound load 0.001 inch, it does 0.01 inch-pound of work. Since the force in each of these devices varies with stroke, we must integrate under the force-stroke curve to determine the available work. With this performance criterion, there is a trade-off between force and stroke. A device which has a fairly constant force over a long stroke may be just as useful as one with a much higher force at some point but which maintains its high force for a relatively short displacement. In actual systems design, however, the engineer may be more interested in the shape of the force-stroke curve. If this shape is not similar to that of the load he is trying to move, he may not be able to extract the total available work from the actuator. Excess available force at some displacements may result in unwanted acceleration and higher than necessary velocities which have to be removed by some kind of dissipator. Inadequate available force at other parts of the stroke may be insufficient to move the load such that an intolerable stall condition develops. In some cases, assuming the actuator available work is greater than the work required by the load, springs and masses may be used to accomplish satisfactory operation of statically mis-matched actuators and loads.

The basic actuator specifications, then, are the following:

(1) Available work.
(2) Maximum force.
(3) Stroke.
(4) Primary condition at maximum force (Γ or Δ).
(5) Primary dissipation at maximum force.

If the actuator is to be applied as a shaker, information regarding its mechanical friction and the dynamic properties of the primary may also be required.

1.10 Transfer Variable Dissipators

From a functional point-of-view, the essence of many systems is based in the component which allows large amounts of power to be controlled by smaller amounts. This is the "valve" function which is carried out by transistors in electronic systems, spool valves in hydraulic systems, and remotely actuated clutches in mechanical systems. As will be shown in Chapter 2, there are other ways to accomplish the control of power. However, in this section, we shall devote our attention exclusively to the kind of device which is a variable dissipator and which allows the amount of dissipation to be controlled by the same medium in which the dissipation occurs. Because there is only one medium, these devices can be built with only three terminals. The basic units are, in fact, constructed that way and we shall further limit this discussion to them.

One rotary mechanical form of such a device is shown at Figure 1.22. There are three rotating bodies of basic interest here [see part (b) of the figure]. One is the shaft at the right which is rigidly connected to a drum, but which otherwise rotates freely in bearings. The second is the shaft at the left which holds the forementioned bearings and terminates in a spider arrangement. At the ends of each lobe on the spider are pivot pins. The third basic element is a hollow cylinder, which we shall call the "runner." The runner is carried on bearings and rotates about the spider shaft. The principle of operation of the unit is related to the friction-type, over-running clutch discussed earlier. Shoes with friction linings are carried by links pivoted at the ends of the spider. These shoes operate on the inside of the drum and the links are arranged so that they are almost aligned with the resultant of the shoe normal force and the perpendicular friction component. Because the resultant force is almost parallel to the link, a small, externally applied torque on the link creates large changes in the magnitude of the resultant. This control torque is applied by a

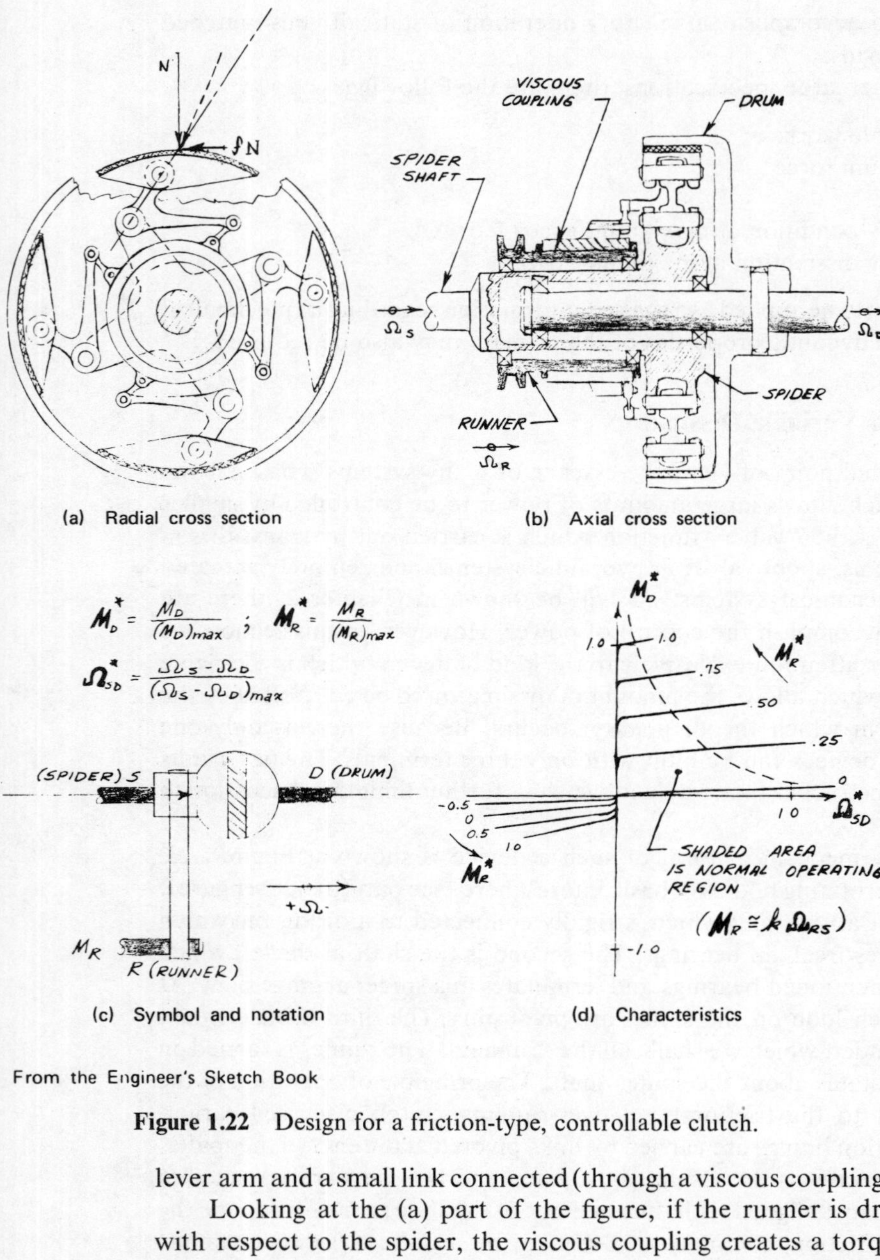

(a) Radial cross section

(b) Axial cross section

(c) Symbol and notation

(d) Characteristics

From the Engineer's Sketch Book

Figure 1.22 Design for a friction-type, controllable clutch.

lever arm and a small link connected (through a viscous coupling) to the runner.

Looking at the (a) part of the figure, if the runner is driven clockwise with respect to the spider, the viscous coupling creates a torque which must be balanced by compression forces in the small links. By the mechanism described above, greatly magnified contact forces between the shoes and the drum are thereby developed. Relatively low torques applied to the runner may then control the large torques in the drum shaft. Depending on the viscous coupling, the relative speed between the runner and the spider may be small relative to that between the drum and the runner. Therefore, large power dissipation between the drum and spider may be controlled by small dissipation between the runner and the spider. In the extreme case, the small links may be pivoted directly on the runner so that there is no speed difference between the runner and spider; this results in the device using zero control power such that it has infinite power gain (the ratio of controlled power to control power). Another extreme case would exist if an inertial actuator were mounted on the runner and used to provide the link torque. This would result in large speed differences between the runner and spider, but smaller runner torques; therefore, the power gain would again approach infinity. In parallel with our grouping of actuators, we could consider these to be "conductive control" and "nonconductive control," respectively.

Another approach to high power gain is to make the links that hold the shoes exactly parallel with the resultant force. This is impractical, however, because lining wear, manufacturing tolerances, and changes in the coefficient of friction with temperature, etc., would make the performance of the clutch totally unpredictable. At some times the runner would lose its influence and the device would simply be an over-running clutch; at others, it would have proper but extremely sensitive operation that would be difficult to control.

A symbol which may be used for this type of clutch is shown at Figure 1.22(c). The symbol is based on that for the over-running clutch—an obviously close relative—and uses a small box to represent the runner cylinder. The arrow on the runner symbol indicates the direction of relative rotation between the runner and spider which tends to engage the clutch. In the use of this symbol, we shall apply the convention that the clutch is disengaged unless some runner-to-spider speed difference exists, or unless a torque is applied to the runner; furthermore, we shall assume that reverse relative rotation can occur and that this insures complete disengagement of the clutch. The general nature of the torque speed characteristics of the drum shaft are shown at (d); the very marked influence of the direction of the relative rotation between the drum and spider is apparent here.

Many commercially available clutches utilize a four-terminal arrangement which includes a housing that is normally fastened to a foundation of some kind. The actuating motion may then be brought in at any orientation, or actuation may be carried out electrically or with fluids. However, the three-terminal construction shown here allows the application of mechanical variable dissipators to be discussed in a very basic way and also allows some

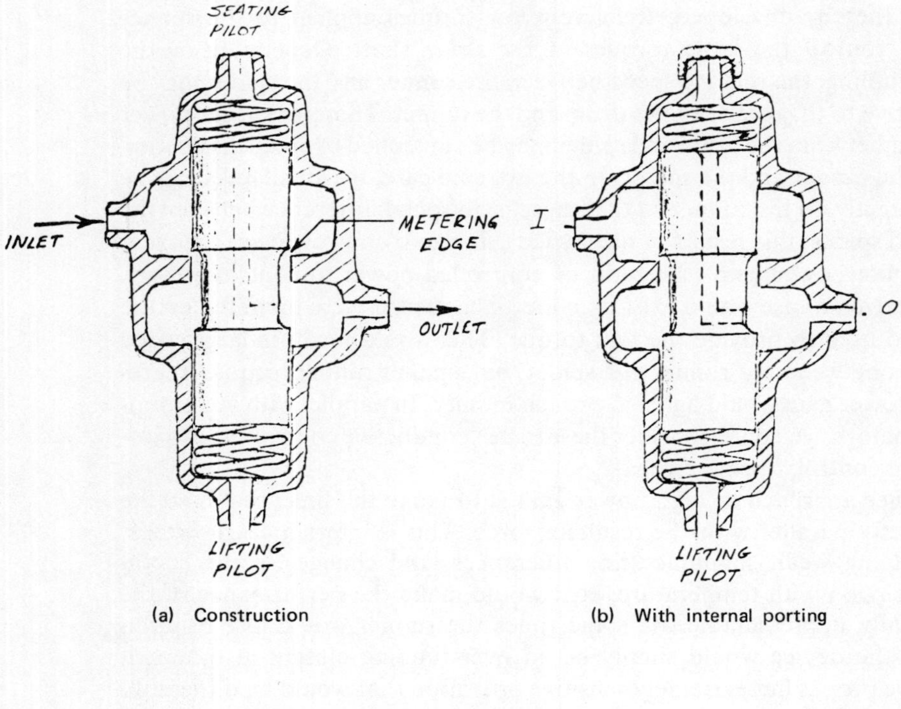

(a) Construction

(b) With internal porting

(c) Symbol and notation

(d) Nature of characteristics

From the Engineer's Sketch Book

Figure 1.23 Basic pilot-operated spool valve.

interesting systems, which would otherwise be difficult to construct, to be built in a straightforward way.

The hydraulic analog of the above clutch is the spool-type valve shown in Figure 1.23. As shown at (a), the spool, which carries a stepped-down section, is held in a central location by springs. A land on the body of the valve acts with the step on the spool to form a flow restriction of variable area. If pressure at the seating pilot end of the spool exceeds that at the lifting pilot, the valve tends to close, and vice versa. In the basic realization of the valve, the seating pilot connection is closed off and the cavity ported to the outlet side of the valve by a hole inside the spool. This is shown at (b). The operation of the valve then depends on the pressure difference between lifting pilot connection and the outlet connection. The symbol for this kind of valve is shown at (c); it is assumed that, unless the pilot pressure exceeds the outlet pressure, the valve is closed. A rough indication of the inlet connection characteristics is shown at (d). Although there is bound to be some leakage from the pilot to the outlet in this valve, since there must be a little clearance around the spool if it is to move freely, the tendency is for this to be a non-conductive control arrangement.

A relatively new form of hydraulic control valve is shown in Figure 1.24. This valve operates on the vortex principle used by the vortex dissipator shown earlier. When there is no control flow, fluid flows relatively smoothly from the inlet to the outlet. However, when a control flow is introduced, a strong vortex is developed because of the tangential velocity of the control fluid. This causes a large pressure difference to develop between the inlet and the outlet, giving the desired valve action. It is clear that this valve utilizes a conductive type of control; as will be seen in Chapter 2, it bears the relationship of "duality" to the spool valve.

One of the most important devices in technology today is the subject of Figure 1.25 and the next several paragraphs. Since this text is written, basically, for non-electrical engineers, the author assumes that the reader is not familiar with the device known as the transistor. Moreover, the author is a mechanical engineer who found himself well advised to learn something of transistors and has benefited greatly from his modest understanding of them. It is with this general background that the following description and "little transistor lectures" in later chapters have been written. The hope is to make the reader aware that there is nothing very mysterious about the application of transistors, even though the physics is somewhat abstract. There are some functions which can be accomplished so much more effectively with semiconductor devices than any other kind of engineering component that the engineer who refuses to work with them is compromising his effectiveness to a very significant extent.

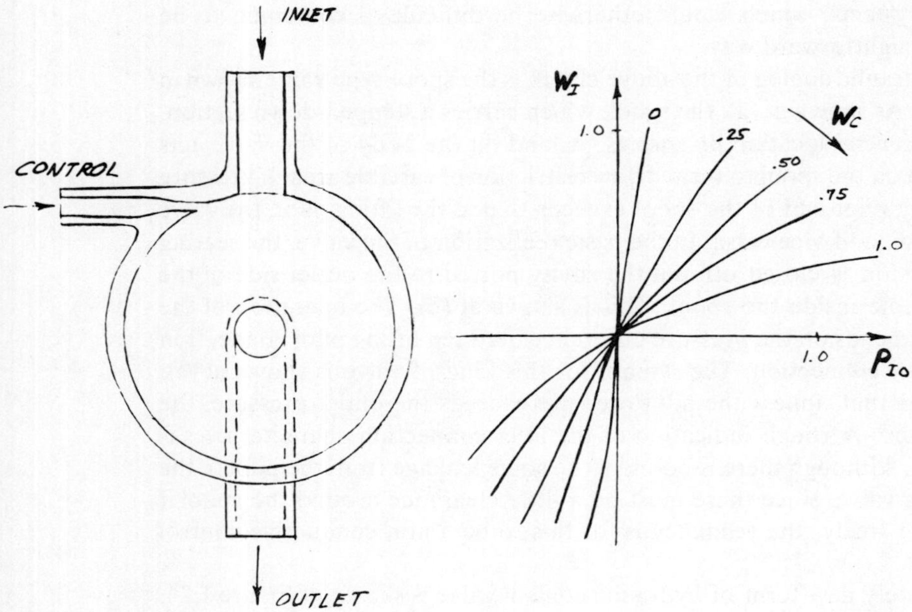

(a) Construction (b) Nature of characteristics

From the Engineer's Sketch Book

Figure 1.24 Vortex amplifier.

The transistor is constructed as shown in Figure 1.25(a). Three wires are held in a glassy insulator which seals off the opening in an airtight metal can. One of these leads terminates at a metal plate sitting on the insulator; the others pierce it without making contact and extend a short distance into the cavity. A small silicon (or germanium) cube, like those from which diodes are constructed, is soldered to the metal plate. Very fine wires connect the silicon chip to the remaining leads. In a typical transistor for general radio and instrument use, where the maximum current is about 1 ampere, the chip may be about 0.050 inch on a side. When we open a transistor for the first time, we may—without close inspection—decide that they left the insides out. The fine wires which connect to the chip may be only one or two thousandths of an inch in diameter. The device is constructed under a microscope with remarkably fine tools. Some firms conduct this part of the manufacturing in Hong Kong or other places where labor is inexpensive. Shipping is not a major consideration with parts of this size. Depending on the transistor specifications, the cost per

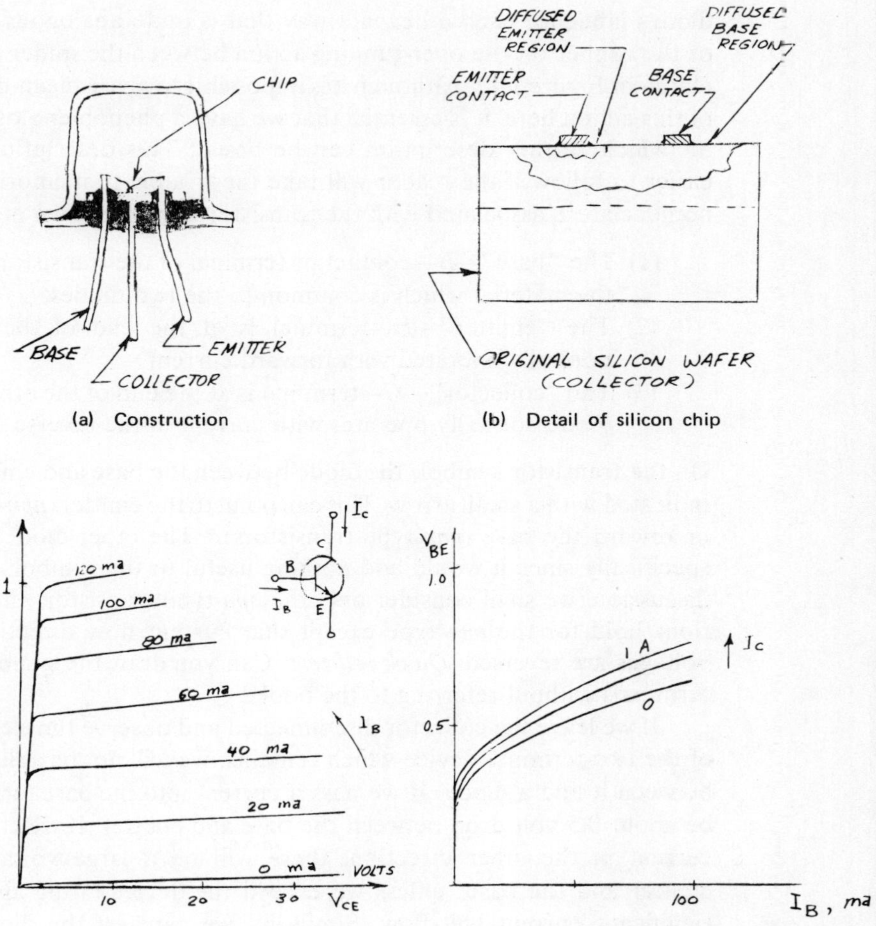

(a) Construction (b) Detail of silicon chip

(c) Collector terminal characteristics (d) Base terminal characteristics

From the Engineer's Sketch Book

Figure 1.25 A typical medium-power, diffused silicon *NPN* transistor.

unit may be as low as 25¢ or as high as several dollars, the average being about 50¢ or less.

A detailed section of one kind of semiconductor chip for use in transistors is shown at Figure 1.25(b). The original material might be *n*-type silicon. Through the use of masks and a process known as diffusion, an area near the top of the chip is converted to *p*-type and then a portion of this is converted back to *n*-type. The result is two diodes in intimate physical contact. These diodes influence each other in a way that is quite analogous to the influence of the runner on the over-running action between the spider and drum of the clutch in Figure 1.22. Although it is impossible to give a meaningful description of this action here, it is essential that we have a phenomenological description on which further description can be based. This description will be much easier to follow if the reader will take the trouble to memorize the following nomenclature associated with the transistor symbol shown at (c):

(1) The "base"—*B*—contact or terminal of the transistor is connected to the material which is common to the two diodes.
(2) The "emitter"—*E*—terminal is at the end of the diode which is normally operated with forward current.
(3) The "collector"—*C*—terminal is at the end of the other diode and this diode normally operates with current in the reverse direction.

On the transistor symbol, the diode between the base and emitter terminals is indicated with a small arrow. This can point to the emitter (*npn*-type transistors) or toward the base (*pnp*-type transistors).* The other diode is not indicated specifically since it would add nothing useful to the symbol. In the following discussion, we shall consider only the *npn*-type transistor; the same explanations hold for the *pnp* type except that current flow directions and relative voltages are reversed. *Quick self-test*: Can you draw the symbol and name the terminals without referring to the book?

If we leave the collector disconnected and observe the electrical behavior of the two-terminal device which remains, we will not be able to differentiate between it and a diode. If we pass a current into the base terminal, there will be about 0.5 volt drop between the base and emitter. If we attempt to force a current in the other direction, there will be a large voltage between the emitter and the base; unless we exceed the device rating and destroy it, no significant current will flow. Similarly, we can test the diode between the collector and base terminals.

Now, if we leave the base disconnected and try to force current to flow from the collector to the emitter, the collector-to-base diode will act to prevent the current; the base-to-emitter diode has no effect since it would permit current to flow in this direction. We may now observe the essential aspect of the operation of the transistor. If we continue to try to force current to flow from the collector to the emitter and, simultaneously, cause a current to flow from the base to the emitter, we find that the base current "disengages" the collector-to-base diode and allows it to pass a certain amount of current, in proportion

* The direction of the little arrowhead in the transistor symbol is easily recalled if we think of it as a compass pointing to the *n* portion of the transistor.

to the amount of base current. These effects are shown by the characteristic graphs at (c) and (d) of Figure 1.25.

In simple terms, we can summarize transistor behavior as follows:

(1) With no base current, a transistor is non-conductive.
(2) With enough base current, a transistor is conductive.
(3) The base-to-emitter voltage in a transistor is never large when the base current flows in the forward direction through the base-emitter diode.
(4) In the normal operating region, the base and collector currents are both inwards for an *npn* transistor and outwards for the *pnp* type.

It has been the policy, so far in this chapter, to discuss device imperfections and specifications by group instead of for each element. Here, however, such an approach is out of the question because of the amount of material. Therefore, we will consider some of these aspects in the transistor—thereby giving the reader a little further familiarity with it—and hope that the generalization to analogous components is obvious.

The heat which must be dissipated by a transistor is proportional, basically, to the greatest product of collector current and collector-to-emitter voltage that will occur steadily in a given application. Although there is some dissipation because of the base current and the base-to-emitter voltage, both are much smaller than the previously mentioned quantities. Therefore, their product is usually completely insignificant with respect to the other product. In many applications, the transistor is operated in such a way that it is either fully conductive or fully non-conductive. In this case, the heating is proportional to the so called "saturation voltage"—the collector-to-emitter voltage when more than enough base current flows—times the current level. Transistors for this application are constructed to have as small a saturation voltage as possible. The basic static specifications for transistors are then the heat dissipation ability, the maximum collector current, the maximum collector-to-emitter voltage, and the amount of base current required to pass the rated collector current. The leakage, under reverse voltage conditions, of both diodes may be important in some applications and is usually specified by manufacturers. Also, the maximum reverse base voltage which may be applied is required for some designs. Transistors tend to be very temperature sensitive and many specifications include information on this matter.

As with diodes, transistors exhibit dynamic aspects which must be known in some applications. The kind of capacitance discussed in the diode description is the major source of this.

The functional dual of a transistor would be a device in which the amount of dissipation was controlled by means of a non-conductive terminal pair.

The vacuum tube is exactly this; the field-effect transistor is its semi-conductor equivalent. These components are very important when a weak physical process must be observed with an electronic instrument since they allow much higher power gain than the common transistor. An example would be in the first stage amplifier of a thermocouple termperature-measuring system. Here, we wish to know the voltage without drawing any current from the thermocouple. There is no way to operate a normal transistor without current so we must adopt something with a non-conductive control. On the other hand, there are some physical processes which produce currents at no significant voltage; a non-conductive control device would be a poor choice here.

Many applications of transfer variable dissipators are discussed in Chapter 2. Therefore, we will not mention any examples here.

1.11 Transfer Exchangers

A common problem in system engineering is to make a certain source and load operate effectively together even though they are not directly compatible. Assuming the power levels and energy forms are the same, a device such as a gearbox, transformer, or hydraulic intensifier (see Figure 1.26) can be applied. In all of these, there is very high efficiency and a fixed ratio between the terminal variables at the two sides of the component. Unfortunately, the transformer is not a perfect analog of the gearbox, since it cannot operate under steady dc conditions. (See the limitation described on the figure; this is a result of the flux saturation of the iron.) Incidentally, the two dots on the transformer symbol indicate the ends of the two windings which simultaneously change in the same direction of relative voltage; also, if the inward current at one dotted terminal is increased, the inward current at the other will increase. Although transformers are most frequently applied to systems with constant frequency sinusoidal excitation, they are also used to change the voltage and current levels in pulse systems and in some audio-frequency systems.

The hydraulic analog of a lever is a two-area ram arrangement such as is found in the brake system of an automobile. This also suffers from an inability to operate continuously. Since there is a need for a continuously operating device, the "intensifier" shown at Figure 1.26(c) and (d) has been developed. In this unit, the piston strokes back and forth, at the end of each stroke moving a "hydraulic limit switch." This switch controls a snap-action, pilot-operated, reversing valve which allows the stroke to reverse. In this unit, the product of the supply pressure, P_1, and the supply flow, W_1, is approximately equal to the product of the outlet pressure and flow ($P_2 W_2$). The check valves steer the delivered flow so that it always flows the same direction in the load. The insightful reader will recognize that a perfectly analogous network could be used to

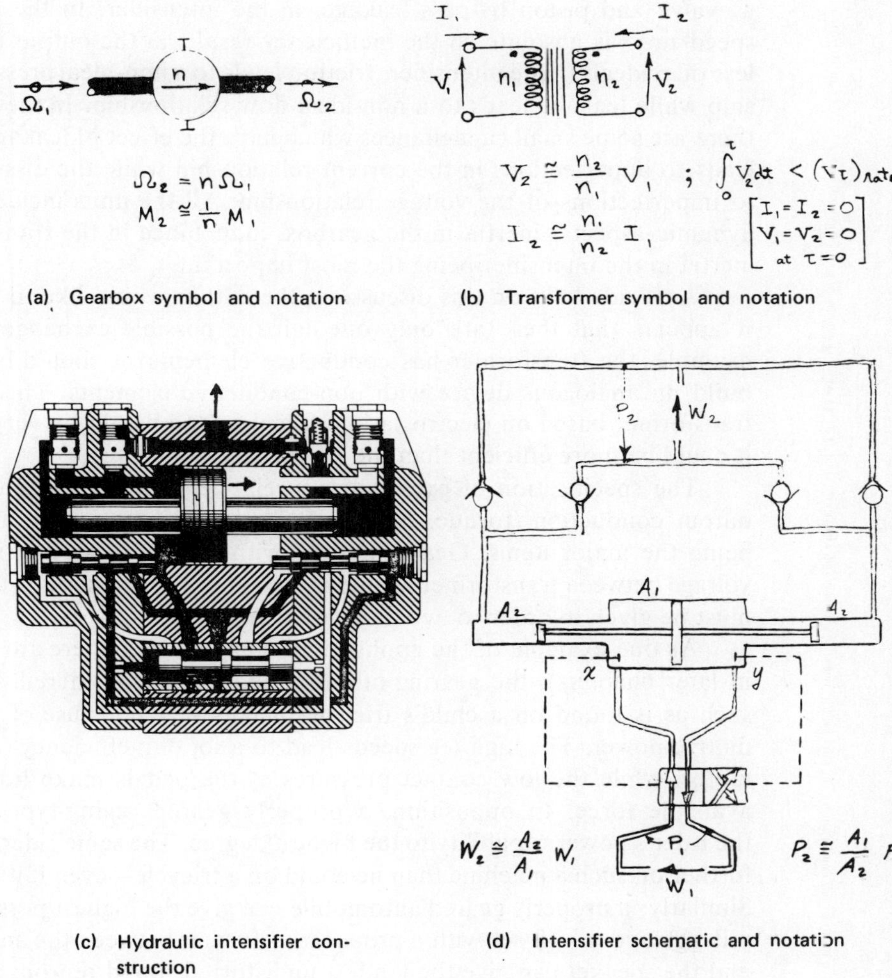

(a) Gearbox symbol and notation

(b) Transformer symbol and notation

(c) Hydraulic intensifier construction

(d) Intensifier schematic and notation

From the Engineer's Sketch Book

Figure 1.26 Transfer exchanger elements.

allow a transformer to operate on dc power. Also, he might note that the teeth on gears carry on a function which is similar to that of the stroking piston; they overcome the inherent inability of a lever system to operate continuously.

The major imperfections in this group of devices are the various inefficiencies: bearing and geartooth friction in the gearbox; coil resistance and eddy current losses in the transformer; mechanical and fluid friction, as well as valve and piston by-pass leakage, in the intensifier. In the gearbox, the speed ratio is absolute so the inefficiency results in the output torque being less than ideal. In the intensifier, friction leads to a non-ideal pressure relationship while leakage leads to a non-ideal flow relationship. In the transformer, there are some small capacitances which have the effect of leakage paths; this leads to imperfections in the current relationship while the dissipation leads to imperfections of the voltage relationship. All the units include undesired dynamic aspects; inertia in the gearbox, inductance in the transformer, and inertia in the intensifier being the most important.

As a side light on this discussion, the author would like to mention that it appears that these are only one-half the possible exchanger forms. For example, the transformer has conductive elements; it should be possible to build an analogous device with non-conductive elements. This would be a transformer based on electrostatic principles, and would be very useful since it could be more efficient than the magnetic kind.

The specification of exchangers is relatively simple, the exchange ratio, output conduction (torque, current, flow rate), output power, and efficiency being the major items. Gearbox speed with respect to an inertial reference, voltage between transformer windings, and absolute pressures in the intensifier must be given in order to avoid destructive conditions.

As one example of the application of exchangers (there are several more in later chapters), the gearing of a bicycle may be considered. Direct drive, such as is found on a child's tricycle, makes very poor use of the available motive power. The high leg speeds lead to a lot of inefficiency of the organic motor, while the low contact pressures at the pedals make little use of the available force. In opposition, a properly geared racing-type bicycle uses the rider's power capability to the highest degree. The same rider can go much further on such a machine than he could on a tricycle—even if it were his size. Similarly, a properly geared automobile can give the highest possible gasoline mileage and a hi-fi set with a proper transformer between the audio amplifier and the speaker can give the loudest undistorted sound reproduction.

1.12 Positive Converters

A class of components which can be particularly useful combine the exchange function described in section 1.11 with the conversion from one energy form to another. We discuss a few examples, very briefly, in this section.

An electromagnetic unit which may be operated either as a motor or as a generator is shown in Figure 1.27. The device is known as a unipolar motor (or generator) and utilizes a purely radial magnetic field as shown at (b). This field may be created with a permanent magnet or with a separate exciting coil;

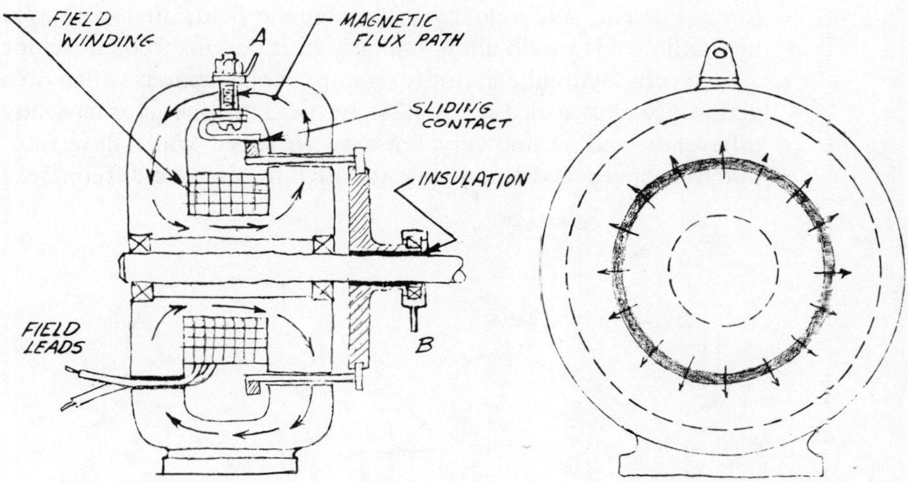

(a) Axial cross section

(b) Front view with rotor removed

From the Engineer's Sketch Book

Figure 1.27 Unipolar generator/motor.

the field structure is toroidal in nature. The air gap, through which the rotor moves, is cylindrical and as short as practical. The rotor is simply a conductive cylinder to which contact is made by slip rings at each end. In this machine, dynamic effects are small due to simplicity and the lack of commutation; the torque is almost perfectly proportional to the rotor current and insensitive to speed. The voltage is almost exactly proportional to speed and very insensitive to current, since the rotor resistance is extremely small (well-designed, heavy slip rings are required on high-performance machines of this nature). Although both torques and voltages are small, speeds and currents can be high and the efficiency good. One application of the generator is in railroad rail-test cars; these pass extremely large currents through the rail as they move along. If there is a rail defect, a voltage change occurs. Since the rail takes high current at very low voltage, this generator is perfectly matched to the application.

The lack of commutation makes this motor unique among the whole group of electrical and hydraulic motors. To the best of the author's knowledge, this is the only device which provides continuous rotation without either internal commutation or external timing such as sinusoidal excitation. The physical fact that makes this possible is the orthogonal relationship between force, current, and velocity in a magnetic field; analogous relationships in hydraulic fields would allow similarly neat designs to be developed.

Several hydraulic motor (or pump) configurations which do have positive action are shown in Figure 1.28. In these, torque goes basically as pressure difference and is not very sensitive to speed while flow rate goes pretty much as speed and is not very sensitive to pressure difference. Friction and

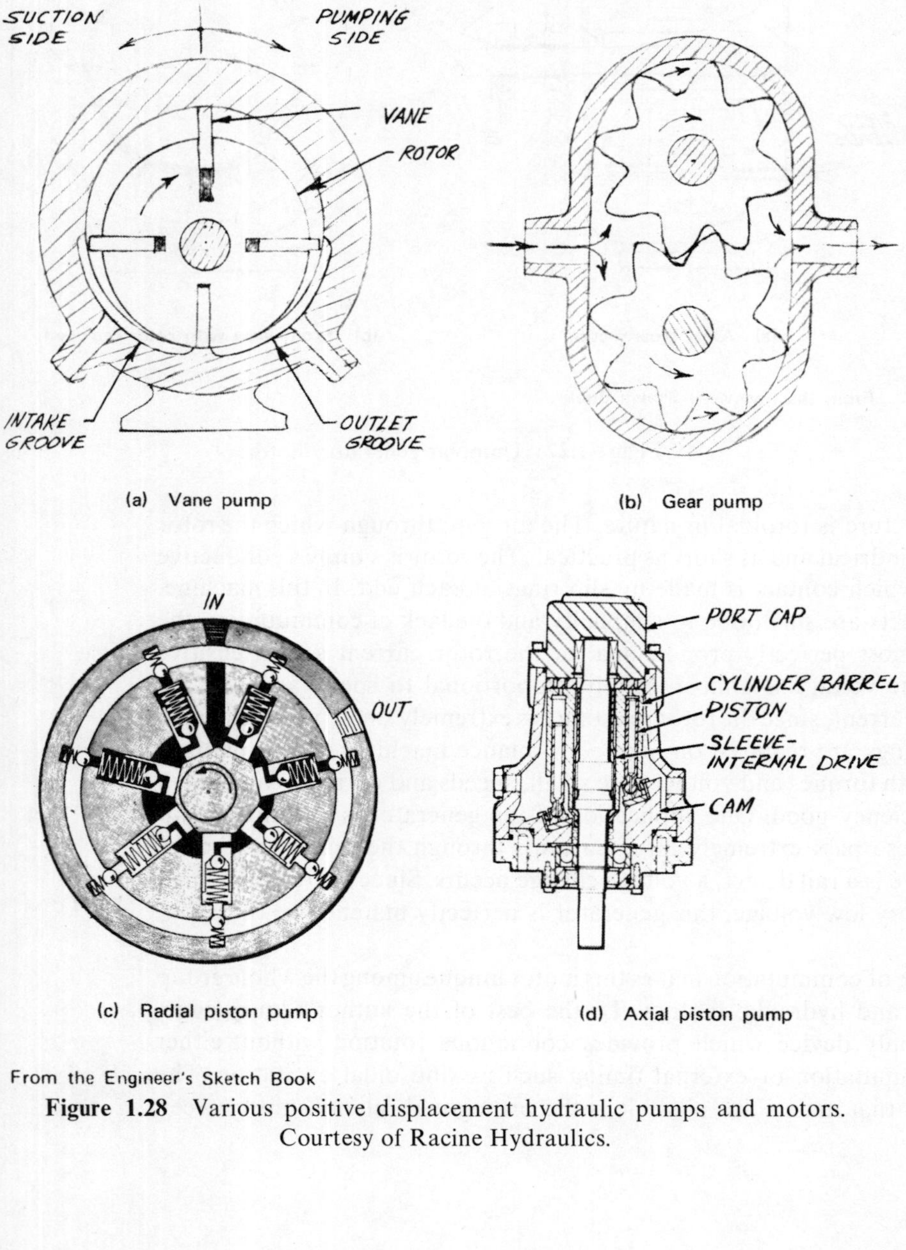

From the Engineer's Sketch Book

Figure 1.28 Various positive displacement hydraulic pumps and motors. Courtesy of Racine Hydraulics.

leakage make both of these idealizations somewhat imperfect, however. Except for the axial piston pump at (d), operation of the units is obvious. In that pump, the cam has a swash plate action which causes the pistons to stroke up and down. Porting in the "port cap" provides the necessary alternating connection of each cylinder to the inlet and outlet lines.

1.13 Free Converters

In many electrical/mechanical and hydraulic/mechanical converters, the exchanger-like performance of the units in section 1.12 is not obtained. For example, in a shunt-would dc motor, the stall torque is proportional to current, but drops off rapidly as the speed is increased.

A useful kind of converter of this nature, for application to hydraulic systems, is shown in Figure 1.29. This is a vane-type pump, but the ring in which the vanes move is supported by a spring. When the pressure in the system is small, the spring moves the ring hard over onto one edge of the rotor. This results in maximum displacement or flow rate. At high pressures, the ring is forced toward the center where the displacement and flow rate are minimized. By using different spring constants, different shapes of flow versus pressure curves may be obtained. It should be noted that this is a non-dissipative approach.

The construction and shaft characteristics, at standard voltage, for a

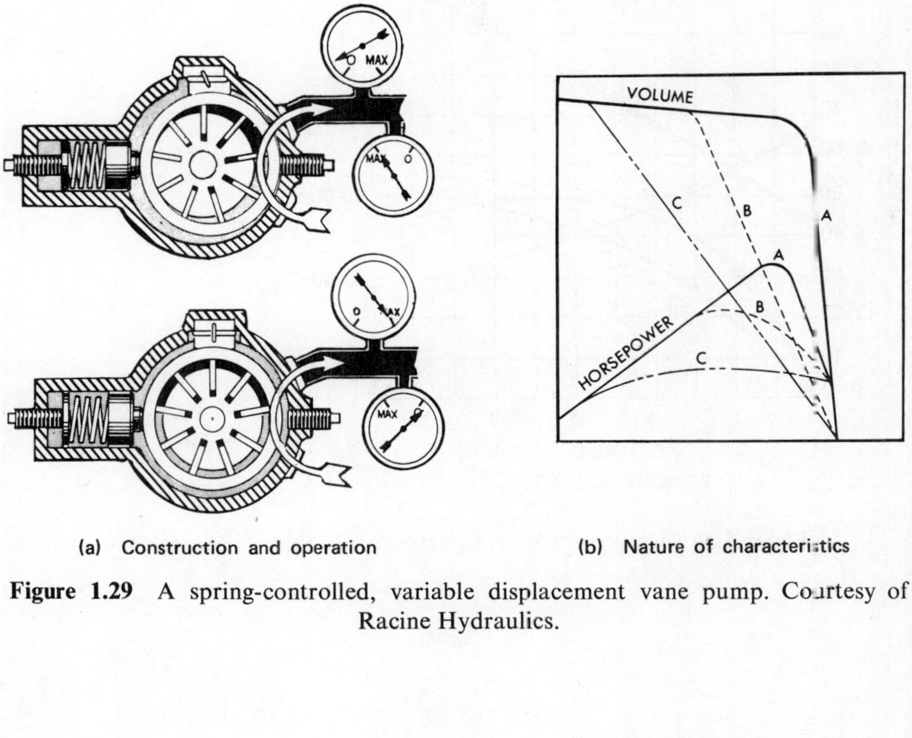

(a) Construction and operation (b) Nature of characteristics

Figure 1.29 A spring-controlled, variable displacement vane pump. Courtesy of Racine Hydraulics.

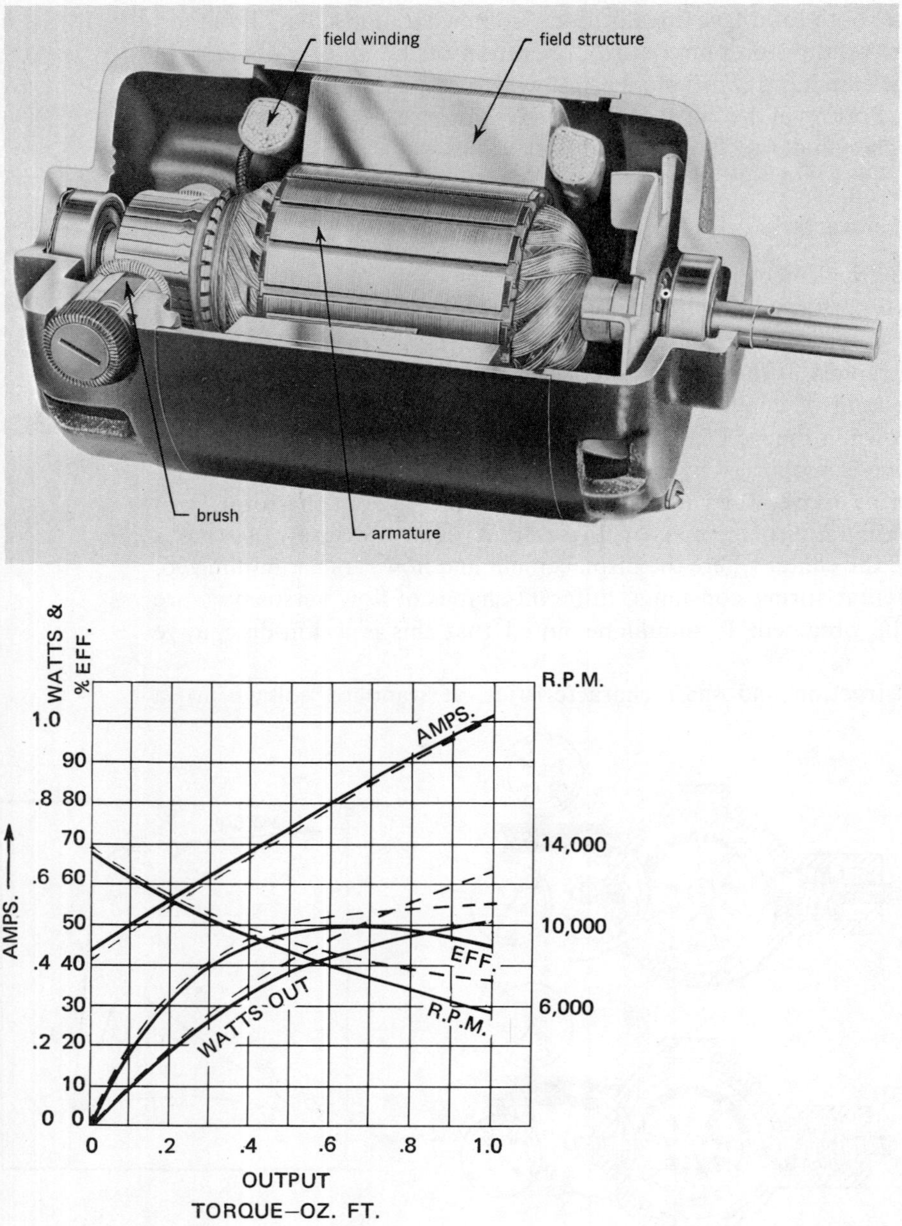

Figure 1.30 Universal motor. Courtesy of Robbins & Meyers, Inc.

typical kind of electrical motor are shown in Figure 1.30. More attention will be paid to this kind of device in Chapter 3.

A group of devices which may be used as the starting point for motor designs is shown in Figure 1.31. Note that the problem is essentially that of timing; all these rotary actuators have dead-center positions and opposed torque directions on each side of dead center. If we provide a switching arrangement which opens and closes the excitation path of the device according to its position relative to dead center, we build a "commutator" type of motor. If we turn the excitation on and off at some periodic rate, we build a "synchronous" machine; we might think of this as externally commutated. In this case, the motor speed relates to the external switching rate.

Other problems of motor design, such as getting maximum output power per unit weight and achieving maximum efficiency, lead to the various motor construction arrangements which we actually see in practice. An interesting exercise is to take all the actuator forms shown in Figures 1.20 and 1.21, convert them into their rotary forms, and then imagine how commutated motors could be constructed from them. The all-mechanical devices which result are particularly instructive.

1.14 A Unified Component Symbolism

In order to simplify the schematic diagrams in the remainder of the book and provide the reader with a summary of the categorization and discussion of this chapter, the comprehensive table of symbols shown in Figure 1.32 has been prepared. Some departures from convention, made in the interest of readability and consistency here, have been made. These are discussed in the notes immediately following the figure. Perhaps the most debatable usage here is in that for the various kinds of conductors (shaft, wire, pipe); a single line is adequate for each and would be used if only a single medium were being considered. Here, the author felt that the diagrams would be clearer with the three media identified in this way.

The reader may be required to use some imagination from time-to-time in later portions of the text; there are combinations and permutations of these symbols which the author uses without further discussion or specific definition. It is hoped that the general scheme of the mechanical system, which is totally non-conventional, will seem reasonable. (No universally accepted convention exists, unfortunately.) In Chapter 2, the emphasis is on the use of analogous components in analogous networks; the categorization of this chapter, as exhibited by the symbol chart, forms the basic background for this work.

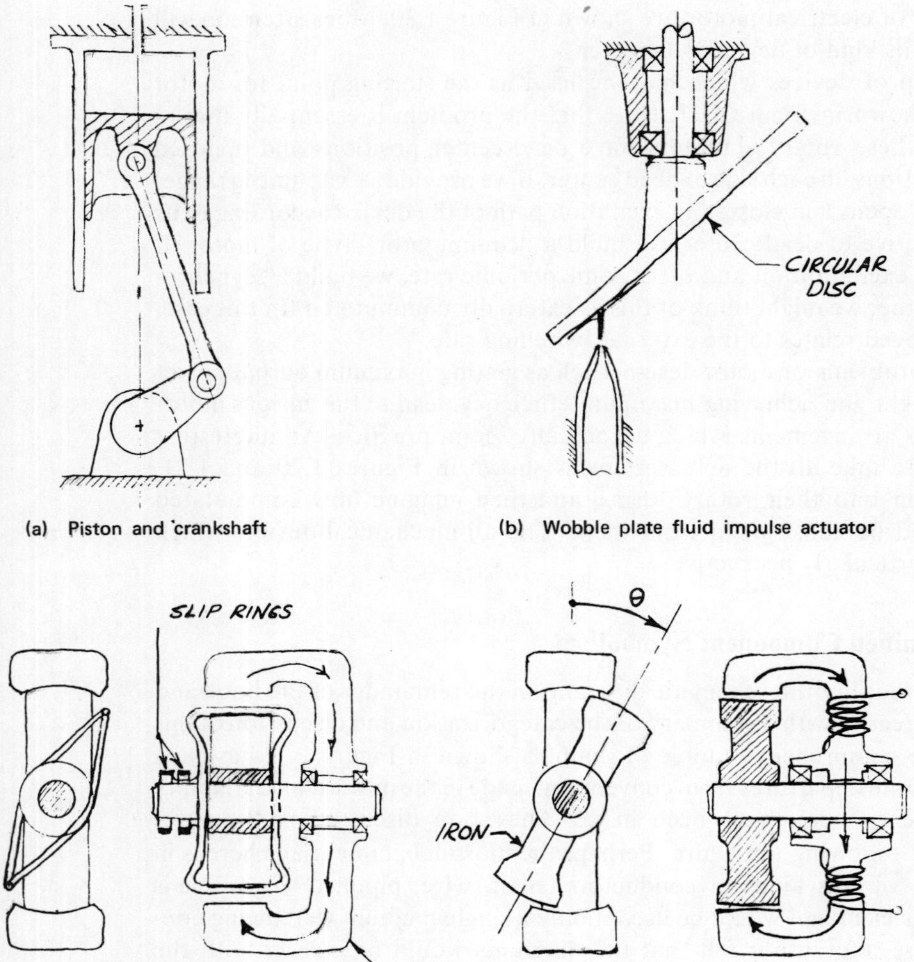

(a) Piston and crankshaft

(b) Wobble plate fluid impulse actuator

(c) Coil type electromagnetic actuator

(d) Variable reluctance type rotary actuator

From the Engineer's Sketch Book

Figure 1.31 Rotary actuators from which motors may be constructed.

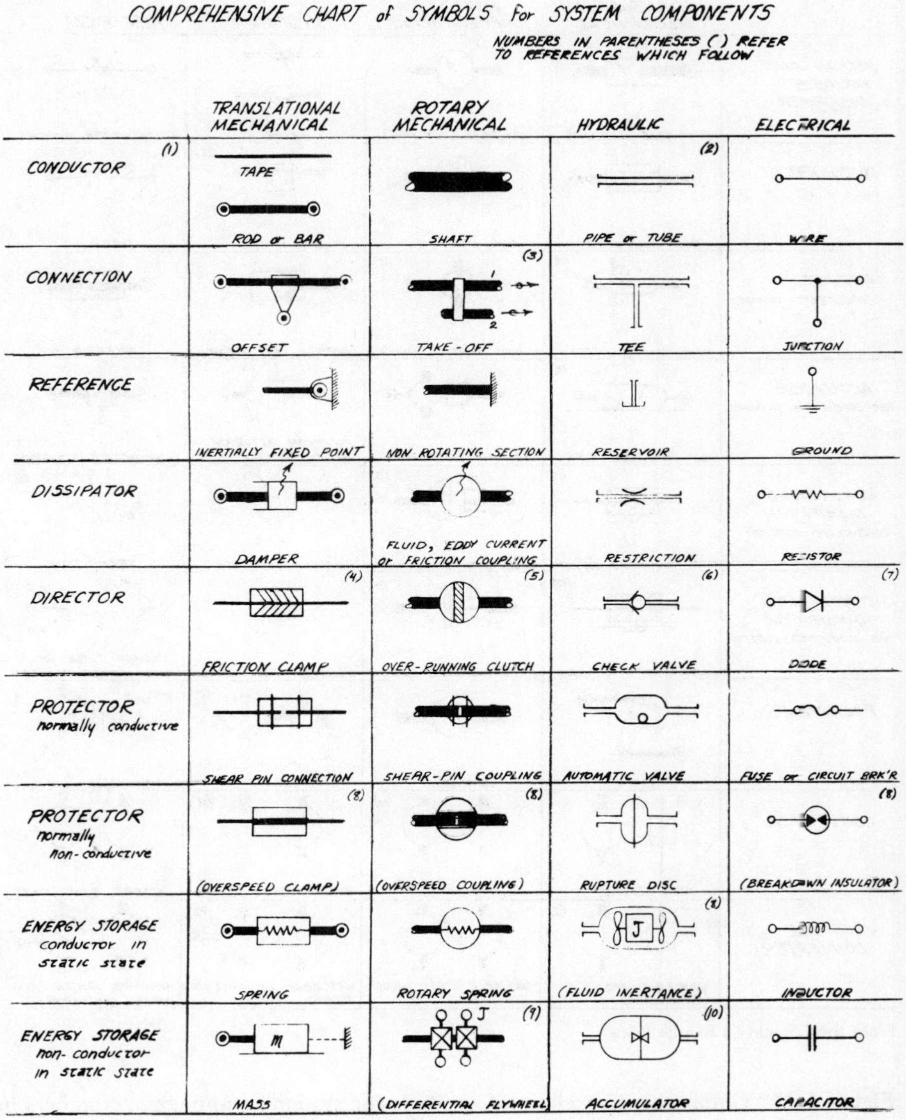

From the Engineer's Sketch Book

Figure 1.32 (*continued overleaf*).

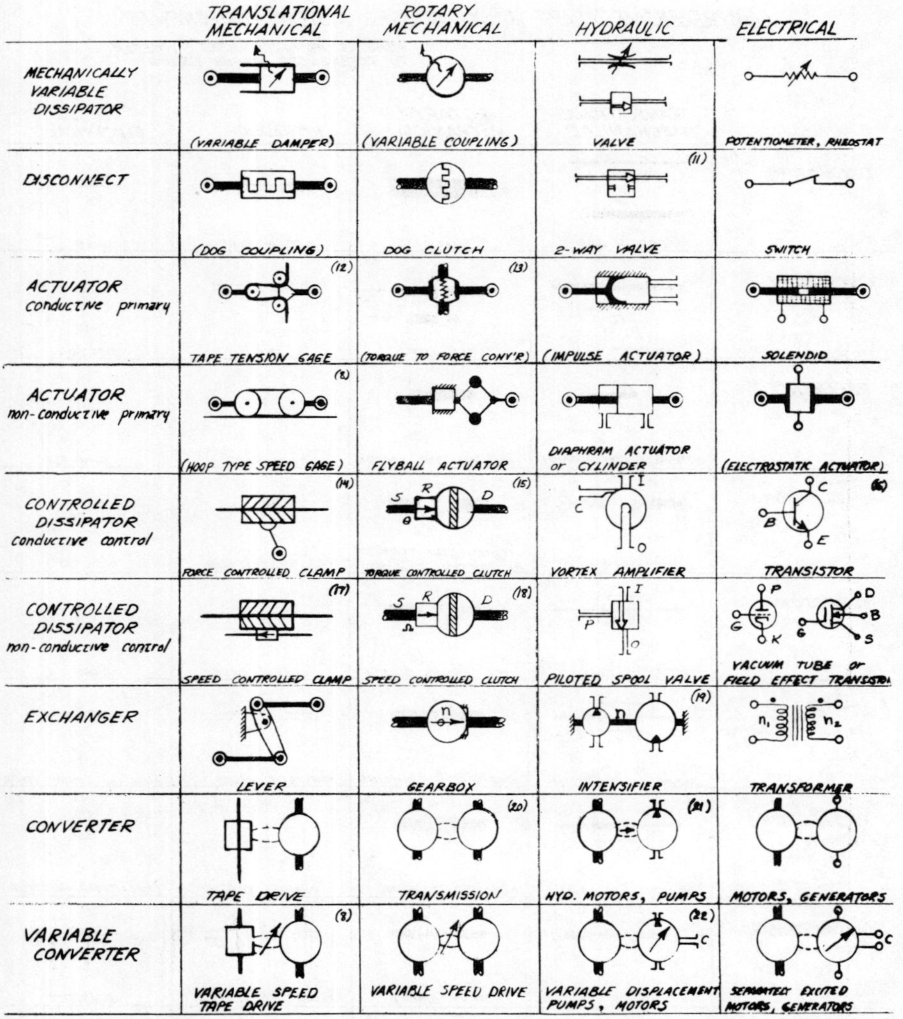

From the Engineer's Sketch Book

Figure 1.32 Comprehensive chart of symbols for system components. Numbers in parentheses refer to notes in the text which follows.

(1) A single, solid line may be used to represent any conductor in a schematic where only one medium appears.

(2) A single line has been adopted as the hydraulics industry standard representation of a pipe or tube; all the symbols shown here differ from the standards in that respect.

(3) The speeds of shafts 1 and 2 are identical, independent of the shaft diameters.
(4) The band may move to the right with respect to the clamp body.
(5) The symbol indicates a clutch capable of transmitting tightening torque; the direction of the hachures is reversed for one which transmits loosening torque.
(6) Fluid flow from left to right is permitted.
(7) Current flow from left to right is permitted.
(8) This device is not commercially available and probably of little value; it is included here in the interest of completeness. The assigned name is arbitrary.
(9) The symbol for a standard flywheel, with inertial reference, shows only a pair of weights on arms connected to the shaft.
(10) The majority of commercially available accumulators have only a single connection which is active in circuits; the symbol is then oriented vertically and a single small triangle indicates liquid level.
(11) Hydraulic valve symbols have been designed so that we may think of the box as moving back and forth perpendicularly to the pipe direction; various kinds of actuators may be indicated. There is some confusion in the system over variable dissipators and valves with only two useful positions. Here, if the valve can act as a variable dissipator as well as a flow-steering device, the wiggly dissipation symbol is added to the standard representation.
(12) The primary is a tape moving in the vertical direction.
(13) The actuator direction has been rotated 90° from that of the construction shown in the text in order to facilitate use of the symbol.
(14) The clamp is normally disengaged; a leftward force on the lever arm tends to brake the tape from moving to the right with respect to the clamp body.
(15) The clutch is normally disengaged. Torques which produce rotational displacements of the runner with respect to the spider, in the direction indicated by the arrow, tend to make it like the tightening torque transmitter type of over-running clutch. Either control or transmission direction can exist, yielding four different symbols.
(16) This symbol is for the *npn*-type transistor; the arrowhead points from the emitter to the base in the symbol for the *pnp* type.
(17) The clamp is normally disengaged; motion of the control tape to the left with respect to the clamp body tends to engage the clamp. Four forms are possible.
(18) Rotation of the runner with respect to the spider in the direction indicated by the arrow tends to engage the clutch. Four versions are possible (see 15 above).
(19) The small triangle inside the circle indicates the flow direction of the higher pressure fluid; one unit acts like a pump and the other like a motor.
(20) This would be a shorthand symbol for hydrostatic or electrical transmissions joining two physically separated mechanical networks.
(21) The small arrow between the units may be used to indicate irreversible machines. In this, the arrow shows that power may be transmitted from left to right, but not vice versa.
(22) A separate pair of connections, which may serve to vary the displacement, has been indicated.

Problems

1. Would the performance of the couplings in Figure 1.1 be improved by addition of thermal insulation over the outside surface?

2. Suppose that an eddy-current coupling operating at 10 lbs-ft torque and 1000 RPM slip speed is undesirably hot. Do you think a clever inventor might be able to change the internal design of the coupling in such a way as to reduce the amount of heat being generated?

3. Design a rotary mechanical coupling based on the hysteresis of flexure in rubber materials; i.e., the friction of rolling of a deformed rubber cylinder.

4. Design a rotary mechanical coupling based on the power loss of a fluid being forced through an orifice.

5. Would the curves of torque versus slip-speed for the couplings of Figure 1.1 go through the origin? If yes, do they go through exactly, or within "engineering accuracy"?

6. Is it possible to revise one of the couplings of Figure 1.1, or devise a new design, which would operate with the reverse function; i.e., when heated, create a relative rotation of the shafts?

7. Calculate the power dissipated by the No. 200K coupling of Figure 1.2 for slip speeds of 200, 400, 800, and 1600 RPM (use horsepower units). Is the torque peak at about 500 RPM also the power peak?

8. Is there any relationship between the hydraulic dissipator of Figure 1.3 and the fluid coupling of Figure 1.1(b)?

9. Determine the power rating required for each of the following resistor applications:

a. 250 ma (milliamps), 1 volt.

b. 300 ma, 2 volts.

c. 300 ma and 10 KΩ(1 KΩ = 1000 ohms)

d. 100 Ω at 3 amps.

e. 10 KΩ at 600 volts.

Note: Resistors are available in the following power ratings in this general range: $\frac{1}{4}$, $\frac{1}{2}$, 1, or integral numbers of watts.

10. Sketch a constant torque coupling based on spring-loaded friction material shoes running on a drum.

11. A group of two terminal devices which would be particularly useful would be the following:

a. Rotary mechanical coupling with constant slip speed.

b. Hydraulic valve with constant flow rate.

c. Electrical device with constant current.

Why do you think constant torque, pressure, and voltage device are so common and these so uncommon, if not actually unrealizable?

12. A kind of director element is used on the pull-cord of venetian blinds. Inspect one and make a sketch of the design. Another translational mechanical director is used on tent ropes and aids in tightening them. It consists, basically, of a stick with holes in the ends. Make a sketch showing how it works.

13. An extremely simple form of rotary mechanical director consists of a close-fitting spring, as the driver, slipped over the driven shaft. Does this transmit tightening or loosening torque if the spring is wound as a right-hand screwthread.

14. Draw the torque versus slip-speed relationship of a perfect rotary director. Also, show how real units differ from the ideal. Does one or both curves go exactly through the origin?

15. In Figure 1.7(d) the manufacturer of a rotary director describes the "sense" (tightening or loosening) of the unit by referring to the allowable motion instead of the torque transmitting ability. Devise a pair of schematic symbols for rotary directors, showing side views exclusively, which clearly differentiate the two kinds of unit, using allowable rotation—instead of torque—as the determinate aspect.

16. Does the human heart contain check valves? If it does, sketch the general construction.

17. Will the diode of Figure 1.9 heat up when conducting 6 amps forward current? If so, what is the power which must be dissipated?

18. A common problem with the application of diodes is insufficient reverse voltage capability. Diodes may be used in series to increase the capability, but do not always share a load equally due to minor internal differences. This can lead to destruction of one, then the other. Show how two diodes can be forced to share a reverse voltage evenly by using two high-valued resistors, at the expense of increased "leakage" current.

19. Design a coupling which protects against overspeed by using friction to melt a tin-lead alloy slug.

20. Can normally conducting and normally non-conducting protectors be used simultaneously to give added assurance against destruction of a single valuable system component? How would this be done for:

a. The electrical portion of a motor?
b. A turbine, to protect against overspeed?
c. An hydraulic actuator, to protect the fluid side?

21. A number of effective protectors rely on indirect mechanisms for the actual accomplishment of switching between the conducting and non-conducting states. For example, current flowing in a coil may move a plunger upon reaching a certain level; the plunger may open a switch and thereby turn off the current. Sketch schematic arrangements for the following indirectly operated protectors:

a. An overspeed protector in which a permanent-magnet field induces voltage in a coil and this voltage triggers the operation of a brake.
b. An over-flow protector in which a propellor generates torque and this torque, upon reaching a predetermined value, throws a toggle valve. [See Figure 1.19(d)].
c. An over-voltage protector where a piezoelectric beam is deflected, in proportion to the voltage, until it closes a contact which, in turn, activates a relay.

22. For each of the energy storing devices of Figure 1.14, what material property sets the limit of static conductivity?

23. For each of the devices of Figure 1.15, what sets the limit of non-conductance?

24. For the devices of Figures 1.14 and 1.15, what is the basic source of unwanted dissipation?

25. Some buses in Switzerland operate from the energy stored in a flywheel, instead of from an engine of some sort. The flywheels are revved up at the end of each (relatively short) run. How many horsepower-hours are stored in a steel flywheel, 3 ft in diameter and 6 ft long (not the specifications of those in the buses) turning at 8000 RPM?

26. Is an automobile brake a "mechanically variable dissipator"?

27. Prepare a characterization of the valve shown in Figure 1.18 so that flow rate is given as a function of pressure drop, with "turns open" as a parameter.

28. Under certain conditions, each of the devices of Figures 1.20 and 1.21 can be used as a gage. List the quantity which would be measured by each. What conditions would have to be enforced on the plunger of each device in order to make the gage accurate?

29. Doing the foregoing problem in reverse, we can ask "what kinds of actuators can be obtained from the most common kinds of gages?" For example, what kind of actuator is:

a. The Bourdon tube?
b. The D'Arsonval movement?
c. The drag-cup of a tachometer?

30. What kind of actuator does the following jobs in your car:

a. Operate brake light switch?
b. Advance spark under power load?
c. Connect generator to battery when needed?
d. Open radiator connection when engine gets warmed up?
e. Operate brakes?
f. Engage starter and ring-gear?

31. The old-fashioned radial engine for aircraft suggests a geometry for a variety of motors. Show how electromagnets could be used in a radial arrangement to produce continuous rotation. Show the analogy between valving of the gas engine and switching (commutation) of the electromagnets.

32. Redesign the following actuators so that they pull instead of push, or vice versa:

a. Figure 1.20(b). c. Figure 1.21(a).
b. Figure 1.20(c). d. Figure 1.21(b).

33. Design a three-terminal controllable clutch based on:

a. The eddy-current dissipator principle.
b The fluid shear principle.
c. The disc brake principle.

34. Most hydraulic spool valves are arranged to carry out several functions at one time. Show how a single spool can open a path between the inlet and one outlet while simultaneously closing a path to another outlet.

35. For the transistor of Figure 1.25,

a. Graph I_C versus I_B for V_{CE} 10 volts.

b. Draw, on the I_C versus V_{CE} characteristic, the parabola which indicates 5 watts dissipated power.
c. Determine the power dissipated due to a base current of 100 ma when the collector-to-emitter voltage is 10.

36. For the hydraulic intensifier of Figure 1.26(c) and (d), indicate by schematic diagrams the valve conditions and flows for the piston:
a. Stroking right.
b. Stroking left.
c. Stationary at full right stroke, accelerating left.

37. How does an impulse turbine relate to the actuator of Figure 1.20(b). Does this type of turbine have the "commutator" action suggested as essential in section 1.13 of the text?

2 Some basic engineering systems

Much of system engineering is devoted to the control of power transfer. In some instances, such as in the selection of a gearbox, the engineering objective is to maximize the power which can be transferred from a motor to a machine. In others, it may be necessary to control the voltage at which electrical power is delivered or the pressure of hydraulic power. Similarly, it may be necessary to convert from reciprocating to steady power transmission, to time the periods during which power is transmitted, or to influence power transmission in such a way that information can be carried. Since these needs arise so frequently, it is useful to be acquainted with the basic techniques for their accomplishment. Beyond this practical objective, the purpose of this chapter is to explore further the very important relationships between electrical, mechanical, and hydraulic systems. Among other examples, the conceptual identity between certain voltage, pressure, and speed regulators will be discussed.

The role of the kind of system to be discussed is indicated schematically in Figure 2.1. Here a "source," which converts fossil fuel, falling water, or some other primary energy form to that required by a "load," is connected

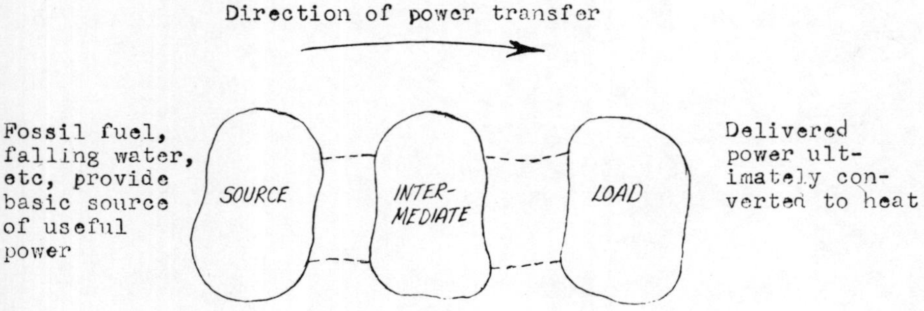

From the Engineer's Sketch Book

Figure 2.1 Showing the role of the "intermediate" system.

to the intermediate system in which we are interested. The load, also connected to the intermediate, carries out the particular objective of the total system and ultimately converts the power it receives to heat or stored energy. Of course, it is possible that other sources and loads may interact with the system; however, consideration of the indicated system will accomplish the objectives of this chapter.

Although the emphasis of this chapter is on the configuration and operating principles of a selection of intermediate systems, an attempt is made to point out the design considerations of most consequence. The kind of knowledge we must have of the source and load characteristics, before we can make a sensible start on the design of the intermediate system, is pointed out. Actual design examples and exercises regarding these systems, however, are reserved until later chapters where the necessary analytical techniques are developed. In particular, Chapter 4, which is concerned with the static analysis of systems, continues the discussion directly.

2.1 Conversion between Periodic and Continuous Power

Typical of the frequently encountered problem of converting from periodic to continuous power is the design of a battery charger. In this it is convenient to utilize the alternating current electrical power delivered to most homes and shops; however, the charger must deliver direct current to the battery. The well-known rectifier network shown in Figure 2.2(a) is often applied in this case, there usually being a transformer between the ac leads and the network in order to establish the proper voltage. Also shown in Figure 2.2 are several analogous networks constructed from *directors* from other power media.

Referring to the check valve arrangement shown at (b), the operation of this network form can be analyzed, for example, as follows: When the piston is forced to the left, so that the pressure in the left-side chamber exceeds that of the right, check valve D_1 opens while D_3 closes so that fluid is forced into the delivery line. The high pressure in the line holds D_2 closed while the suction in the right-side chamber causes D_4 to open, thereby supplying fluid to the chamber from the reservoir below. On the right-hand stroke, the roles of the check valves are reversed and fluid is delivered whenever the piston is in motion. Note that, at any instant, two of the valves are in the conducting state while the other two are non-conducting. The reader should analyze the action of the networks of (c) and (d) and note the high degree of analogy which exists between the four. Although reciprocating motion is normally converted to steady rotary motion by the use of an inverse slider-crank, the network of (c) is more generally applicable since reciprocating motion of any amplitude can be accomodated.

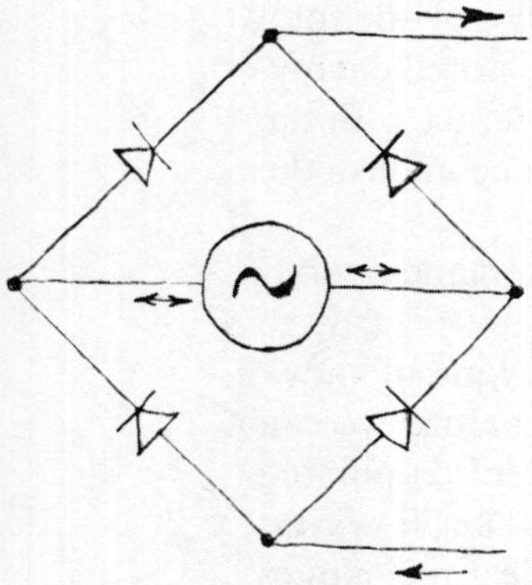

(a) The well-known full-wave rectifier

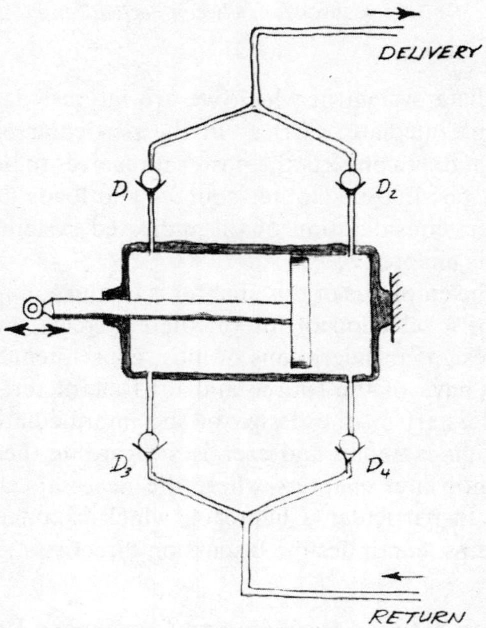

(b) Check valve arrangement for a double acting reciprocating pump

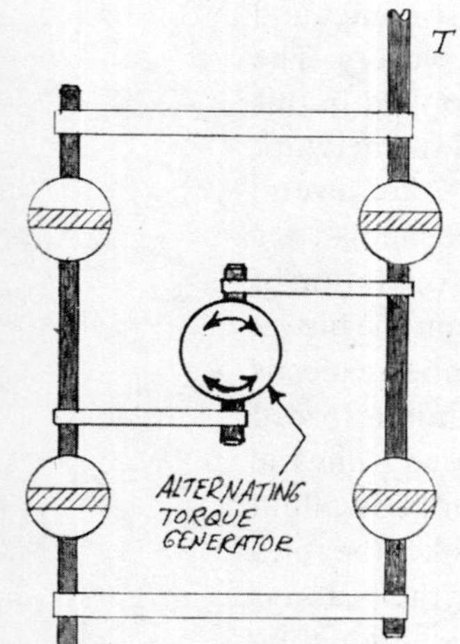

(c) Over-running clutch bridge which converts alternating torque to uni-directional torque

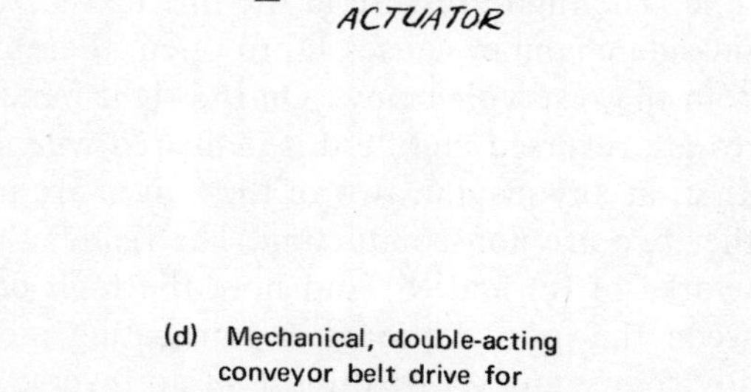

(d) Mechanical, double-acting conveyor belt drive for one-directional, intermittent motion situation

From the Engineer's Sketch Book

Figure 2.2 Some director bridge networks.

Selecting the correct size of director element for this kind of network is straightforward. We must insure that the element can withstand the greatest voltage, pressure, torque, or force to which it will be subjected, and that it will not overheat at the required levels of current, flow rate, or relative speed. Note that one specification arises from the conducting state while the other comes from the non-conducting condition. The efficiency of this conversion can be very high since director elements, in either state, have very small dissipation with respect to the product of their Γ and Δ ratings.

The inverse conversion, from continuous to periodic power, is seldom so readily or efficiently accomplished. A relatively straightforward approach can be taken where it is sufficient to simply turn the continuous power on and off at regular intervals. An example is the jackhammer; networks giving this kind of action are described later in section 2.2 which discusses timing systems. However, when more or less sinusoidal excitation is required, such as when a variable frequency, variable amplitude vibration test table is to be built, an "oscillator" network must be utilized. This group of networks is based on the fact that the differential equation

$$\frac{d^2X}{dt^2} + \omega^2 X = 0 \tag{2.1}$$

has a sinusoid as its solution. (If this is unfamiliar, it should be verified by substitution of sin ωt for X.)

The implication of the equation is that if we can build a network which is described by such an equation, the network will be capable of sinusoidal action. One very straightforward way of doing this—using computer components—is indicated in Figure 2.3(a). Here we use components which have a voltage output that is equal to the time derivative of the input, in volts per second. Two such devices have been connected in such a way that the voltage representing X is forced to be equal to the voltage representing d^2X/dt^2. Therefore, the circuit is described by an equation similar to (2.1); if it is started by the application of the correct initial conditions, it will continue to oscillate as predicted by the solution of the equation. There are some practical problems to resolve in the application of this approach (the amplitude will drift, due to computer imperfections, if it is not artificially stabilized; frequency adjustment is inconvenient), but it is representative of the concept on which most oscillators are built.

In order to apply this technique where large amounts of power are being handled, we would operate the oscillator itself at low power and use one of the amplifiers described in section 2.3 to provide the required power level. A somewhat more practical approach to oscillator construction is indicated in

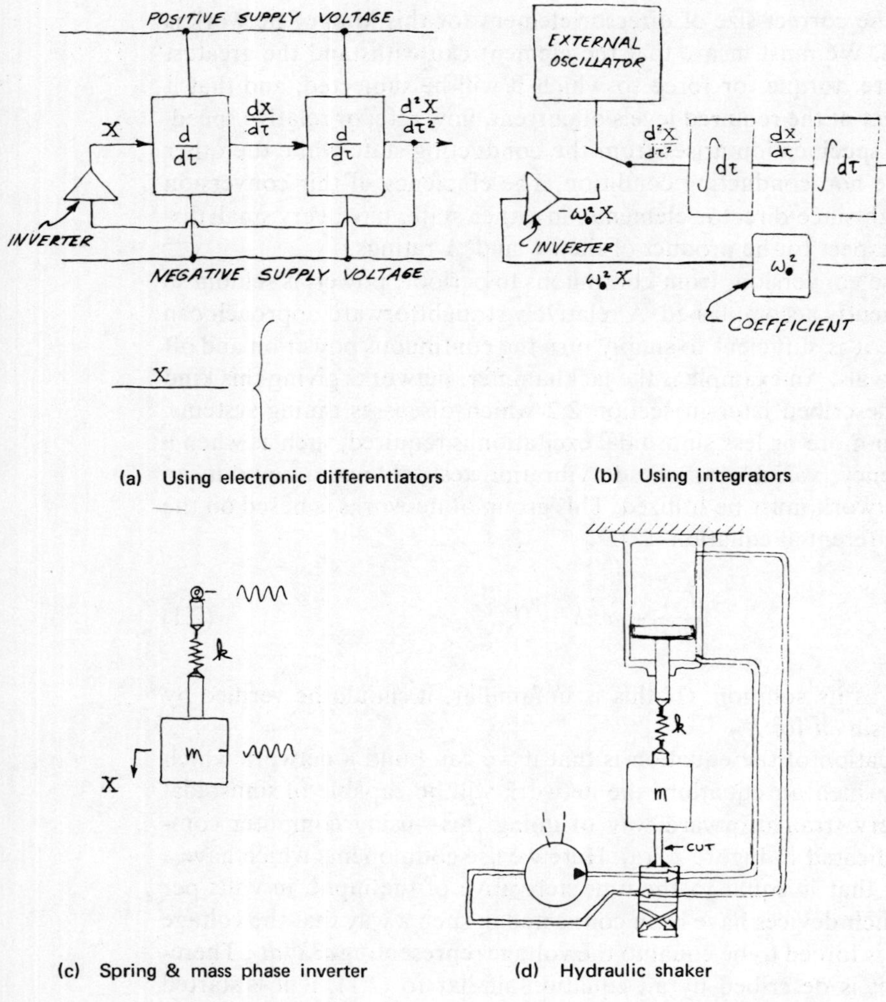

From the Engineer's Sketch Book

Figure 2.3 Oscillator principles and construction.

Figure 2.3(b) where the differentiators have been replaced by integrators—which are more readily constructed. The operation of the circuit can be explained on a rather intuitive basis as follows: Suppose that a separate oscillator is used to generate a sinusoid and this is made the input of the first integrator, the original circuit having been opened as indicated. If the frequency of the applied signal is greater than ω_0, the output of the first integrator will be of lower amplitude than that with ω_0 since

$$\int \sin \omega t dt = -\frac{1}{\omega} \cos \omega t \qquad (2.2)$$

Similarly, the output of the second integrator will be even smaller and it is obvious that, at this frequency, it is impossible for the signal $\omega_0^2 X$ to regenerate the oscillations, as it must if the circuit is to operate. By considering an applied signal of lower frequency than ω_0, it is easy to see that it is also impossible for this to be an operating frequency for the system. This leads to a simple criterion for the operating frequency of a system: If a system is capable of oscillation, the frequency is that at which, when one of the system's connectors is cut and then sinusoidally excited by an external source, the sinusoidal signal on the free portion of the cut connector is the same as the excitation. In control systems language, we say that "the oscillation frequency is that at which the open loop gain is exactly unity."

It is clear from study of these examples that the magnitude of the feedback signal ($\omega_0^2 X$ here) must change with frequency and the phase must be constant if we are to build an oscillator based on this principle. (Note that differentiators and integrators have a constant 90° phase shift; in the circuits of (a) and (b) the invertor adds another 180°, bringing the total to a constant 360° as required.) As will be shown later in Chapters 7 and 8, oscillation can arise in systems of a different nature than those described here. However, this is an effective and generally applicable route to the design of an oscillator, and it is the only one we will pursue here. Now let us try to design one, using only simple engineering components.

In Figure 2.3(c) is shown the familiar spring and mass combination from the theory of vibrations. The differential equation for this combination is

$$\ddot{X} + \frac{k}{m} X = 0 \qquad (2.3)$$

This is obviously the same form as (2.1) and must have a solution of $X = \sin \sqrt{k/m}\, t$. Of course, it cannot serve as an oscillator of the kind under consideration since it is not a power converter. It does, though, provide a starting point for the design of a proper oscillator.

Note that, if the top of the spring is oscillated below the natural frequency of the device, the mass moves in phase with the oscillation (when the top of the spring is going up, so is the mass, and vice versa) while excitation above the natural frequency yields an exact phase reversal. Simple experiments with a large rubber band and some kind of slug will verify this if the statement is found questionable. It is also known that, above the natural frequency, as we increase the excitation frequency the amplitude of the response decreases. It is possible then that, if we installed this combination in a system so that the spring was oscillated by the system output and the motion of the mass was coupled back to the controller of the output, an oscillator might be constructed. Obviously, such a system would require a power amplifier.

The system shown in Figure 2.3(d) has been developed around the foregoing argument. The hydraulic valve serves as the power amplifier and the actuator simply converts hydraulic power to mechanical. Suppose we cut the system where the mass is connected to the valve spool (we would probably hook the mass to a dummy valve operating a similar system, so as to retain the influence of the friction and other forces). By oscillating the valve spool at different frequencies, we could verify that the necessary phase relationship between spool motion and motion of the mass had been achieved and can find the proper supply pressure to get equal amplitudes. Because this is a relatively unsophisticated system, there would probably be some irregularities in the phase versus frequency graph, but it will average zero degrees over the range of interest if everything is in order. Finally, when the system is reassembled, we can expect the desired oscillations when the supply pressure is set as determined. Incidentally, this would not be an unreasonable approach to the design of a hydraulic shaker.

In evaluation of the system, we should first note that the efficiency does not compare to that of the inverse conversion. Here we have had to apply a variable dissipator (the hydraulic valve) and, in doing its work, it will waste considerable power as it throttles the flow between the full-open and full-closed positions. At maximum power transfer, when the cylinder pressure varies all the way from zero to supply pressure, the maximum theoretical efficiency is on the order of 50 percent. When higher efficiencies are required, a "kinematic" approach is usually adopted. For example, the foregoing problem would have been solved with a variable stroke piston pump while an electrical conversion might be accomplished by using a dc motor to drive an ac generator. These, however, can get complex and expensive. In some cases, the overall efficiency is not markedly greater.

2.2 Filters

Having accomplished the conversion of periodic power to continuous, it is frequently desired to smooth or filter the result, thereby removing—to some required extent—vestiges of the periodic action. It turns out that the networks used for this purpose transmit power well at low frequencies, but transmit poorly at high frequencies. Therefore, they are often referred to as "low pass filters." They are also applied in information transmitting systems and in the stabilization of systems.

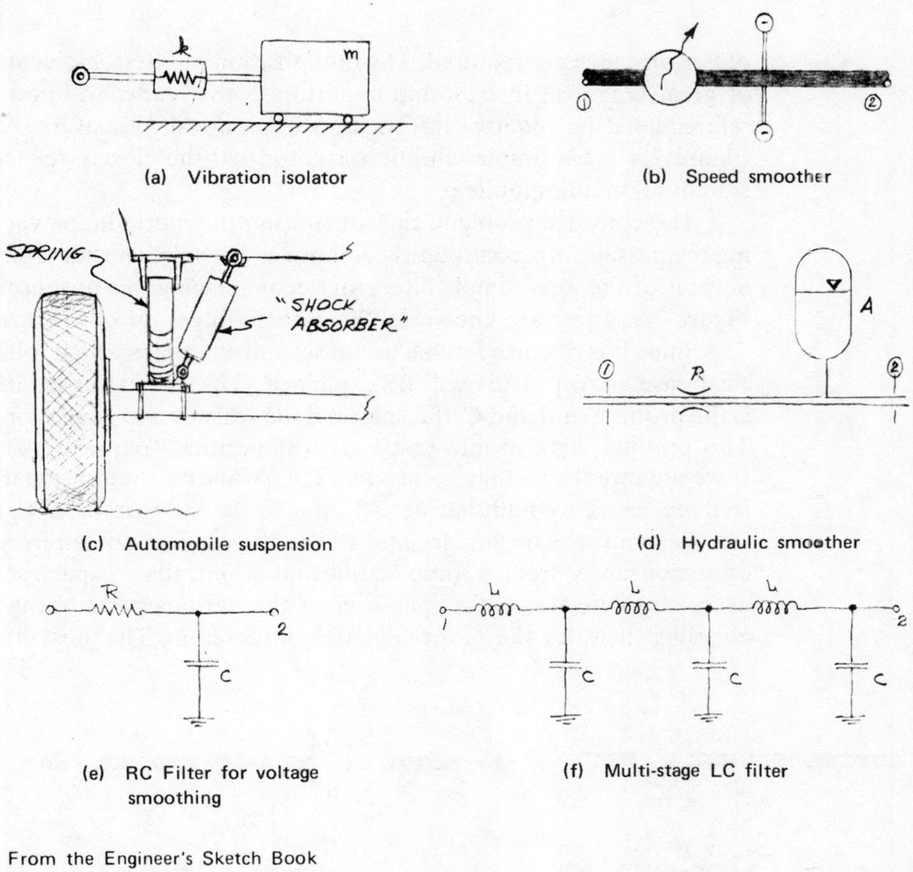

Figure 2.4 Basic configurations for reducing the pulsations of a power source (Source connected at terminal 1, load at 2.)

Figure 2.4 presents a variety of these networks. Taking the hydraulic example shown at (d) for explanation, we note that a sudden rise in pressure at terminal 1 would not be transmitted to terminal 2 because of the combined action of the restriction R and the accumulator A. The rise would create an increased flow through R and the portion of this which is not taken out at 2 will flow into the accumulator. This will create a gradual rise in the pressure at 2, but the downstream elements will not have suffered the shock that they would have otherwise. The fluid-coupling, flywheel combination at (b) and the RC filter at (e) serve similar functions in rotary mechanical and electrical systems respectively. At (f) is shown a multi-stage electrical filter which can give far superior performance to that of (e). As would be expected, multi-stage filters are applied in both mechanical and hydraulic systems when high levels of performance are required. The optimization of filters, incidentally, is a topic of great technical interest and importance; the reader will find innumerable references if he requires information of an advanced nature. Also shown in Figure 2.4 are a simple vibration isolator and the closely related suspension system of an automobile.

The converse problem, that of transmitting periodic power while blocking the passage of a continuous component, is also of considerable importance. Several of the most simple filters for accomplishing this purpose are shown in Figure 2.5. These are known as "high-pass" filters for obvious reasons.

Since it is of considerable use in several examples which follow, the electrical case shown at (d) will be explained. The critical figure in such a filter is the product of R and C, the values of the resistor and capacitor respectively. This product, divided into unity, gives the critical frequency ω_c for the filter. If we observe the voltage at terminal 2 (V_2) while exciting terminal 1 at different frequencies ω, we find that $V_2 < V_1$ if $\omega < \omega_c$ and that $V_2 \cong V_1$ if $\omega > \omega_c$.

In many electronic circuits, where the quantity of interest is often the excursion away from a static equilibrium point, this "capacitor coupling" is used to isolate the static operation of the various parts of the circuit while coupling them for the proper dynamic interaction. The most effective way to

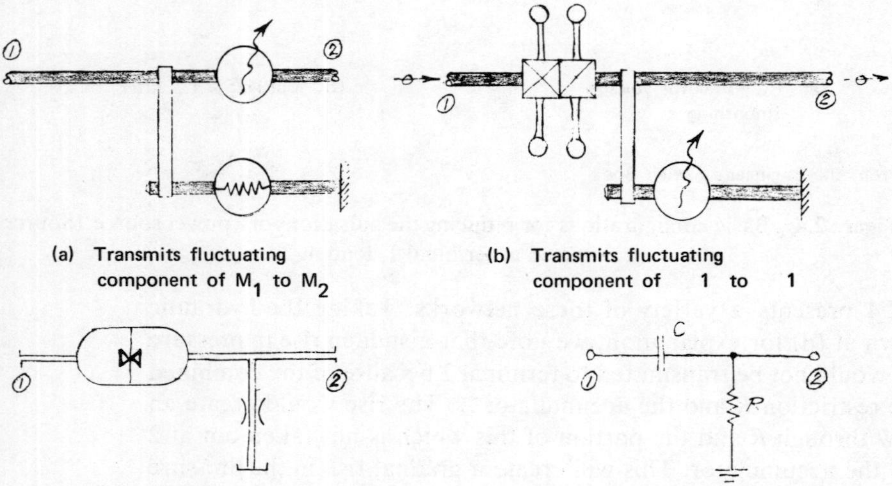

(a) Transmits fluctuating component of M_1 to M_2

(b) Transmits fluctuating component of 1 to 1

(c) Coupling which transmits pressure pulsations from terminal 1 to 2

(d) Capacitor coupling which transmits ac but not dc

From the Engineer's Sketch Book

Figure 2.5 Some filters which transmit the dynamic and block the steady components of power.

look at such a device when attempting to understand a circuit is to assume the capacitor is a short circuit for things which happen rapidly and an open circuit for those which are slow. In the way of a physical explanation, we can see that any current which flows through the capacitor—thereby charging it and changing the voltage across it—must flow through the resistor. This is a restrictive element, and therefore will not allow the capacitor voltage to change very rapidly. Therefore, a fast change in V_1 is bound to be transmitted to terminal 2. If V_1 has a dc component, the capacitor has plenty of time to charge up to this value. Since the capacitor cannot conduct direct current, the average value of the voltage V_2 must be precisely zero.

In designing this form of filter, we first select a capacitor with the required voltage rating; i.e., the maximum magnitude of the average value of V_1. Combinations of R and C are considered, and a selection is made based on the desired critical frequency. Finally, the wattage of R is selected on the basis of the expected ac component of V_1.

2.3 Timing Networks

The essence of the timing problem to be considered here is the control of how often something happens or the duration of its action when it does occur. Representative of the first possibility is the flashing of a light used to warn motorists of excavations or other road hazards, while the second is represented by the shutter-speed control in a camera. Although related timing problems are frequently solved by reference to 60-cycle ac power, the networks to be discussed here are self-contained. In cases where the rate is high compared to 60 cycles per second, the duration much less than 1 second, or ac power is unavailable, this is the only alternative. Otherwise, the use of a synchronous timing motor or stepping device may be the most rational choice.

Considered very broadly, timers are based on the rate at which energy is stored in one of the basic two-terminal storage elements. All timers contain at least one such element, be it spring, mass, capacitor, inductor, accumulator, or the equivalent thing appearing as a secondary aspect of some other system element. A second universal aspect of timers is their relationship to bistable networks; this will be made apparent in the following discussions.

The fundamental nature of timers is well exhibited by the doorbell buzzer network shown in Figure 2.6. In this device, the electromagnet of the relay is arranged so as to open the contact when it is energized; i.e., it is the "normally closed" relay configuration. Also attached to the contactor is the clapper—which has no essential role in the network function. Suppose that a visitor pushes the doorbell button S_1. The contact is closed, so the electromagnet becomes energized. However, as soon as the force overcomes the inertia of the

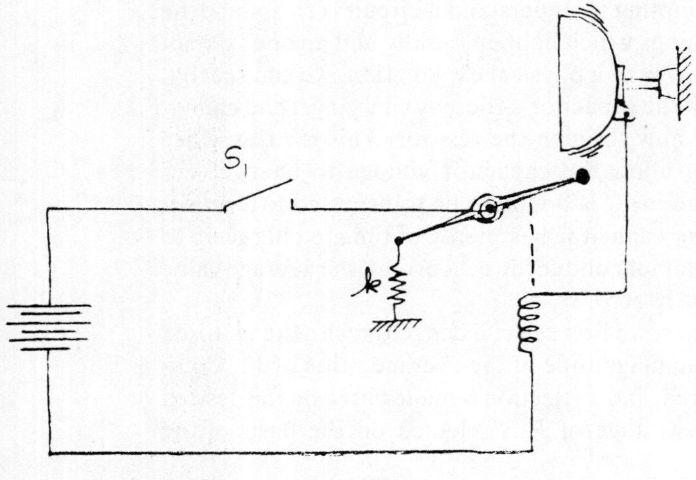

Figure 2.6 Doorbell buzzer.

moving parts of the buzzer, the contact opens and the electromagnet is de-energized. The spring which holds the relay contact closed now takes over and returns the moving parts to their original position. So the cycle repeats, and continues at the visitor's discretion. In this example the forementioned storage elements are the mass of the moving parts and the spring; the values of these, and of the electromagnet force, determine the rate at which the bell is struck. The relationship to a bistable device is that the wiring of the relay is just opposite to that of a bistable relay, as will be shown next.

Several bistable networks are shown in Figure 2.7. In each of these, the action is similar to that of the flyball actuated clutch shown at (b). In this, a spring is arranged to hold the clutch normally open. Suppose that the lower shaft is rotating at some constant speed. Then it is easy to see that there are two stable conditions which the network can take on. With the clutch open, the flyball actuator does not rotate and there is no tendency to close the clutch. However, if someone comes along and tweaks the clutch closed, the flyballs will rotate and the actuator force (assuming the network has been designed properly) will overcome the spring force, thereby holding the clutch closed and keeping the flyballs rotating.

All the networks shown in Figure 2.7 have the memory property which is now so important in the fabrication of digital computers; many of the random access memories in use utilize precisely analogous networks. Note that simple modifications can be made to each of these networks so that they become "astable"; i.e., so that they have *no* stable steady-state operating condition. A

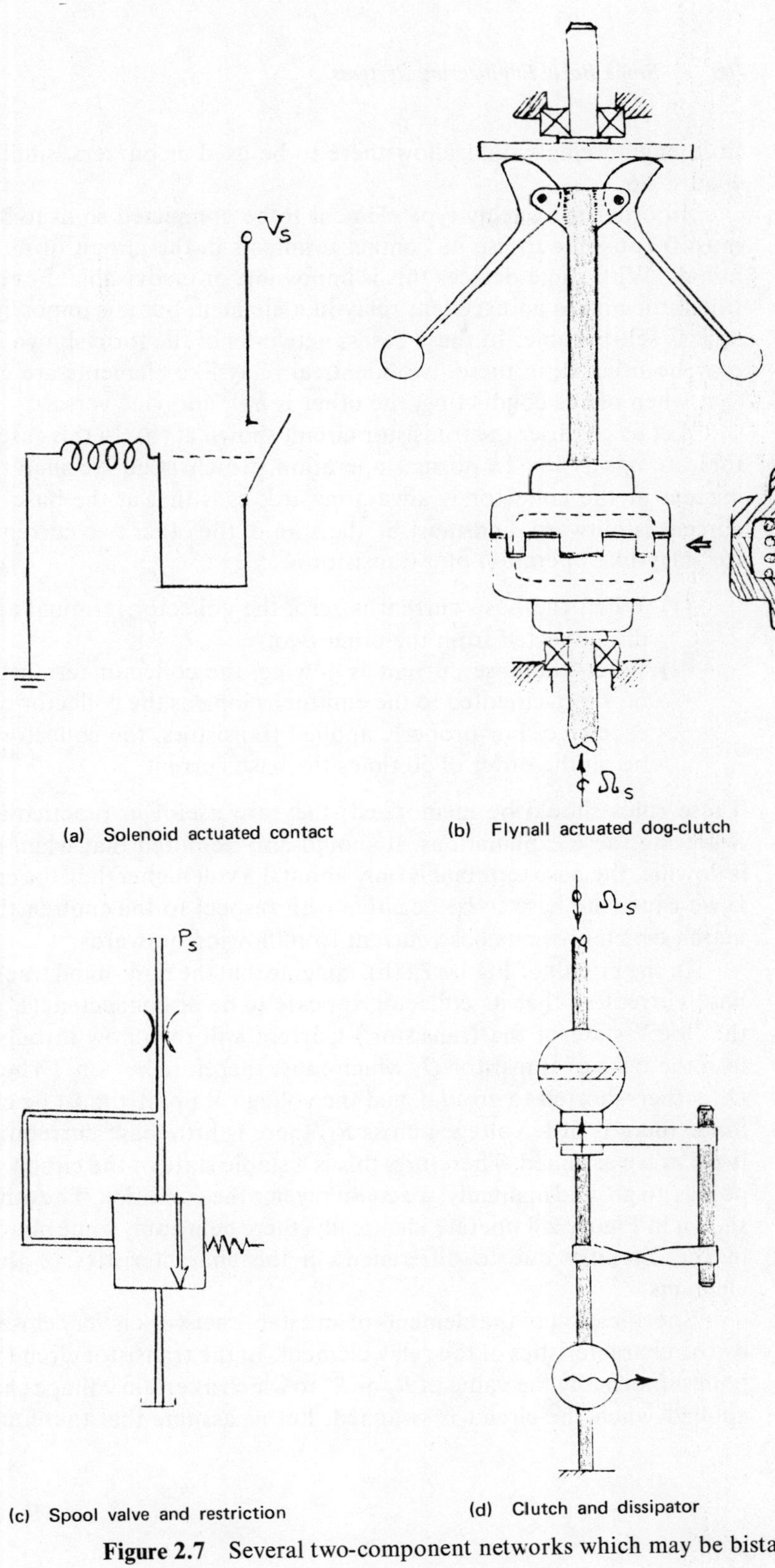

(a) Solenoid actuated contact

(b) Flynall actuated dog-clutch

(c) Spool valve and restriction

(d) Clutch and dissipator

Figure 2.7 Several two-component networks which may be bistable.

little engineering would allow these to be used as buzzers, similar to that of Figure 2.6.

In order for a relay-type element to be connected so as to be bistable, it must be possible to use its contact terminals in the circuit of its primary terminals. With some devices this is impossible or inadvisable. For example, the transistor makes an excellent relay-like element, but it is impossible to wire it so it is self-bistable. In these cases, networks of the form shown in Figure 2.8 may be utilized; in these, two identical relay-like elements are connected so that when one is conducting, the other is not, and vice versa.

Let us condiser the transistor circuit shown at (b). In this case, the transistors are *npn* types. In normal operation, which is all we shall consider, the current at the collector is always inward, as is that at the base. The emitter current is outward, and must be the sum of the other two currents. The rules for relay-like operation of a transistor are:

(1) When the base current is zero, the collector terminal appears to be disconnected from the other two.
(2) When the base current is flowing, the collector terminal appears to be short-circuited to the emitter as long as the collector current is not excessive. For properly applied transistors, the collector current can be on the order of 50 times the base current.

These rules should be memorized; they are useful in practice as well as in following these explanations. It should also be noted that when base current is flowing, the base terminal is only about 0.5 volt higher than the emitter. Also, if we cause the base to be negative with respect to the emitter, there is diode action which prevents base current from flowing outwards.

In the circuit of Figure 2.8(b), imagine that the right-hand transistor has no base current so that its collector appears to be disconnected. (Let us call this the "off" state of the transistor.) Current will then flow through R_2 and R_3 into the base of transistor Q_1 which must therefore be "on." The collector of Q_1 is then shorted to ground, and the voltage at point x must be close to zero. Since there is little voltage across R_4, there is little base current in Q_2, and it is off as we assumed. Therefore, this is a stable state of the circuit; by shorting point y to ground manually, we could reverse the condition. The other networks shown in Figure 2.8 operate identically, there being only some minor variations in the circuitry due to differences in the characteristics of the relay-like elements.

Specification of the elements of an astable network is very closely governed by the characteristics of the relay element. In the transistor circuit, the starting point is normally the value of R_1 or R_2 to which a certain voltage change is to be applied when the circuit is switched. Let us assume that these are the same.

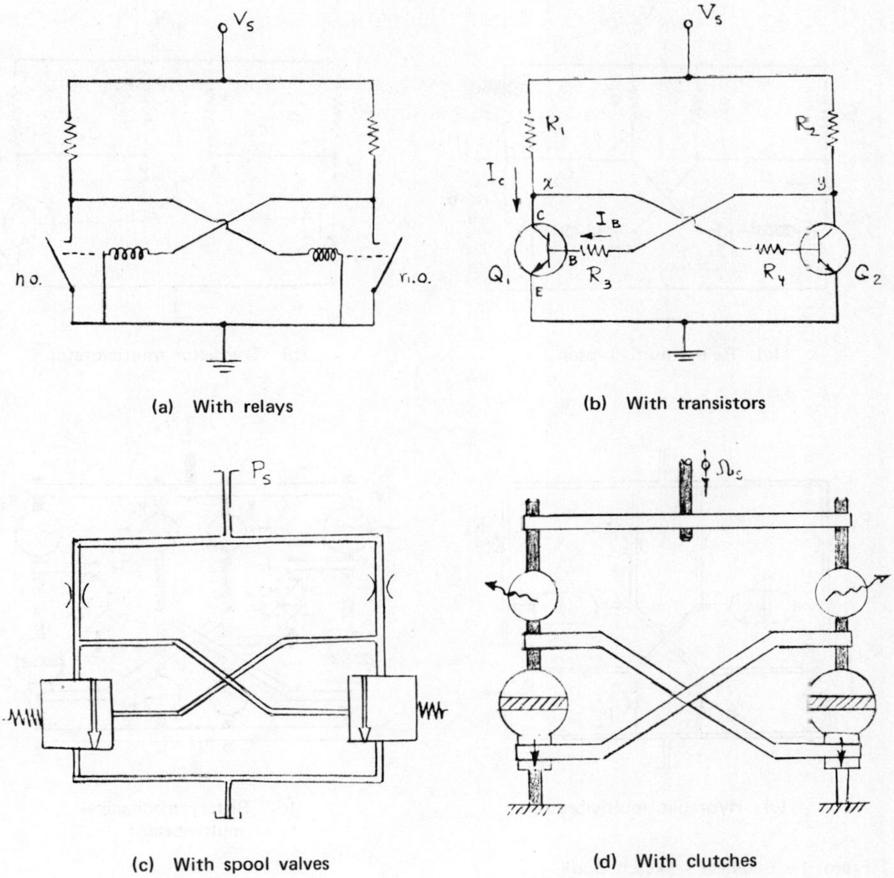

From the Engineer's Sketch Book

Figure 2.8 Some bistable networks which contain two identical valves.

Taking the desired voltage change as V, the maximum transistor current can be estimated to be V/R_1 plus an allowance for the fact that the supply voltage V_s will have to be larger than V because of the influence of R_3 and R_4. A tentative choice for the transistor would be made, using an appropriate safety factor for both the voltage and current ratings. The amount of base current necessary to turn the transistor on with the estimated collector current would then be determined from the transistor characteristics. R_3 and R_4 would be chosen to pass at least this much current at a voltage of V. Because of the amplification effect in the transistor, these will be quite a bit larger than R_1 and R_2, a factor

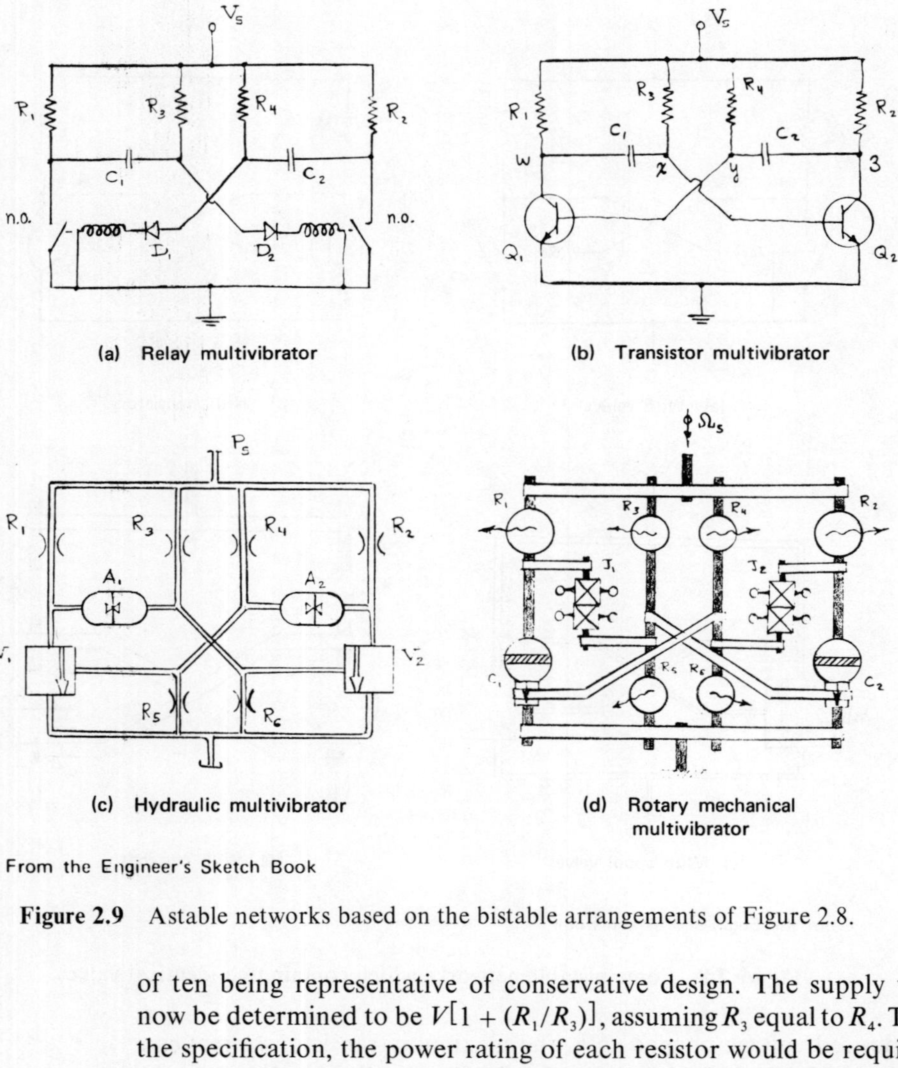

From the Engineer's Sketch Book

Figure 2.9 Astable networks based on the bistable arrangements of Figure 2.8.

of ten being representative of conservative design. The supply voltage can now be determined to be $V[1 + (R_1/R_3)]$, assuming R_3 equal to R_4. To complete the specification, the power rating of each resistor would be required.

A group of networks having the nature of buzzers and based on the bistable networks of Figure 2.8 are shown in Figure 2.9. These are utilized when it is desired to transmit power at evenly spaced intervals of time, and are most commonly referred to as "vibrators." Usually, the load to which power is transmitted is one or both of the dissipators labeled R_1 and R_2 in the figure, or the signals appearing in the network are used to control a separate relay element. Both the rate and the duration of the power pluses can be controlled by design, as will be shown later.

Continuing with the transistor case, which is of the greatest practical utility, let us assume that the voltage at point w is a square wave, Q_1 turning on and off as demanded by the circuit. Also, assume that when Q_1 is on, Q_2 is off, and vice versa. When Q_2 is on, base current flowing to it through R_3, the voltage at point x must be about 0.5 volt; at the same time, Q_1 is off and the voltage at point w is at its positive value. Therefore, capacitor C_1 is charged. Now, suppose that Q_1 turns on. The voltage at point w drops very rapidly to zero; the capacitor voltage in the high-pass filter formed by R_3 and C_1 cannot change rapidly, so the voltage at point x drops just as much as that at w. This makes x quite negative, and means that Q_2 must be off, which is consistent. Gradually, C_1 will charge through R_3, the voltage at x increasing exponentially toward V_S. However, as this voltage gets to about 0.5 volt, Q_2 turns on, thereby turning Q_1 off. Hence the cycle will repeat indefinitely as we assumed.

The other circuits shown in Figure 2.9 operate in an analogous fashion; the reader may find it instructive to analyze them. In particular, the odd mechanical vibrator network at (d) is stimulating, since we must think in terms of rotary speeds in the place of voltages for the electronic example.

Returning to the circuit of (b), note that the off duration of Q_1 is determined by the charging rate of C_2 supplied through R_4, while the off time of Q_2 is governed by C_1 and R_3. These times are roughly equal to the RC products; it is a simple matter to design the circuit so as to achieve a certain desired non-symmetric waveform and over-all pulse rate. In the component specification for the circuit, the basic approach is not dissimilar to that for the bistable circuit. Here, because the base voltage goes negative, we must be careful about the reverse voltage rating of the base terminal of the transistor.

The relay circuit shown at (a) is a bit peculiar because it is not self-starting. This is due to the fact that there is no transmission of small signals through the feedback circuits, since the contacts are absolutely open or closed, and an essential aspect of self-starting is missing. In the transistor circuit, self-starting can be insured by making R_2 and R_3 somewhat larger than they should be to turn the transistors fully on. This way, when the power is first applied, there is some voltage on each transistor. A small, fast noise signal will trigger the vibration from this state, as follows: Suppose w is jostled upward. C_1 transmits the pulse, and Q_2 is sent more toward full conduction. This pulses z downward, tending to turn Q_2 off and regenerating the original positive pulse at w. Hence, the circuit snaps into the Q_1 off, Q_2 on condition and the vibration is begun.

An important variation on the theme is indicated in Figure 2.10. The networks shown operate as event timers; when actuated, they allow power to be conducted for a preset interval of time. In the transistor circuit of (a), Q_1 is normally on while Q_2 is normally off. If switch S_1 is closed, or point y driven

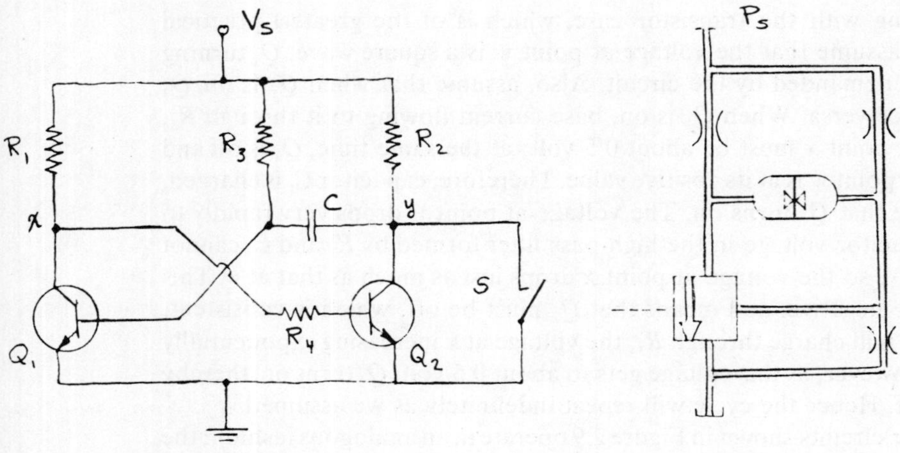

(a) Electronic monostable multivibrator

(b) Hydraulic timer built from a single valve, astable network

From the Engineer's Sketch Book

Figure 2.10 Event timers based on bistable networks.

to ground in some other way, Q_1 will turn off so that Q_2 must simultaneously go on. This condition will remain until the capacitor charges up enough to allow Q_1 back on, Q_2 going off at this time, and the circuit then remains in this condition until actuated again. This allows an event of some fixed duration to be carried out at random intervals. A possible application of the circuit would be as a precision timer for a photographic enlarger; switch S would be the operator's control, and should be arranged to give a short grounding pulse.

2.4 Amplifiers

In many cases of power transfer, it is either necessary or significantly more economical to utilize relatively low-power devices and networks to control the main power level. For example, in controlling the speed of a car descending a steep hill, the brake power—which may approach 100 HP—must be controlled by the driver whose horsepower rating is less than one. In this case, the brake is the valve by which the necessary power amplification is achieved. In other systems, hydraulic valves, transistors, and various other controlled dissipators serve this function.

It should be noted that the level of power transfer can be controlled by means other than variable dissipation. In later sections, switching-type and source or load modification-type control is discussed. Here, we consider only the valve-type amplifiers. An important aspect of this kind of control is that it is always wasteful of power delivered by the source. In systems where the power levels are high, this approach is usually avoided even though it is technically feasable. In others, such as electronic devices intended mainly for information transmittal, the wasted power is of little consequence and the simplicity of the method merits its use.

One amplifier problem that has a long history is that of the "follower." In steering gear for ships, reversing gear for steam locomotives, and power steering for automobiles, increasing vehicle size eventually required help for the human operator. The basic solution to these problems is indicated schematically at Figure 2.11(a). In this, we assume that a linkage has been provided so that the operator can move the valve spool, which is part of a four-way valve. If the operator moves the spool upward, high-pressure fluid is admitted to the bottom of the piston and the upper chamber is connected to the reservoir. Therefore, the load is forced upward until it carries the valve body into alignment with the spool. Downward motion occurs similarly. The power lost because of throttling in the valve is not important here because the system is normally at rest with the spool centered; there is little flow and little power loss during this time. (Spool dimensions greatly exaggerated).

A related device for controlling the pressure in a fluid line is shown in Figure 2.11(b). Here, a pilot-operated check valve is used as the amplifier valve; we assume that a relatively small positive pressure difference between the pilot and opening side connections of the valve will unseat the plug. Suppose that the pressures were as shown on the drawing; if P_c were increased to 700 psi, the plug would open wide. This would allow more flow through the restriction R and the pressure P_o would rise, probably to the vicinity of 650 psi. Since the valve must pass more fluid at this condition, the plug must be more open than previously and increased pilot-pressure differential is required. Basically, however, P_o follows P_c. The error depends on the sensitivity of the valve, and can be made very small if necessary. The basic usefulness of the circuit is that the pilot pressure can be taken from a device, such as a hydraulic computing element, which cannot deliver large flows while the supply pressure may come from a very husky source.

The transistor follower circuit shown in Figure 2.11(c) is a very useful arrangement and also allows us to continue our exploration of transistor applications. In considering the relay-like use of a transistor, it was pointed out that the transistor looks like a short circuit so long as the collector current is not too large. It turns out that, for a fixed amount of base current, once we draw the maximum collector current, the device "chokes" and collector

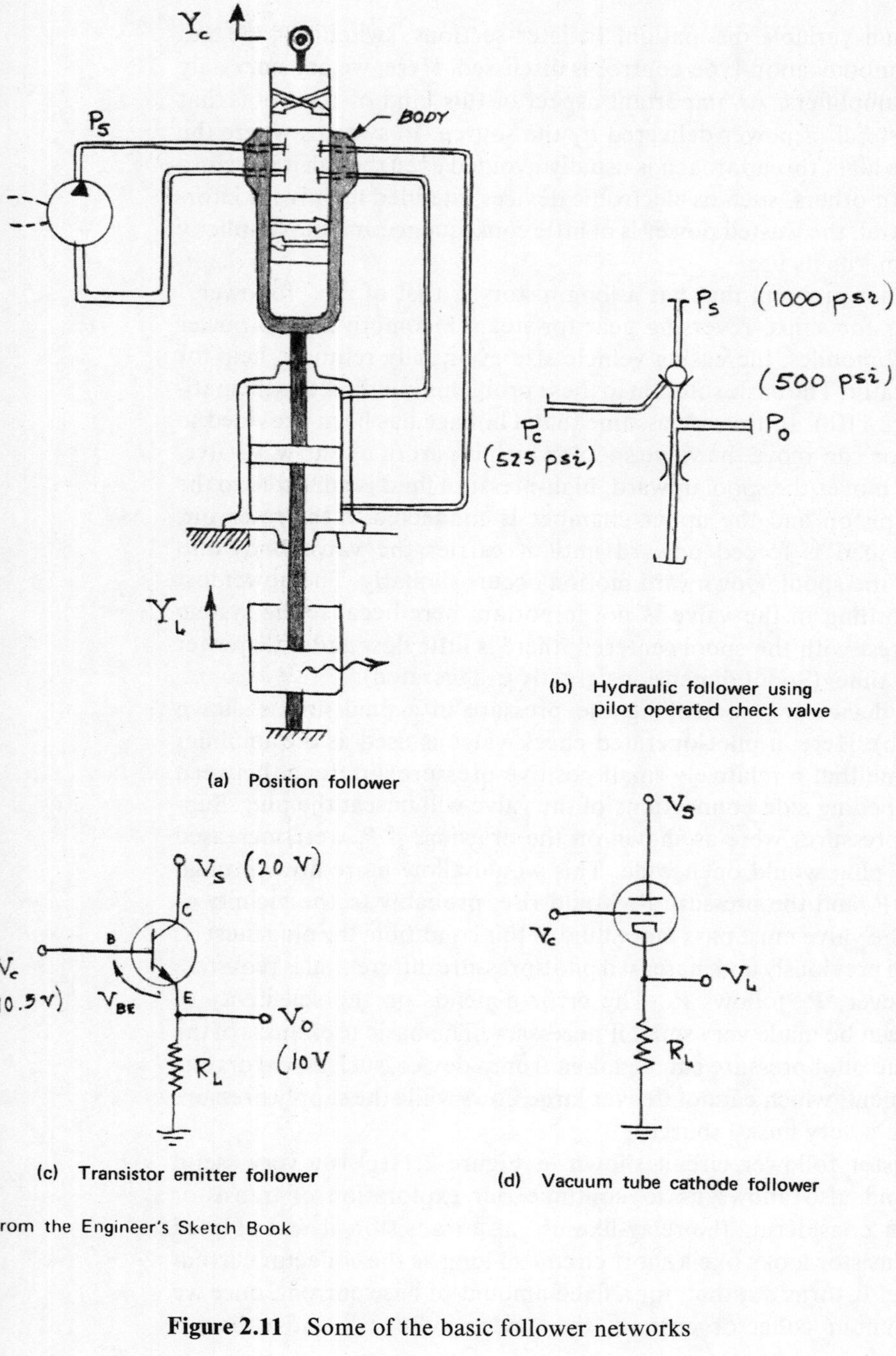

(a) Position follower

(b) Hydraulic follower using pilot operated check valve

(c) Transistor emitter follower

(d) Vacuum tube cathode follower

From the Engineer's Sketch Book

Figure 2.11 Some of the basic follower networks

current cannot be increased significantly by increasing the collector to emitter voltage. If we increase the base current, the base to emitter voltage rises only a very small amount, such as 0.1 volt, and the level at which the collector current is choked increases by 10 to 100 times the amount of the increase in base current.

Suppose that the circuit at (c) is operating at the voltages shown. If we increase the base voltage, the base current would tend to increase and the collector current would tend to increase by a much greater amount. Since this current has to flow through R_L, V_o is forced to increase proportionally. The transistor is so sensitive that the equilibrium is at a point where the base to emitter voltage is only slightly increased from its original value. Therefore, the emitter voltage follows the base voltage and the circuit is exactly analogous to the hydraulic example discussed previously. The old-fashioned cathode follower circuit, using a vacuum tube, performs similarly.

In one sense, this group of amplifiers may not seem very impressive; they have, in electronics terminology, a gain of only unity. However, the power gain is quite respectable and their usefulness in overcoming practical network design problems is widely recognized by experienced engineers.

A group of high torque, pressure, and voltage gain amplifiers is shown in Figure 2.12. The electronic amplifier shown in Figure 2.12(c) gives both current and voltage gain if designed properly. Similarly, the hydraulic circuits shown at (b) and (d) can be made to give both pressure and flow gain.

Since the design problems associated with each of these amplifiers are similar, we will again consider the transistor case for the usual reason. In Figure 2.12(c), the base current can be estimated to be V_I/R_B so long as V_I is quite a bit larger than 0.5 volt, the transistor base-to-emitter voltage. As noted earlier, increases in base current lead to magnified increases in collector current (and both are zero when the base-to-emitter voltage is zero or negative). A very crude, but useful, mathematical representation for the transistor is

$$I_C = \beta \; I_B \tag{2.4}$$

so long as the collector-to-emitter voltage is larger than 2 (or thereabouts, depending on the transistor) and β is the current gain of the transistor. More formally, this is the "dc forward current-transfer ratio" and is represented by h_{FE} instead of β, but our usage here is widely accepted and should be familiar. In the given circuit, the collector current must flow through R_C so that $V_S - V_O = I_C R_C$ and, using (2.4), we have

$$V_0 = -\frac{V_I}{R_B}\beta R_C + V_S \tag{2.5}$$

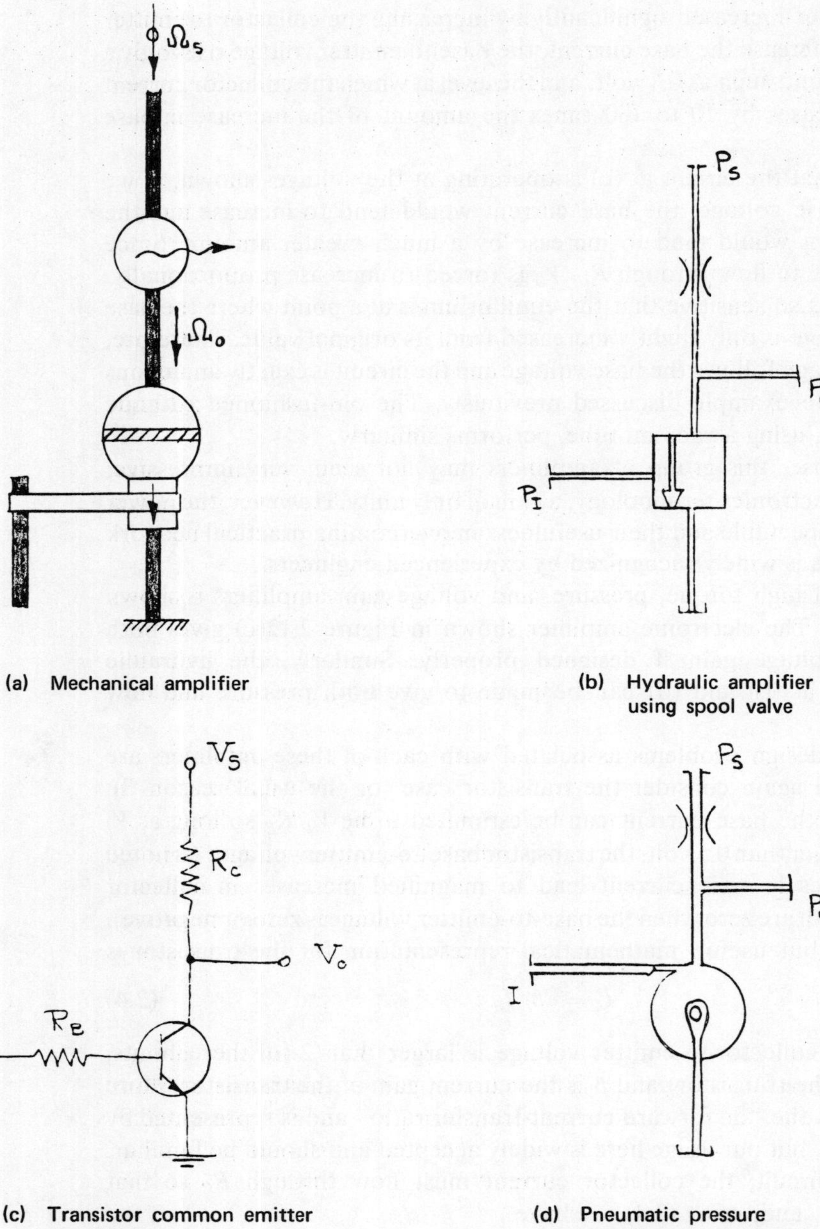

(a) Mechanical amplifier

(b) Hydraulic amplifier using spool valve

(c) Transistor common emitter amplifier

(d) Pneumatic pressure amplifier using a vortex amplifier

From the Engineer's Sketch Book

Figure 2.12 Some basic single-stage high-power-gain amplifiers.

Therefore, the voltage gain of the circuit is approximately $(R_C/R_B)\beta$. Usually, V_S and R_C are known, and the specification of components simply amounts to finding a transistor with a collector-to-emitter voltage rating of at least V_S, a current rating of at least (V_S/R_C), a power rating (for the worst case) of $\frac{1}{2}V_S{}^2R_C$, and an adequate β. Incidentally, the power rating arises because the transistor dissipation is greatest when V_O is $\frac{1}{2}V_S$. The two resistors would be chosen for adequate power dissipation, assuming that the value of R_B had been selected to give the desired gain. In the way of warning, this circuit can exhibit a thermal runaway problem in certain cases and should not be applied in practice until that problem is understood.

In Figure 2.13, the subject is the so-called operational amplifier, an advanced form of the high gain voltage amplifier. The main feature of the amplifier is that its gain is so high, as shown at (b), that the input voltage is negligible in magnitude with respect to the output. By grounding one of the input terminals and using the appropriate external circuitry, we may build power-gain circuits of accurately predictable performance. At (c) is shown one example. Here, the amplifier input voltage at x is always very small. Since the amplifier input current is also made extremely small by careful design, all the current through R_1 also flows through R_2. It is easy to then show the input/output relationship indicated on the figure. At (d) is shown a related transistor amplifier configuration. If we assume the base current is negligible and that the signal voltages are large compared to the base-to-emitter voltage, the indicated input/output relationship may be obtained. The circuit has the useful property that the gain is rather insensitive to changes in the transistor characteristics or to transistor replacement. Note that the circuit is sort of half follower and half high-gain amplifier.

When many orders of magnitude of power gain are required, it is usually necessary to use a cascade of amplifiers. Although this is a very simple matter from a conceptual point of view, the practical aspects can be somewhat challenging. Most valve-like devices are somewhat erratic in their performance; temperature, age, and general environmental conditions all having some influence on their operation. When the valve in the first stage of a high-gain, cascade amplifier changes, the change is reflected on down the line and, due to amplification, can send the system into a completely inoperative condition. The nature of the coupling between stages has a marked influence on this problem, amplifiers which respond to non-periodic inputs being notoriously difficult to design.

2.5 Dissipative Regulators

One of the most frequently occurring problems in the design of systems is to provide for the transmission of power at a constant level of one of the variables.

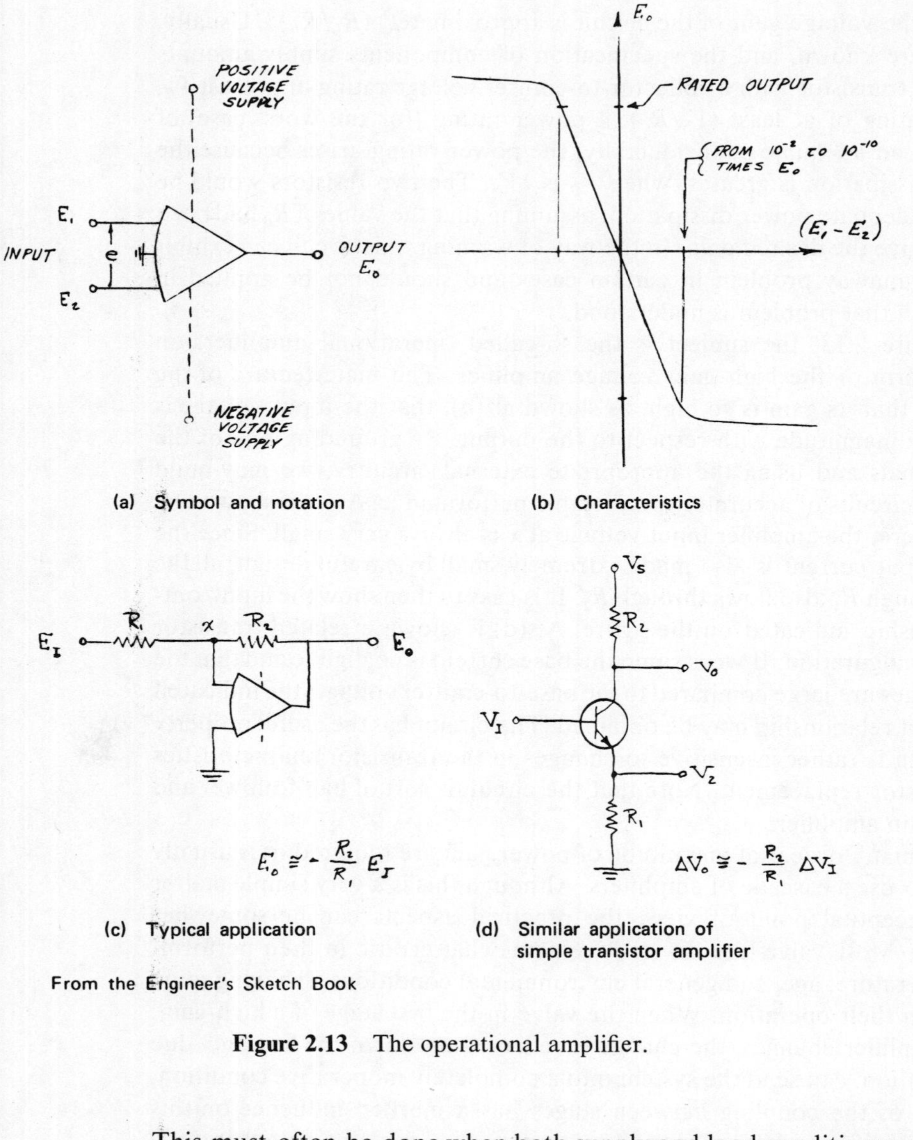

(a) Symbol and notation

(b) Characteristics

(c) Typical application

(d) Similar application of simple transistor amplifier

From the Engineer's Sketch Book

Figure 2.13 The operational amplifier.

This must often be done when both supply and load conditions are changing. For example, in supplying gas to an acetylene torch, the regulator must hold the supply line pressure more or less constant while the bottle pressure drops and the torch valves are adjusted to suit the working conditions. As with amplification, regulation can be accomplished by variable dissipation, by pluse switching in the power path, or by modification of the source or load characteristics.

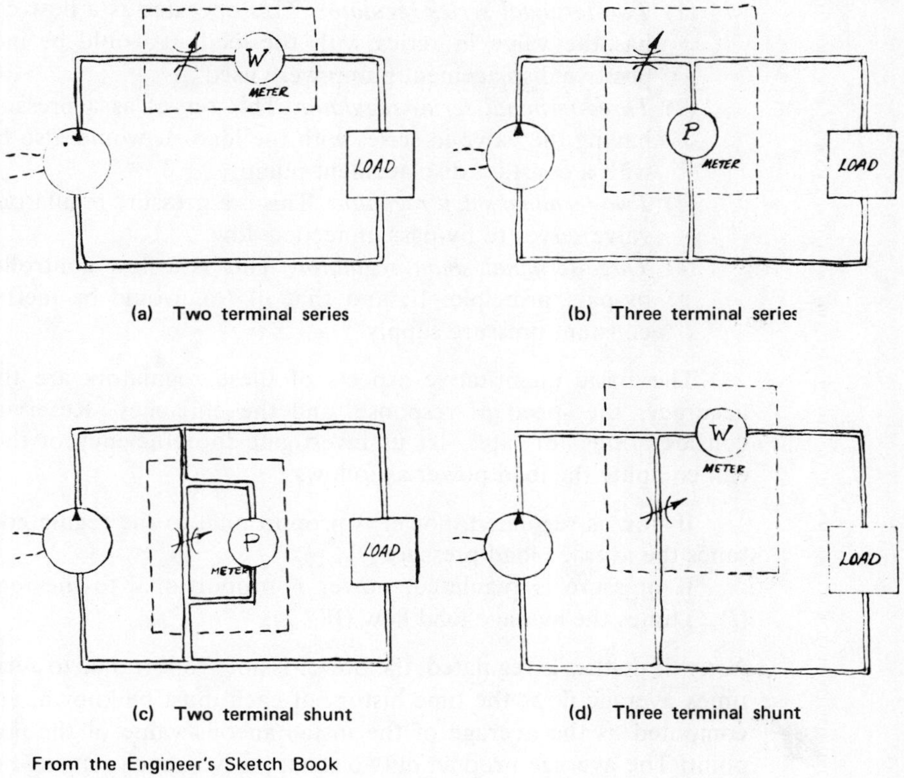

Figure 2.14 Various forms of manual regulators for hydraulic systems.

Dissipative regulators utilize the amplifiers discussed in section 2.3, some sort of reference device, and the network components required for using the two together. It must be mentioned that all of the constant Δ-variable dissipators appearing in Chapter 1 are, of themselves, regulators. Here, we set out to achieve improved performance by unburdening these devices, which are often sensitive and expensive, and do the high-power work of the regulator with a more appropriate element.

In order to point out the basic regulator configurations which are possible, Figure 2.14 has been prepared for the case of hydraulic systems. Here, it is assumed that regulation is to be accomplished by manual adjustment of a valve, using the reading on a pressure or flow meter for guidance. The cases shown are, referring to the figure subtitles:

(1) *Two-terminal series regulator.* This operates as a flow controller and has the valve in series with the load. It would be ineffective if a positive displacement pump were used.
(2) *Three-terminal series regulator.* This serves as a pressure controller having the valve in series with the load. It would also be ineffective with a constant displacement pump.
(3) *Two-terminal shunt regulator.* This is a pressure regulator in which the valve serves to by-pass unneeded flow.
(4) *Three-terminal shunt regulator.* This is a flow controller, using the by-pass principle. It, and that of (c), would be ineffective with a constant pressure supply.

The basic quantitative aspects of these regulators are the regulation accuracy, the speed of response, and the efficiency. Reserving the other considerations for later, let us investigate the efficiency of the group. We can compute the load power as follows:

If flow is regulated, power is proportional to the regulated value (W_{reg}) times the average load pressure ($P_{\text{av},L}$).

If pressure is regulated, power is proportional to the regulated value (P_{reg}) times the average load flow ($W_{\text{av},L}$).

Note: If neither is regulated, the power is not proportional to average pressure times average flow; the time history of each must be known, and the power computed as the average of the instantaneous value of the power. Case in point: The average product of two square waves, each ranging between 0 and 1 so that each has an average of $\frac{1}{2}$, is zero if the waves are exactly out of phase.

The source power may be computed similarly if either the pressure or flow is constant. With the two-terminal regulators, this occurs automatically. With the three-terminal regulators, unless the pump is inherently constant flow or constant pressure, the power has to be computed from time averages, utilizing information on the pump characteristics and the load variations.

Let us consider the case of a load which is to be operated at 1000 psi, the characteristics of the load changing in such a way that the average flow is 50 gpm but occasional maxima of 120 gpm are required. Using a two-terminal shunt regulator, the pump would always work against 1000 psi and would have to deliver at least 120 gpm at this pressure. Therefore, the system efficiency will be:

$$\frac{\text{Load Power}}{\text{Source Power}} = \frac{(P_{\text{reg},L})(W_{\text{av},L})}{(P_{\text{reg},S})(W_{\text{av},S})} = \frac{50}{120} = 41.6\%$$

If the three-terminal shunt regulator were chosen, supposing that a variable volume pump were available, higher efficiencies are theoretically possible. In the extreme case, a constant-pressure pump which would operate at 1000 psi would give 100% efficiency, no regulator being required. If the load flow stays constant at 1000 psi, but the pump pressure is sporadic having an average of 2400 psi, the efficiency falls back to 41.6%. This could happen if the source were, for example, a discharging accumulator or a reservoir of fluctuating surface elevation.

The schematic diagrams of automatic two-terminal series regulators are shown in Figure 2.15. Note that the valves in these networks are utilized in a follower configuration; the reference device provides that the follower enforce a particular operating level on a metering dissipator. In the mechanical example shown at (d), the constant speed dissipator R_R is driven by the excess speed difference between shafts 1 and 2, the supply dissipator R_S taking up the difference and providing the necessary reaction torque. If the load, connected at shaft 2, somehow starts to take too much torque, the metering dissipator, R_M, will increase its slip rate, since it must transmit the torque. The relative speed between shafts x and y will increase by the same amount since the relative speed across the reference device will not change. With the clutch form shown, the difference tends to unclamp the clutch, thereby increasing its slip and reducing the load torque. As in all these regulators, the accuracy depends on the sensitivity of the valve (clutch) to the (speed) error, marked e on these schematics. Analogous explanations pertain to the other regulators shown in the figure.

Several three-terminal series regulators are shown in Figure 2.16. In the hydraulic example at (c), when the load pressure at 2 exceeds the reference pressure at x, the valve tends to close, thereby reducing the load pressure.

Several two-terminal shunt regulators are shown in Figure 2.17. In the electronic example at (b), if the voltage V_{12} becomes too large, the base-to-emitter voltage of the transistor is increased, thereby causing it to be more conductive. This increases the current drain on the source and decreases the voltage via the source internal resistance. This circuit would be ineffective if the source were of low internal impedance such as a battery.

Figure 2.18 shows a group of three-terminal shunt regulators. In the mechanical network at (d), an excess of load torque increases the clutch control speed difference, thereby creating a braking torque increase at the drive shaft and a reduction of load torque. The component selection for this regulator exhibits most of the aspects which are universal to regulators, and goes roughly as follows: The source and load characteristics are examined to determine the torque, speed, and dissipation requirements that would have to be met by the clutch. When a suitable clutch has been located (or designed), the torque

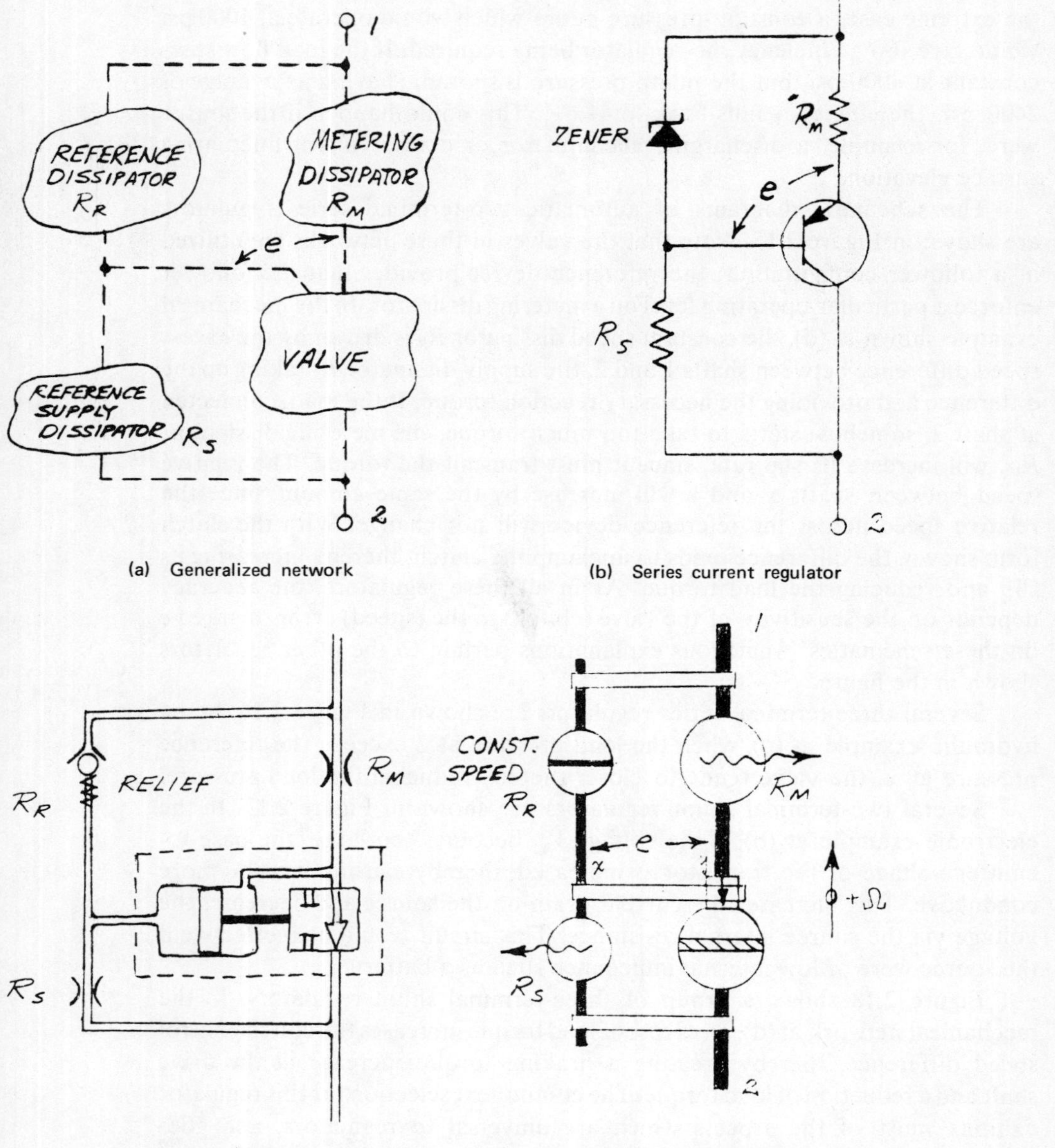

(a) Generalized network

(b) Series current regulator

(c) Inline flow regulator

(d) Clutching torque regulator

From the Engineer's Sketch Book

Figure 2.15 Two-terminal series regulators.

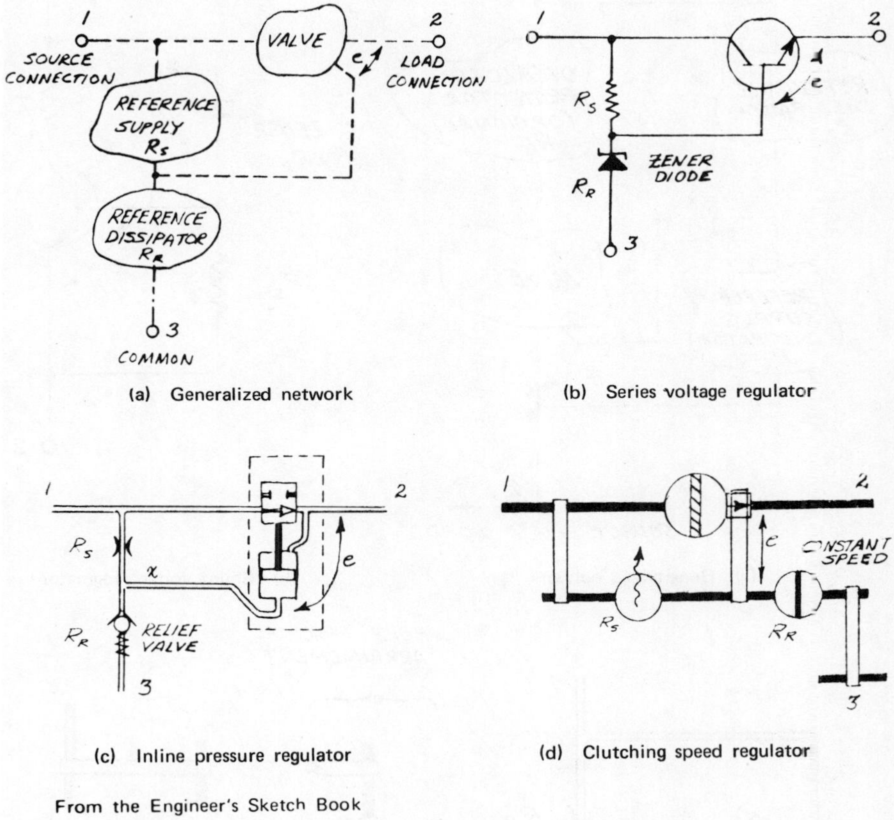

Figure 2.16 Three-terminal series regulators.

and speed differences required to operate it would be known. Since the operating torque must be derived from R_R and R_S, these elements must have at least that rating. Since the control speed difference arises from the *changes* in the slip of R_M, the element would be chosen soft enough to give the necessary sensitivity but as stiff as possible, otherwise, in order to minimize dissipation. This specification forces the use of a particular speed difference rating at R_R, hence its dissipation is known. Finally, R_B can be chosen so that the slip between shafts 1 and x is adequate to drive R_S when maximum clutch control torque is required. As is always the case, the characteristics of the valve determine the design; the other elements simply service it.

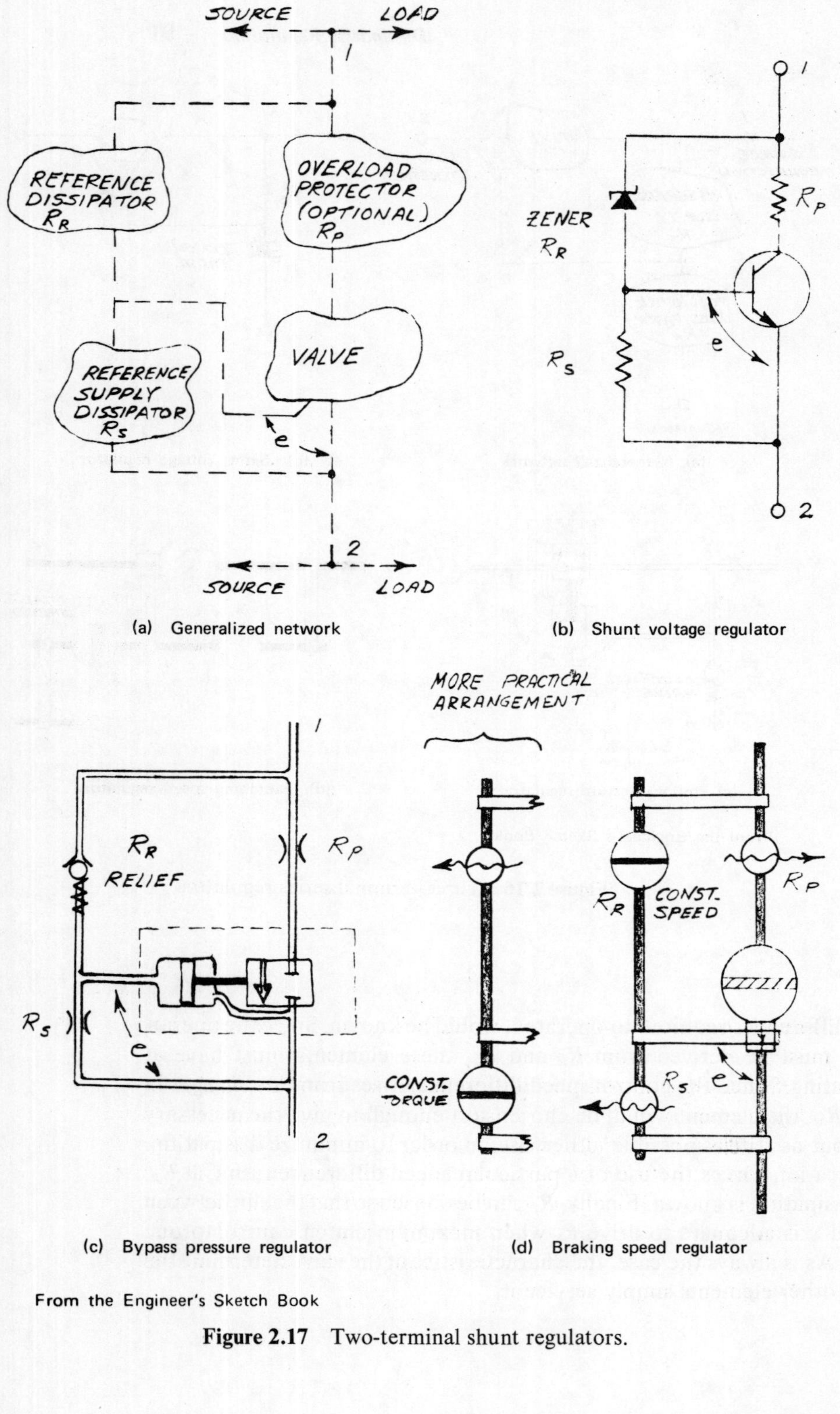

(a) Generalized network

(b) Shunt voltage regulator

(c) Bypass pressure regulator

(d) Braking speed regulator

From the Engineer's Sketch Book

Figure 2.17 Two-terminal shunt regulators.

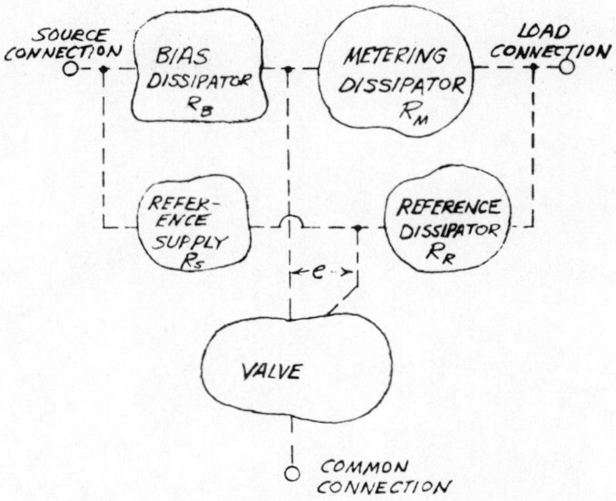

(a) Generalized network

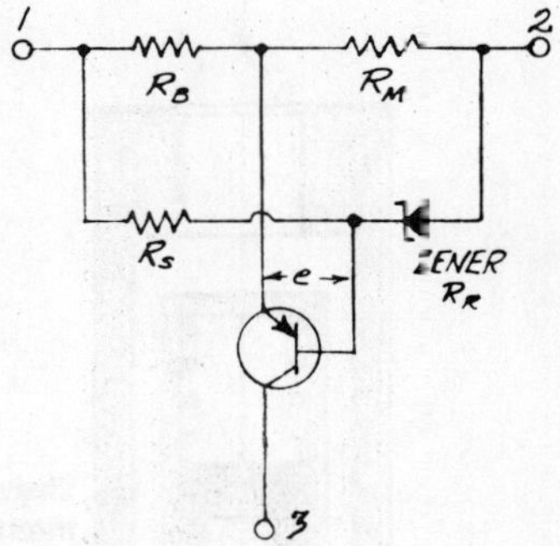

(b) Shunt current regulator

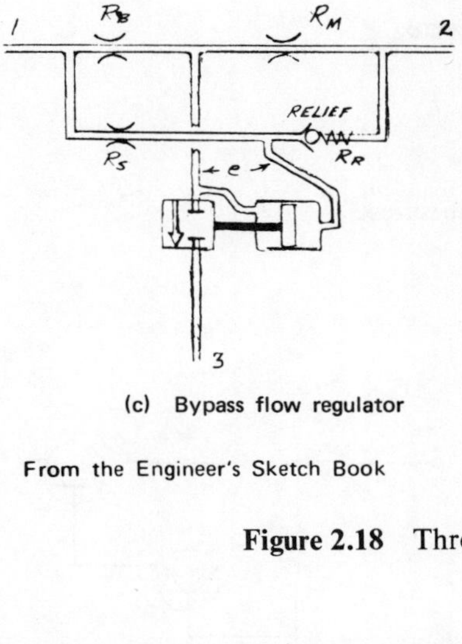

(c) Bypass flow regulator

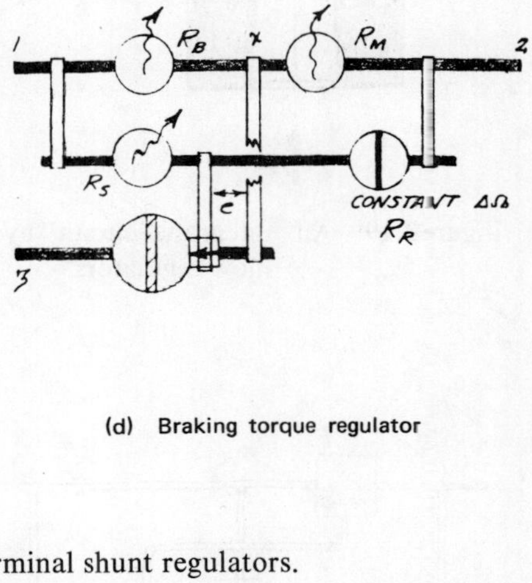

(d) Braking torque regulator

From the Engineer's Sketch Book

Figure 2.18 Three-terminal shunt regulators.

To bring us back to the completely practical world, Figure 2.19 shows, in cross section, the construction of a commercially available series, hydraulic flow regulator. This device is often applied to fork lift trucks in order to control the speed at which loads are lowered.

By using cascaded amplifier valves in a regulator, we may improve the overall performance of the device. Two examples are shown in Figure 2.20. It is not unusual to see a two-stage regulator applied to gas dispensing from bottles; oxygen for hospital use is often handled this way.

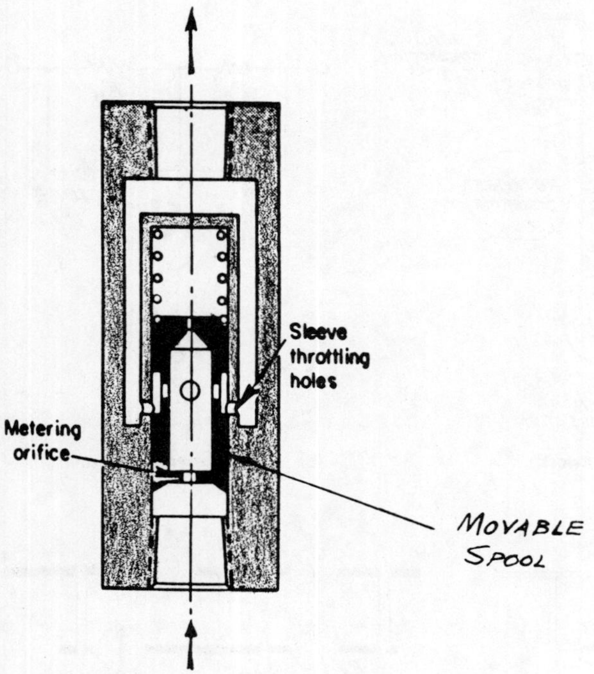

Figure 2.19 An "integrated circuit" hydraulic series flow regulator.

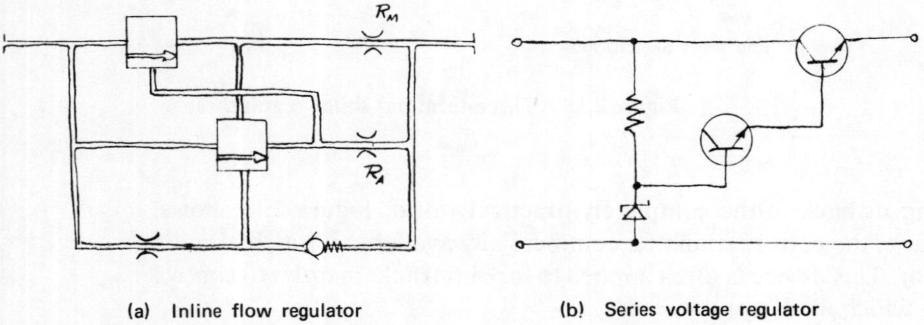

Figure 2.20 Regulators in which two valves are employed to obtain improved performances.

2.6 Switching Regulators

Regulators of high inherent efficiency can be constructed with switching devices, the basic concept being that we either allow, or do not allow, the transmission of power from a source to a load on a pulsed basis. For example, we might simply place a relay in an electrical line and excite it with a variable ratio vibrator. If the load were a heater, the "on" time of the relay, as a proportion of the "off" time, would determine the heating power and, therefore, the temperature of the enclosure in which the heater is located. In all regulators of this nature, we must be able to tolerate the pulsed nature of the signals in the system. We either don't care, or we use filters to reduce the pulsing to a tolerable level. The fundamental reason for the high efficiency of these regulators is that the regulator, theoretically, takes no power.

A simple system of this general form is shown in Figure 2.21. Here, a pilot-operated relief valve is used in the by-pass configuration to control load pressure. A positive displacement pump is used; when the valve is open there

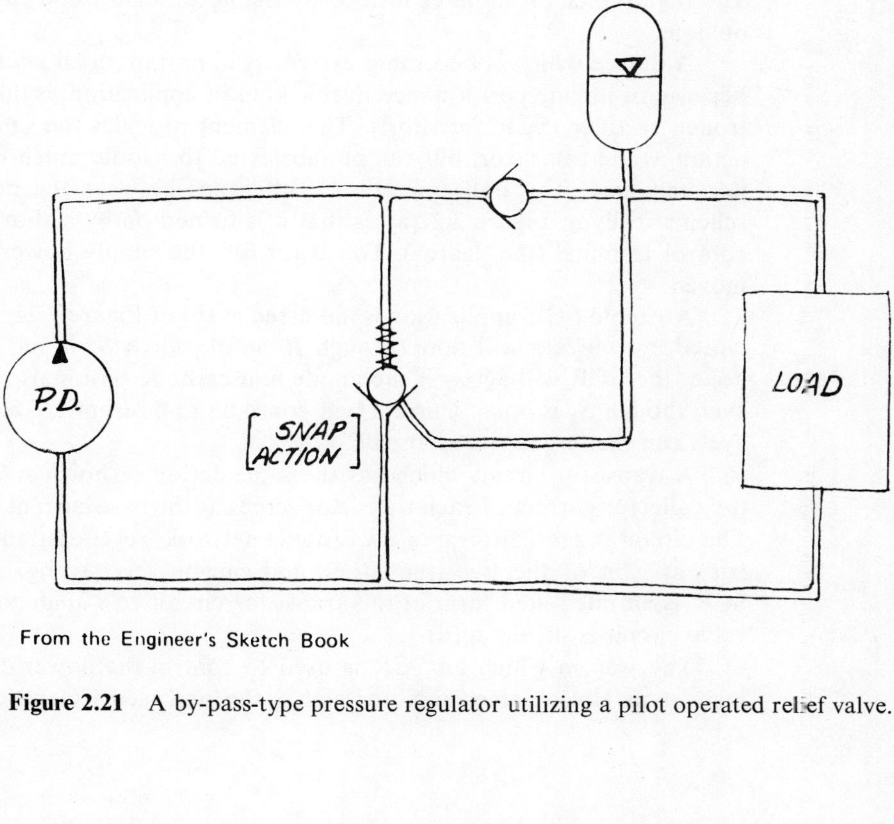

From the Engineer's Sketch Book

Figure 2.21 A by-pass-type pressure regulator utilizing a pilot operated relief valve.

is no pressure on the pump. Therefore, the motor which drives it is taking no power (other than that due to friction and other system imperfections). When the valve is closed, all the flow passes through the upper check valve and is available to do work. Therefore, no pump power is wasted. In order to make the system work properly, the pilot-operated valve should be a snap action device, and should close at a lower pressure difference than that which opens it. The system operates as follows: If the load pressure is below the desired amount, the pilot pressure is insufficient to open the valve (a spring bias, equivalent to the desired pressure level, holds the valve plug closed). Therefore, the pump flow enters the load and the accumulator. The portion going into the accumulator charges it, thereby raising the load pressure. Obviously, the pump must be able to deliver the flow required by the load at the desired pressure plus an increment satisfactory for charging the accumulator at a reasonable rate. When the desired pressure has been reached, the valve will open and the pump flow will by-pass freely through it. The check valve will prevent the accumulator from discharging. The system will retain this condition until the load has drawn enough flow from the accumulator to lower the pressure to the point where the pilot valve again closes, and the cycle will repeat. The importance of the filter formed by the accumulator and the load is also obvious.

A device which is becoming extremely important in all kinds of systems, because of its fine performance in this kind of application, is the silicon controlled rectifier (SCR for short). This element provides the same relay-type action as the transistor, but can be fabricated to handle much higher power per unit cost. The only operational difference between the device, shown schematically at Figure 2.22(a), is that it is turned on by a short pulse at the control terminal (the "gate"). To turn it off, the supply power must be removed.

A simple SCR application is indicated at (b) of Figure 2.22. Assuming S_1 closed, no current will flow through R_L until switch S_2 is hit. Once this is done, the SCR will act as if the anode and cathode terminals were shorted, even though S_2 is open. Current will continue to flow until S_1 is opened. The cycle can then be repeated at will.

A transistor circuit which has the same action is shown at (c). Note that the collector current of each transistor serves as the base current for the other. The circuit is another form of the bistable network; regeneration is due to the current gain of the two transistors, and can be very strong. Actually, the SCR is an integrated form of this transistor circuit. An analogous hydraulic valve circuit is shown at (d).

The way in which an SCR is used to control the power delivered to a load—typically, a universal motor such as the kind used in electric drills, etc.—

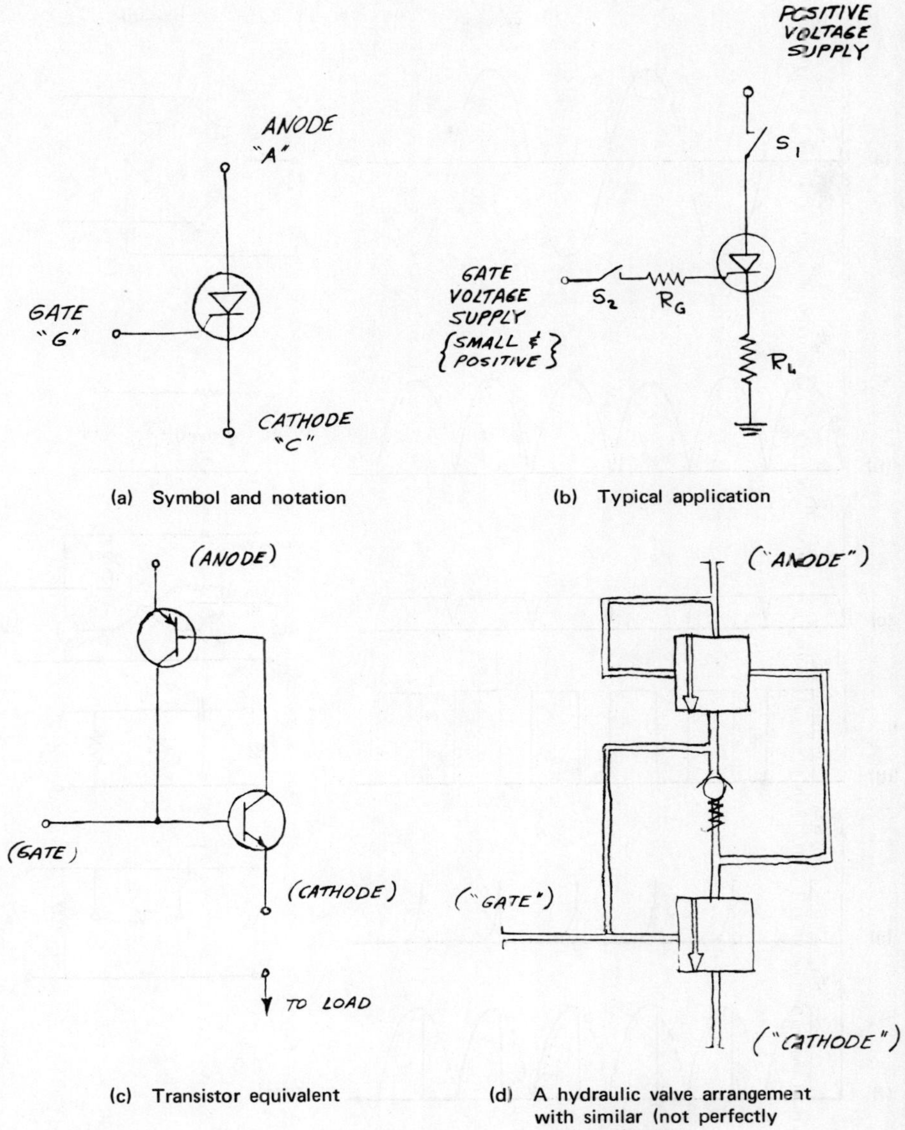

(a) Symbol and notation

(b) Typical application

(c) Transistor equivalent

(d) A hydraulic valve arrangement with similar (not perfectly analogous) action

From the Engineer's Sketch Book

Figure 2.22 The silicon-controlled rectifier.

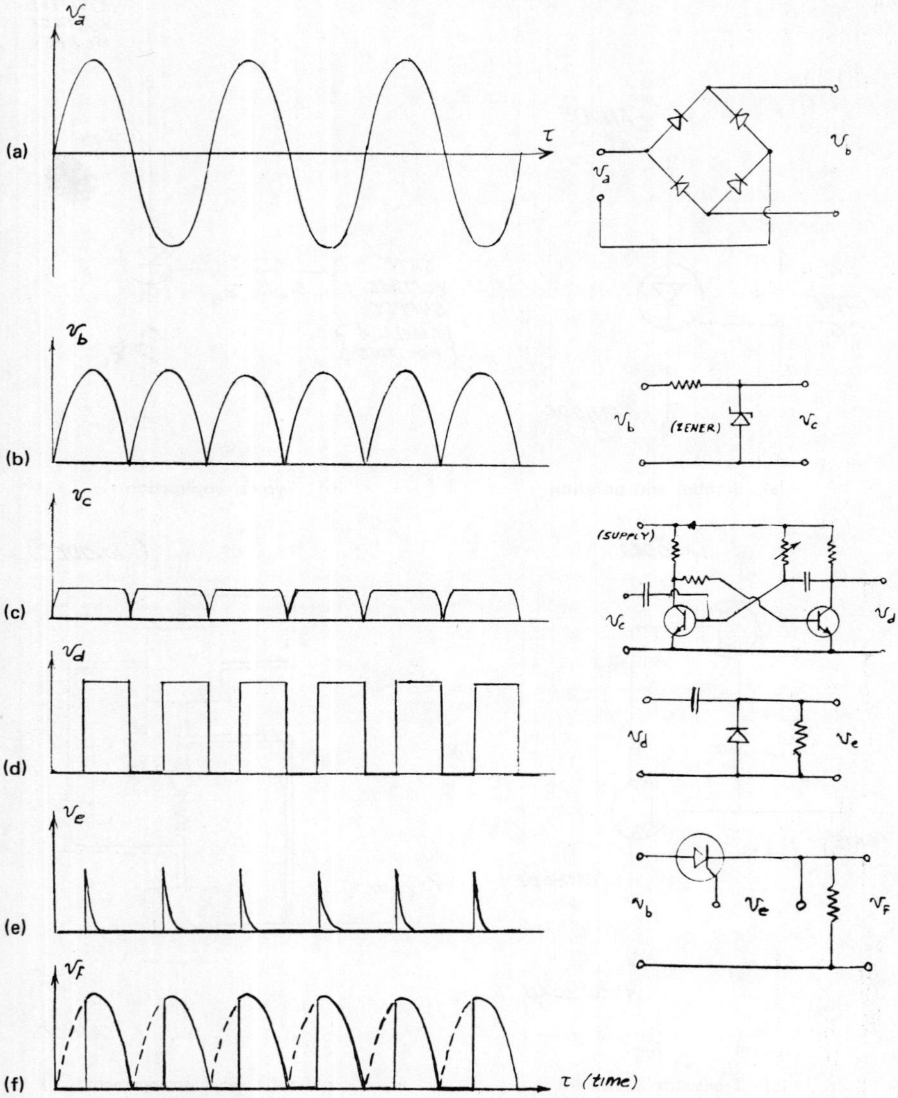

From the Engineer's Sketch Book

Figure 2.23 Voltage waveforms in one kind of SCR control network.

is indicated by Figure 2.23. Sinusoidal line voltage (a) is applied to a rectifier bridge resulting in the full wave rectified voltage at (b). This is clipped by using a simple Zener diode circuit to provide an upside-down string of pulses (c). These pulses are coupled, through a capacitor, to a monostable multivibrator having an adjustable *on* time. The pulse fires the vibrator-timer, which then stays on for up to 1/120 second. The output of this circuit goes through a high-pass filter in which the time constant has been made small for negative outputs by the inclusion of a diode. This produces a string of positive pulses, delayed from the zero-crossing of the sinusoid, by the on time of the vibrator-timer. These pulses fire the SCR, thereby imposing the voltage waveform shown at (f) on the load. By changing the timer period, the delivered power may be controlled between 0 and 100%. Except for that required by the control circuitry, which can be quite small, there is no wasted power.

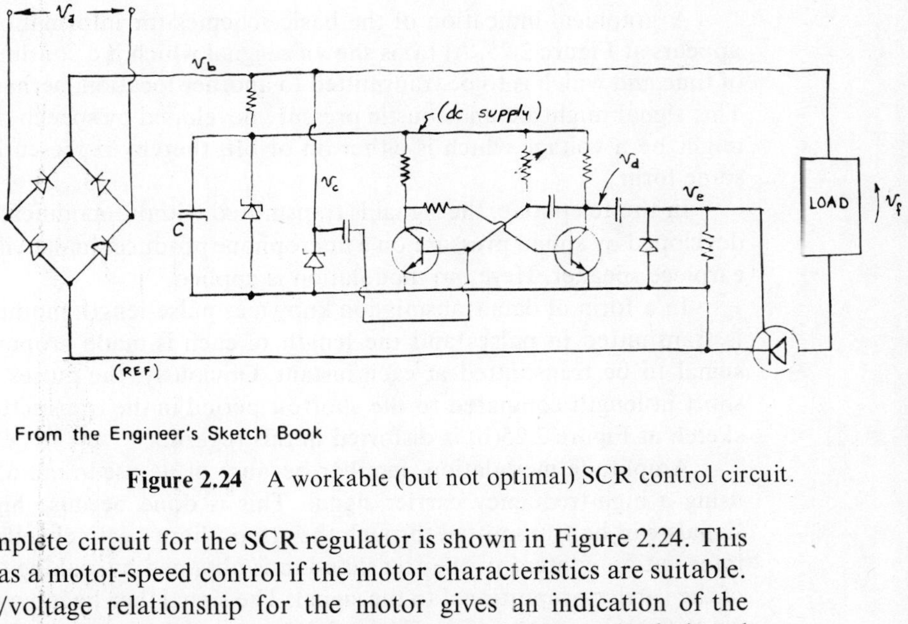

From the Engineer's Sketch Book

Figure 2.24 A workable (but not optimal) SCR control circuit.

The complete circuit for the SCR regulator is shown in Figure 2.24. This can be used as a motor-speed control if the motor characteristics are suitable. The current/voltage relationship for the motor gives an indication of the speed; a network which utilizes this to control the timer period can be designed so that speed control is achieved.

The switching-type regulator is applicable to many media and the reader should not be misled by our emphasis here. It is just that we saw this as a nice opportunity to introduce an important device. The cable-car system of San Francisco, created in a different age of technology, utilizes just this principle to regulate the average speed of the cars. It is a full-go or full-stop system;

the grip-man who attempts to slip with respect to the cable finds the error in his concept exemplified by fire at his wooden grip-block. Andrew Hallidie was too smart an engineer to apply a dissipative regulator in this high-power application!

2.7 Modulation for Information Transfer

Although we shall not be able to consider many information transmitting systems in this book—because of space limitations and because the field has become so sophisticated that even a basic coverage must be lengthy—it is appropriate to be acquainted with the basic information transmitting processes. For one thing, this pretty well rounds out the study of the ways in which power transmittal between a source and load is controlled for engineering purposes. The obvious examples of this are telegraphy, the telephone, radio, and television.

A graphical indication of the basic schemes for information transmittal appears at Figure 2.25. At (a) is shown a signal which is a continuous function of time and which is to be transmitted to another location, perhaps far distant. This signal might be an acoustic pressure developed by speech or music, or it might be a voltage which is either on or off, thereby representing a code of some form.

In the telephone, the signal is transmitted without modification. Voltages developed by sound pressure on a microphone produce similar vibrations in an earpiece speaker. Here, no modulation is applied.

In a form of data transmission known as pulse-length modulation, power is transmitted in pulses and the length of each is made proportional to the signal to be transmitted at each instant. Obviously, the pulses must be very short in length compared to the shortest period in the transmitted signal; the sketch at Figure 2.25(b) is distorted in this regard.

Amplitude modulation, familiar because of its use in radio, transmits by using a high-frequency carrier signal. This is done because high-frequency signals can be transmitted through the atmosphere relatively efficienctly. The information content of the transmission is obtained by making the amplitude of the carrier proportional to the signal. The signal then becomes the envelope for the carrier, as shown in Figure 2.25(c).

Frequency modulation also utilizes a carrier signal, but the amplitude is held constant and the base frequency is normally much higher than that for AM transmission. Here, the carrier frequency is changed, between limits, such that its excursion away from the center represents the signal to be transmitted. This system is capable of high quality performance in the atmosphere because interference has more influence on the amplitude than the frequency of radio

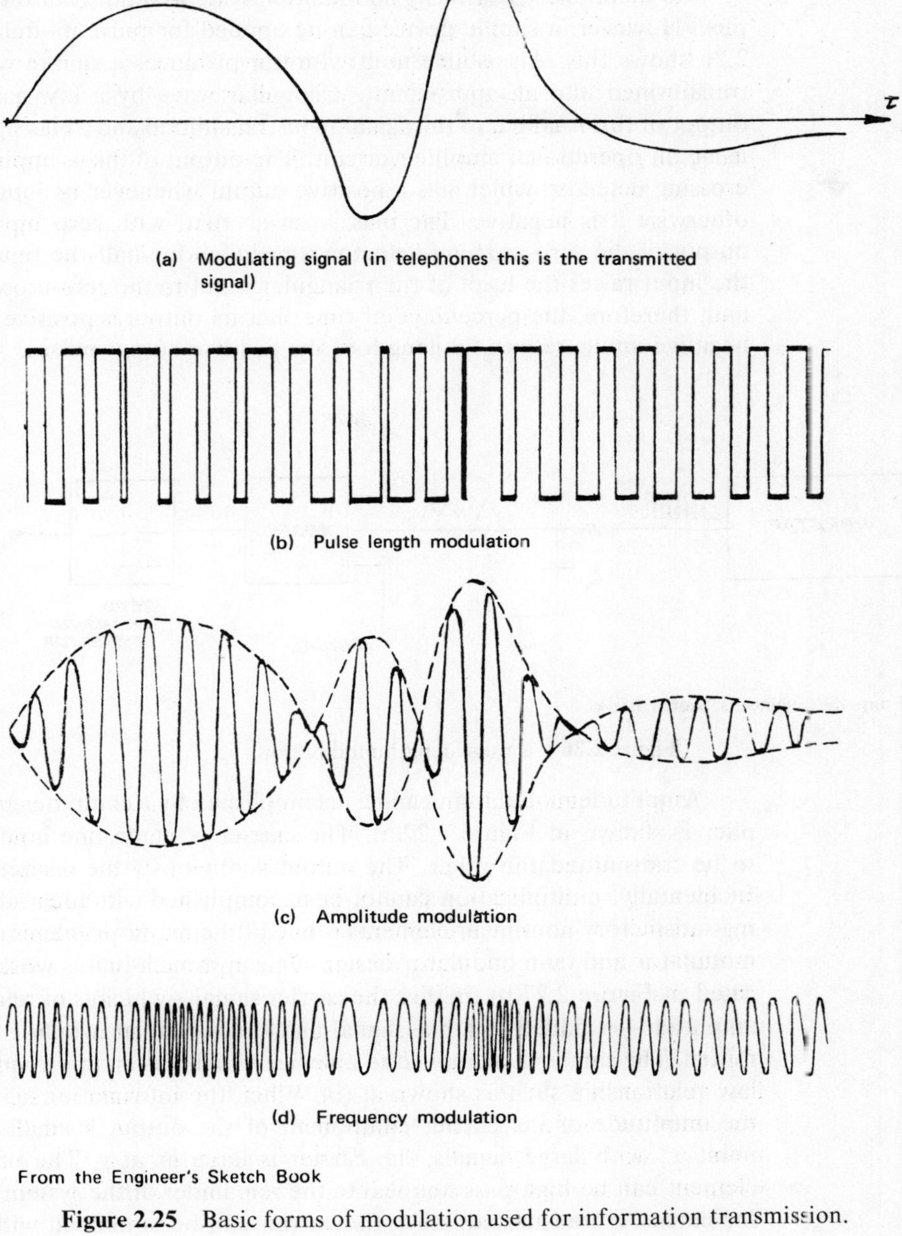

From the Engineer's Sketch Book

Figure 2.25 Basic forms of modulation used for information transmission.

waves. This system is also used in quite a bit of instrumentation work; for example, in the tape recording of experimental data.

As mentioned previously, modulation systems tend toward the very complex. However, a simple device can be applied for pulse modulation. Figure 2.26 shows this. An astable multi-vibrator produces a square wave which is transformed into an approximate triangular wave by a low-pass filter. The output of this is added to the signal to be transmitted and a bias signal, perhaps using an operational amplifier circuit. The output of this is applied to a zero-crossing detector, which has a positive output whenever its input is positive; otherwise it is negative. The bias is set so that, with zero input signal, the output of the zero-crossing detector is positive one-half the time. Increasing the input raises the level of the triangular input to the zero-crossing detector and, therefore, the percentage of time that its output is positive. Conversely, negative inputs reduce the length of the positive output pulses.

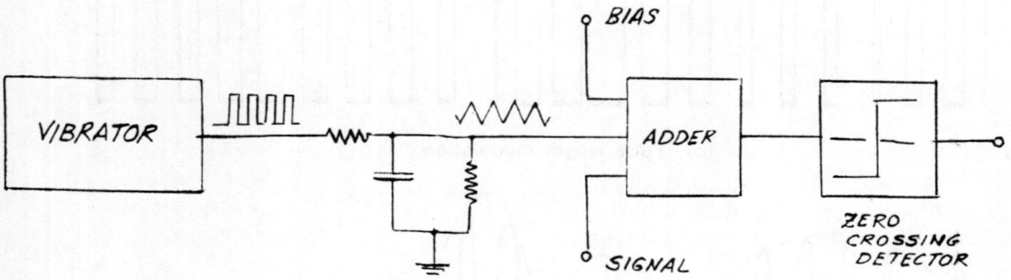

From the Engineer's Sketch Book

Figure 2.26 A pulse-length modulation.

Amplitude modulation can be accomplished by a computer-type of multiplier as shown in Figure 2.27(a). The carrier is made one input; the signal to be transmitted the other. The output is obviously the desired waveform. Incidentally, multiplication cannot be accomplished with linear devices; finding satisfactory non-linear elements is one of the major problems of amplitude modulator and de-mondulator design. One approach that is workable is indicated in Figure 2.27(b). In this, the carrier signal—which is of relatively small amplitude—is added to a bias signal and the waveform which is to be transmitted. The sum is then applied to a device which has an approximate square-law relationship such as shown at (c). When the information signal is small, the amplitude of the carrier component of the output is small as shown at point x; with large signals, the carrier is large as at y. The output of the element can be high-pass coupled to the remainder of the system so that only the carrier is transmitted. Also, the carrier can be combined with an out-of-

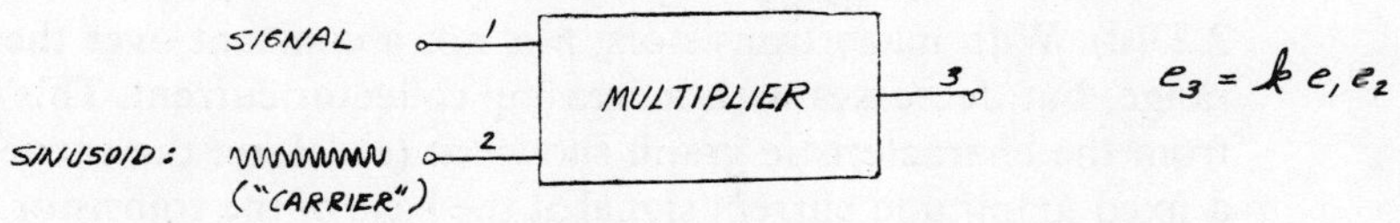

(a) Multiplier used for amplitude modulation

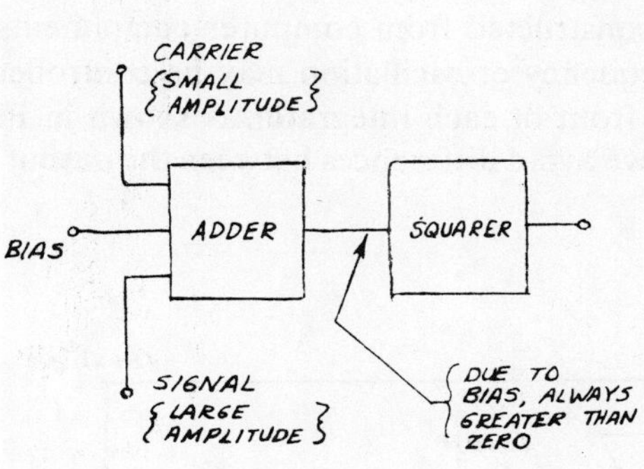

(b) Squarer used to accomplish modulation

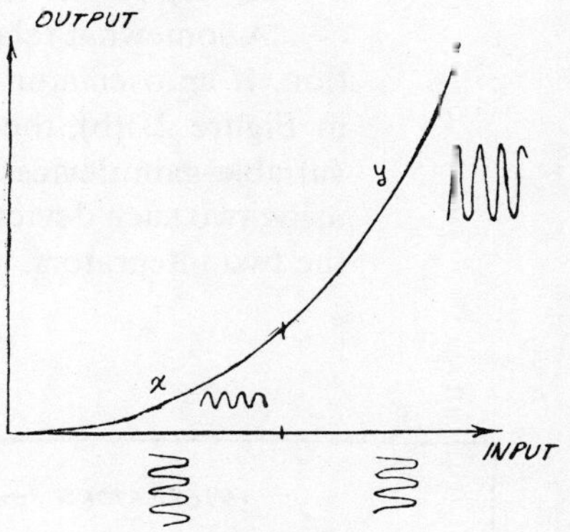

(c) Squarer modulator analysis

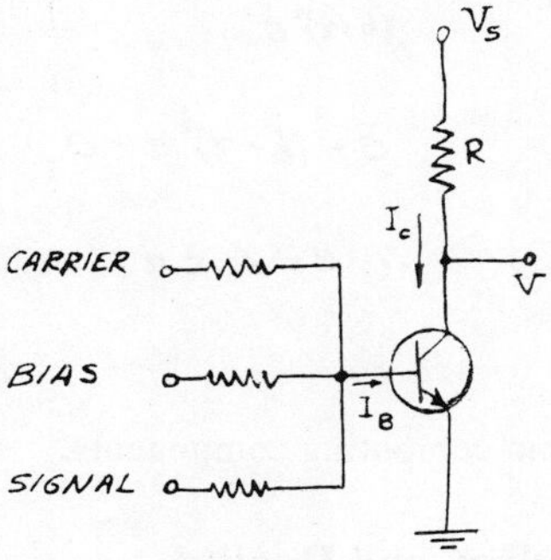

(d) A modulator which utilizes the non-uniform amplifier characteristics of a transistor

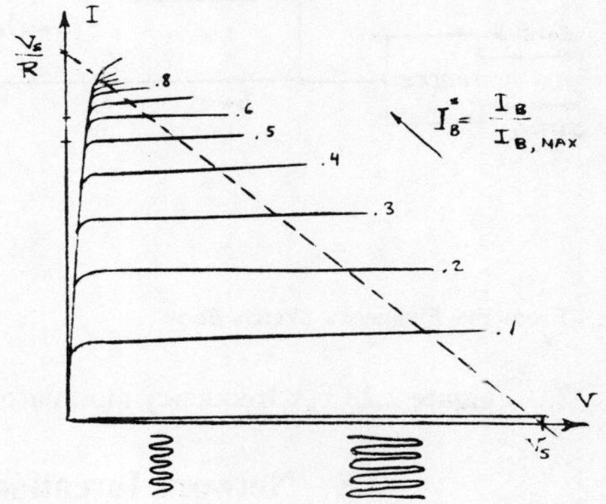

(e) Transistor modulator analysis

From the Engineer's Sketch Book

Figure 2.27 Basic aspects of amplitude modulation.

phase reference carrier so that the resulting amplitude is zero when the information signal is zero.

A practical way of carrying out the above approach is indicated in Figure 2.27(d). With many transistors, β is not a constant over the whole operating range, but decreases with increasing collector current. This effect is apparent from the characteristic graph shown at (e). Here, the carrier is introduced as a fixed amplitude current signal at the base of the transistor. The information signal causes large excursions of the operating point, such that the changes in β cause the carrier to change in amplitude as required.

A somewhat related approach may be used to achieve frequency modulation. If an oscillator is constructed from computer components, as was done in Figure 2.3(b), the frequency of oscillation may be controlled by inserting variable-gain devices in front ot each integrator as shown in Figure 2.28. By using two such devices, we avoid differences between the output amplitudes of the two integrators.

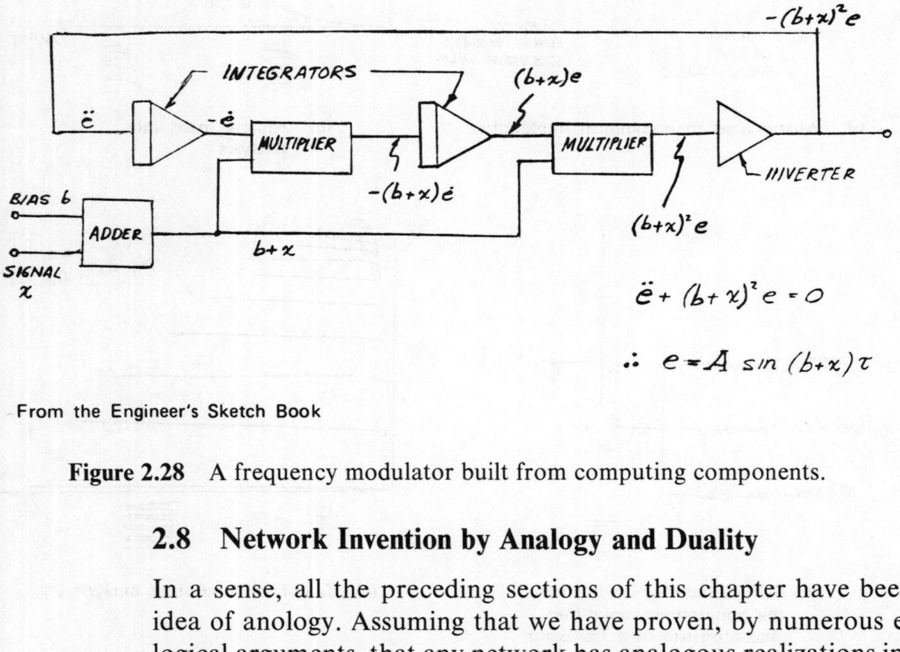

Figure 2.28 A frequency modulator built from computing components.

2.8 Network Invention by Analogy and Duality

In a sense, all the preceding sections of this chapter have been devoted to idea of anology. Assuming that we have proven, by numerous examples and logical arguments, that any network has analogous realizations in mechanical, hydraulic, and electrical forms, we devote this section to a lesser known, but somewhat similar concept. Because it is not well known, it offers a very rich area for academic exploration and has many interesting practical implications.

This concept is known as "duality." In a crude sense, it is that which can be ascertained by the interchange of paired quantities and relationships. In geometry, duality appears as a pattern relating the quantities *point* and *line* to the various ideas such as *betweenness*, *greater than*, and *less than*. In system engineering, the relationships are between the two quantities which are multiplied together to get power (Γ and Δ) and the topology of networks. If the reader will bear with us, we will develop the subject abstractly at first, reserving the practical interpretations for awhile. We begin by showing how two system components can be dually related, and then proceed to dually related networks.

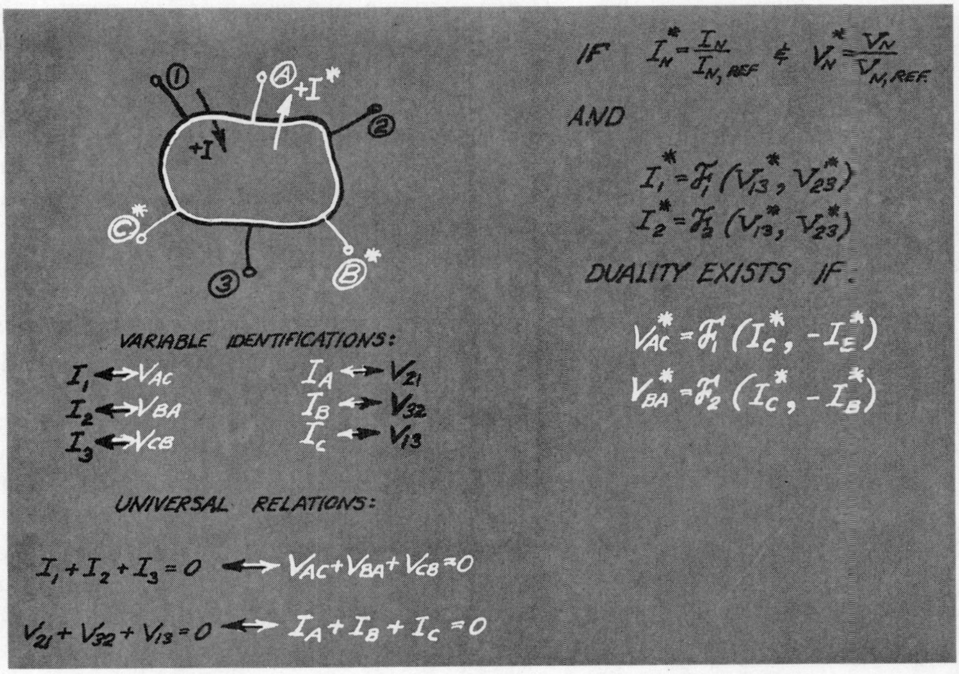

From the Engineer's Sketch Book

Figure 2.29 Duality between electrical components.

At Figure 2.29 is shown the procedure for determining the relationships between two three-terminal, dual electrical components. The basic idea is that each current in the original component is dually related to some voltage difference in the dual component, and some current in the dual corresponds to a voltage difference in the original. It is convenient to make a sketch which shows the terminals of the original alternatingly arranged with those of the dual (see Figure key for reference to the duals). In this way, we may identify which currents go with which voltage differences; specifically, the current at any terminal is understood to correspond to the voltage difference between the adjacent dual component terminals. Here, we have arbitrarily chosen to make clockwise voltage increases positive; in the figure, for example, V_{21} is positive if V_2 is greater than V_1 and this is taken to correspond to positive I_A, which we take as positive outward.

A somewhat mysterious aspect of this business is that we must define the positive currents of the original component and those of the dual component in opposite directions; otherwise, physically unrealizable dual components are developed. If this is not done, the dual of a dissipator is a generator, the dual of a storage device increases in stored energy as energy is removed from it, and other weird things happen. Although these are interesting, they are not useful in engineering and we use the convention metioned earlier. In this, the dual of a dissipator is another dissipator and the dual of a statically conductive storage device is a statically non-conductive storage device. This leads to devices and networks of practical interest. With regard to the example, inward currents at terminal 1, then, correspond to positive voltage rises V_{AC}, while outward currents at terminal A correspond to positive voltage rises V_{21}. Below the component symbol are shown the corresponding variables. Below this are shown the relationships which may be written without knowledge of the nature of the component (these come from universal conservation laws and definitions). It should be noted that we may replace the variables in any of these with the dually corresponding variables and develop another true relationship.

In order to formally establish duality, we must normalize the variables involved. This is indicated at the upper right-hand corner of the figure. In this case, the reference quantities need not be maximum values; any quantity that will satisfy the necessary conditions will do. Finally, duality is established by inspection of the characteristics of the original component. If some device can be found which has the same characteristic *functions* ($\mathscr{F}_1$ and $\mathscr{F}_2$ here), with the corresponding variables interchanged, the device is the dual of the original. As a simple example, it is easy to see that a resistance is the dual of a conductance, since

$$E_{12} = I_1 R$$

$$I_B = E_{AB} G \qquad (G = 1/R).$$

Here, we take the order of the terminals to be 1, A, 2, B in the clockwise direction.

Determining the duals of rotary mechanical components is a bit more complicated, as indicated in Figure 2.30, because of the way that the conserva-

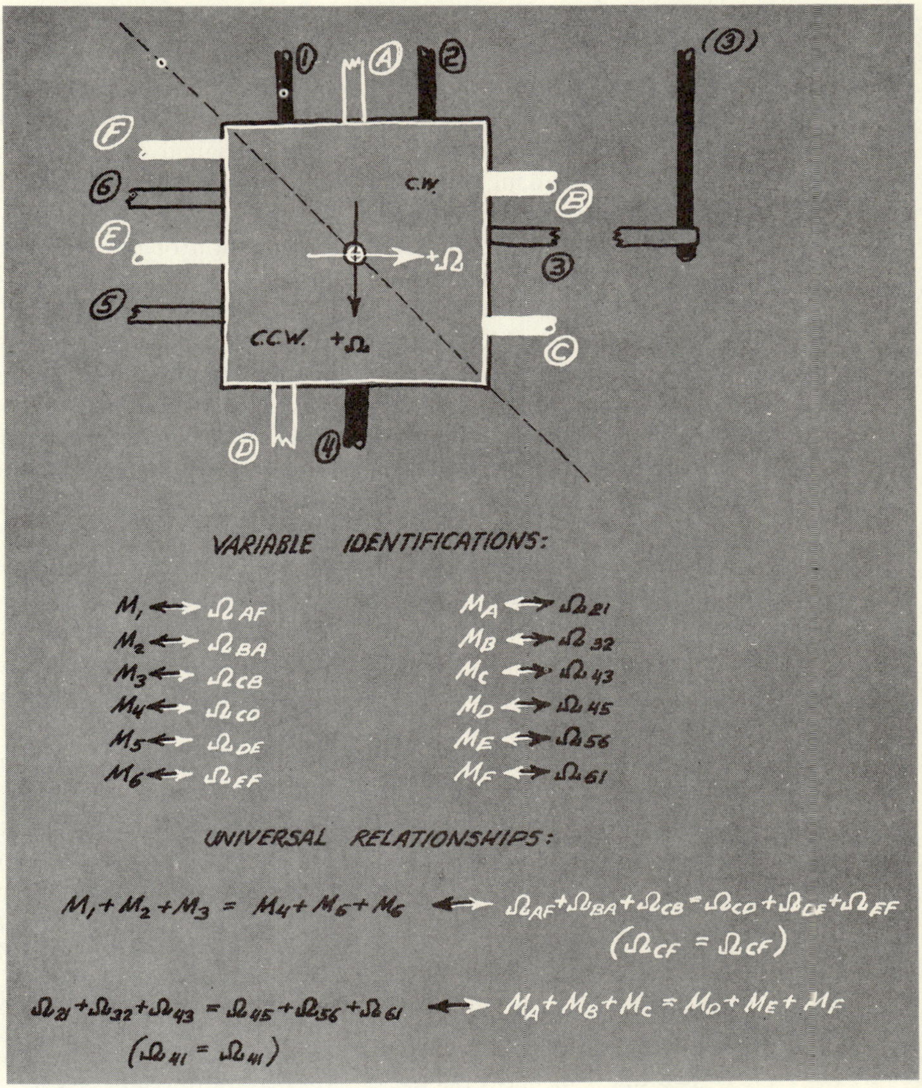

From the Engineer's Sketch Book

Figure 2.30 Dual relationships for rotary mechanical components.

tion of momentum is written. In the example shown in the figure, the positive Ω direction of the original component is downward and the positive direction for the dual component is selected toward the right. The component to be dualized is enclosed in a rectangular box, for convenience. A diagonal, running between the two positive rotation vectors, is drawn. In one-half the diagram, clockwise increases in speed are related to positive (tightening) torques, whereas in the other half, counterclockwise increases correspond to tightening. For example, in the figure, M_1 corresponds to Ω_{AF} while M_D corresponds to Ω_{45}. This rule provides that dissipation in the original component always yields dissipation in the daul; it does not matter which half of the box is selected for the "clockwise positive" direction.

In Figure 2.31 is shown the determination of the dual for a diode. At the top left is shown the original component, at the top center is the original and its dual superposed, and at the top right is the dual alone. The variable identifications are below these sketches and then the universal relations follow. Finally, the characteristics of the diode are described in a simple way, the interchange of corresponding variables is made, and it is determined that a diode is its own dual. It might be noted that the small amounts of dissipation (or imperfection) in the diode are carried through the determination although the exact form of the characteristics is not; in all of this, as in analogy, there is some room for approximation.

The dual of a transfer clutch (with a conductive control) is established in Figure 2.32. In this case, the procedure is straightforward, but the result is not. In fact, the dual turns out to be an interesting engineering component which is quite constructable, but which has not become commercially available. The dual is another kind of clutch. It is essentially a constant-speed dissipator where the speed at which power is dissipated is determined by the relative speed between the runner and the spider shafts. The unit could be built quite readily if we would be willing to hold the spider shaft at zero speed relative to the earth; the runner control would then be a flyball actuator and the drum-to-spider friction would be developed by centrifugal forces on the shoes or other friction members. The flyball actuator would provide a bias working against the centrifugal force; increases in Ω_{ZY} would tend to increase the force on the frictional elements and, therefore, the torque transmitted from the spider to the drum. It should be noticed that the original device is a constant-torque dissipator with a conductive control; the dual is a constant-speed dissipator with a non-conductive control.

The duals of networks are determined quite readily as shown in Figure 2.33. In any network schematic (with some restrictions), the paper is divided into a number of separate areas. Each component borders on as many areas as it has terminals; the dual of each component has terminals in each of these areas. The dual of a network is simply the network which is indicated when we connect all the terminals of the dual components, drawn as before, which enter a particular area of the schematic of the original network. For example, in Figure 2.33 the duals of both components II and IV have terminals in area

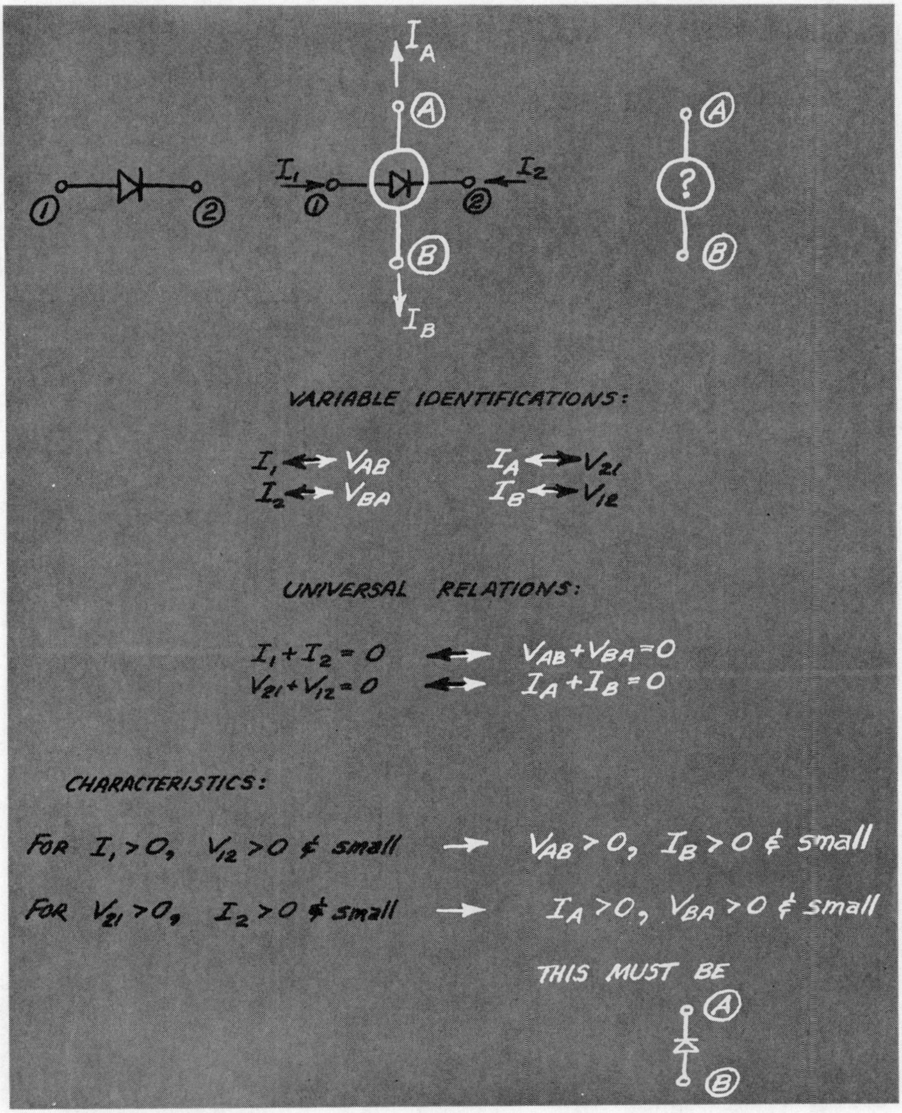

From the Engineer's Sketch Book

Figure 2.31 Determination of the dual of diode.

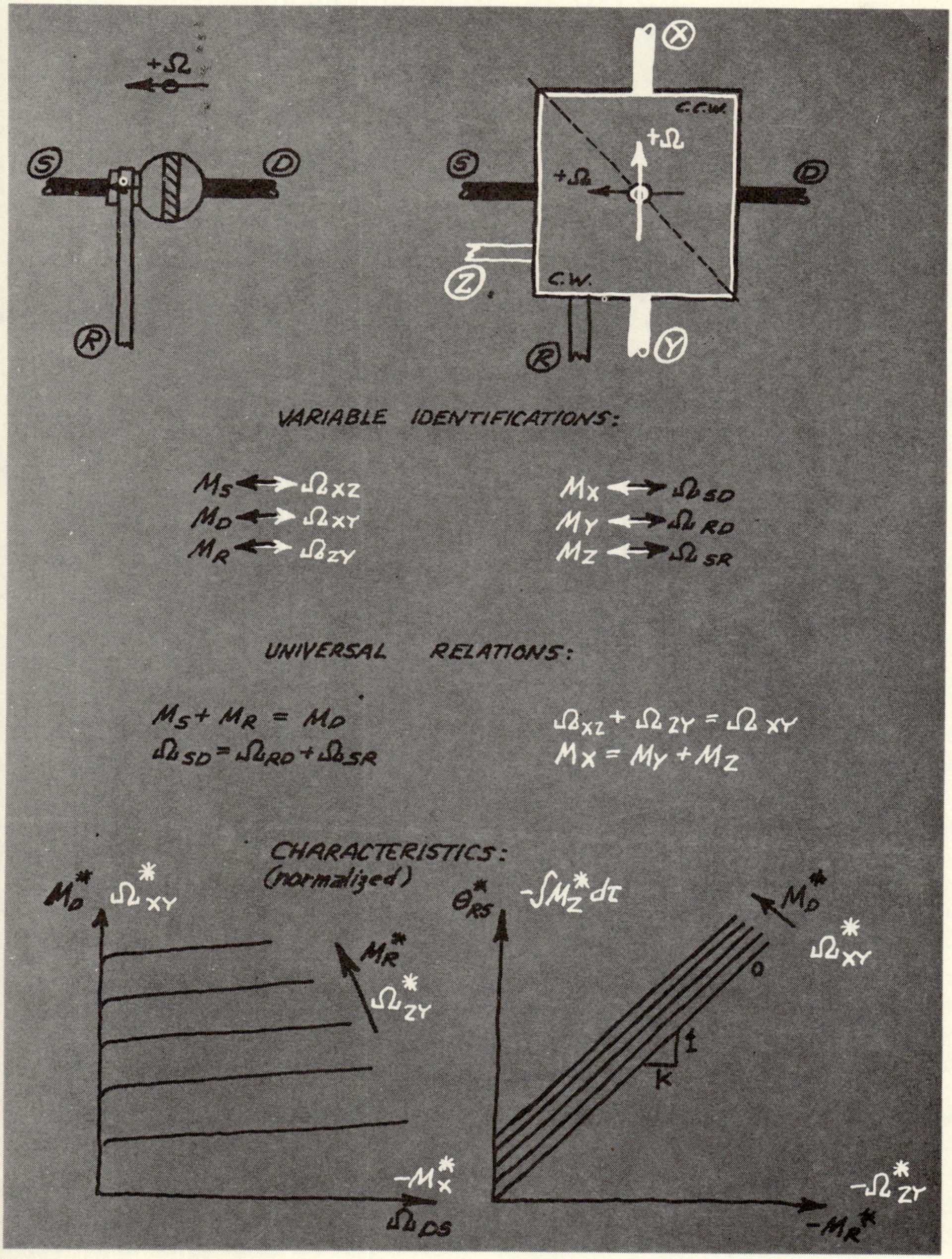

From the Engineer's Sketch Book

Figure 2.32 Determination of the dual of a clutch.

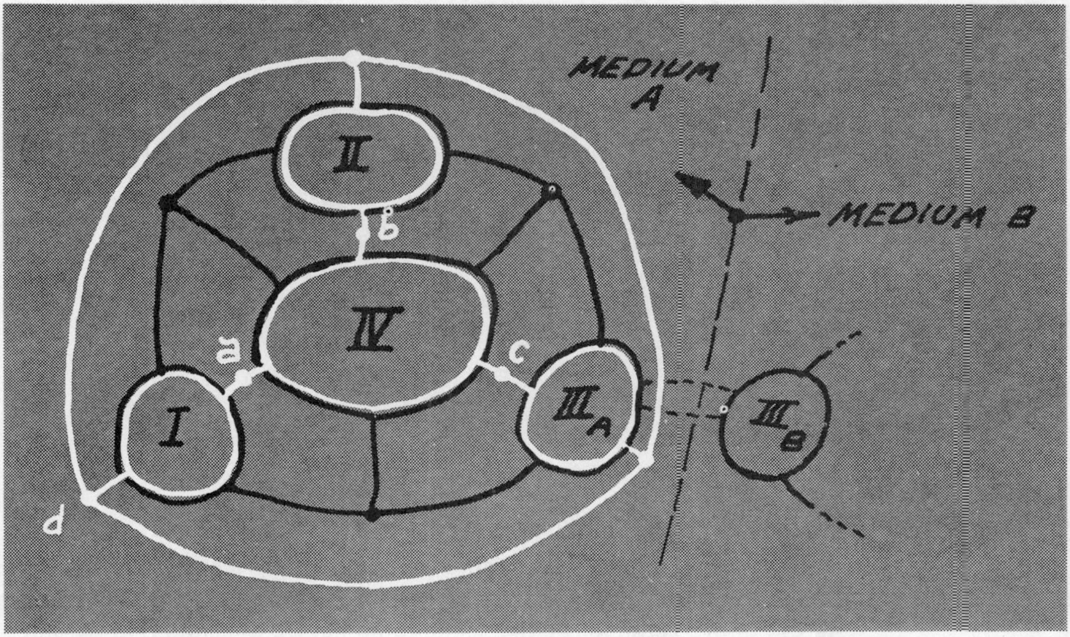

From the Engineer's Sketch Book

Figure 2.33 Dual networks.

b; in the dual network, these are connected. In the case that more than one medium is involved (such as A and B in this example), the dualization may be confined to a single medium if desired. That is, we may dualize the network of medium A and leave that of medium B undisturbed. The component symbols suggested in Chapter 1 have been developed to accomodate this process.

In Figure 2.34 is shown an example of the usefulness of the dualizing process. The original network is an upper selector ($P_3 = P_1$ or P_2, whichever is greatest), drawn in a somewhat unusual configuration. The dual network is developed in the center and isolated at the right. As might be expected, the pressure generators P_1 and P_2 become flow generators (W_1 and W_2) in the dual; the check valves are self-dual, as may be anticipated from the dualization of the diode. The dual network has the interesting property that the flow through the restriction is the greatest of the two flows W_1 and W_2. The deployment of flow, in the case that $W_1 > W_2$, is shown in the right-hand sketch. From a designer's point of view, this scheme is quite nice because, if he knows of some network which does the dual of what he wants to do, he may be able to derive the network he needs in a methodical instead of happenstance way.

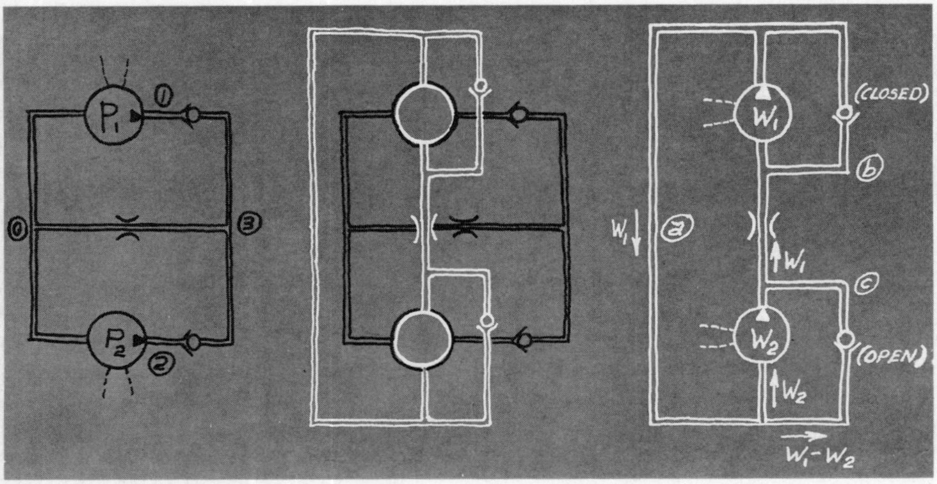

From the Engineer's Sketch Book

Figure 2.34 The hydraulic upper-pressure selector and its dual.

Also, by understanding the relationships of duality, his overall grasp of system concepts is markedly advanced.

As one more example of the remarkable nature of duality, a hydraulic bistable arrangement constructed from spool valves is considered in Figure 2.35. The original network is shown at (a); here, the valves are oriented similarly so as to obviate the necessity of determining their dual twice. This makes the network appear unfamiliar, but it is the same as considered previously. The general form of the dual network is determined at (b). Notice that the crossing of lines does not divide the paper into areas in a proper vay; a component labeled "X" has been put in place of the crossing in order to allow the method to be applied. We will take the dual of this component just as if it were something of more interest.

At (c) in the figure are shown the duals of the crossing and the valves. Surprisingly, the dual of a simple crossing is not very simple: it must have equal flows of alternating directions at each terminal, and the opposing pressures must be equal. Fortunately, this kind of device can be built. In hydraulics, it has the nature of a cavity which is divided into four regions by a rotor and specially shaped stator; the symbol indicates this roughly. Continuous-acting units can be constructed from gear-type pumps. The dual of a spool-type valve is approximately a vortex valve. The first has a non-conductive control while the second is conductive; in the bistable circuit here, we don't care about the characteristics other than for "off" and "on"; this makes the duality quite good.

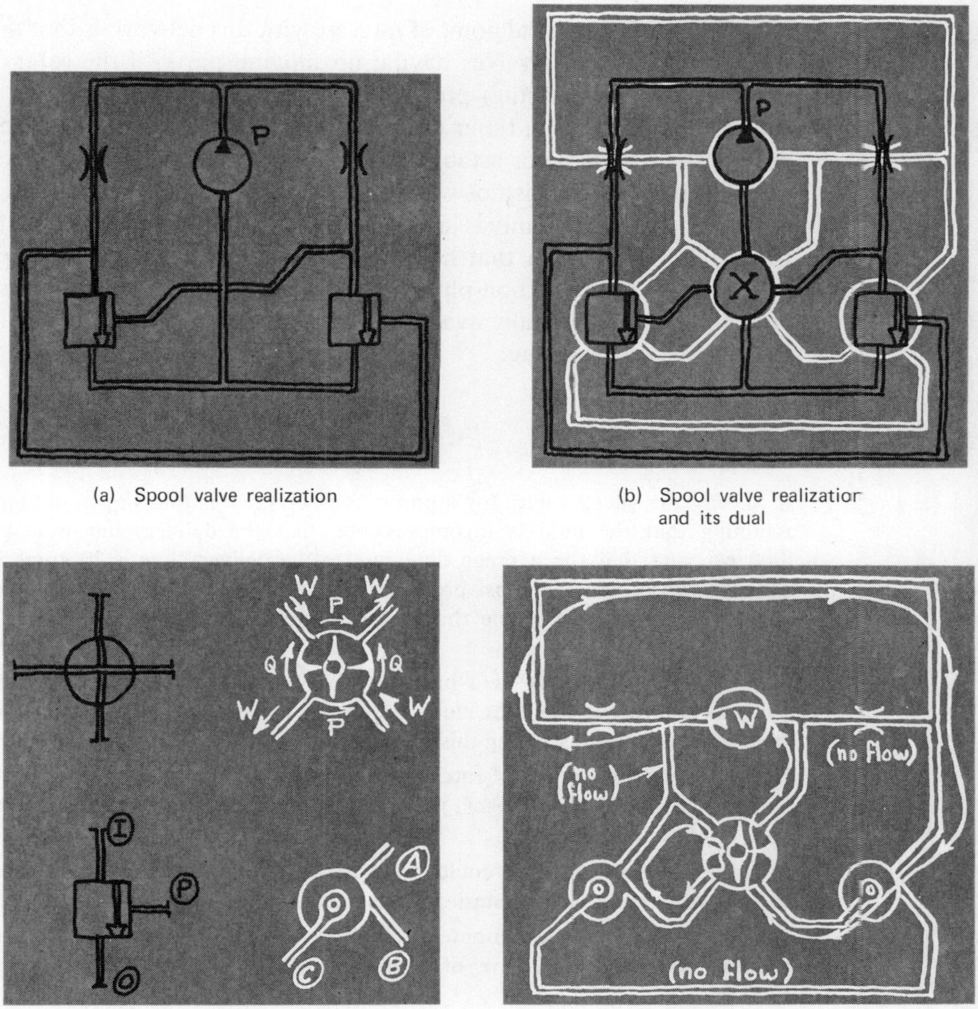

(a) Spool valve realization

(b) Spool valve realization and its dual

(c) Duals of line crossing and valve

(d) Dual network in one of the stable states

From the Engineer's Sketch Book

Figure 2.35 Bistable hydraulic spool-valve network and its dual.

The final dual network is shown at (d); the flows for one of the two stable states are indicated. Here, it has been assumed that the vortex valve is completely conductive when there is no control flow, and that the control flow may be sucked out the inlet when the outlet pressure is high with respect to the inlet. The main practical point of interest with this network is that the control is accomplished with valves having no moving parts; if the rotary actuator (dual of the line crossing) provides mechanical work to a system, it can be caused to operate as a timer or vibrator by providing the usual facilities required of monostable or astable vibrators.

Although duality is not widely recognized as a means for engineering innovation, it most certainly is just that. In the way of encouragement, the author would like to mention that he has never been able to find a component or network which had a non-physical dual. Since very few dual pairs of components are commercially available, there *must* be opportunity for invention via this line of reasoning.

Problems

1. Specify the check valves for a pump rectifier such as that shown in Figure 2.2(b), assuming that the fluid is incompressible, that the delivery line is at a constant 1000 psi, and that the average flow rate in the delivery line is 20 gpm. Supposing that the valves have a 50-psi pressure drop across them when passing fluid in the forward direction, determine the wasted power (use HP units) and the efficiency of the system.

2. What is the efficiency of a bridge of over-running clutches, as in Figure 2.2(c), if the friction torque when the clutch is running in reverse is 5% of the torque at which the bridge operates, assuming this to be constant?

3. Draw the arrangement of ratchets which allows a double acting mechanical jack to operate. (In this kind of jack, the load is raised on both the upward and the downward strokes of the handle.)

4. Show how an oscillator could be constructed by using one computer integrator and one differentiator, plus static elements as necessary.

5. Suppose there was a computer element which could directly produce the double integral, with respect to time, of its input. Could an oscillator be constructed from it? How?

6. Explain what must be happening when, in a public-address system having the microphone fairly close to the speakers, a high-pitched whine develops. Assuming that there is negligible phase shift in the system electronics, predict the frequency of the whine if the microphone and speakers are 110 feet apart.

7. In some record players, where the speaker is mounted in the cabinet with the turntable, an objectionable noise develops when the volume control is turned up too high. Explain what must be occurring.

8. If the stem-and-washer assembly on a kitchen faucet becomes badly worn and loose, it is possible for the pipes throughout the house to whine at certain openings of the valve. The frequency of the whine is found to be related to the length of pipe between the faucet and the first major flow obstruction, such as the water meter or pressure regulator. Can you explain what is occurring?

9. Devise low-pass filters built with the following:

(1) A rotary spring and rotary dissipator.

(2) A rotary spring, dissipator and inertia, where the dissipator does not act when the power transmission is smooth.

(3) A resistor and inductor.

(4) A resistor and capacitor.

In each case, specify the variable which would be smoothed by the filter and show how sources and loads would be connected to it.

10. Devise a flyball actuator, dog clutch bistable network in which the actuator is not rotating when the clutch is engaged.

11. Devise a bistable relay arrangement in which the electromagnet is not energized when the contact is closed. (*Hint*: Use a power supply that can withstand short-circuiting to ground.)

12. Devise an hydraulic valve, fluid restriction arrangement in which there is no flow through the restriction when the valve is open, but which is bistable.

13. Devise a bistable transfer-clutch, rotary dissipator arrangement in which the elements are in parallel.

14. Show how two flyball actuators can be used to construct a bistable network in which one rotates when the other does not.

15. Suppose that, in the relay network shown at Figure 2.8(a), 100 ma is required to hold the relay closed and the coil resistance is 50 ohms. If a supply voltage of 20 is to be used, what should be the resistance and power ratings of the resistors?

16. For the network shown at Figure 2.8(d), develop a general expression for the dissipator value R (RPM/in.-lb) as a function of the supply speed Ω_S, the clutch runner slip Ω_{RQ} to maintain a locked spider, and the clutch runner to spider torque ratio (assume this is much greater than unity) K_{PR}. This quantity is the ratio of the spider torque to the minimum runner torque that will maintain lock, and may be assumed constant for this problem.

17. Devise an astable multi-vibrator based on Figure 2.7(a), (b), (c), (d). Be sure to include specific elements for determining the vibration rate.

18. Make a sketch of the voltage waveforms which exist in the circuit of Figure 2.9(b) at points w, x, y, and z. (*Hints*: The voltages at x and y never go over half a volt positive; the voltages at w and z are out-of-phase square waves.)

(1) Same, for (a) in the figure.

(2) Same, for (c) in the figure.

(3) Same, for (d) in the figure.

19. Devise a rotary mechanical monostable multi-vibrator.

20. Devise an event timer based on an astable relay circuit.

21. The judicious use of director elements allows a bistable network, constructed from a single valve-like element, to be adapted to event timing. The director elements serve to bring the timing components into play when the circuit is in its temporary state. Devise such a circuit which is

(1) electronic. (3) rotary mechanical.
(2) hydraulic. (4) electromagnetic.

22. Specify all the components for an asymmetric astable multi-vibrator where [Figure 2.9(b)] Q_2 is to conduct for 100 msec and then be off for 200 msec. The supply voltage is 25, the transistors are to have 1ma of base current when on, and the collector current level is to be 50 ma. In addition to the R and C values, give resistor power and capacitor voltage ratings.

23. Devise a rotary mechanical follower network, using a transfer clutch.

24. Show how two valve-like elements could be used in each of the networks shown in Figure 2.11. Use a minimum of dissipative elements—one only in (c) and (b), for example.

25. Devise an electronic and an hydraulic high-gain amplifier where the valve-like element is in parallel with the dissipator. Assume that a constant current or a positive displacement source is available.

26. Show four configurations of manual, electrical regulators where the variable dissipator is a potentiometer. Specify any condition under which the regulator would be ineffective.

27. Show four configurations of manual mechanical regulators and specify any conditions under which they would be ineffective.

28. Show a revision of the regulator of Figure 2.15 which would allow a constant-torque dissipator to serve as the reference element.

29. In order to get the highest possible accuracy from the regulator of Figure 2.15(c), should R_S be made as restrictive to flow as possible?

30. Suppose that, in the regulator at Figure 2.15(b), the Zener diode voltage does not stay perfectly constant, but tends to increase as increased current flows through it. Does this tend to compensate, or strengthen, the regulation error due to R_S?

31. What happens in each of the regulators shown in Figure 2.16, if the reference dissipator destructs so that

(1) The **Δ** variable goes to zero.
(2) The **Γ** variable goes to zero.

32. What is the maximum possible efficiency of the regulators of Figure 2.17 if the maximum load power is N times the average?

33. In the regulator of Figure 2.17(d), what should be the value of R_P (RPM/in.-lb) if the clutch maximum torque rating is 100 in.-lbs and the supply speed is 1200 RPM?

34. Suppose that R_B were left out of each of the circuits of Figure 2.18. Would the valve-like device be rendered inoperative in every case? If not, which ones and why?

35. Design, and make carefully prepared cross-sectional drawings of, a speed-difference regulator which bears the same relationship to Figure 2.15(d)—with the constant-torque reference modification as suggested for Figure 2.17(d)—that Figure 2.19 bears to Figure 2.15(c).

36. Show for the (a), (b), (c), or (d) portions of Figures 2.15 through 2.18 how the use of two valve-like elements is possible.

37. Devise a series-type pressure regulator utilizing a snap action pilot operated relief valve. Refer to Figure 2.21 for background information.

38. Devise an arrangement of transfer-clutches which is analogous to the transistor regenerative switch, and therefore the SCR, of Figure 2.22. Note that the runner torque will be considerable in this situation; we would probably elect clutches having zero runner slip speed, at lock, in order to keep heating and wear to a minimum.

39. Suppose the SCR control circuit of Figure 2.24 is to operate from 60-cycle, 115-volt power and that the load draws 10 amperes maximum. Specify all the system components, giving value, power rating, and other necessary quantities, if the SCR requires a 1-millisecond gate current pulse, of at least 1 ma and 1 volt, to turn it on. Take $V_S = 10$ volts.

40. Devise a transistor circuit which would act as a pulse-length modulator. (*Hint*: Both the adder and the zero-crossing detector can be built from simple high-gain transistor amplifiers.)

41. Derive the equation of the square law amplitude modulator of Figure 2.27(b) and (c). Then compare the result with the ideal input/output relationship that would be achieved by the system in (a).

42. Show how a rectifier and filter combination might be used to "undo" (formally, "detect") amplitude modulation. How should the critical frequency of the filter relate to the carrier frequency?

43. Would an integrator or differentiator be of any help in detecting frequency modulation? If yes, show how.

44. If only one variable-gain device is used in the circuit of Figure 2.28, amplitude modulation will also occur. Show why.

3 Static characteristics of components

Probably the most taken-for-granted aspect in all system engineering is the static characterization of components. Eloquent testimony to the misfortune of this situation is given by the multitude of incomplete, uninterpretable, clumsy, and outright incorrect characterizations which we may obtain by writing for manufacturers' literature on system components. In the modern disciplines in which system engineering concepts are well established, semiconductor electronics being a notable example, the matter of component characterization is admirably well done. In some of the more traditional disciplines, the mechanical area being notable, the opposite is true. A fundamental objective of this chapter is to show *exactly* what a proper static characterization should be; it is hoped that the reader will join the author in promoting higher standards in this area.

The static characteristics of components serve several basic functions in system engineering. The first of these is in the component selection process where the shape and levels of the characteristic curves are used to ascertain that proposed combinations of components will indeed operate together as desired. Because the characteristics show how component variables change with respect to one another, they provide the basic information for error analysis and for small excursion dynamic analysis. Characteristics also indicate the levels at which components can be destroyed, both by overheating and by overstressing. Finally, characteristics give information on the temperature sensitivity of the component, as well as other information on how it relates to its environment.

The aspect of characterization that is not widely recognized is that there is a set of universally applicable principles which apply to the process. If we are acquainted with these principles, we can prepare proper characterizations of the components we have designed, we can determine possibly confusing imperfections in characterizations prepared by others, and we can understand completely unfamiliar components more easily and rapidly. The derivation and explanation of these principles makes up the subject matter of this chapter. We begin by examining several (unusually well-done) characterizations prepared by manufacturers. These include examples from electromechanical, hydraulic, and electronic power areas. We point out a few questions that might be asked about the why and wherefore of these examples, and then proceed to answer these in the remainder of the chapter. Toward the end of the discussion we review the examples, looking to see what kind of agreement our principles and their practice exhibit.

The kinds of components we shall consider are the following:

(1) *Translational mechanical* components in which the motion occurs along parallel lines that do not move with respect to an inertial reference, in which power is transmitted by bodies which may be considered rigid, and in which there is no other form of power transfer except the dissipation of heat.
(2) *Rotational mechanical* components having inertially fixed parallel axes of rotation, rigid shafts, and no other form of power transfer except heat dissipation.
(3) *Hydraulic* components having non-accelerating cases, parallel-inlet and outlet lines, operating on incompressible fluids, and having no other form of power transfer except heat dissipation.
(4) *Electrical* components operating on direct current and having no other power transfer except heat dissipation.
(5) *Combined* components which include two or more of the above kinds of power transfer.

Note that this list includes practically all the components described in Chapter 1 so long as the application is normal. The characterization of certain other kinds of mechanical components, fluid components which utilize compressible fluids, and alternating current electrical components is closely related to the cases considered here, but does not yield too well to a generalized discussion; therefore, these cases have been reserved for Chapter 8. The reader may wish to look that material over upon completing this chapter; he should have no trouble with it. Actually, the methods we shall develop here apply to a broader range of components than has been listed above; the author is simply wary of claiming applicability for cases which he has not considered in some detail.

It is a convenience in this discussion to be able to discuss the component types without reference to the power-transfer medium involved. This is facilitated by the use of the following defined quantities:

The Γ variable

Tension or compression *force* in translational mechanical systems.
Tightening or loosening *torque* in rotary systems.

Flow rate in hydraulic systems.
Current in electrical systems.
In a two-terminal dissipator, the magnitude of the **Γ** variable is the same at each terminal.

The Δ variable

Translational *speed.*
Rotational *speed.*
Pressure.
Voltage.
In a two-terminal dissipator in which **Γ** is not zero there is some difference, or drop, in the magnitudes of the **Δ** variables.
Also, the product of the magnitude of **Γ** with that of the **Δ** drop is proportional to the power being dissipated.

Note that conservation relationships, such as those for momentum, mass, and charge, are written in terms of the **Γ** variables. Also, **Δ** variables are always measured on a relative basis; absolute definition of them is impossible.

The **Γ** and **Δ** variables have been used by many authors and referred to in several ways, the most common being as the "through" and "across" variables. In this usage, reference is made to the way in which the variable is to be measured. For example, in an electrical network, voltage is measured by connecting a voltmeter *across* the component at which voltage is of interest while current is measured by connecting an ammeter so that the current in the component also goes *through* the meter. This is quite a workable usage, but seems a bit obtuse to this author. Therefore, I have adopted the **Γ** and **Δ** notation and suggest that these symbols can be read in any way the reader likes; i.e., read:

Γ as "gamma," "through," "conducted," or "conserved."
Δ as "delta," "across," "difference," or "dropping."

3.1 The Characterization Problem

The proper characterization of a component must allow the engineer to predict exactly how the device will perform when installed in a system. This prediction takes the form of a determination of all the **Γ** and **Δ** variables which appear at the component terminals during static equilibrium operation of a system. The form taken by the characterization may be tabular, graphical, or functional; each method may be used to specify the relationships between the several terminal variables. For use in computer simulation of systems it is necessary to utilize the functional form of characterizations. For most paper-and-pencil engineering work, the graphical presentation is most convenient. Here, we shall exhibit examples of each form, although our discussion shall be phrased in terms of the functional presentation.

Before going further, it might be appropriate to say exactly what we mean by a "static equilibrium" condition. First, this does not mean the case where there is no motion, no flow, or no current. These are trivial and of little engineering consequence. Static equilibrium refers to the situation in which all the terminal variables have no rate of change with respect to time. There may be considerable power transfer under these conditions; a truck ascending a hill of constant grade at constant speed represents a system which is in static equilibrium. This condition is sometimes called the "quiescent" point of a system, especially in electronics work. Another point of definition is our use of the word "terminal." Generally, this refers to the bars, belts, shafts, pipes, and wires which come out of the casing of a component and which are used to connect the component to the system in which it operates; all the power transfer of the components we consider here occurs at these connections (except the dissipative heat transfer, which occurs over the whole surface of the casing). In order to achieve consistency in our presentation it is necessary to consider the place where a mechanical component is bolted down, if it is, as a terminal. This is just as important as thinking of all the wires coming out of a transistor as being terminals, even though one might be soldered to a ground connection. As a practical point, we can always take the casing of a mechanical device, such as a motor, and mount it on a rotating shaft; this brings out the full versatility of unit, which might not be considered if the casing is discounted in the identification of terminals.

Now let us look over a few examples of component characterizations and try to see what the whole business is about. The first example is a magnetically operated clutch coupling, the manufacturer's characterization being reproduced in Figure 3.1. (Incidentally, this represents a much more thorough presentation than is conventionally used in the clutch-brake field.) The clutch characterization consists basically of a graph which shows how torque varies as coil current and slip speed change. Listed in the specifications is the value of the coil resistance; this implies that we can get the coil voltage as a function of current. This characterization has two functions, one for torque and the other for voltage. There are four terminals (two electrical and two rotary mechanical); there are four useful variables (slip speed, torque, current, and voltage); given any two of the four variables, except voltage and current together, we can determine the others from the characterization. Some questions that we might ask about this device are: (1) Can it operate in reverse? (2) Is the coil voltage actually insensitive to slip speed, as implied? (3) Can it run indefinitely at any of the conditions shown? (4) What is the reason that the characteristics are not shown beyond certain values?

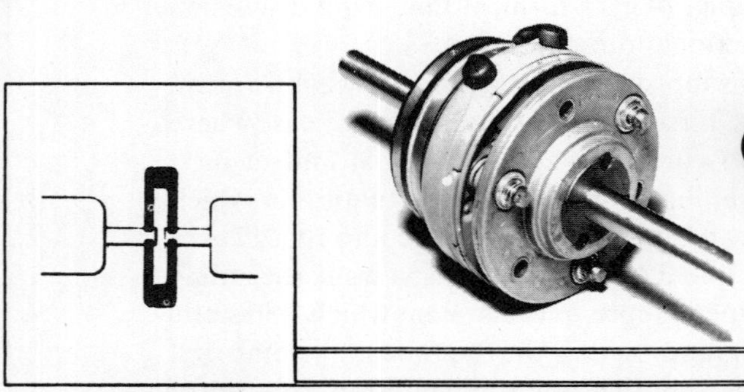

PCC-500
I-25542

**CLUTCH COUPLING
NORMAL DUTY**

TORQUE & HEAT DISSIPATION CURVES

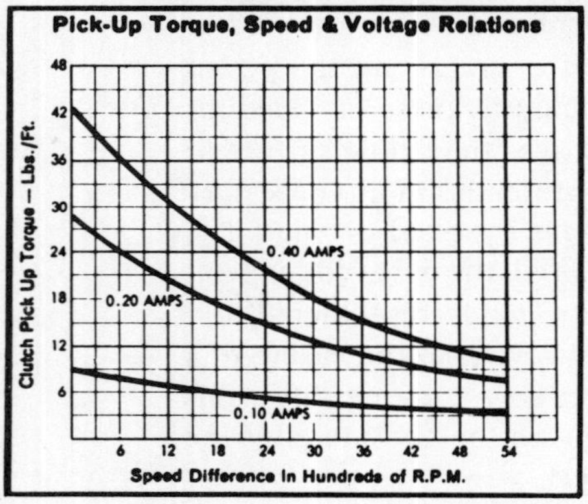

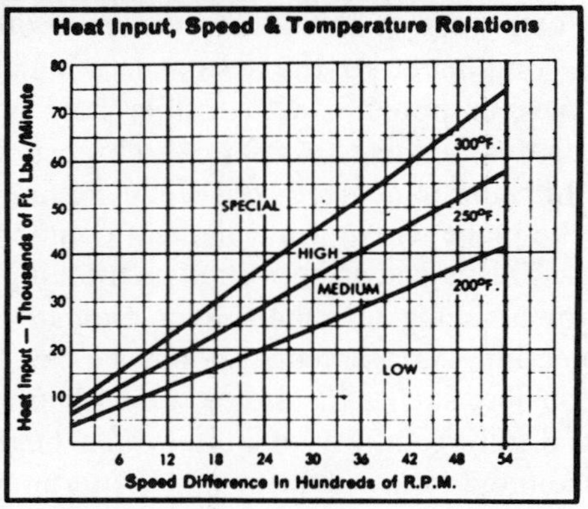

TECHNICAL DATA

MECHANICAL

Static Torque 40 lb. ft.
Maximum Speed 5,400 rpm

DESCRIPTION	Average Weight Lbs.	Inertia of Rotating Components Lbs. ft. ²
Magnet	2.800	.074
Magnet Hub Assembly	1.224	.031
Armature	1.017	.029
Armature Hub	.941	.018
Bushing	.436 Max. Bore to .842 Min. Bore	.002 to .003

ELECTRICAL

Standard Voltages - D.C.	6	90
Resistance @ 20°C - Ohms	1.76	218
Current - amperes	3.42	.413
Watts	21	37
Coil Build-up milliseconds	58	50
Coil decay - milliseconds	38	30

WARNER ELECTRIC

WARNER ELECTRIC BRAKE & CLUTCH COMPANY

193

BELOIT, WISCONSIN.

Figure 3.1 Manufacturer's characterization of electromagnetic clutch. Courtesy of Warner Electric Brake and Clutch Co.

A second, somewhat differently organized, example of characterization is shown in Figure 3.2. Here, the manufacturer of a positive-acting hydraulic motor, normally used within the wheel of large vehicles, has elected to present the performance of a whole line of products. The various curves shown are labeled with the model number of the unit represented; the arrangement of the presentation is helpful in the rapid selection of an appropriate component, given the system requirements. For any particular model of the motor, the curves give maximum power and torque versus speed, torque as a function of pressure, and flow rate as a function of speed. Since power can be computed from torque and speed, the set of power curves may be considered to be just a convenience to the designer. It appears then, that we have a four-terminal device in which four meaningful variables occur; the characterization is in the form of two independent relationships. Again, there are some questions which might be asked: (1) Is the torque actually independent of speed as implied by the second graph? (2) Is the flow rate actually independent of pressure, as implied by the third graph? (3) How about reverse operation, variable limits, and the possibility of continuous operation, as in the preceding example?

The third example, that of a widely used, relatively high voltage, Silicon *npn* transistor is shown in Figure 3.3. Here, the basic characteristic presentations are in the form of graphs showing, first, collector current as a function of collector-to-emitter voltage and base current (the region of small voltages is shown magnified in one graph); and, second, collector current as a function of base-to-emitter voltage and base current. This is a three-terminal device in which four useful variables appear. Two independent functions (graphs) are used to describe the relationships between the variables. Notice that the manufacturer gives a thorough list of the maximum allowable values for the variables as well as some information on thermal conditions under which the component operates. Some of the questions which might be asked are: (1) What does the component do if we try to operate it in reverse? (2) Why are there two graphs which give collector current?

Instead of trying to draw any conclusions from these examples, let us simply try to define the questions that must be answered before a component characterization can be carried out. These will reflect, mainly, the things which might appear arbitrary about the foregoing examples. We should keep in mind that the examples have been carefully chosen to represent responsible work; questions which the author wishes to discuss here have been answered properly by the engineers who prepared the examples.

Here are the questions which appear most important:

(1) How many variables appear in the relationships of a proper characterization?
(2) How many functions (or graphs) must be used to completely specify the performance of a component?

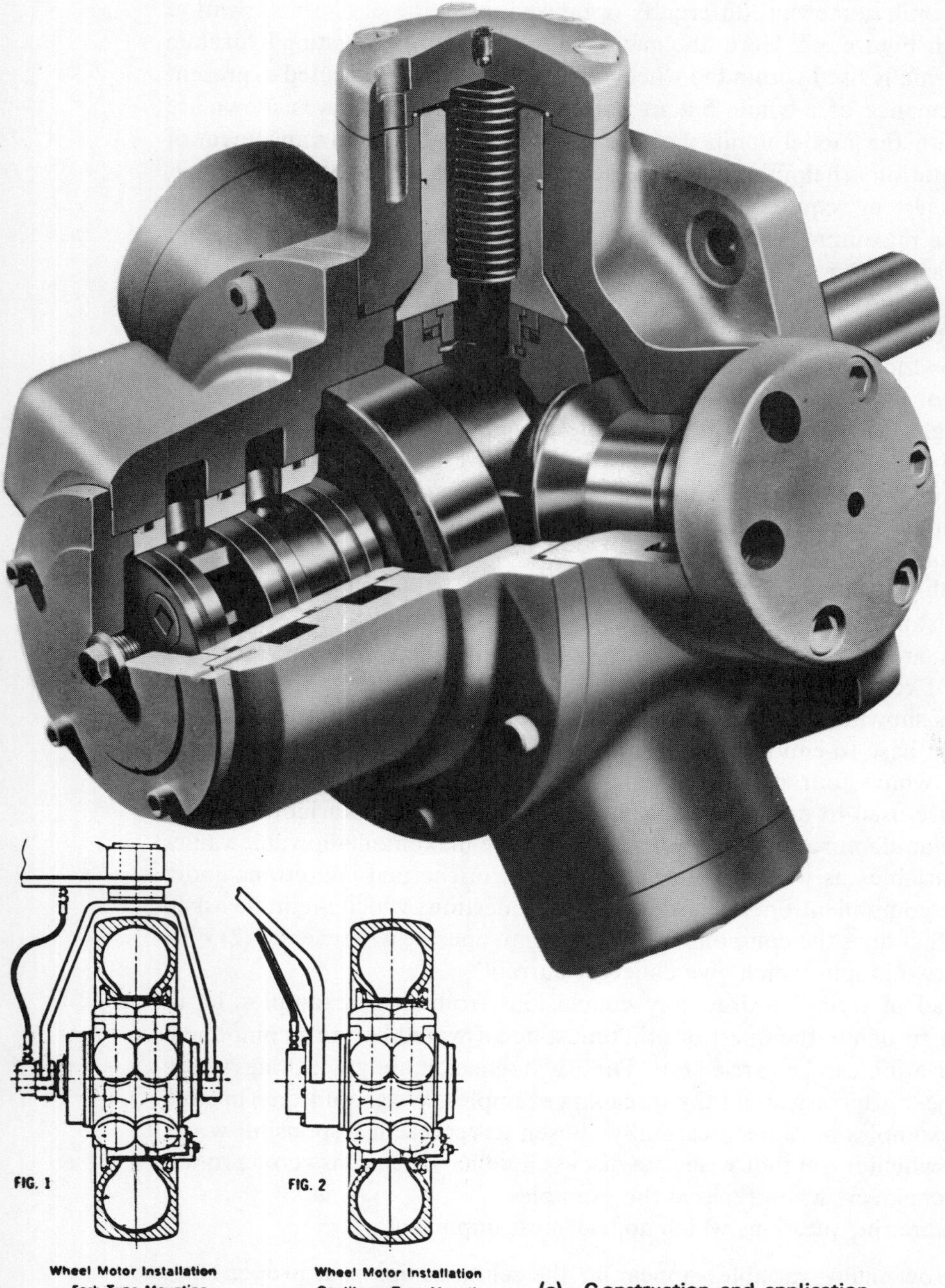

(a) Construction and application

Figure 3.2 (*continued*)

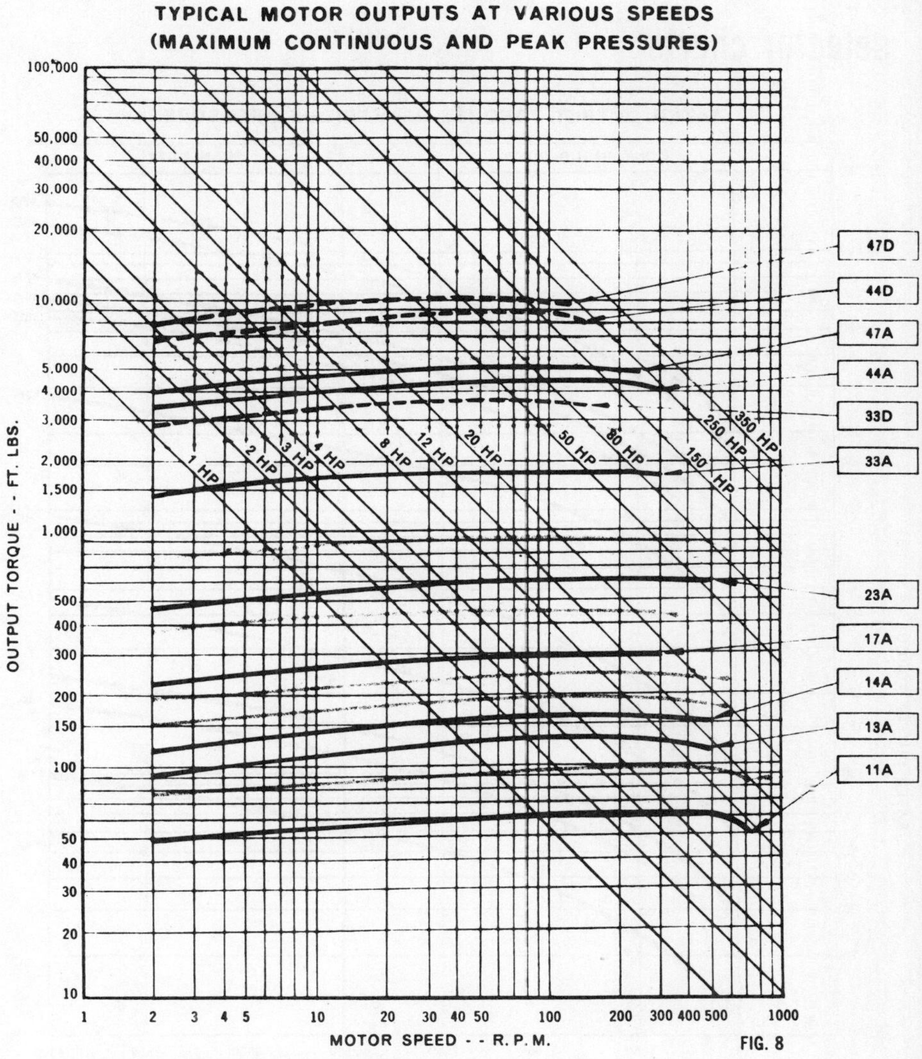

Notes: Continuous torques (at 2500 PSI) are shown in black.
Peak torques (at 4000 PSI) are shown in blue.
23A Wheel Motor maximum speed is 300 RPM.

(b) Torque and power versus speed characteristic

Figure 3.2 (*continued*)

selector charts

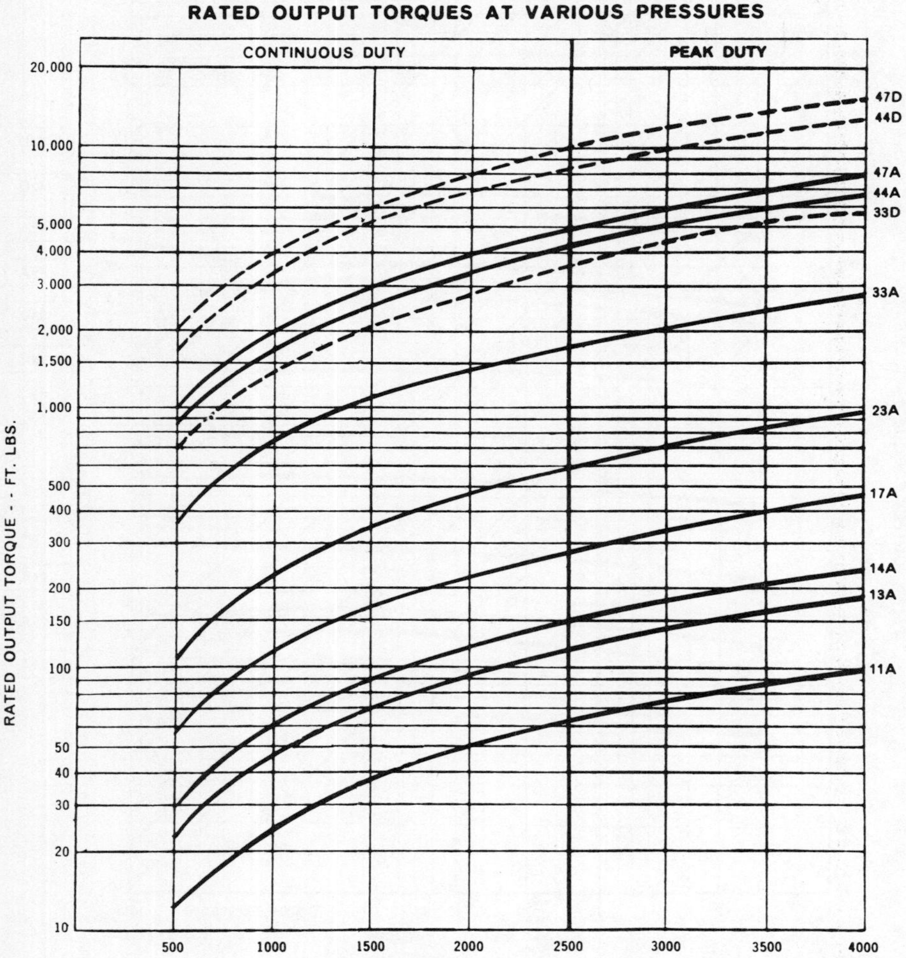

FIG. 9

(c) Torque versus pressure curves

Figure 3.2 (*continued*)

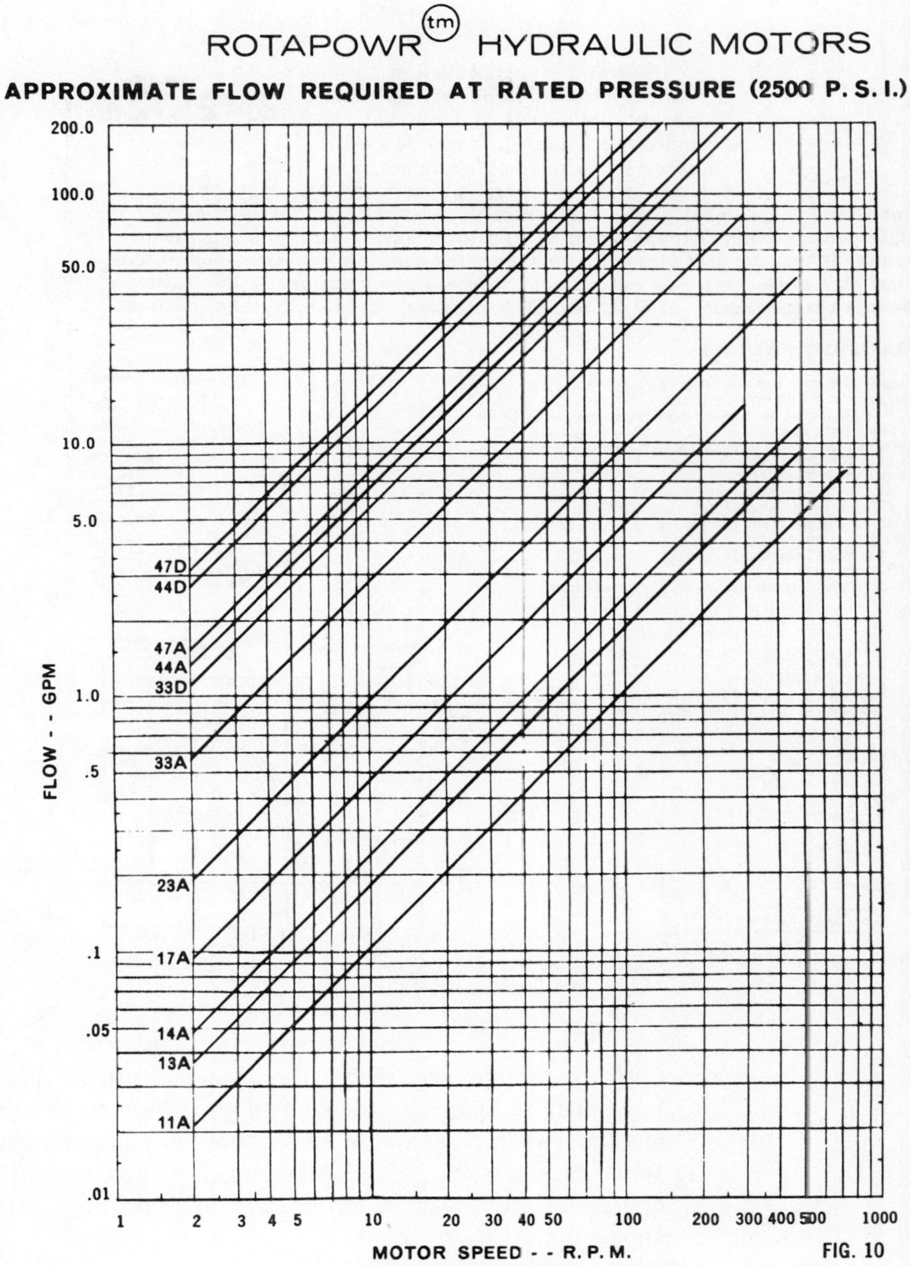

(d) Flowrate versus speed curves

Figure 3.2 Manufacturer's characterization of hydraulic motor. Courtesy of Houdaille Hydraulic for (a).

TRANSISTOR

2N2102

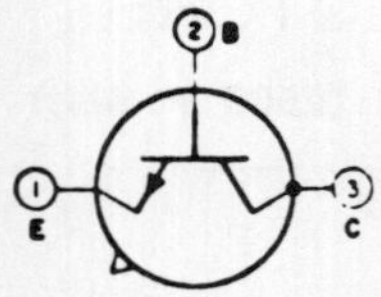

Silicon n-p-n type used in a wide variety of small-signal and medium-power applications in industrial and military equipment. It can be used in rf service as an amplifier, mixer, oscillator, and converter; in af service for small- and large-signal driver and power applications; in switching service for high-speed switching circuits requiring transistors having high voltage and current values. It features low saturation voltage, high sustaining voltage, high dissipation, high pulse beta, low output capacitance, and exceptionally low noise and leakage characteristics. JEDEC No. TO-5 package; outline 6, Outlines Section.

MAXIMUM RATINGS

COLLECTOR-TO-BASE VOLTAGE (with emitter open)	120 *max*	volts
COLLECTOR-TO-EMITTER VOLTAGE:		
With external base-to-emitter resistance = 10 ohms or less	80 *max*	volts
With base open	65 *max*	volts
EMITTER-TO-BASE VOLTAGE (with collector open)	7 *max*	volts
COLLECTOR CURRENT	1 *max*	ampere
TRANSISTOR DISSIPATION:		
At case temperatures up to 25°C	5 *max*	watts
At ambient temperatures up to 25°C	1 *max*	watt
At case or ambient temperatures above 25°C	See curve page 68	
TEMPERATURE RANGE:		
Operating (junction)	−65 to 200	°C
Storage	−65 to 300	°C
LEAD TEMPERATURE (for 10 seconds maximum)	300 *max*	°C

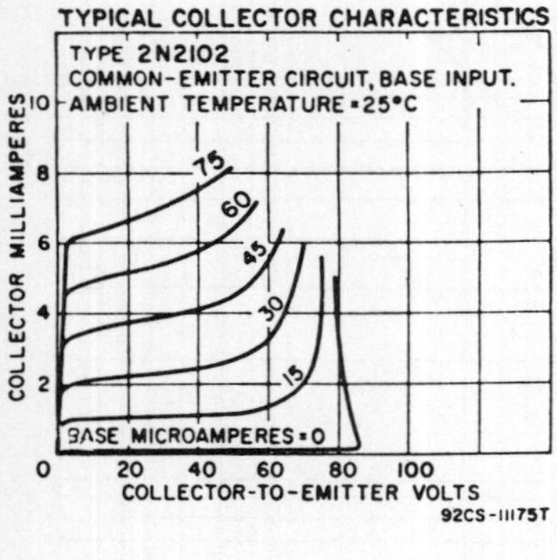

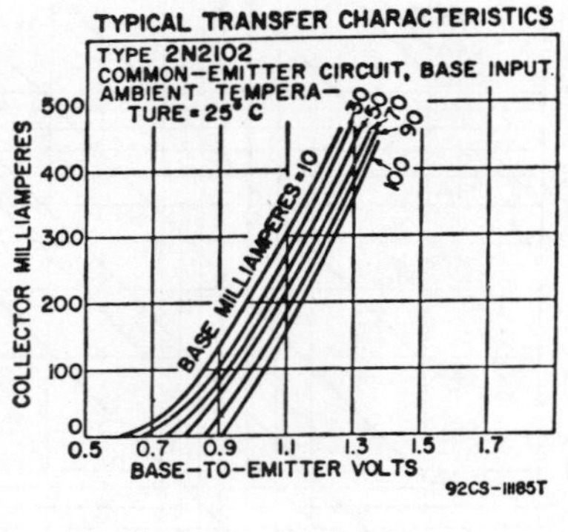

Figure 3.3 (*continued*)

(3) Can any variables whatever, of those appearing in the characterization functions, be taken as independent; i.e., used to determine the remaining variables, as would be required for computer simulation.

(4) What details, such as variable limits, thermal conditions, reverse operation, etc., must be referred to in order to cover all possible applications?

In the following sections, the answers to these and some related questions are developed.

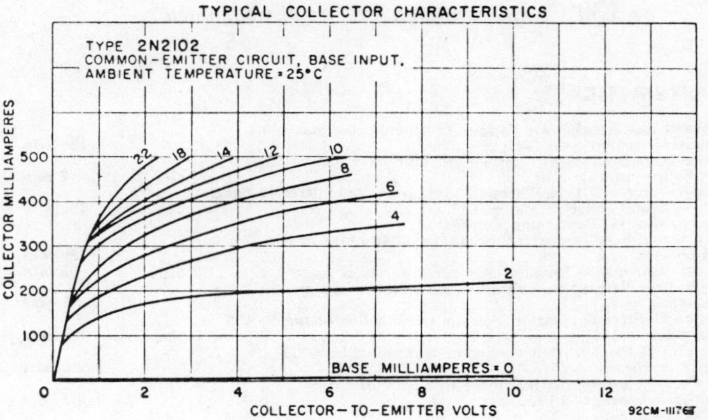

In Common-Emitter Circuit

DC-Pulse Forward Current-Transfer Ratio:*		
With collector-to-emitter volts = 10 and collector ma = 150	40 to 120	
With collector-to-emitter volts = 10 and collector ma = 500	25 *min*	
With collector-to-emitter volts = 10 and collector ma = 1000	10 *min*	
DC Forward Current-Transfer Ratio:		
With collector-to-emitter volts = 10 and collector ma = 0.01	10 *min*	
With collector-to-emitter volts = 10 and collector ma = 0.1	20 *min*	
With collector-to-emitter volts = 10 and collector ma = 10	35 *min*	
Small-Signal Forward Current-Transfer Ratio:		
With collector-to-emitter volts = 5, collector ma = 1, and frequency = 1 kilocycle	30 to 100	
With collector-to-emitter volts = 10, collector ma = 5, and frequency = 1 kilocycle	35 to 150	
With collector-to-emitter volts = 10, collector ma = 50, and frequency = 20 Mc	3 *min*	
Noise Figure (with collector-to-emitter volts = 10, collector ma = 0.3, generator resistance = 1000 ohms, circuit bandwidth = 15 kilocycles, and signal frequency = 1 kilocycle)	6 *max*	db
Total Switching Time† (delay time plus rise time plus fall time)	30 *max*	nsec

* Pulse duration = 300 μsec, duty factor = 0.018.

† See *Total-Switching-Time Measurement Circuit below.*

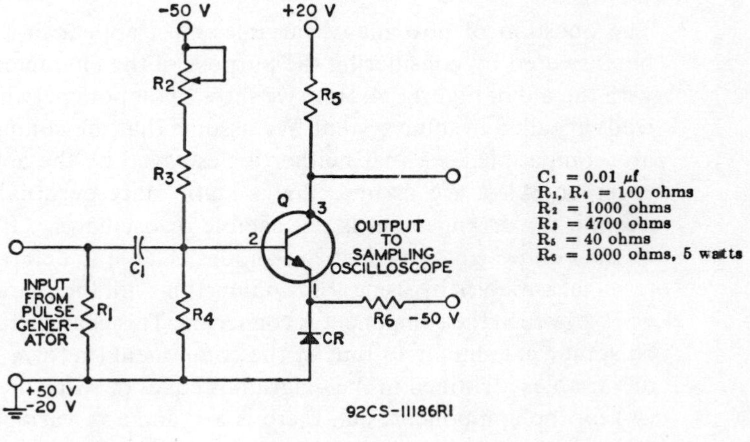

Figure 3.3 (*continued*)

CHARACTERISTICS

Collector-to-Base Breakdown Voltage (with collector ma = 0.1 and emitter current = 0)	120 *min*	volts
Emitter-to-Base Breakdown Voltage (with emitter ma = 0.1 and collector current = 0)	7 *min*	volts
Collector-to-Emitter Reach-Through Voltage (with emitter-to-base volts = 1.5 and collector ma = 0.1)	120 *min*	volts
Collector-to-Emitter Sustaining Voltage:		
With external base-to-emitter resistance = 10 ohms or less and collector ma = 100	80 *min*	volts
With collector ma = 100 and base current = 0	65 *min*	volts
Base-to-Emitter Saturation Voltage (with collector ma = 150 and base ma = 15)	1.1 *max*	volts
Collector-to-Emitter Saturation Voltage (with collector ma = 150 and base ma = 15)	0.5 *max*	volt
Collector-Cutoff Current (with collector-to-base volts = 60 and emitter current = 0)	0.002 *max*	μa
Emitter-Cutoff Current (with emitter-to-base volts = 5 and collector current = 0)	0.002 *max*	μa
Thermal Resistance:		
Junction-to-case	35 *max*	°C/watt
Junction-to-ambient	175 *max*	°C/watt

In Common-Base Circuit

Input Resistance at 1 kilocycle:		
With collector-to-base volts = 5 and collector ma = 1	24 to 34	ohms
With collector-to-base volts = 10 and collector ma = 5	4 to 8	ohms
Input Capacitance (with emitter-to-base volts = 0.5 and collector current = 0)	80 *max*	pf
Output Capacitance (with collector-to-base volts = 10 and emitter current = 0)	15 *max*	pf
Output Conductance at 1 kilocycle:		
With collector-to-base volts = 5 and collector ma = 1	0.1 to 0.5	μmho
With collector-to-base volts = 10 and collector ma = 5	0.1 to 1	μmho
Small-Signal Open-Circuit Reverse Voltage-Transfer Ratio at 1 kilocycle:		
With collector-to-base volts = 5 and collector ma = 1	0.0003 *max*	
With collector-to-base volts = 10 and collector ma = 5	0.0003 *max*	

Figure 3.3 Manufacturer's characterization of transistor. Courtesy of R.C.A.

3.2 Basic Principles of Characterization

The question of how many variables must appear in a characterization can be answered by considering the purpose of the characterization, which we do with the aid of Figure 3.4. Here we show a component (which is to be characterized) installed in some system. We assume that the component and the system are compatible; i.e., that neither is destroyed by the act of connecting them together. Also, we assume that a static state eventually develops; i.e., we assume the arrangement is not astable or oscillatory. If the characterization of the component is proper, we should be able to determine the values of the terminal variables by using it in conjunction with the characterization of the network to which the component is connected. This characterization will, obviously, be somewhat similar to that of the component (in form, that is). The number of variables identified in the diagram is equal to twice the number of terminals (N) on the component; i.e., there is a Γ and a Δ variable at each connection between the component and the system. The number of variables with which we must work, however, is less than this because of some conservation relation-

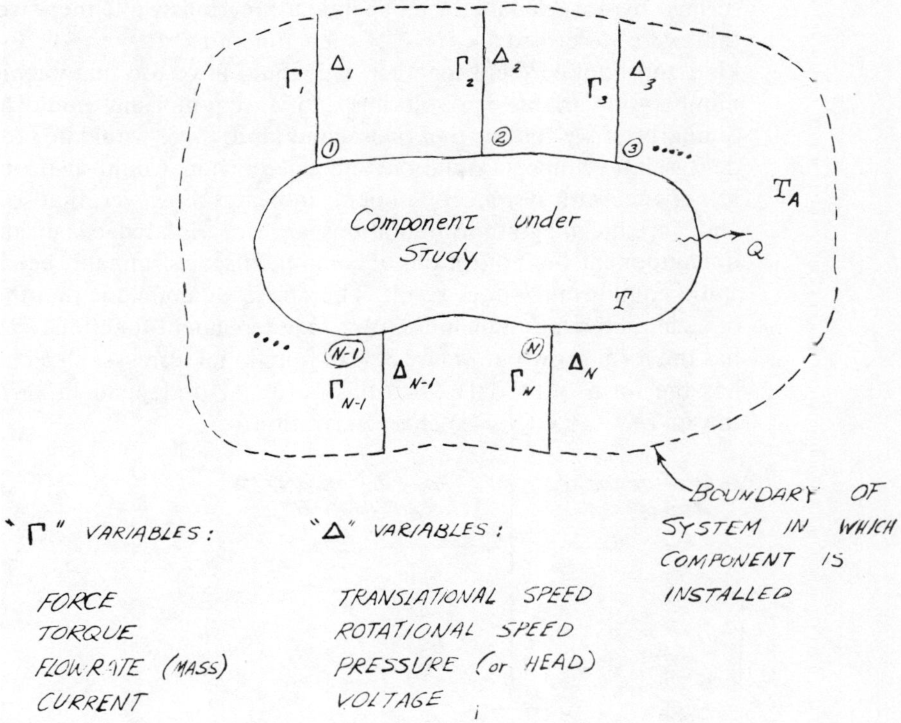

From the Engineer's Sketch Book

Figure 3.4 Generalized component in static equilibrium with a system.

ship and because the $\mathbf{\Delta}$ variables are relative quantities. From some conservation equation, we may write

$$\mathbf{\Gamma}_1 \pm \mathbf{\Gamma}_2 \pm \mathbf{\Gamma}_3 \pm \cdots \mathbf{\Gamma}_N = 0$$

Because the $\mathbf{\Delta}$ variables are relative, we may arbitrarily take $\mathbf{\Delta}_N = 0$ for this analysis; the other $\mathbf{\Delta}$ variables may then be measured with respect to $\mathbf{\Delta}_N$. The number of variables which are actually of engineering significance is then $2(N - 1)$; in order to solve for this many variables we must have an equal number of independent functional relationships between them. Thus, the characterizations of the component and the external system must include a total of $2(N - 1)$ independent functions or graphs.

There is a great temptation to say that each characterization includes the same number of independent functions; i.e., that each has $(N - 1)$. This is true for almost all useful components, for the following reason. Suppose it were not true. There would be some N terminal component that was described by, for example, N independent functions while there would be others that were described by $(N - 2)$ such functions. If we took two of the first kind and hooked them together, we would have too many equations for the number of variables; a solution and an equilibrium would be impossible. Similarly, if we took two of the second kind, there would be too few independent relationships; a solution and equilibrium would again be impossible. Experience with system components indicates, however, that so long as power and variable magnitude limitations are not violated by an interconnection of components, or collections of components, determinable behavior—usually static equilibrium—does result. Therefore, we conclude that the proper form of a characterization includes $(N - 1)$ independent functions if the component has the form shown in Figure 3.4. In functional terms, such a characterization has the form of $(N - 1)$ equations in $(N - 1)$ independent variables; we call this an $(N - 1) \times (N - 1)$ characterization.

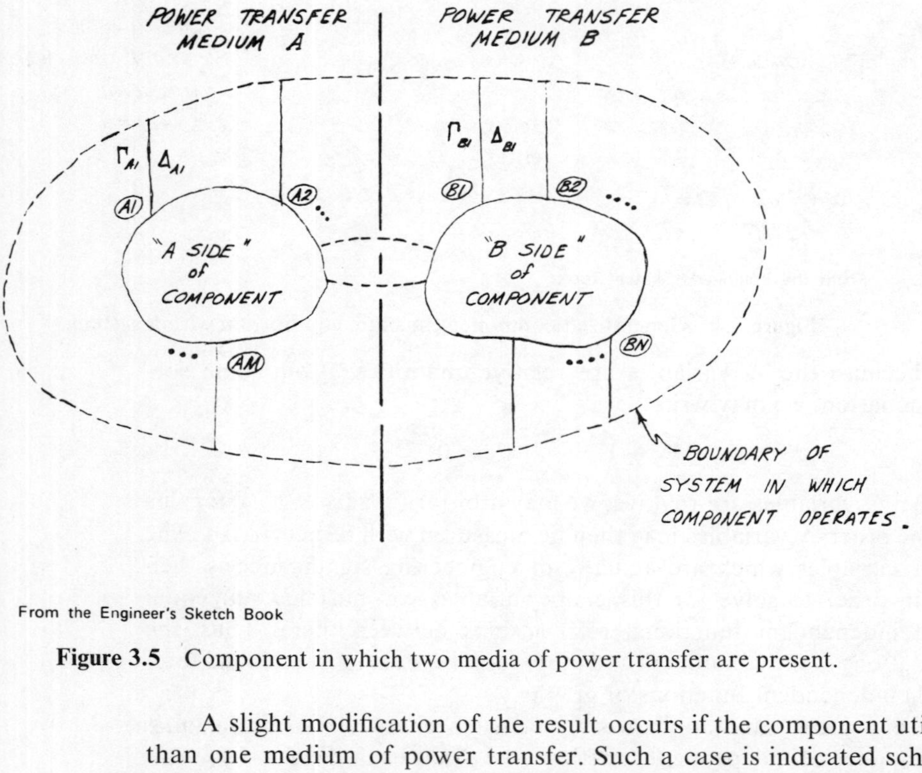

From the Engineer's Sketch Book

Figure 3.5 Component in which two media of power transfer are present.

A slight modification of the result occurs if the component utilizes more than one medium of power transfer. Such a case is indicated schematically in Figure 3.5. Here the number of variables is $2(M + N)$ where M is the number of terminals of the "A" type power transfer side of the component and N

is the number on the "B" side. In this case, there are two conservation-type relations which may be written

$$\mathbf{\Gamma}_{A1} \pm \mathbf{\Gamma}_{A2} \pm \cdots \mathbf{\Gamma}_{AM} = 0 \quad \text{and} \quad \mathbf{\Gamma}_{B1} \pm \mathbf{\Gamma}_{B2} \pm \cdots \mathbf{\Gamma}_{BN} = 0.$$

Also, there are two possibilities for defining relative measurements. For example, we may arbitrarily take $\mathbf{\Delta}_{AM} = 0$ and $\mathbf{\Delta}_{BN} = 0$. Thus, the characterization need contain only $(M + N - 2)$ functions. A similar result can occur if a component has only one kind of power transfer, but contains completely separated conduction paths for two or more groups of terminals. In this case, as many conservation relations may be written as there are separated paths; an equal number of relative measurement references may be defined. Thus, it is generally true that, if R is the number of conservation-type relationships which may be written for a component and N is the number of its terminals, the number C of independent functions in its characterization is given by

$$C = N - R. \tag{3.1}$$

Let us return, momentarily, to the simple component of Figure 3.4. We might argue that the first law of thermodynamics applies to any system, and why didn't it enter the characterization argument? Actually, it does, as follows: An expression of the form

$$\pm\mathbf{\Gamma}_1\mathbf{\Delta}_1 \pm \mathbf{\Gamma}_2\mathbf{\Delta}_2 \pm \cdots \pm \mathbf{\Gamma}_{N-1}\mathbf{\Delta}_{N-1} \qquad (\mathbf{\Delta}_N = 0)$$

gives the power which is being dissipated in the static state by the component (provided the signs are chosen correctly). This is the power that would have to be dissipated by $(N - 1)$ separate dissipators connected from the various terminals to terminal N. This power must be dissipated as heat in order to satisfy the first law; this is shown as Q in the figure. Now, the heat transfer Q will be some function of the temperature of the surrounds in which the component is embedded (this is normally referred to as the ambient temperature T_A) as well as the temperature of the component. (Let us assume that the fluid or gas flow about the component is specified so that its nature does not enter the discussion.) Actually, the temperature of the component is undefinable; there will be a continuous distribution of temperature throughout the device, and this distribution will be dependent on the static operating condition. For engineering purposes, we often utilize the temperature of the hottest spot in the component as an indication of the thermal condition; we may then say that the heat transfer is a function of this and the ambient temperature. This heat transfer relationship is a separate part of the characterization of the component; in most cases it does not have to be known with much precision. Let us suppose that we did know it, however. The first law would give us the amount of heat to be transferred; i.e., Q. The ambient temperature would have to be specified independently. With these values, we could use the heat-transfer characteristic to determine the operating temperature of the component. Therefore, we have used all our equations and have specified the operating conditions of the component. Fortunately, we may usually use gross estimates of the thermal aspect of the characterization; our earlier result on the relationships between terminal variables remains unchanged.

In determining the possible functional form for the characterizing relationships for a component there is one restriction that must be observed. Let us suppose that a particular N terminal component is to be characterized in the form.

$$\begin{aligned} Y_1 &= \mathcal{F}_1 \quad (X_1, X_2, \ldots, X_{N-1}) \\ Y_2 &= \mathcal{F}_2 \quad (X_1, X_2, \ldots, X_{N-1}) \\ &\vdots \\ Y_{N-1} &= \mathcal{F}_{N-1} \quad (X_1, X_2, \ldots, X_{N-1}) \end{aligned}$$

Here the list of independent variables X is a random mixture of $\mathbf{\Gamma}$ and $\mathbf{\Delta}$ variables, as is the list of dependent variables Y. The selection of these would be based on the user's anticipated application of the characterization or, perhaps, strictly on whim. However, if the formulation is to be utilized in conjunction with a computer analysis, it is necessary that for any small change in one of the variables X the corresponding changes in the values Y must be computable. This requires that all the partial derivatives $\partial \mathcal{F}/\partial X$ be defined (non-infinite). In many physical devices, one terminal variable remains constant (such as at zero value) over an extensive portion of the operating region of the device. The partial derivative of such a variable with respect to several other variables is zero; the inverse derivative is infinite. It is always bad practice to select such a variable as independent in a characterization. Unsatisfactory formulations can also arise when two variables associated with a device are proportionally related over the operating region. If both are chosen to be independent, or both dependent, trouble ensues. This may be seen as follows, in the case of a three-terminal device. We know there are four variables related by two equations; this may be represented as follows without making any assumptions about independent and dependent variables:

$$\mathcal{F}(U, V, W, X) = 0$$
$$\mathcal{G}(U, V, W, X) = 0.$$

For small excursions, the variables will be related as follows:

$$\frac{\partial \mathcal{F}}{\partial U} u + \frac{\partial \mathcal{F}}{\partial V} v + \frac{\partial \mathcal{F}}{\partial W} w + \frac{\partial \mathcal{F}}{\partial X} x = 0$$

$$\frac{\partial \mathcal{G}}{\partial U} u + \frac{\partial \mathcal{G}}{\partial V} v + \frac{\partial \mathcal{G}}{\partial W} w + \frac{\partial \mathcal{G}}{\partial X} x = 0.$$

(If the concept of small excursions is unfamiliar to the reader, Section 3.7 may be studied at this time and before attempting to follow this argument.) Solving for the u and v excursion variables in terms of the w and x variables, we get

$$u = \frac{\begin{vmatrix} \quad & \quad \\ \quad & \quad \end{vmatrix}}{\begin{vmatrix} \frac{\partial \mathcal{F}}{\partial U} & \frac{\partial \mathcal{F}}{\partial V} \\ \frac{\partial \mathcal{G}}{\partial U} & \frac{\partial \mathcal{G}}{\partial V} \end{vmatrix}} \qquad v = \frac{\begin{vmatrix} \quad & \quad \\ \quad & \quad \end{vmatrix}}{\begin{vmatrix} \frac{\partial \mathcal{F}}{\partial U} & \frac{\partial \mathcal{F}}{\partial V} \\ \frac{\partial \mathcal{G}}{\partial U} & \frac{\partial \mathcal{G}}{\partial V} \end{vmatrix}}$$

where the entries in the numerator determinants are not relevant, and thus not written out. The denominator automatically becomes zero if the entries of the columns are proportional or if the entries of the rows are proportional. These will be the cases if U is proportional to V or if W is proportional to X, respectively. By similar developments, it can be shown that *if two variables are proportional in some portion of the characterization of a component, one must be taken as independent and one as dependent in any proper characterization.*

A basic condition which must be met in the selection of independent and dependent variables is that none of the dependent variables be infinitely sensitive to any of the independent variables. The matter is illustrated by the simple case shown in Figure 3.6 where V may be computed as a function of U, but the opposite is impossible. Another illustration arises in the simulation of director elements where a real quandary develops; there are regions of operation for these devices where either possible formulation of the characteristics is unallowable; in fact, the formulation must be changed, in computation, depending on the state of conduction of the component.

Example 3.1 The schematic symbol and notation for a vacuum tube appears in Figure 3.7; this is a simple three-element tube (triode) in which the conductivity between the plate (P) and the cathode (K) is controlled by the grid (G); the heater, with terminals H_1 and H_2, brings the cathode temperature up to the level at which electrons can escape into the vacuum inside the tube and be collected by the plate. Let us attempt to formulate the proper characterization of the tube, without going into any further detail on its operation. We note that two current conservation relations can be written for the device since the heater circuit is separate from the other. The total number of terminals is five; therefore, the characterization should have three independent

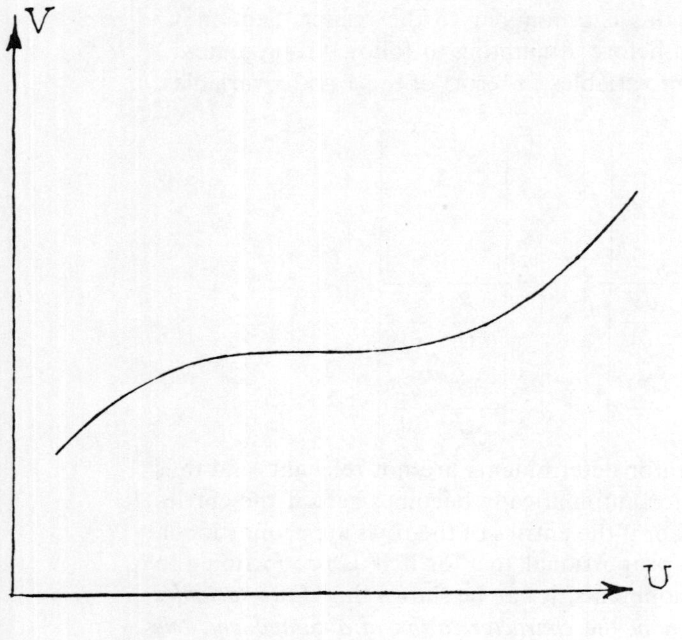

From the Engineer's Sketch Book

Figure 3.6 A function with only one usable section of the independent variable.

functions. Conventionally, the cathode terminal is used as a voltage reference; either heater terminal may be. Therefore, the list of variables appearing in the characterization will

$$E_{PK}, E_{GK}, E_H, I_P, I_G, I_H$$

In selecting the independent variables, some knowledge of the tube behavior is required. For example, in normal operation,* the grid current I_G is zero. Therefore,

$$\frac{\partial I_G}{\partial E_{PK}} = \frac{\partial I_G}{\partial E_{GK}} = \frac{\partial I_G}{\partial I_P} = 0$$

and I_G should not be selected as an independent variable. It also turns out that the relationship between heater current and voltage is quite insensitive

* In the region $E_{GK} \leq 0$.

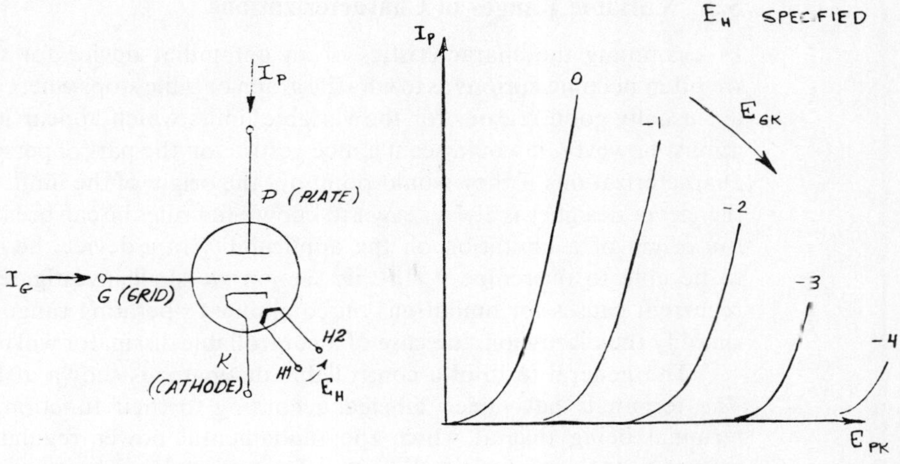

(a) Symbol and notation

(b) Nature of characteristics

From the Engineer's Sketch Book

Figure 3.7 Characterization of the vacuum triode.

to the other aspects of tube operation; it may be represented rather well by $E_H = I_H R_H$. This indicates that any formulation in which both I_H and E_H were supposed to be independent, or both were supposed to be dependent, would be faulty. Of the possible choices of formulation available under these constraints, one of the most useful is that which gives the currents in terms of the voltages:

$$
\begin{aligned}
I_P &= \mathscr{F}_1(E_{PK}, E_{GK}, E_H) \\
I_G &= \mathscr{F}_2(E_{PK}, E_{GK}, E_H) \\
I_H &= \mathscr{F}_3(E_{PK}, E_{GK}, E_H).
\end{aligned}
$$

This is proper, but may be simplified by taking account of some of our earlier observations and also by noting that the heater voltage is normally set at a preselected value for any particular tube. The practical form for the characteristics is then

$$
\begin{aligned}
I_P &= \mathscr{F}_1(E_{PK}, E_{GK}) \quad \text{for } E_H = \text{rated value} \\
I_G &= 0 \qquad\qquad\qquad\qquad \text{and } E_{GK} \leqq 0 \\
I_H &= E_H/R_H
\end{aligned}
$$

3.3 Variable Ranges in Characterizations

In examining the characteristics of an unfamiliar device for the first time, we often become curious as to why the graph or table stops where it does. There are usually good reasons for the variable limits which appear in characterizations; however, it would seem a nice gesture on the part of persons preparing characterizations if they would point out the origin of the limit. In particular, the clever designer is always eager to know what rules he can break; if he knows the origin of a limitation on the applicability of a device, he is more likely to be able to overcome it. In this section we shall investigate some of the recurrent causes for limitations on component operating ranges. In order to simplify the discussion, the case of a controllable dissipator will be considered.

The general form of a controllable dissipator is shown at Figure 3.8(a). The terminals have been labeled according to their function, the "valve" terminal being that at which the fundamental power regulation action is carried out. Control of the dissipator is associated with terminal number one; we assume that the device has considerable power gain so that

$$|\mathbf{\Gamma}_1\mathbf{\Delta}_1| \ll |\mathbf{\Gamma}_2\mathbf{\Delta}_2|$$

One of the fundamental limitations on device performance is maximum allowable temperature. The general nature of temperature versus power (Q) is shown at (b). This shows that, since the control power is much less than the valve power, the basic limitation is that the product $\mathbf{\Gamma}_2\mathbf{\Delta}_2$ not exceed the value at which this temperature occurs. It may be seen that this is very strongly related to the ambient temperature in which the device operates; high temperatures require appropriate power derating of the component.

A convenient set of characteristics for this kind of device is shown in Figure 3.8(c) and (d). The forementioned $\mathbf{\Gamma}_2\mathbf{\Delta}_2$ product limitation appears as a hyperbola on the graph at (d); this is labeled the "thermal limit." Of course, the hyperbola would have to be moved inward as ambient temperature was increased, resulting in a restriction of the operating range. At both (c) and (d) are shown maximum magnitudes for the various $\mathbf{\Gamma}$ and $\mathbf{\Delta}$ variables; these arise from material limitations such as yield stress, breakdown voltage, etc., and from phenomena such as centrifugal force and local overheating because of excessive current. These limitations may, or may not, override the thermal limitation described earlier; at (c) they do while at (d) they do so only in some of the operating region. At both (c) and (d), the actual operating region is shown shaded; this is less than the region specified by the previously mentioned constraints because of the nature of the component characteristics. For example, from (d) it is clear that $\mathbf{\Gamma}_2$ cannot exceed a certain limit; let us call this 1. Referring back to (c), we see that the boundary

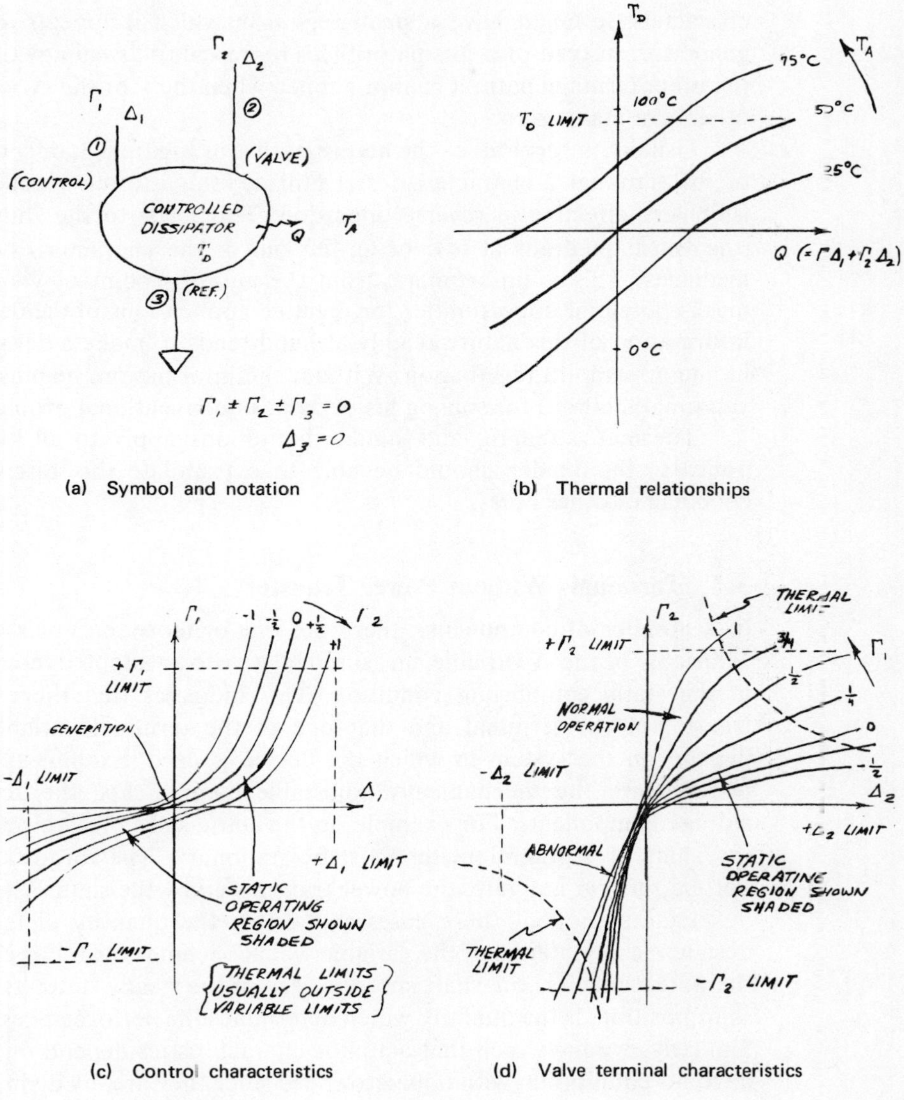

(a) Symbol and notation (b) Thermal relationships

(c) Control characteristics (d) Valve terminal characteristics

From the Engineer's Sketch Book

Figure 3.8 Variable ranges for controlled dissipator.

of the shaded region is given by the curve for $\Gamma_2 = 1$. By similar means, the complete boundary of the operating region is established. In these figures, the curves have purposely been drawn through the second and fourth quadrants, in a small region near the origin, to indicate the possibility that either characteristic might have a small region in which it appears to represent a generator, instead of a dissipator. This represents a "stealing" of power from the other terminal path; it cannot happen when the Γ or the Δ variable of the other terminal is zero.

Usually, in devices of the nature of that in Figure 3.8, only one quadrant of the terminal 2 characteristics is utilized; this is often because the device isn't very effective in reverse operation. This leads to the abnormal region [the third quadrant at (d)] being left out of the characterization in many instances. This is unfortunate, from the author's point of view, because it masks potential opportunities for creative applications of the device. Having information of this nature readily at hand tends to make a designer ask how he might exploit the situation. Without the information, he must utilize conventional methods for solving his perhaps unconventional problem.

Thermal, strength, and similar limitations apply to all kinds of components; the reader should be able to extrapolate this discussion into a coverage of other cases.

3.4 Terminals Without Power Transfer

In a number of components, there are one or more terminals in which the Γ variable or the Δ variable, measured relative to an adopted reference, is zero in the static equilibrium condition. This indicates that there is no power transfer at the terminal and that one of the terminal variables does not depend on the system in which the device is used. Examples of such components are the mechanically adjustable components, the actuators, and storage components. For example, in the static state of an electronic system the shaft of a potentiometer must be stationary. The torque may, or may not, be zero; at any rate, the power transmitted by the shaft is zero.

Examination of these cases shows that the quantity of interest to the designer is the integral of the variable which becomes zero. For example, with the potentiometer, the shaft speed is zero in the steady state; its integral, the shaft position, is the quantity which determines the performance of the device. Similarly, we have seen that actuator characteristics depend on stroke; they have no equilibrium with non-zero speed since they are, by definition, limited travel devices. With energy-storing elements the result is the same. For example, in a spring, the force is a function of the stretch; i.e., the integral of the stretching speed. In a capacitor, the voltage is a function of the integral of current. In both of these, and energy-storing elements in general, there is no change in the stored energy for operation at the static equilibrium of a system. The general result—that the integral of the zero-valued variable is the quantity of interest—actually coincides with our approach here in quite a reasonable way. Up to this point we have emphasized the variables which represent the power transfer conditions in a system; it turns out that the integrals in question represent the total work done, or energy stored, in the development of the resultant static equilibrium. Since the amount of energy stored in a component, or the amount of work which can be done by an actuator, or the amount of work required to bring about the adjustment of a mechanically variable machine, is the measure of the status and nature of those machines having no power transfer, the energy-oriented characterization is the natural adjunct to the power-oriented approach used earlier.

A typical example of the zero-power-transfer condition, at one terminal of a component in static equilibrium, arises from the variable displacement vane pump shown in Figure 3.9(a). By applying the usual reasoning, it can be determined that the following is a reasonable formulation for the characterization of this component:

$$
\begin{aligned}
M &= \mathscr{F}_1(P, \Omega, X) \\
W &= \mathscr{F}_2(P, \Omega, X) \\
F_{CA} &= \mathscr{F}_3(P, \Omega, X)
\end{aligned}
$$

Here we have written the integral of the control arm velocity by the direct measure of it; i.e., the displacement variable X. The last function in the characterization is somewhat different from most we have observed so far; its general nature is shown at (b) in the figure. Here, it has been assumed that there is quite a significant amount of stiction in the control arm packing. At any particular equilibrium of the unit, no force is required to hold the arm fixed. However, if it is necessary either to increase or decrease X, a significant amount of force is required. As indicated by the form of the characterization, this force will probably be dependent on both the pressure and speed of the unit. Since it is impossible to draw a graph with three independent variables (without getting very complicated), only the variation with pressure has been indicated; a particular speed of operation is assumed. It may be noted that the work required to increase X from value a to value b is given by the integral under the curve of F versus X; this would be useful if an actuator were being selected to control the unit. Of course, the control characteristics of a particular pump might be completely different, as to levels, from those sketched here. For example, the stiction might be very small compared to the force due to

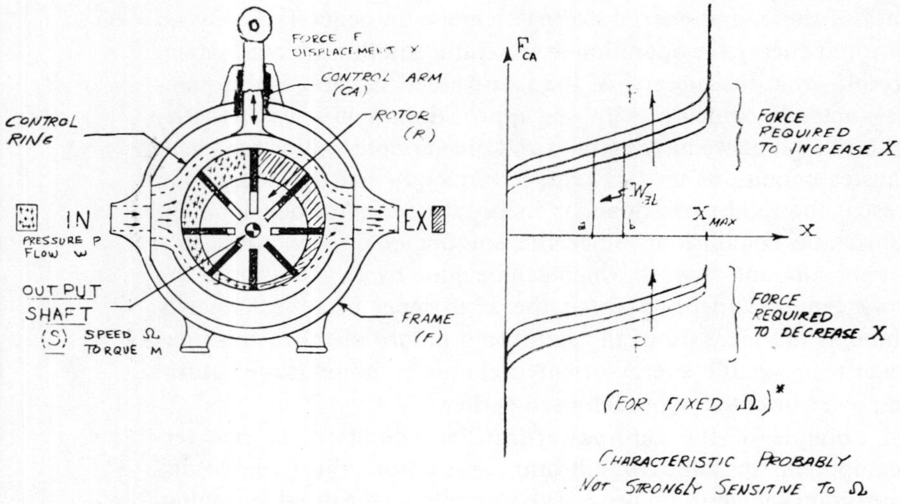

(a) Construction and notation (b) Control characteristics

From the Engineer's Sketch Book

Figure 3.9 Variable displacement vane pump.

pump pressure. This would result in the control characteristic having the nature of a spring; the spring constant might be either negative or positive. Perhaps the reader has observed a similar phenomena in water-faucet valves which are badly worn; when opened a tiny bit, such a valve can proceed to turn itself completely on.

Another example of a zero-power terminal component is the hydraulic accumulator shown in Figure 3.10. This is constructed with two hemispherical shells and an impermeable bladder. Before admitting liquid, the air side of the unit is charged with highly compressed air and then sealed off. This forces the bladder down along the lower hemisphere and actually closes off the fluid inlet. The characteristics, shown at (b), indicate that the hydraulic pressure (with respect to an absolute reference because of the involvement of a compressible fluid) must exceed the charging value before any flow is admitted to the accumulator. As predicted, the characteristics are properly given by the integral of flow rate into the accumulator since the flow must be zero in any static equilibrium. The integral of flow rate gives information about the volume taken up by the liquid and, therefore, the volume of the trapped gas. This indicates the physical origin of the component characteristics. It should be noted that the energy stored in the accumulator by the flow of fluid into it is given by the integral under the characteristic curve, taken to the value of the

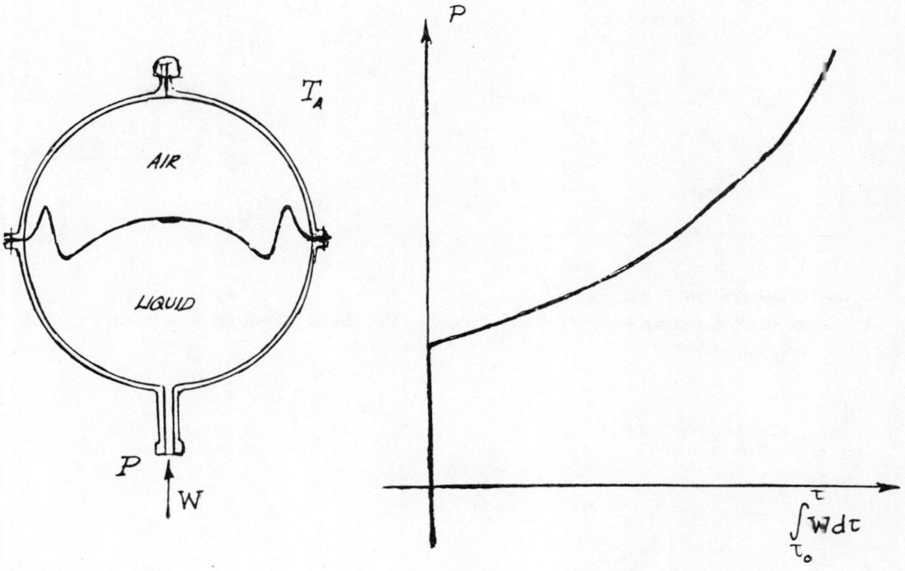

(a) Construction (b) Form of characteristic

From the Engineer's Sketch Book

Figure 3.10 Hydraulic accumulator.

displacement for the static condition in question. As an incidental issue, we might note that quite some time would be required for this accumulator to reach its static equilibrium after a sudden change; the temperature of the air in the vessel would have to adjust to that of the ambient.

3.5 Usefulness of Alternative Characterization Formulations

As has been shown, there are a number of proper, alternative formulations for the characterization of any particular component. The selection of one formulation over others is based mainly on the usage of the characterization in analysis procedures. Again, the controlled dissipator provides some instructive lessons in this respect.

A typical controlled dissipator application is indicated in Figure 3.11(a). Device I may be supposed to provide the control action (therefore, it must have the characteristics of a generator), while device II is the load controlled by III. Two possible formulations for the characteristics of III are:

$$\mathbf{\Gamma}_1 = \mathscr{F}_1(\mathbf{\Delta}_1, \mathbf{\Delta}_2) \qquad \mathbf{\Gamma}_1 = \mathscr{F}_3(\mathbf{\Gamma}_2, \mathbf{\Delta}_2)$$
$$\mathbf{\Gamma}_2 = \mathscr{F}_2(\mathbf{\Delta}_2, \mathbf{\Delta}_2) \qquad \mathbf{\Delta}_1 = \mathscr{F}_4(\mathbf{\Gamma}_2, \mathbf{\Delta}_2)$$

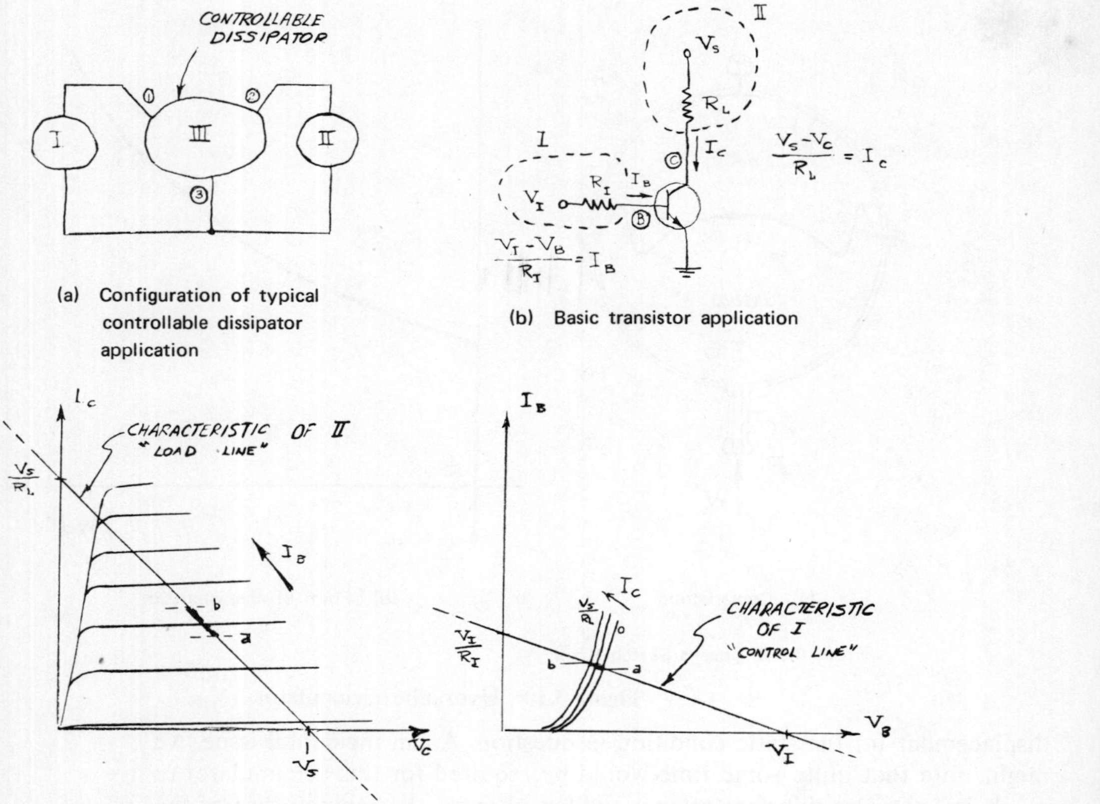

(a) Configuration of typical controllable dissipator application

(b) Basic transistor application

(c) Collector terminal analysis

(d) Base terminal analysis

From the Engineer's Sketch Book

Figure 3.11 Characterization form and static analysis.

We assume that either one of these satisfies all the conditions established earlier; i.e., that either is a proper characterization. In the application shown, which may not be the only possible arrangement, static analysis would probably consist of determining $\mathbf{\Gamma}_1$, $\mathbf{\Delta}_1$, $\mathbf{\Gamma}_2$, $\mathbf{\Delta}_2$, and taking terminal 3 as the reference for $\mathbf{\Delta}$. Of course, $\mathbf{\Gamma}_3$ would come from a conservation relationship if it were of interest. Since devices I and II each have only two terminals, there is little choice for the formulation of their characteristics, especially if a graphical presentation is utilized. Thus, it is clear that we will start with one relationship between $\boldsymbol{\Gamma}_1$ and $\boldsymbol{\Delta}_1$ and another between $\boldsymbol{\Gamma}_2$ and $\boldsymbol{\Delta}_2$. Obviously, if we elect the first characterization form above, there is an almost perfect match; a simple process of superposition will indicate the basic aspect of the static equilibrium. Alternatively, if the second form for the characterization of III is utilized, a trial-and-error or an elaborate manipulation procedure will be required to establish the static equilibrium. In this case, the first form is markedly superior and would be utilized if the indicated network were the fundamental application of III.

This discussion can be put on a somewhat more substantial basis by considering a specific example; the circuit shown at Figure 3.11(b) is satisfactory. In this, the control device consists of the input voltage V_I and the resistor R_I. The load consists of the supply voltage V_S and the resistor R_L. The equations which describe these components are shown in the figure. At (c) and (d) is shown how the best selection of a characterization form simplifies static analysis. The choice for the transistor characterization formulation is:

$$I_C = \mathscr{F}_1 (V_{CE}, I_B)$$
$$V_{BE} = \mathscr{F}_2 (V_{CE}, I_B)$$

In the graphical presentation, the axes have been chosen such that terminal power is given by the product of the ordinate and abcissa; the second independent variable is utilized parametrically. With this arrangement, the characteristics of the control and load components can be drawn, quite readily, on the same set of coordinates. When this is done, each graph represents a relationship which must exist between terminal variables in the static state. Any static state must, of course, lie on the static characteristics of all the components. As shown by the darkened regions of the control and load lines, the general location of the static equilibrium is almost obvious. The small possible range of I_B is found from the graph at (d) and runs between a and b. This range is picked off the graph at (c), thereby defining the operating region. By a simple procedure to be shown in the next chapter, or by straightforward trial and error, the exact location of the equilibrium can then be found. If a transistor characterization form had been used that did not allow the superposition step, the analysis would have been much longer. *Thus, the ideal formulation of a characterization depends on the way a component is used in networks.*

It is of some discomfort that there are often several effective ways in which a particular component can be utilized. Keeping our attention on the controlled dissipator, we examine one case in Figure 3.12. In this, all the possible arrangements of a controlled dissipator, a control source, and a power load are displayed. (This is not the only set of components which might be used with the controlled dissipator.) Several of the arrangements shown are absolutely

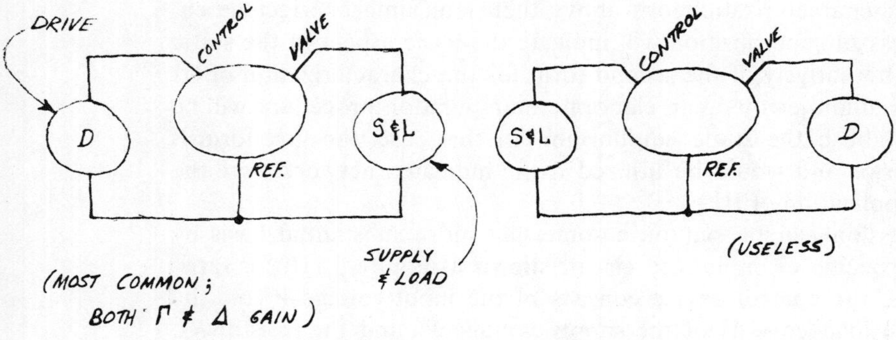

(a) Possible "common reference" connections

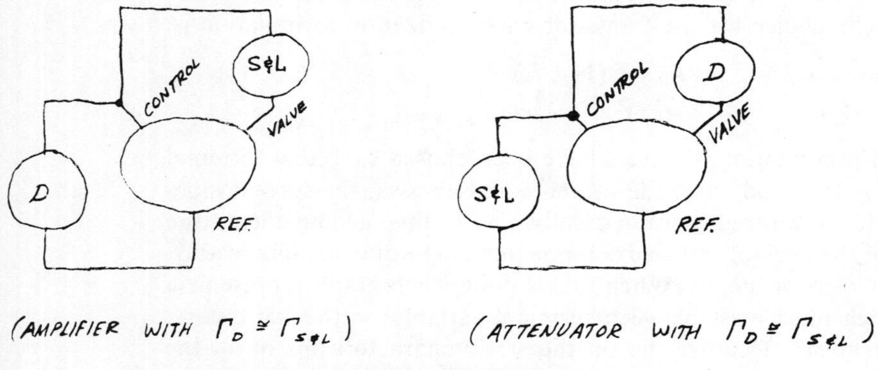

(b) Possible "common control" connections

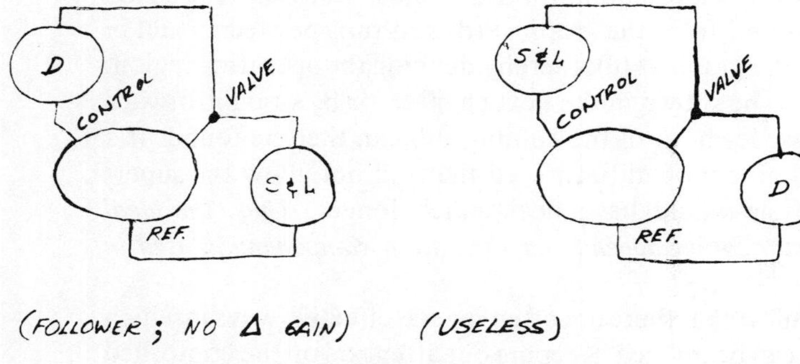

(c) Possible "common valve" connections

From the Engineer's Sketch Book

Figure 3.12 Controlled dissipator application configurations.

useless, as indicated, because they require the large load power to be handled by the control terminal of the dissipator while the small power-control unit operates into the valve terminal of the dissipator. However, all three of the arrangements shown at the left of the figure are potentially useful; that at (a) is the usual high-gain amplifier while that at (c) is the follower arrangement. The arrangement at (b) has power gain, but the Γ variable in the drive component must be about equal to that in the load; this is usually not as useful as other arrangements and is not very frequently applied. It is clear, however, that there are two formulations for the device characteristics that would be useful. In this kind of situation, the manufacturer would have to provide both or elect one over the other, leaving some engineers to derive the necessary information for static analysis from a formulation which does not suit their needs exactly. However, the manufacturer might made his product more attractive by taking the trouble to provide both formulations of the characteristics.

3.6 Summary of the Characterization of the Basic Components

In this section we shall quickly review the nature of proper characterizations for the basic components.

The characterization formulations for the basic two-terminal components are shown in Figure 3.13 in graphical presentations. For the dissipator, the operation is confined strictly to the two quadrants representing dissipation; the curves go through the origin exactly. The temperature sensitivity of the device is shown by the family of curves and the maximum allowable dissipation by the dashed lines. Director characteristics, indicated at (b), are similar to those for the dissipator since the director has a somewhat dissipative nature; the main difference in the characteristics is the knee at the origin and the fact that variable limits are usually set by material properties instead of dissipation limits. Characteristics for the two kinds of storage components are shown at (c) and (d); these would generally be temperature sensitive, as the dissipator.

A typical actuator characterization is given at Figure 3.14. This is an acceptable presentation, but does not show the current versus voltage and stroke relationship explicitely. Also, the origin of the variable range limits is not explained, although it is relatively obvious. A hypothetical set of characteristics for a typical converter, the dc motor, is shown in Figure 3.15. These are relatively self explanatory, except that it might be said they represent the author's opinion of how motors should be characterized, not how it is actually done in practice. In the formulation given, both mechanical and electrical network analysis is greatly facilitated. Also, the full range of possible applications for the device is rather clearly exhibited. The odd effect of stiction, yielding the break in the curves at (a) as the speed goes through zero, is an

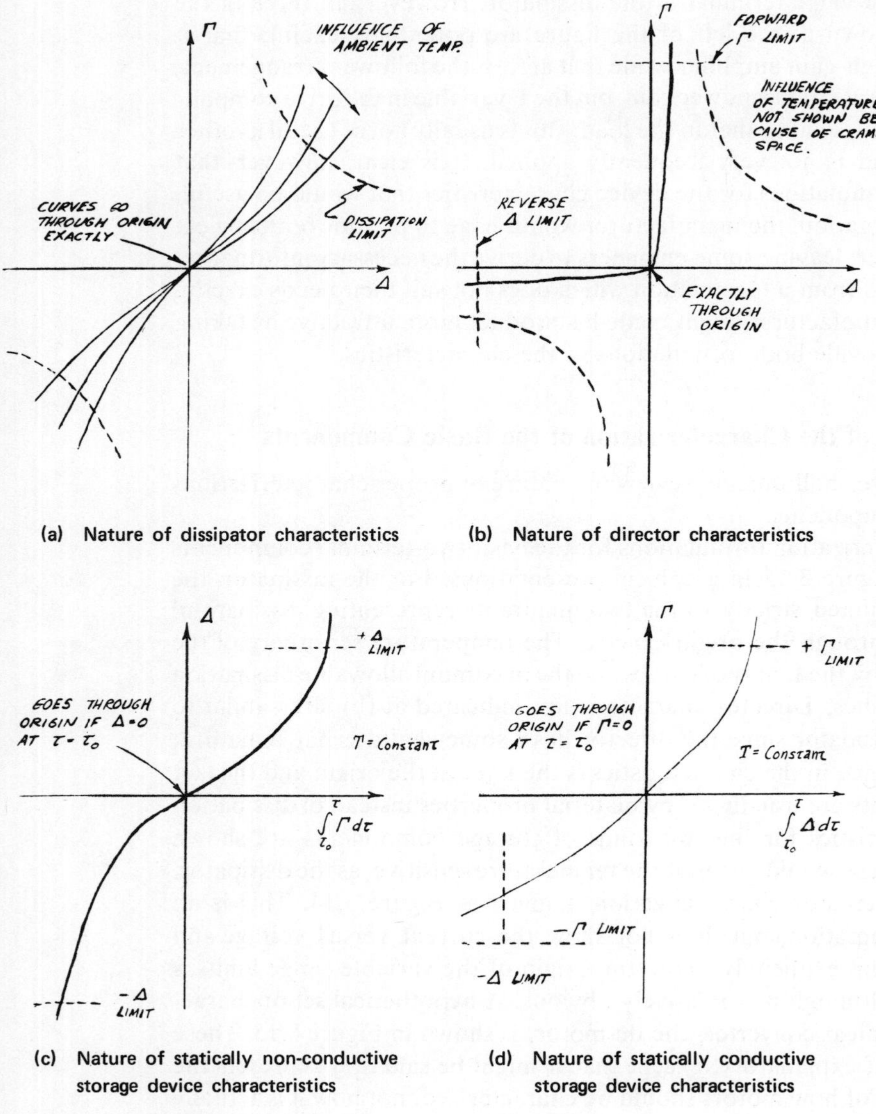

(a) Nature of dissipator characteristics

(b) Nature of director characteristics

(c) Nature of statically non-conductive storage device characteristics

(d) Nature of statically conductive storage device characteristics

From the Engineer's Sketch Book

Figure 3.13 Characteristics formulations for 2-terminal components.

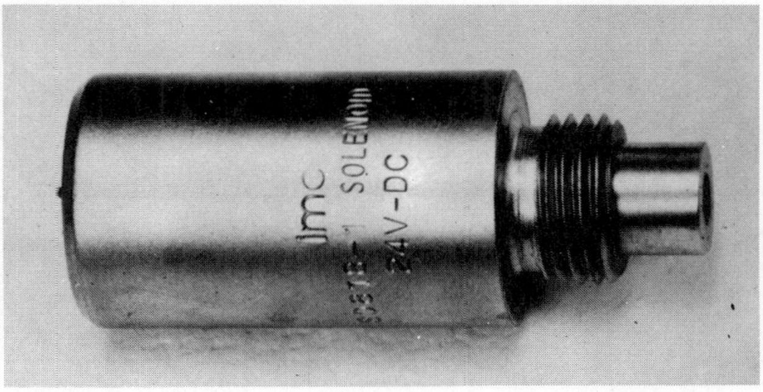

Factory Stocked Smallest two-coil solenoid available
High force (3.5 lbs at 0.120-inch stroke)
Light weight (under 2 ounces) Low current drain.

Solenoid design takes advantage of the fact that less power is required to hold a load than to pull it in. In two-coil construction, an actuating coil pulls the load in (3.5-pound minimum force at 0.120-inch stroke), and is then switched out, leaving a low drain (0.20 Amp) coil to hold the load. This technique decreases heat and increases work output. Because of its small size and light weight, it is particularly suited for avionic systems. Its high force and low drain are valuable characteristics for computer peripherals, control systems, instruments, and test equipment.

Voltage range	VDC	18 to 30
Max pull current	Amps	5.0
Max hold current	Amps	0.20
Min. pull force	lbs	3.5 at 0.120-in stroke
Min. hold force	lbs	6.0 at 0 stroke
Temp range	deg F	−65 to +160
Weight, unit	grams	53
Weight, plunger	grams	8

Duty cycle: continuous at 18 to 30 VDC.
*at 24 VDC, 78 degrees F.

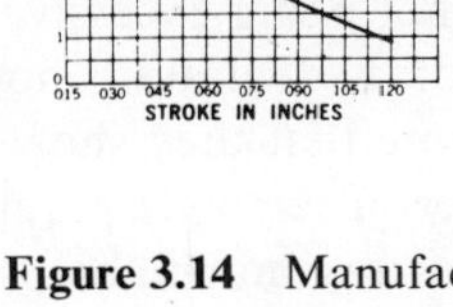

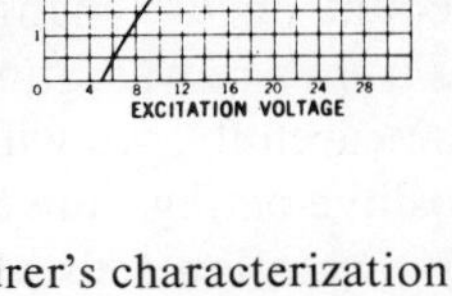

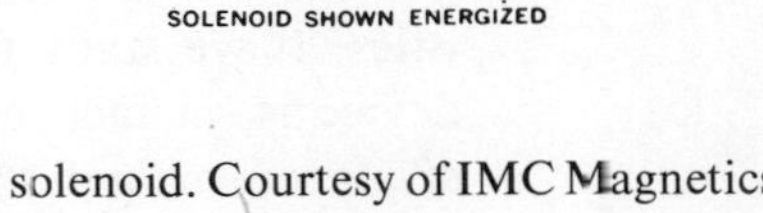

Figure 3.14 Manufacturer's characterization of solenoid. Courtesy of IMC Magnetics Corp.

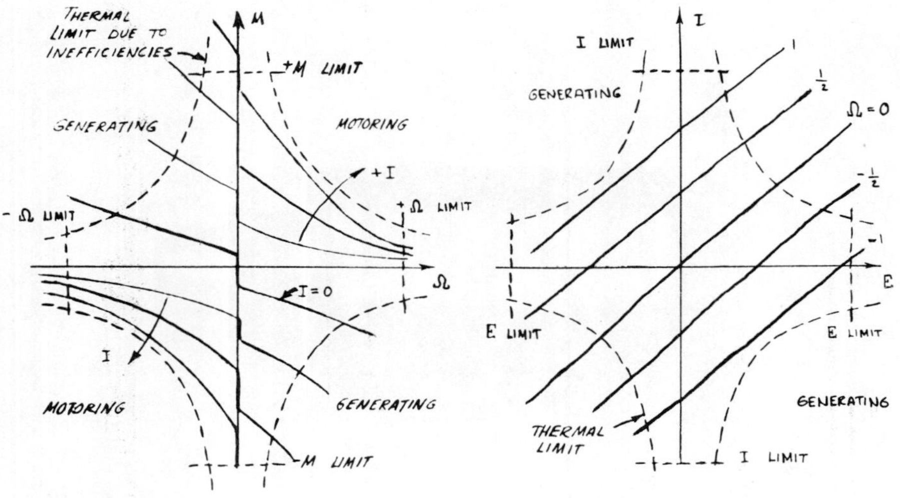

(a) Mechanical characteristics (b) Electrical characteristics

From the Engineer's Sketch Book

Figure 3.15 Typical converter characterization (dc motor).

effect that must be anticipated if a motor is to be bidirectionally operated in any kind of system. The presentation here brings this out most clearly. Also, the use of the motor as a tachometer is quite apparent from these curves.

To return briefly to the examples with which we began the chapter, it may be pointed out that each was a generally correct characterization. A check of the terminal number and conservation relationship number verifies this in a general sense. Some of the presentations seem a bit over-simplified and some over-done in small respects. With the clutch characterization, a more thorough job on the electrical side might be in order; since the component temperature changes with orperating point, the coil resistance, and therefore the current-voltage relationship, will change. This could be shown easily on a graphical presentation; even if it is very small, the sophisticated engineer will be grateful to know that he can count on the change not being excessive. Perhaps the greatest surprise comes from a reevaluation of the hydraulic motor characteristics. If we study these carefully, we will note that they show only that the device is, in fact, a positive-acting converter. The single parameter of fluid displacement per shaft revolution tells almost as much about the device as all the curves. This is not to detract from the value of the manufacturer's presentation, which is certainly a well organized, convenient design guide; it is simply to point out how misleading a first impression can be.

3.7 Linearization of Characteristics

The linearization of device characteristics accomplishes an extremely important simplification in the error and dynamic analysis of a system, and is used frequently in the remainder of the text. A typical situation is indicated in Figure 3.16 where one portion of a two-by-two characterization is indicated. The point Q is assumed to be the static operating point of the characterized component. If the operating point is never greatly different from Q, the characteristic of the device may be represented by the equation

$$\mathbf{\Gamma}_2 - \mathbf{\Gamma}_{2Q} \cong \frac{\partial \mathbf{\Gamma}_2}{\partial \mathbf{\Delta}_1}(\mathbf{\Delta}_1 - \mathbf{\Delta}_{1Q}) + \frac{\partial \mathbf{\Gamma}_2}{\partial \mathbf{\Delta}_2}(\mathbf{\Delta}_2 - \mathbf{\Delta}_{2Q})$$

where the partial derivatives are evaluated at Q. The representation may be simplified by the use of "excursion variables" which represent the difference between the value of a variable and its value at the quiescent point, as follows:

$$\gamma_2 = \mathbf{\Gamma}_2 - \mathbf{\Gamma}_{2Q}, \qquad \delta_1 = \mathbf{\Delta}_1 - \mathbf{\Delta}_{1Q}, \qquad \delta_2 = \mathbf{\Delta}_2 - \mathbf{\Delta}_{2Q}.$$

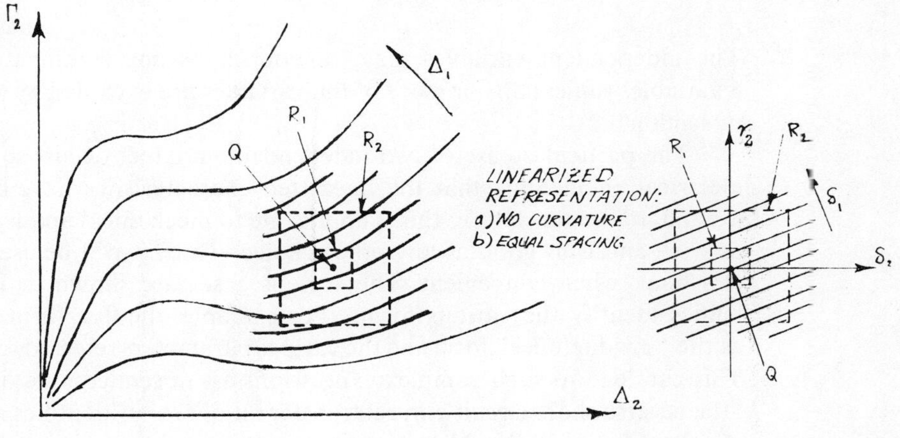

From the Engineer's Sketch Book

Figure 3.16 Partial linearization of 2 × 2 characterization.

Thus, the linearization becomes

$$\gamma_2 = \frac{\partial \mathbf{\Gamma}_2}{\partial \mathbf{\Delta}_1}\,\delta_1 + \frac{\partial \mathbf{\Gamma}_2}{\partial \mathbf{\Delta}_2}\,\delta_2 .$$

The graphical interpretation of this is shown at the extreme right of the figure. It may be noted that the linear representation is relatively accurate in the small region R_1, poorer in the larger region R_2, and totally misleading if applied over the whole first quadrant of the characteristic. The partial derivatives in such linearizations are frequently given special symbols to provide convenience in lengthy analytical procedures. Thus, the above expression might be indicated by

$$\gamma_2 = g_{11}\delta_1 + g_{12}\delta_2$$

The frequency of use of linearized characteristics justifies an efficient, standardized form of representation. For multi-terminal devices a standard matrix form, as follows, is used:

$$\begin{aligned}
\gamma_1 &= g_{11}\delta_1 + g_{12}\delta_2 + \ldots + g_{1m}\delta_m \\
\gamma_2 &= g_{21}\delta_1 + g_{22}\delta_2 + \ldots + g_{2m}\delta_m \\
&\vdots \\
\gamma_m &= g_{m1}\delta_1 + g_{m2}\delta_2 + \ldots + g_{mm}\delta_m
\end{aligned}$$

The independent variables may, of course, be any combination of γ, and δ variables (unless one or more of the variables are excluded by the conditions of section 3.2).

The particular case of two independent variables occurs so frequently in electrical engineering that the linearized characteristics have been assigned a standard nomenclature; this is applicable to mechanical and hydraulic work as well, and is as good as any other choice. Thus, it will be used throughout this book when convenient. The various cases are shown in Table 3.1; the names identify the various forms. For example, the first form is recognized as the "conductance" form and the various parameters referred to accordingly. This can lead to such complex expressions as, in semiconductor technology, "the reverse short-circuit current transfer ratio" which identifies a_{22} in the third standard form. It should be noted that the six forms given include all the possibilities which exist within the convention that one-subscripted variables are written before or above two-subscripted variables, then δ variables are written before or above γ variables.

Table 3.1

Conductance	Hybrid	Reverse Transfer*
$\gamma_1 = g_{11}\delta_1 + g_{12}\delta_2$	$\delta_1 = h_{11}\gamma_1 + h_{12}\delta_2$	$\delta_1 = a_{11}\delta_2 + a_{12}\gamma_2$
$\gamma_2 = g_{21}\delta_1 + g_{22}\delta_2$	$\gamma_2 = h_{21}\gamma_1 + h_{22}\delta_2$	$\gamma_1 = a_{21}\delta_2 + a_{22}\gamma_2$
Resistance	Inverse Hybrid	Forward Transfer
$\delta_1 = r_{11}\gamma_1 + r_{12}\gamma_2$	$\gamma_1 = m_{11}\delta_1 + m_{12}\gamma_2$	$\delta_2 = b_{11}\delta_1 + b_{12}\gamma_1$
$\delta_2 = r_{21}\gamma_1 + r_{22}\gamma_2$	$\delta_2 = m_{21}\delta_1 + m_{22}\gamma_2$	$\gamma_2 = b_{21}\delta_1 + b_{22}\gamma_1$

* The reverse transfer form given here differs from electrical engineering convention in the sign of a_{12} and a_{22}, which are oddly made negative therein.

Since any one representation completely specifies the linearized behavior of a component, the coefficients in the above forms are not independent but may be determined from one another. For example, if we happen to have the conductance representation for a device but would prefer to use the hybrid form in a particular problem, we may determine the h's in terms of the g's by the following procedure:

$$\gamma_1 = g_{11}\delta_1 + g_{12}\delta_2 \longrightarrow \delta_1 = \frac{1}{g_{11}}\gamma_1 - \frac{g_{12}}{g_{11}}\delta_2$$

$$\gamma_2 = g_{21}\delta_1 + g_{22}\delta_2 \longrightarrow \gamma_2 = g_{21}\left(\frac{1}{g_{11}}\gamma_1 - \frac{g_{12}}{g_{11}}\right)\delta_2 + g_{22}\delta_2$$

$$= \frac{g_{21}}{g_{11}}\gamma_1 + \left(g_{22} - \frac{g_{21}\,g_{12}}{g_{11}}\right)\delta_2$$

thus,

$$h_{11} = \frac{1}{g_{11}}, \quad h_{12} = -\frac{g_{12}}{g_{11}}$$

$$h_{21} = \frac{g_{21}}{g_{11}}, \quad h_{22} = g_{22} - \frac{g_{21}g_{12}}{g_{11}}$$

Problems

1. Calculate the heating power (ft-lb/min) due to friction in the clutch of Figure 3.1 at 3000 RPM slip speed and 0.40 amp actuating current. Enter this value on the graph at the right. Is the meaning of the curves at the right completely clear?

2. Do you think the clutch of Figure 3.1 can be operated by-directionally? What would be the appearance of the left-hand graph if this case were to be shown explicitly? What happens if the coil current is reversed? Graph "pick-up torque" versus current for 1200 RPM slip speed. Estimate, from the graph, the torque at 0.60 amp. Why do you suppose there are no higher actuating currents given?

3. What category of device, according to the material of Chapter 1, is the unit of Figure 3.2? Theoretically, in a device of this nature, torque is strictly a function of pressure. What, then, is the source of the "droop" in the curves at (b)? Why aren't the curves at (c) straight lines? Judging from the curves at (d), can we conclude that these motors have very little leakage?

4. For the transistor of Figure 3.3:
(1) What is I_C when $V_{CE} = 40$ volts and $I_B = 45$ microamps?
(2) What is I_C when $V_{BE} = 1.0$ volt and $I_B = 50$ ma?
(3) What is V_{CE} when $I_C = 300$ ma and $I_B = 22$ ma?
(4) Show how a 120-volt battery could safely be connected to the transistor.
(5) What is the largest collector current which can be handled by the device?
(6) Why do you suppose the curves of the first graph stop in "mid-air"?

5. For the three devices and characterizations of Figures 3.1, 3.2, and 3.3:
(1) Estimate the prices of each device from this list:
$1000, $100, $10, $1, 10¢.
(2) Which characterization, in your opinion, is the most
(i) thorough?
(ii) devoid of undefined terminology?
(iii) redundant?
(v) lacking in essential details?

6. Identify the **Γ** and **Δ** variables for the three units discussed in Figures 3.1, 3.2, and 3.3. Write the conservation relationships which apply to each. Point out the relative measurements used in describing the characteristics of the devices.

7. In the characterization of the following devices, by means of sets of equations, are there any variables which should not be taken as independent?
(1) Transistor
(2) eddy current coupling
(3) impulse turbine
(4) relief valve

8. Carry out, as in the discussion associated with Figures 3.7, an analysis of the proper characterization form for a vacuum tube with two grids, G_1 and G_2 (a "tetrode"). Normally, such tubes are used with one of the grids at constant voltage with respect to the cathode. How would this influence the characterization "for normal applications"?

9. On a tracing of the transistor characteristics of Figure 3.3, indicate the thermal and other limits of operation. Shade in the total usable area appearing on the diagram. Are there any usable operating conditions which do not appear on the graphs?

10. Sketch the primary and mechanical characteristics of the following adjustable components:
(1) inductor.
(2) globe valve.
(3) automobile emergency brake.

11. For the following energy-storing devices, sketch the characterizations, with all graph axes specified, to show an ideal characteristic, the effects of saturation, the effects of hysteresis, and variable limits:
(1) inductor.
(2) spring.
(3) capacitor.
(4) air charge accumulator.

12. Estimate the quiescent operating point of a 2N2102 transistor (see Figure 3.3) used in the circuit of Figure 3.11(b) if

$V_S = 60$ volts, $V_I = 5$ volts,
$R_L = 10\ \text{K}\Omega$, $R_I = 1000$ ohms

13. Make up schematics for all the ways that three two-terminal devices may be connected to one three-terminal controlled dissipator. Supposing the devices have the following nature:

I operates over large ranges of **Γ** and **Δ**. *II* and *III* are limited to relatively small **Γ** and **Δ**. The "valve" terminal of the controlled dissipator is the only one which can take large **Γ**.

There must never be large **Δ** between the "control" and "reference" terminals of the controlled dissipator.

(1) Put an X through the schematics which indicate useless networks.
(2) Circle the network which appears most useful as an amplifier.

14. Prepare a presentation, similar to those of Figure 3.13, showing the nature of a proper characterization of an actuator.

15. Prepare linearized representations for the following characteristics, about the specified points. Retain the units of the original graphs and use lower-case notation to indicate excursion variables.

(1) The M, Ω characteristic for the clutch of Figure 3.1, at $\Omega = 1800$ RPM and $I = 0.20$ amp.
(2) As in (1), but at $\Omega = 600$ RPM and $I = 0.40$ amp.
(3) The I_C, V_{CE} characteristic for the transistor of Figure 3.3, at $V_{CE} = 40$ volts and $I_B = 45$ microamps.
(4) The I_C, V_{BE} characteristic of Figure 3.3, at $V_{BE} = 1.1$ and $I_B = 50$ ma.
(5) The F versus X characteristic of the solenoid of Figure 3.14, at $X = 0.60''$ and 18 volts at the coil.
(6) The eddy current coupling characteristic at Figure 1.2 (No. 100K) at 200, 400, and 600 RPM.
(7) The needle-valve characteristic of Figure 1.18, at 30 psi and 8 turns open (use "N" for turns open). Arrange this so flow rate is the dependent variable.

16. Repeat Problem 15(1) above, for the complete range of slip speed (600 RPM increments). Graph the "incremental damping coefficient", $\partial M/\partial \Omega$, over the range. Note that this is a negative quantity; it represents a generator in small excursion analysis and can account for extremely unusual behavior of certain networks.

NOTE: In the following problems, to "discuss" means to:

(1) Make a neat sketch of the device in question, identify each of its terminals with a name and letter (preferably the first letter of the name, to be used as a subscript where necessary), and specify positive directions for the measurement of variables. Use F, V, X; M, Ω, θ; I, V; P, W for the terminal variables, as in the text.
(2) Write all the conservation of current, flow, and momentum (sums of forces, torques) relationships which pertain to the device in static operation.
(3) Show, by use of a schematic diagram, the most common application arrangement for the unit.

(4) Specify the terminals which may be used as reference points for the measurement of Λ variables, in normal applications.
(5) Identify any variables which cannot be independent in a functional characterization because they are unchanging over some region of operation.
(6) Identify any pairs of variables which cannot both be independent or dependent since they are proportionally related in all, or a portion of, the operating region.
(7) Write the functional form of the device characterization that you feel would be most useful.
(8) Sketch some reasonable estimates of the nature of the curves in the graphical form of the characterization.
(9) List the physical effects most likely to set magnitude limits on the terminal variables.
(10) Indicate, on the graphs of (8), the nature of the limits on the operating region due to thermal considerations.
(11) Use a colored pencil to shade in, on the graphs of (8), the complete region on each set of curves over which the device may be operated continuously.

In the following problems, *discuss* (as just outlined) the characterization of the devices shown in the specified figures:

17. Figure 1.1(a).
18. Figure 1.4(a).
19. Figure 1.7(d).
20. Figure 1.8(a).
21. Figure 1.10(a).
22. Figure 1.12.
23. Figure 1.17(a).
24. Figure 1.17(c).
25. Figure 1.20(b).
26. Figure 1.21(a).
27. Figure 1.21(d).
28. Figure 1.22.
29. Figure 1.23.
30. Figure 1.25.
31. Figure 1.26.
32. Figure 1.27.
33. Figure 1.29.
34. Figure 1.30.
35. Figure 3.1.
36. Figure 3.2.
37. Figure 3.9.

4 Static analysis of systems

Static analysis provides information necessary to preliminary component selection and, in many cases, to dynamic analysis. For example, static analysis is used in determining the average power which must be dissipated by each of the components of a system and thus provides the required power ratings. In dynamic analysis where non-linear components are involved, we must determine the static operating point before we can develop linearized representations to facilitate the work. The term "static" is used here in a somewhat different sense from that in mechanics where "static" implies "non-moving"; in systems work, the static operation of a system is understood to be the case in which the terminal variables (such as voltage, current, shaft speed, torque, etc.) are not changing with respect to time. The term "steady state," used in thermodynamics for this case, has been usurped by electrical engineers to identify the case where the system variables are sinusoidal with unchanging magnitudes, frequencies, and phases; there is no choice but to follow this well established usage. Thus, in a mechanical system the static state is that in which the parts have velocities but no accelerations.

In many cases static analysis is a simple matter which can be carried out by inspection and a small amount of algebra. In others, especially where the components are non-linear and where the network has feedback loops, static analysis may become quite challenging. In this chapter we develop universally applicable techniques for both situations. Also, we consider some design problems which relate to the static performance of a system.

4.1 Nature of the Static Analysis Problem

The objective of static analysis is to determine the operating conditions, in the static state, of each of the components of a system. This is equivalent to a determination of the conditions at each of the connectors of the system. A typical case is presented in Figure 4.1 where a clutching-type torque regulator is shown delivering power from an induction motor to a load. This kind of

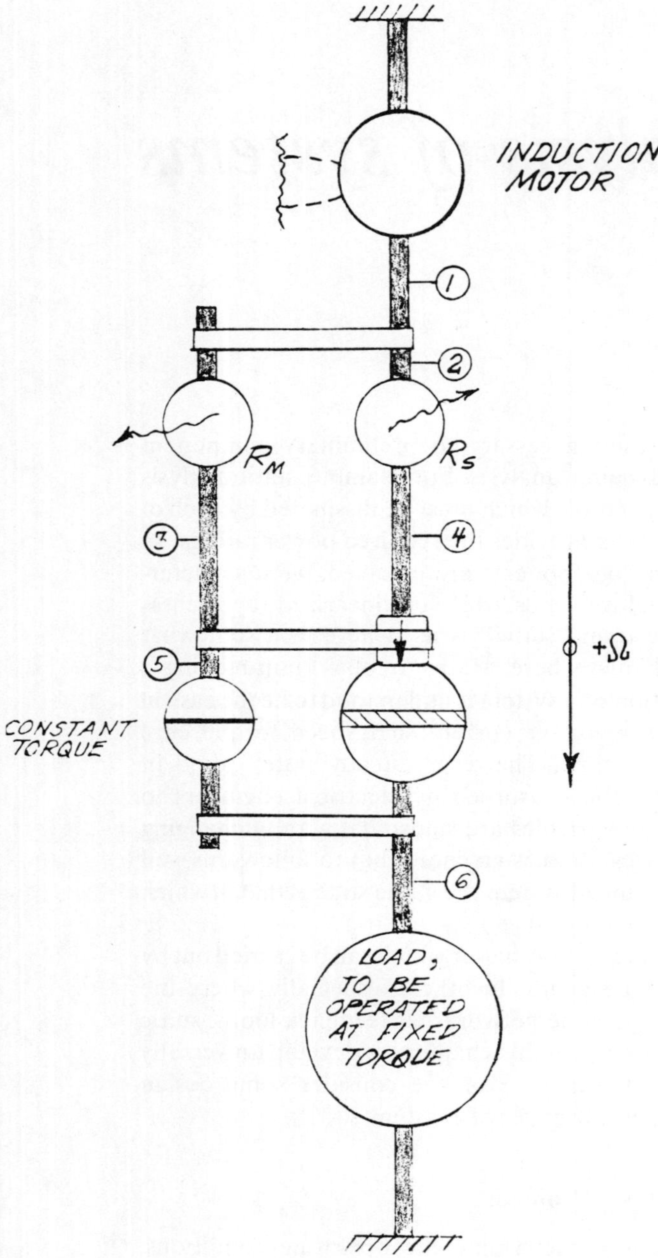

From the Engineer's Sketch Book

Figure 4.1 Showing the nature of a problem in static analysis.

system would be used when the load might be destroyed by excess torque or when constant torque operation is required; one specific example would be a tap being driven by a drill press. The system shown would be a refinement on the normal tapping-head arrangement. Six particular shaft sections have been numbered on the figure; knowledge of the torque and speed at each of these sections would completely specify the operation of each component. Thus, the static analysis must determine the value of twelve variables. Let us verify that this is possible by identifying twelve independent relationships between these variables. We note:

(a) The speeds at sections 1 and 2, 3 and 5 are equal. This gives two independent relationships.
(b) The torque at sections 2 and 4 is the same. This is one relationship.
(c) There are five two-terminal components which must each have one-by-one characterizations. (We consider only the mechanical half of the induction motor characteristic, assuming that the motor is operated at constant voltage.) This gives five independent relationships.
(d) There is one three-terminal component and this will have a two-by-two characterization. This gives two independent relationships.
(e) From a summation of moments, we may determine that the torque at section 1 must equal the sum of the torques at sections 3 and 4; and that at 6 must also equal the sum at 3 and 4. Therefore we have two more independent relationships.

Thus, there are twelve independent relationships. This parity between the number of variables to be determined and the number of independent relationships available is, of course, essential to any successful analysis. In troublesome problems it is frequently worthwhile to check this parity as we have done here; oversights in analysis are quickly exposed in this way.

It should be noted that two kinds of relationship are used in analysis. One kind depends only on the geometry (to be precise, "topology") of the network; i.e., how the components are hooked together. Typical of these relationships are (a), (b), and (e) above. The other kind depends on the components themselves; i.e., the characterizations. As another way of checking on the necessary parity of variables and equations, we may examine our equations to see that all the geometric conditions of the network and all the component characterizations appear. If they do not, a solution is impossible. Naturally, in static analysis it is desirable to have the most efficient possible formulation of the problem. The widely used "node" (or "junction") and "loop" techniques can provide that a minimum number of variables enter the problem, thereby yielding the maximum convenience to the analyst. These techniques take advantage of the simplicity of the geometric conditions as much as possible. We shall consider them in section 4.3.

4.2 Preliminary Aspects of Static Analysis

In all kinds of system analysis, one of the most troublesome aspects is that of sign. When a "plus" somehow gets replaced by a "minus" where it shouldn't, the analysis almost always indicates a kind of performance which is totally different in nature, not just magnitude, from the real thing. Few errors can cause the engineer greater grief. Therefore, one of our major objectives here is to establish, as carefully as possible, a set of sign conventions which reduce the possibility of confusion as much as possible. We try to make such a fuss over this matter that the reader will become particularly sensitive to it. Also, in this section, we review the important role of junctions and idealized generators in analysis techniques.

The static analysis of electrical circuits is well known and is readily extended to hydraulic networks. Unfortunately, mechanical networks have not been studied so systematically and analysis techniques for them tend to be haphazard in form. This probably reduces many engineers' awareness of the utility of mechanical networks. To continue our "media-independent" philosophy here, let us begin by developing some ideas which allow mechanical networks to be considered as readily as others. Some fundamental aspects of this approach are shown in Figure 4.2. At (a) is shown the rotary mechanical equivalent of an electrical junction (wherein a number of conductors are simply brought together and connected metallically). Here, three shafts are shown "banded" together. This is simply a symbolic way of indicating that the shafts (there can be any number) are somehow all caused to revolve at the same speed by some means which does not dissipate power or otherwise influence it. Physically, this might be approximated by a system of pulleys and belts, sprockets and roller chains, gears, kinematic linkages, or welding if the shafts can be coaxial by means of one or more being hollow. (It's just the importance of geometry to mechanical systems that makes this junction seem like a more complex thing than its equivalent in other media.) Mathematically, this junction is described by the equations:

$$\Omega_1 = \Omega_2 = \Omega_3 = \Omega_4 = \Omega_5 = \Omega_6$$
$$M_1 + M_2 + M_3 = M_4 + M_5 + M_6$$

It should be noted that our convention of tightening torques being positive, loosening torques negative, obviates the need for little arc-shaped direction indicators on the drawing. Once we are familiar with this convention we can write system equations much more rapidly than if the traditional approach,

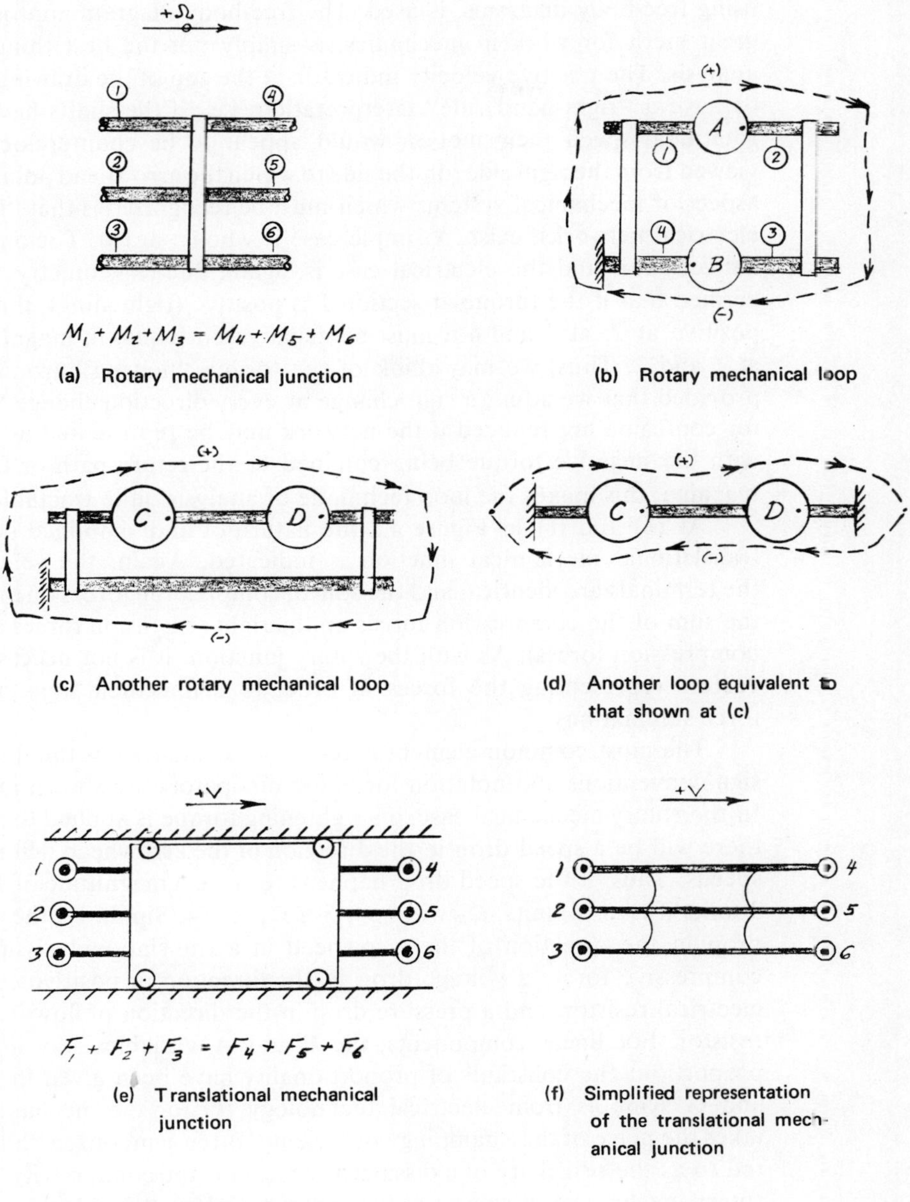

From the Engineer's Sketch Book

Figure 4.2 Junctions and loops in mechanical systems.

using free-body diagrams, is used. The free-body diagram approach, with its great merit for work in mechanics, is simply not the best thing for system analysis. The positive velocity indicator at the top of the drawing is intended to have a "right-hand rule" interpretation; i.e., if the shafts have a positive rotational speed their motion would appear to be counterclockwise when viewed from the right side (or, the side to which the arrowhead points). Another aspect of mechanical systems which must be recognized is that "loops," as in electrical networks, exist. A simple case is whown at (b). The one difference between this and the electrical case is, again, due to geometry. It is readily verified that if the torque at section 1 is positive (tightening), it must also be positive at 2; at 3 and 4 it must be negative but equal in magnitude to that at 1 and 2. Thus, we may think of the torque "flowing" around the loop, provided that we admit a sign change at every direction change. Possibilities for confusion are reduced if the network may be represented as at (c) or (d) with the negative torque being confined to the return path or frame of the machine; this makes the loop technique of analysis quite tractable.

At (e) and (f) in Figure 4.2 the nature of and simplified symbol for a translational mechanical junction is indicated. Again, the velocities of all the terminals are identical and the sum of compression forces on the left equals the sum of the compression forces on the right (extension forces are negative compression forces). As with the rotary junction, it is not necessary to draw vectors representing the forces in order to define their signs or carry out force summations.

The most common element entering static analysis is the dissipator; the sign conventions and notation for linear dissipators are shown in Figure 4.3. In the rotary mechanical case, if a tightening torque is applied to a dissipator, there will be a speed drop in the direction of the arrowhead defining positive speeds. Thus, if the speed drop happens to have a magnitude of 1, Ω_B will be 3 units if Ω_A is 4 units; Ω_B will be -5 if Ω_A is -4. Similarly, there is a speed drop in the direction of the arrowhead in a translational dissipator under compressive force, a voltage drop in the direction of positive current in an electrical resistor, and a pressure drop in the direction of flow in a hydraulic resistor. For linear components, the $\boldsymbol{\Gamma}$ and $\boldsymbol{\Delta}$ variables stay in a constant proportion; the constants of proportionality have been given the familiar R and G symbols from electrical technology (G for the mechanical devices takes the place of the "damping coefficient," often symbolized "b"). R is referred to as the resistivity of a dissipator while G is the conductivity. Note that a rotary mechanical dissipator (coupling) with a high "G" would be quite stiff; i.e., it would transmit large torques with relatively little speed differential (slip).

As will be shown, every static analysis problem can be reduced to one in which only dissipators and ideal generators appear. The symbols and per-

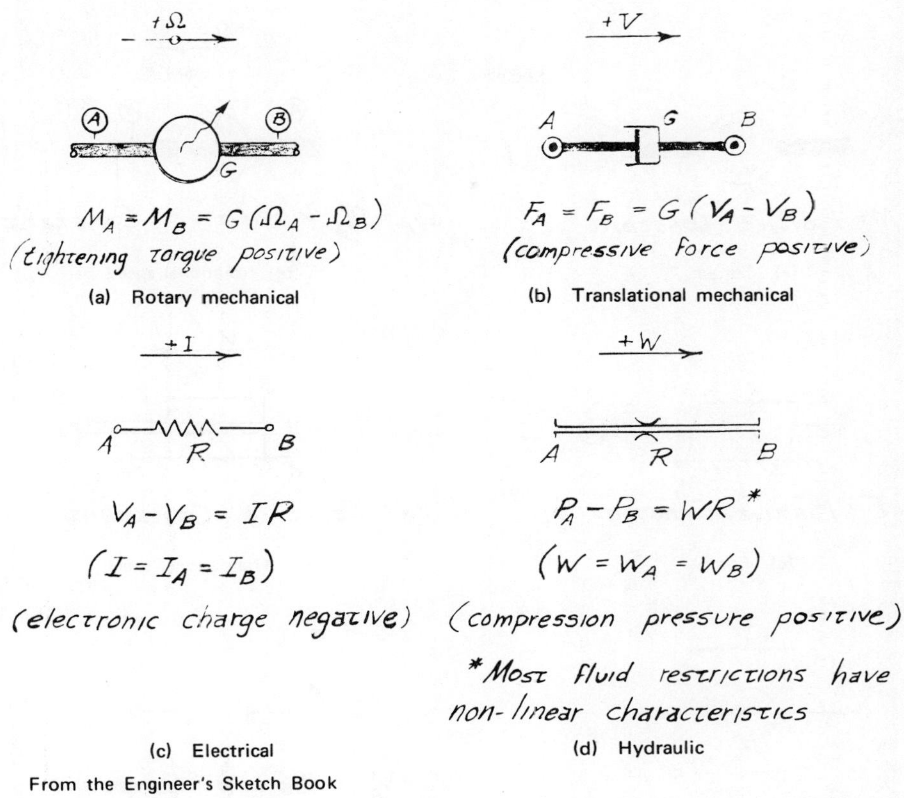

Figure 4.3 Notation and sign conventions for linear dissipative components.

formance of these generators are given in Figure 4.4. In most cases it is difficult to build such generators, but that is immaterial here. (Although the synchronous motor is rather simple, its speed regulation comes from the regularity of ac frequency and this is achieved by an elaborate control system.) Typical of the ideal generators is the current source shown at (e); this is imagined to deliver the current I no matter what sign or magnitude of voltage is applied across it. Symbols for the mechanical sources have been contrived by the author since there is no standard for them. With the rotational speed source, the double-headed arrow (inward or outward, depending on the rotational direction) may seem odd, but is useful. If we imagine either shaft to be held stationary and cover the adjacent arrowhead, the remaining arrowhead indicates the direction of rotation of the free shaft, by the right-hand rule. At (i) and (j) are given examples of the kinematic behavior of such devices. The translational generators are shown fixed to ground because they are

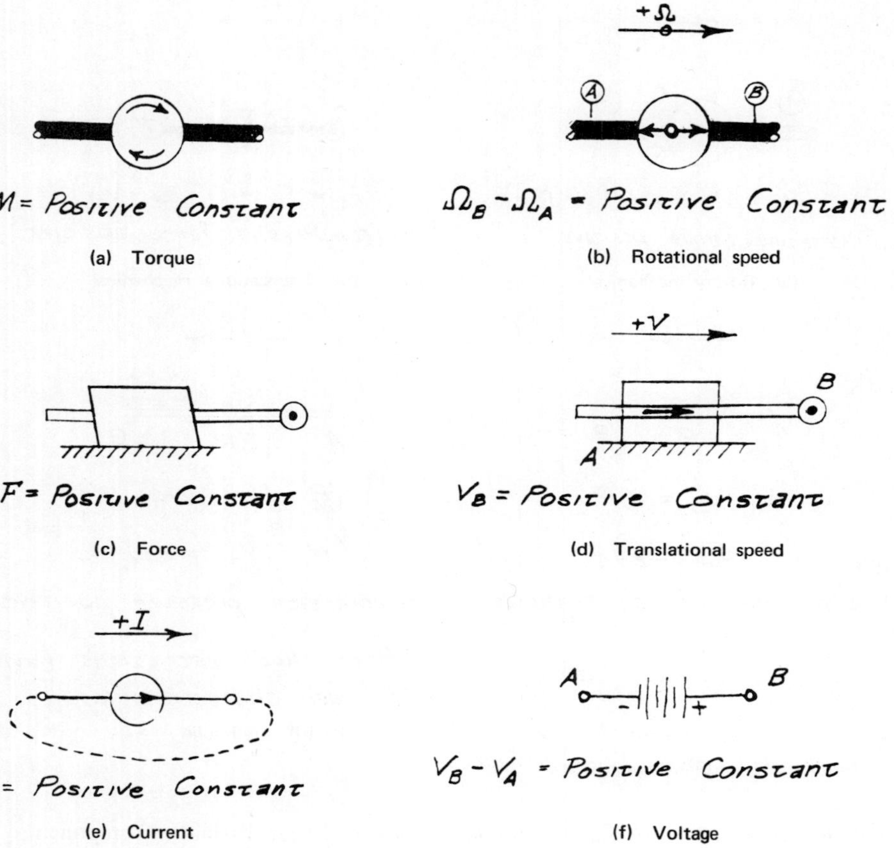

Figure 4.4 (*continued*)

difficult to visualize otherwise; the direction in which the case on the forcer leans indicates the sense.

The use of ideal generators is indicated briefly by Figure 4.5. Here, two devices, constructed from generators and dissipators only and having the same characteristic as a real battery, are shown. The replacement of all real power sources by such equivalents leads to a network schematic having only dissipators and generators. This can facilitate analysis and help clarify some aspects of system performance. In the example, the dissipator would be the same for either equivalent; the use of G and R notation is conventional and arises from the following rules:

(1) The series equivalent of a real electrical source includes a voltage

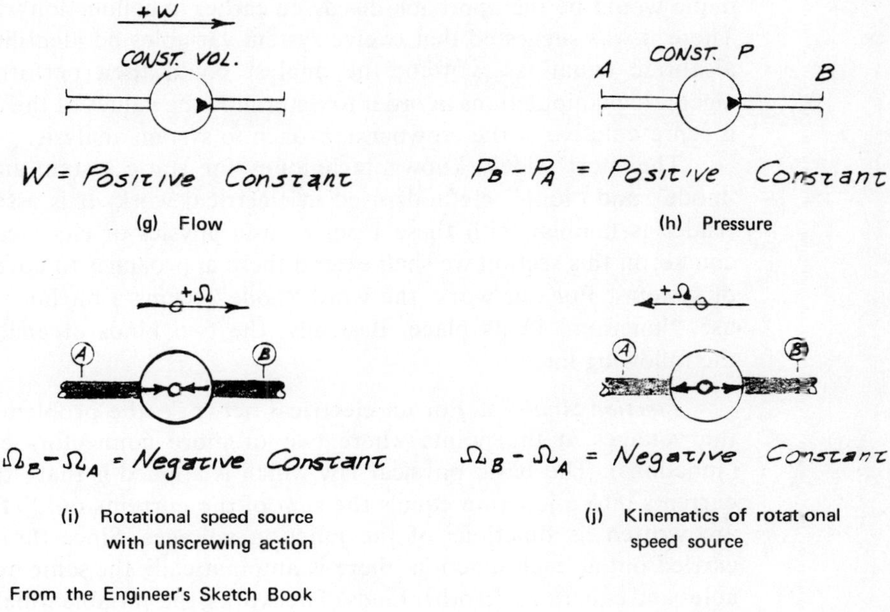

Figure 4.4 Symbols, notation, and sign conventions for idealized power sources.

generator with output equal to the open circuit voltage of the source and a resistor with a value dV/dI.

(2) The parallel equivalent of a real electrical source includes a current generator with output equal to the short circuit current of the source and a conductance with a value dI/dV.

These rules are readily modified to include the other forms of power transfer media. Such equivalent devices represent important building blocks for small excursion equivalent networks for three terminal components, found to be most helpful in dynamic analysis.

4.3 Techniques for the Static Analysis of Linear Networks

The application of a little grace and style—i.e., technique—to the solution of any problem can save time and increase the likelihood of success; this is naturally true with static analysis of systems. In contrast to an efficient technique would be the approach discussed earlier in connection with Figure 4.1. There it was suggested that twelve system variables be identified and twelve algebraic equations written; the analyst could then perform a series of algebraic manipulations in order to determine the values of the variables. This is representative of the crowbar approach to system analysis.

The most widely known techniques for static system analysis are the "node" and "loop" methods used in electrical work. It is assumed that the reader is familiar with these from a basic physics or electrical engineering course; in this section we shall extend these approaches to cover other kinds of systems. For our work, the word "node" seems a bit fancy and we shall use "junction" in its place. Basically, the two kinds of analysis go along the following lines:

Junction Analysis. For an electrical network, the problem variables are the voltages at the points where two or more connectors come together (junctions). The basic physical law which is applied is that "the sum of the currents into a junction equals the sum of the currents out." These currents are written as functions of the junction voltages. Since the summation is carried out at each junction, there is automatically the same number of variables and equations. In other kinds of networks, the variable which is analogous to voltage is used (the Δ variable) and torque, force, or flow-rate summations are carried out.

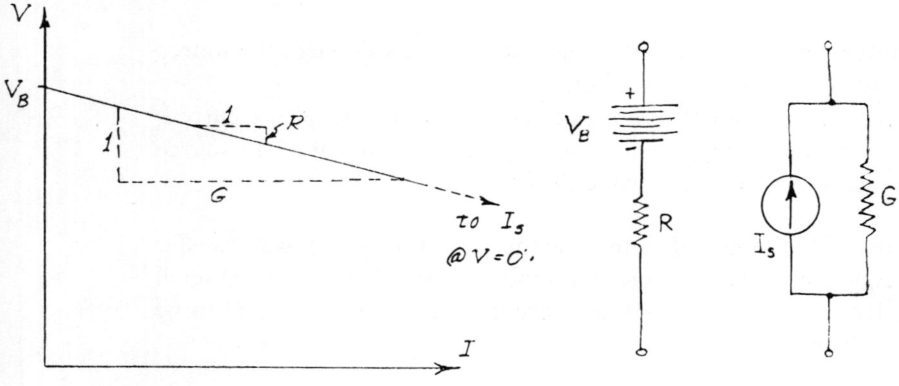

(a) Nature of the characteristic of a real battery.

Two arrangements which are equivalent to the real battery

From the Engineer's Sketch Book

Figure 4.5 Reducing a network to one in which only dissipators and ideal sources appear.

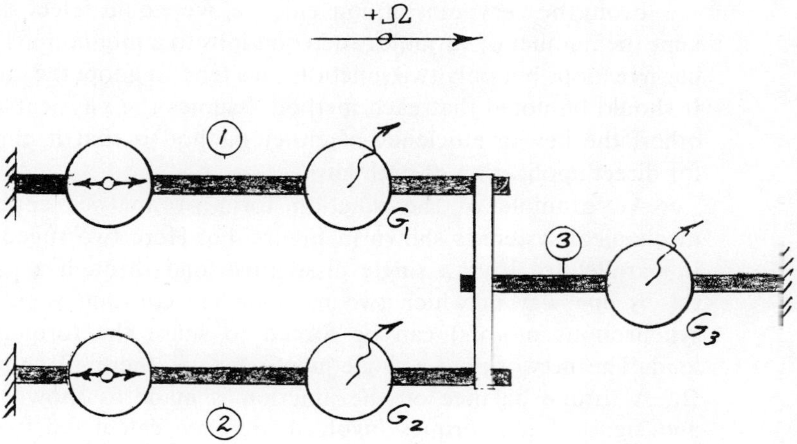

$$M_1 + M_2 = M_3$$

$$G_1(\Omega_1 - \Omega_3) + G_2(\Omega_2 - \Omega_3) = G_3\,\Omega_3$$

$$\Omega_3 = \frac{G_1\Omega_1 + G_2\Omega_2}{G_1 + G_2 + G_3}$$

From the Engineer's Sketch Book

Figure 4.6 An example of the "junction" form of analysis.

Loop Analysis. For an electrical network, several closed paths through the network are selected so that each component is on at least one path, and so that the number of paths is equal to the number of areas into which the network schematic divides the paper on which it is drawn, less one. (This assumes the schematic can be drawn, without lines crossing over one another, on a sheet of paper; i.e., that the network is "mapable" on a plane.) The problem variables are the currents which circulate in the paths. The basic physical law which is applied is that "the sum of voltage drops (or rises) around any path must be zero." These drops are computed in terms of the loop currents. The parallel technique for rotary mechanical systems utilizes torque loops and summations of speed changes. For other kinds of system the technique is readily adapted by means of analogy.

From the viewpoint of convenience, we would select the method which kept the number of variables and equations to a minimum. Thus, if a network has five loops but only two junctions, we tend to adopt the junction technique. It should be noted that each method assumes the physical law utilized in the other; the key to efficiency of either method is that it eliminates the need for direct application of both laws.

An example of the junction form of analysis applied to a rotary mechanical system is shown in Figure 4.6. Here two speed sources operate in parallel to drive a single dissipative load through a pair of couplings; this is one way in which two more-or-less constant speed drives (such as synchronous motors) can be forced to share the torque developed by a load. The network has a single junction; thus, there is one problem variable, Ω_3. A torque balance on the junction is made as shown. The magnitudes and signs of the torques involved are then calculated from the dissipator characteristics and the applied and unknown speeds. Simple algebra produces the value of the junction speed.

Two examples of loop analysis are shown in Figure 4.7. An electrical case is shown at (a). Note that the voltage drop across the battery, taken while summing drops on the loop threaded by I_1, is negative since this is a rise in the direction of positive current. After the resulting equations are put in a standard form, determinants may be used to obtain solutions, as shown. The rotary mechanical case, shown at (b), has been set up so that we work only with positive torques, the negative return paths not threading any components. Note that if the return paths are indicated as shafts, instead of by a surrounding frame, the number of areas within the schematic is equal to the number of loops, as it should be. As with the battery in the electrical case, the speed source here is a negative drop; i.e., a rise. The specific law applied here is that the sum of speed drops on any loop must be zero. Note that the equations are in terms of unknown torques only; in this kind of analysis, unknown speeds never appear explicitly.

The perfect analogy between all kinds of linear dissipative networks allows us to utilize a generalized form of analysis if we are so inclined. Two simple examples are shown in Figure 4.8 where the single component equivalent to both series- and parallel-connected linear dissipators is determined. So long as we are careful to maintain perfect accuracy in variable identification and sign, we can apply the results of analysis in one media to problems in another by using this approach.

In order to achieve obvious analogy between mechanical and electrical networks in this work, all the mechanical systems have been drawn as if all the bars of translating systems or all the shafts of rotary systems were parallel. This eleminates some of the geometric considerations peculiar to mechanical

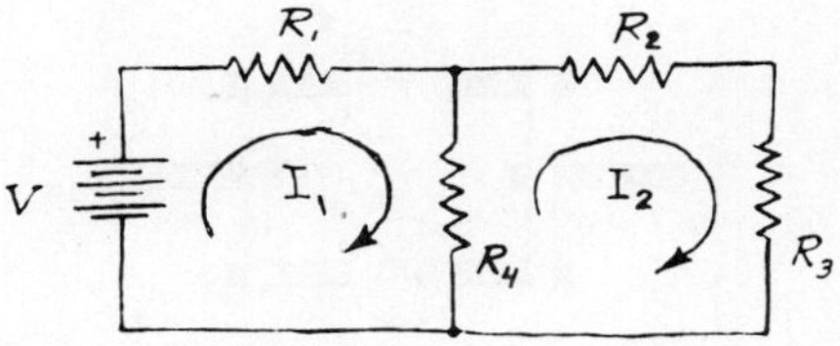

$$I_1R_1 + (I_1 - I_2)R_4 - V = 0$$
$$I_2R_2 + I_2R_3 + (I_2 - I_1)R_4 = 0$$

$$(R_1 + R_4)I_1 - R_4 I_2 = V$$
$$-R_4 I_1 + (R_2 + R_3 + R_4)I_2 = 0$$

$$I_1 = \frac{(R_2 + R_3 + R_4)V}{\begin{vmatrix} (R_1 + R_4) & (-R_4) \\ (-R_4) & (R_2 + R_3 + R_4) \end{vmatrix}}$$

(a) An electrical case

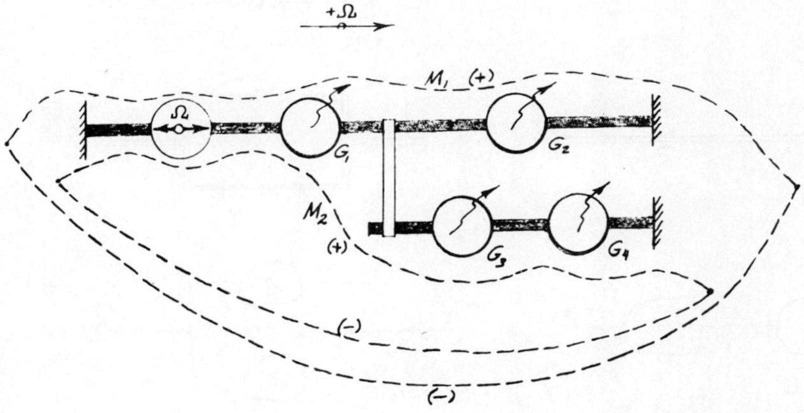

$$-\Omega + \frac{M_1 + M_2}{G_1} + \frac{M_1}{G_2} = 0$$

$$-\Omega + \frac{M_1 + M_2}{G_1} + \frac{M_2}{G_3} + \frac{M_2}{G_4} = 0$$

(b) A rotary mechanical case

From the Engineer's Sketch Book

Figure 4.7 Applications of the "loop" form of analysis.

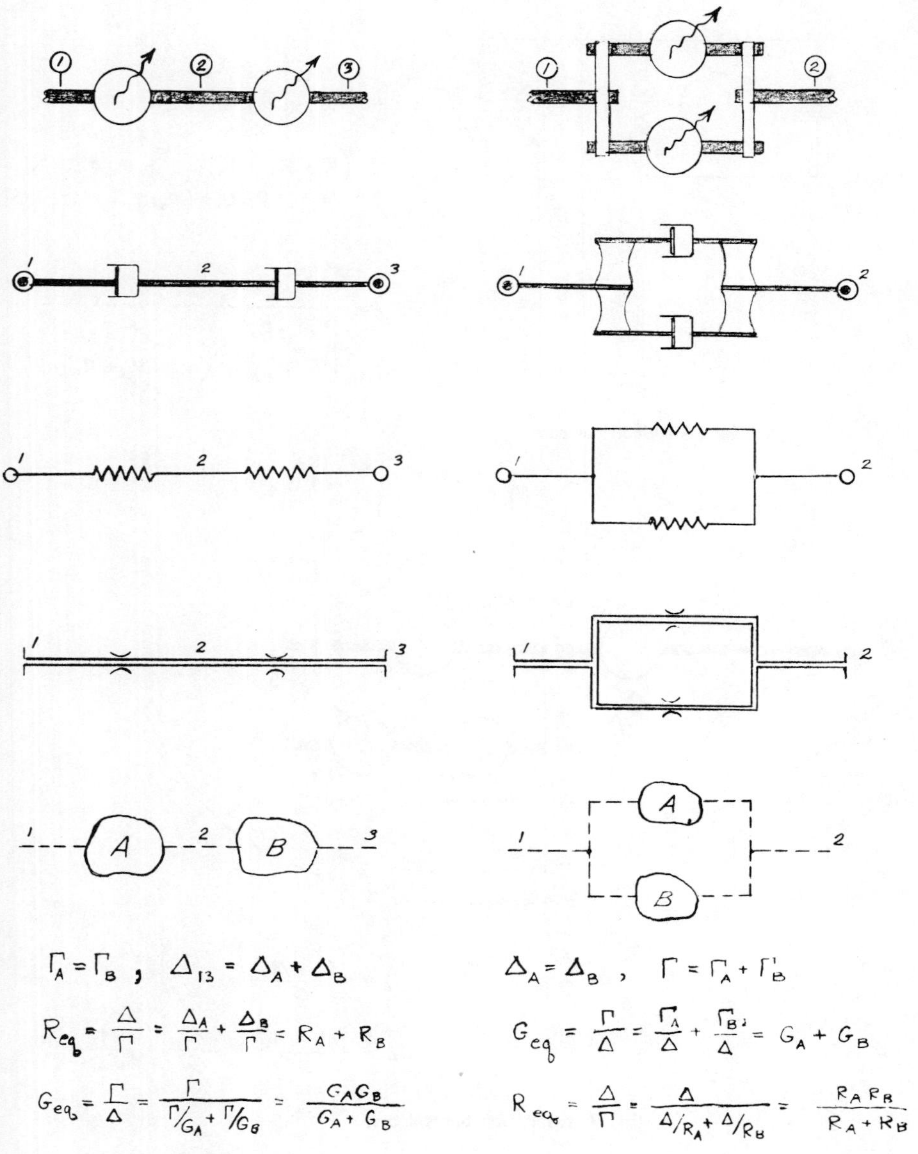

$\Gamma_A = \Gamma_B$, $\Delta_{13} = \Delta_A + \Delta_B$

$$R_{eq} = \frac{\Delta}{\Gamma} = \frac{\Delta_A}{\Gamma} + \frac{\Delta_B}{\Gamma} = R_A + R_B$$

$$G_{eq} = \frac{\Gamma}{\Delta} = \frac{\Gamma}{\Gamma/G_A + \Gamma/G_B} = \frac{G_A G_B}{G_A + G_B}$$

(a) Series connections

$\Delta_A = \Delta_B$, $\Gamma = \Gamma_A + \Gamma_B$

$$G_{eq} = \frac{\Gamma}{\Delta} = \frac{\Gamma_A}{\Delta} + \frac{\Gamma_B}{\Delta} = G_A + G_B$$

$$R_{eq} = \frac{\Delta}{\Gamma} = \frac{\Delta}{\Delta/R_A + \Delta/R_B} = \frac{R_A R_B}{R_A + R_B}$$

(b) Parallel connections

From the Engineer's Sketch Book

Figure 4.8 Applications of the use of generalized network analysis.

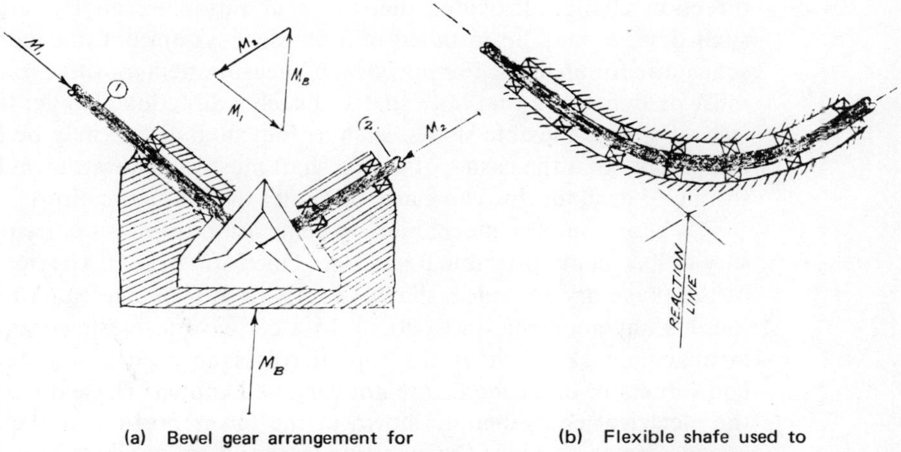

(a) Bevel gear arrangement for changing shaft direction

(b) Flexible shafe used to change directions

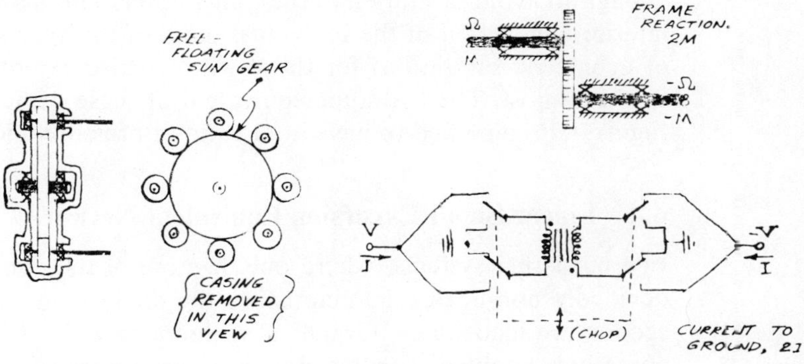

(c) One form of multi-terminal rotary mechanical junction

(d) Inverting gear arrangement and electrical equivalent

From the Engineer's Sketch Book

Figure 4.9 Special considerations for rotary mechanical systems.

(and hydraulic) systems, and helps to bring out the salient features of system operation. Physically, however, a mechanical system may be built with axes in many different directions. So long as certain precautions are taken, the system schematic may be set up as if the system were a parallel shaft or parallel motion arrangement. This allows the present analysis techniques to be applied without modification. Some of the pertinent issues are brought out by Figure 4.9. For example, at (a) is shown a bevel gear arrangement for changing the direction of a shaft. The torque reaction M_B is due only to the direction change. Provided that the gear ratios are all 1:1, any number of such devices may be installed in a physical system but not indicated on the schematic for analysis. For purposes of measurement, positive rotational speed must be defined so that both shafts of such a direction changer have the same sign of speed. Flexible shafts, such as indicated at (b), may be handled similarly. Note that the casing of such a shaft must somehow transmit the reaction torque to ground; this can cause difficulty in real applications.

Where complex mechanical systems are required it is frequently necessary to have a multi-terminal junction. One form of such a device is a variation of the planetary gear idea, shown in (c); this may be made up of bevel gears if there is any good reason to do so. In a certain sense, the common spur-gear arrangement sketched at the top of (d) is an oddity in systems work; its equivalents in other media are not very well known. These do exist, however; the electrical equivalent is shown in the lower portion of the sketch. Here a transformer provides the inverting action. A set of vibrating switches (these can be built from semiconductor devices) reverse the connections rapidly enough to avoid saturation of the transformer. This action is analogous to the intermittent action of the individual teeth of the gear set (preventing a kind of geometric saturation) for those who wish to explore the idea of analogy to the utmost. The hydraulic equivalent of these devices is the intensifier of Figure 1.26, modified to have a 1:1 ratio of pressure and flow rate.

4.4 Linear Small Excursion Equivalent Networks

For most real systems, where one or more of the components usually have decidedly non-linear characteristics, the analysis methods of the preceding section are inadequate. Graphical techniques, to be described later, are most commonly adopted. The linear analysis techniques do have two very important applications, however. Both are associated with the analysis of system behavior when the terminal variables have values which are relatively close to those at static equilibrium. We shall refer to this condition by the qualification "small excursion" and understand such excursions shall be limited in such a way that linearized representations of the characteristics of all the devices in the system are sufficiently accurate. One of the important applications is dynamic analysis; this will be considered in Chapter 5. The other is simply static small excursion analysis, wherein we attempt to find the result of a small change enforced upon one of the system variables or components while the system is in static equilibrium. For example, this method could be used to predict the temperature change of a shower upon the opening of a hot water tap in the kitchen if we knew the characteristics of the pressure regulator at the water company connection, the various pipes and fittings of the system, and the valves, shower head, etc. The most important reason for studying these methods, though, is to prepare for dynamic analysis; we shall refer back to this section at many points in the dynamic analysis study in Chapter 6.

The basic idea of small excursion analysis by equivalent networks is that the real system is replaced by an imaginary one constituted of linear devices exclusively and set up so that the terminal variables are the excursions of the variables of the real system away from their equilibrium values. We then use common linear system methods to carry out the analysis. An important part of the procedure, the determination of an equivalent device, is shown by Figure 4.10. Here, we have considered a general three-terminal device as noted at (a); upper-case letters represent the total terminal variables while lower-case letters stand for excursions. The subscript Q denotes the quiescent (static equilibrium) condition of the system. Thus Γ_{1Q} is the value of Γ_1 in the quiescent state of the system. The usual procedure for the linearization of device characteristics is shown at (b); this was developed earlier (section 3.7). We have arbitrarily chosen the "conductance" form of the linearized representation for this example. In this case, the γ variable at terminals 1 and 2 is the sum of two terms; this implies that parallel devices, each generating one of the terms, could be used to generate the desired γ. Clearly, one of the devices can simply be a linear dissipator since the expression $g_{11}\delta_1$ is the γ in a dissipator of conductance g_{11} and having a δ of δ_1. As shown at (c), the parallel arrangement requires that both devices have the same δ. The second term also has the form of a dissipator characteristic, but the δ is that appearing at the other terminal of the device. Therefore, a dissipator cannot be used; a γ generator must be applied in order to create the needed component of γ_1. This generator must be "controllable" so that its output is proportional to δ_2; although physically complex, such an artifice causes no difficulty in analysis procedures. The full equivalent network is shown at (d) and consists of two dissipators and two generators, properly interconnected and sized. It is typical of the procedure that there are as many devices in the equivalent network as there are terms in the linearized characterization. Thus, the equivalent network for a device with a three-by-three characterization would involve nine two-terminal linear components. Specific examples of the foregoing equivalent network are shown at (e) and (f); note that the sign conventions must be specified carefully.

The four most common forms for three-terminal linear equivalent networks are shown in Figure 4.11. These come from the first four linearization forms of section 3.7; the other two forms, although useful in algebraic analysis, indicate non-physical equivalent networks and are not applied in this method. Each of the forms shown utilizes two dissipators and two generators.

From time to time, we encounter other forms of equivalent networks.

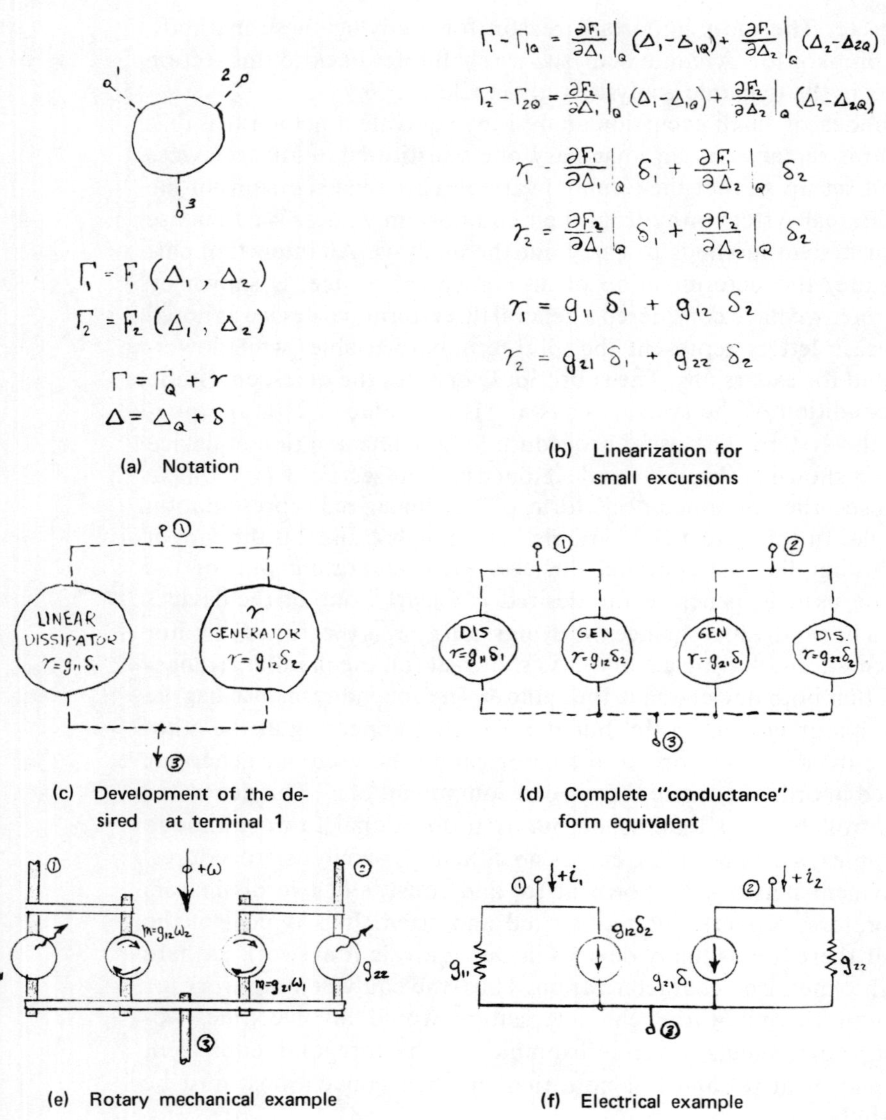

From the Engineer's Sketch Book

Figure 4.10 The small excursion linear equivalent for a 3-terminal component.

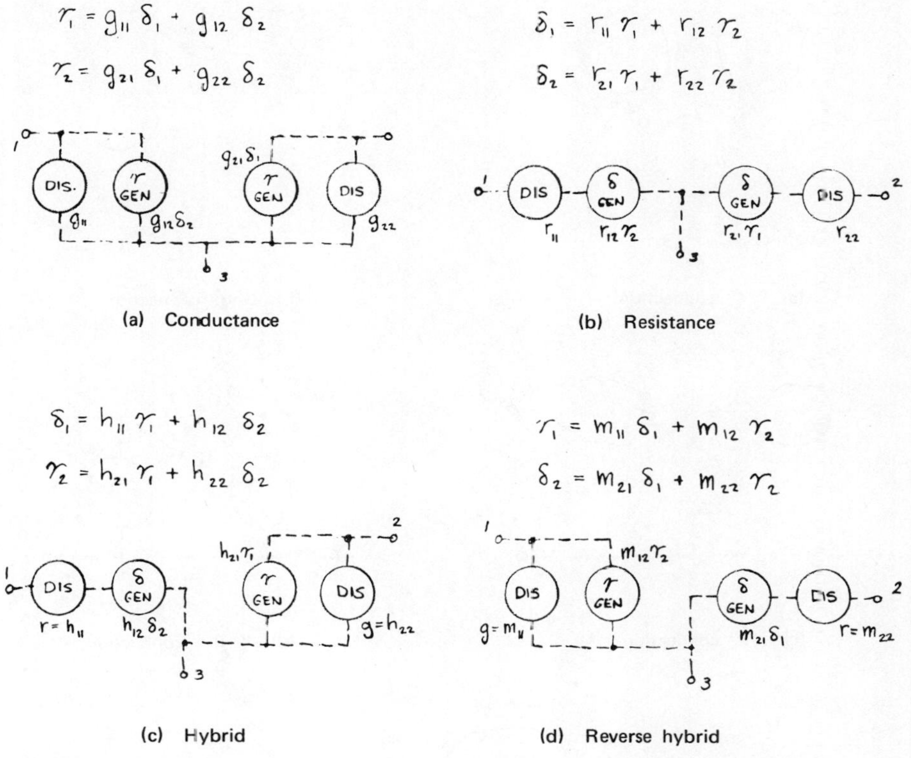

From the Engineer's Sketch Book

Figure 4.11 Linear equivalent representations of 3-terminal component.

Figure 4.12 has been prepared so that these forms would not be completely unfamiliar to the reader. They are based on the "Y" or "star" and "delta" connections of three components, as shown at (a) and (b); for network analysis convenience these are usually drawn in the "T" and "π" forms shown at (c) and (d). Through the addition of one generator to such a network (a γ generator in parallel with one of the dissipators or a δ generator in series with a dissipator) a four-component arrangement capable of representing a general three-terminal component is created. Since there are four coefficients to be determined, and four terms in the linearized characterization, it is possible to determine the coefficient values required. In some cases these representations exhibit the physical nature of the original device more effectively than the two-generator equivalents of Figure 4.11; adoption of one form over another is often a matter of personal preference. The algebraic manipulations

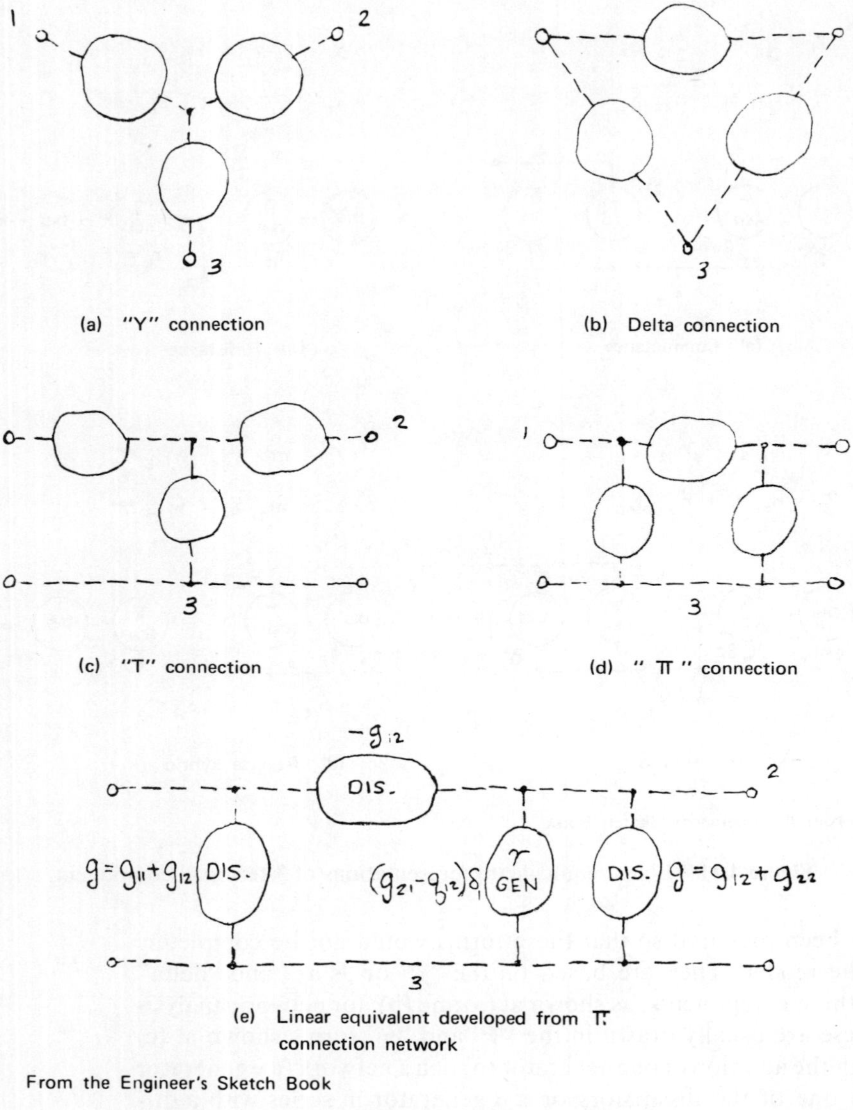

Figure 4.12 Linear equivalents having a single generator.

required to determine the needed coefficients in terms of the commonly accepted ones can be tedious although they are not difficult. Note that dissipators with negative coefficients may occur, as in the example at (e). Depending on the sign of the partial derivative in the characteristic of the original device, this may, or may not, indicate a device which produces power from

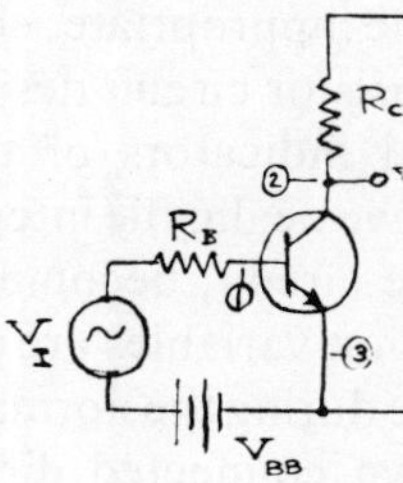

(a) Amplifier schematic

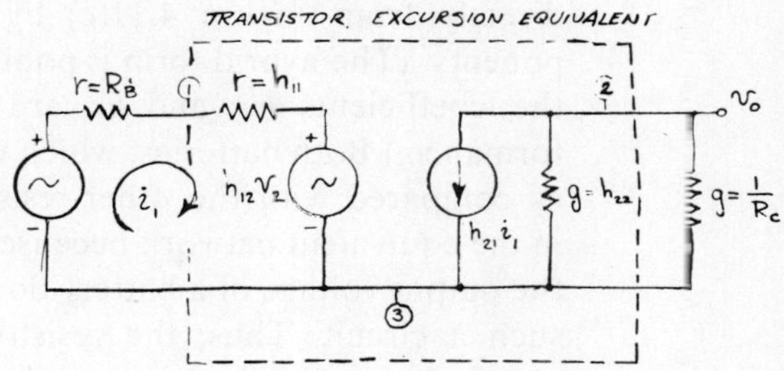

(b) Schematic formed by replacing components by their linear equivalents

$$-v_I + i_1 R_B + i_1 h_{11} + h_{12} v_o = 0$$

$$h_{21} i_1 + h_{22} v_o + \frac{v_o}{R_c} = 0 \quad \rightarrow \quad i_1 = \frac{-\left(h_{22} + \frac{1}{R_c}\right) v_o}{h_{21}}$$

$$v_I + (R_B + h_{11})\left(\frac{h_{22}}{h_{21}} + \frac{1}{R_c h_{21}}\right) v_o - h_{12} v_o = 0$$

$$\frac{v_o}{v_I} = K = -\frac{1}{(R_B + h_{11})\left(\frac{h_{22}}{h_{21}} + \frac{1}{R_c h_{21}}\right) - h_{12}}$$

normally: $h_{11} \ll R_B$, $\frac{h_{22}}{h_{21}} \ll \frac{1}{R_c h_{21}}$, $h_{12} \ll \frac{R_B}{R_c h_{21}}$

$$K \cong -h_{21} \frac{R_c}{R_B}$$

(c) Static analysis of the linear equivalent network for small excursions

From the Engineer's Sketch Book

Figure 4.13 Application of small excursion linear equivalent networks.

heat. Since the equivalent network is conceptual, not real, this does not really matter.

A typical application of small excursion linear equivalent networks is shown in Figure 4.13. Here, the "gain," or amplification factor, of a simple transistor circuit is determined. Note that we need to know nothing about electronics, specifically, in order to carry out this process. The actual circuit is shown at (a) and the small excursion linear equivalent at (b). The hybrid form of equivalent for the transistor has been selected and may be taken directly from Figure 4.11(c) by inserting the appropriate electrical components. (The hybrid form is popular with transistor circuit designers because the coefficients h_{11} and h_{21} are fundamental indicators of transistor performance.) Both batteries, which we assume have negligible internal resistance as compared with the other resistances in the circuit, become short circuits in the equivalent network because their excursion variables are precisely zero; the output voltage of a battery does not change during the normal operation of such a circuit. Thus, the resistor R_C is shown connected directly between terminals 2 and 3 in the equivalent circuit. The voltage generator which provides the input to the amplifier (V_I) would have no dc output; thus it goes directly into the equivalent (and can be neglected in static analysis). Small excursion analysis of the network is carried out in the usual way at (c). A loop analysis is chosen for the left side of the equivalent, while a junction analysis is most efficient on the right. The two resulting equations are manipulated so that the output-to-input ratio K is determined. It may be noted that this can all be done prior to carrying out the actual static analysis of the original network; we cannot, however, determine the values of the h coefficients unless the quiescent operating point of the circuit is determined. The final expression for the amplifier gain comes from some simplifications arising from the general nature of transistor action. For most applications, the coefficient h_{21} will be between 10 and 50 and the amplification of the circuit can be estimated rather well by inspection of the transistor characteristics. Of fundamental importance is that this is the ratio between the *excursions* of the output and input voltages; it has nothing whatsoever to do with the quiescent values of these variables.

4.5 Graphical Static Analysis

Since the characteristics of many important system components are complicated non-linear functions, graphical presentations tend to be the most practical and widely adopted form for their presentation. For similar reasons, and because so many characterizations are graphical, the most efficient and practical technique for static analysis is usually graphical. Generally, device characteristics are approximate in that no two units, even from the same run of one production line, will ever be exactly identical. Thus, graphical accuracy is normally quite sufficient for the purposes of static analysis. Frequently, the engineer is as much interested in the nature of the static condition of a system as in the exact values of the variables; graphical techniques, if properly conceived, can give considerable insight in this regard.

The general nature of static analysis is not changed by the adoption of graphical techniques. It is still necessary to have as many independent relationships between variables as there are unknowns in the problem. It is still necessary to apply the junction conservation laws and the loop summations of Δ variables, as in linear cases. The fundamental difference is that we manipulate graphs instead of equations; the solution of a statics problem always appears as the intersection of two curves on some graph. In this section we develop a technique which is applicable to a wide variety of problems in a uniform way. The author hopes that this approach will clarify the essense of static analysis in such a way that the reader will feel confidence in approaching a problem in an unfamiliar field.

A very simple case of graphical static analysis procedures is indicated in Figure 4.14 where the overall characteristic of a series combination of two non-linear components is determined. The schematic and notation are shown at (a), the individual characteristics at (b) and (c), and the procedure and result at (d) and (e), respectively. The procedure amounts to a simple graphical addition as expressed in the Δ variable summation.

$$\Delta_{20} = \Delta_{21} + \Delta_{10}.$$

Since the two components are in series, the Γ variable must be the same for each (recognition of this amounts to an application of a conservation type relationship; i.e., a geometrical aspect of the network). Thus, by reading the Δ variable for each component at one value of Γ and entering the sum on the final chart, a complete graph of the characteristic may be developed. A pair of dividers may be used to facilitate the work if desired.

The procedure for a parallel connection, the dual of the series connection, would be similar except the roles of the Γ and Δ variables would be interchanged. As usual, the matter of signs must be handled perfectly. It is assumed in Figure 4.14 that both devices are dissipators; this influences the appearance of the procedural graph.

A series combination which appears very frequently in electronics work is shown in Figure 4.15. This network involves a power source as well as a dissipator. Although the graphical procedure just given may be used here, it is much simpler to recognize that the network is linear and, therefore, must be represented by a straight line graph. When the current is equal to zero, there is no drop across the resistor and the voltage must be the battery voltage; the line must go through this point. If the voltage is to become zero, the resistor voltage drop must equal the battery voltage; the current that flows is the same as if the resistor were short-circuited to ground; i.e., V_B/R. These two observations make it quite simple to prepare the graphical characterization. This graph is frequently superposed on vacuum tube or transistor characteristics during static analysis; the line is referred to as the "load line" and is one example of a whole class of related load characteristic curves.

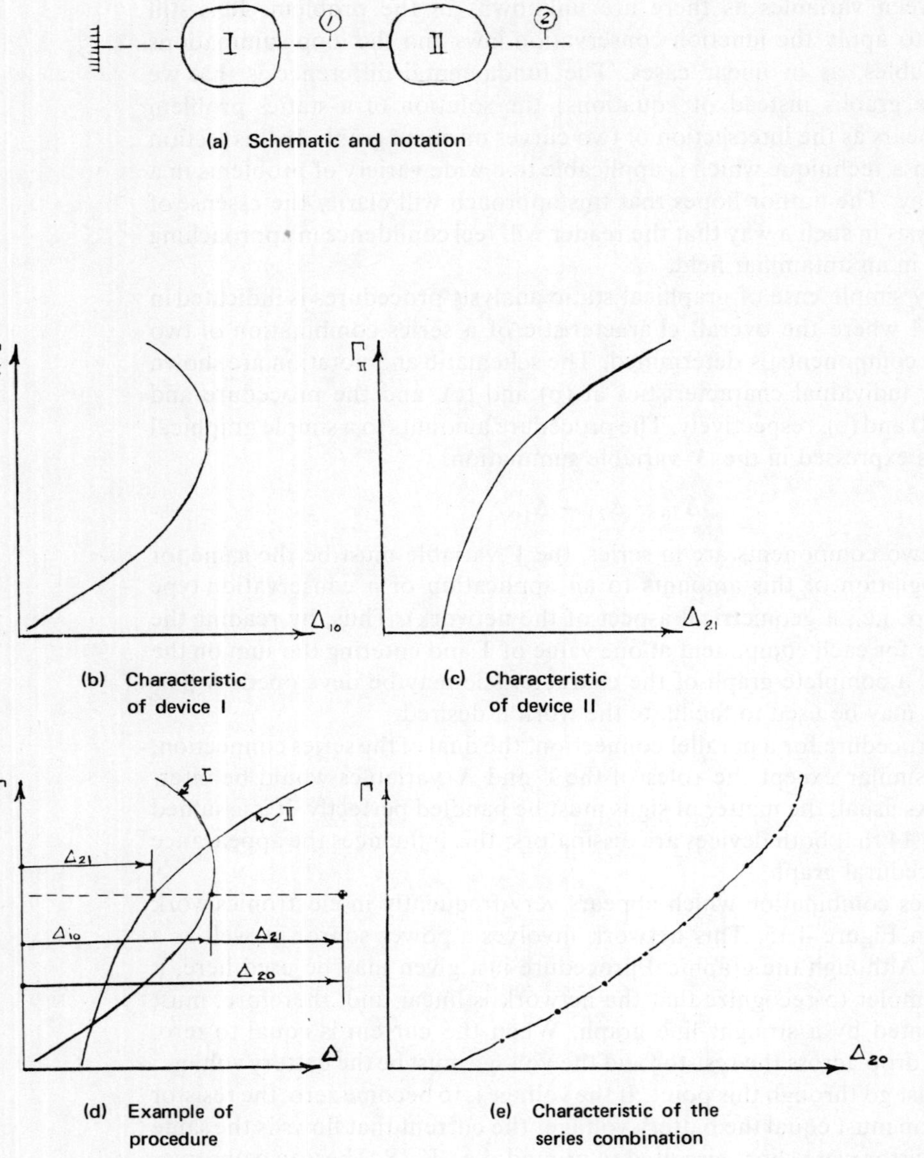

From the Engineer's Sketch Book

Figure 4.14 Methods used to find the characteristic of series connected dissipators.

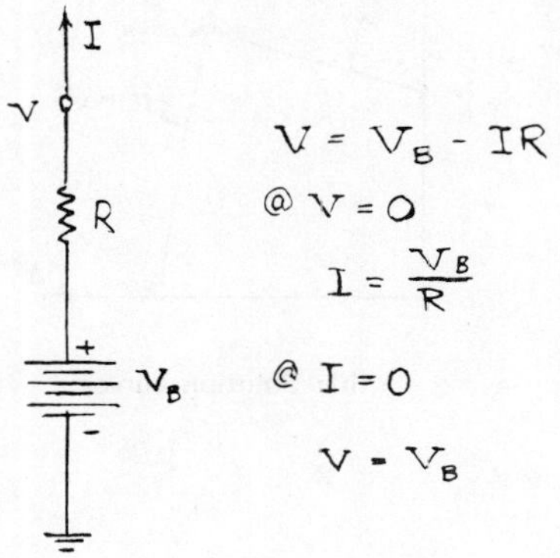

(a) Schematic and notation

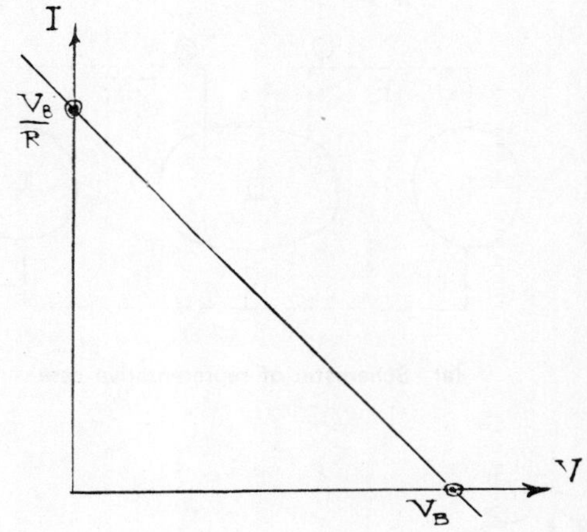

(b) Graphical characteristic

From the Engineer's Sketch Book

Figure 4.15 Characterization of a widely used series connection.

The basic nature of graphical static analysis is brought out in the series of graphs of Figure 4.16. The case considered here may arise when two two-terminal devices are connected to a three-terminal device as shown at (a). There are, however, a number of ways in which a problem involving two one-by-one and one two-by-two characterizations may arise. The procedure is the same for each. The characteristics of device III are shown at (b) and (e), while those of I and II appear at (c) and (f); the arrangement has been contrived to exhibit the procedure as effectively as possible. It should be noted that normalized variables have been adopted here in order to simplify the discussion; these are denoted by the superstar. Since the graphs at (b) and (c) both represent relationships between the variables $\Gamma_1{}^*$ and $\Delta_1{}^*$, they may be used to "eliminate" one variable. This is done by superposing the two graphs. The result, shown at (d), defines a relationship between the three variables $\Gamma_1{}^*$, $\Delta_1{}^*$, and $\Delta_2{}^*$, in which any one may be taken as independent. Similarly, graph (g) gives a relationship between the three variables $\Gamma_2{}^*$, $\Delta_2{}^*$, and $\Delta_1{}^*$. In examining (d) and (g), note that they both specify a relationship between $\Delta_1{}^*$ and $\Delta_2{}^*$; if we examine these relationships simultaneously, we can determine the solution to the problem. This is done at (h) and it is found that only one combination of the two satisfies both relationships. Therefore, the solution has been found. (In order to save space the figure has been arranged out of

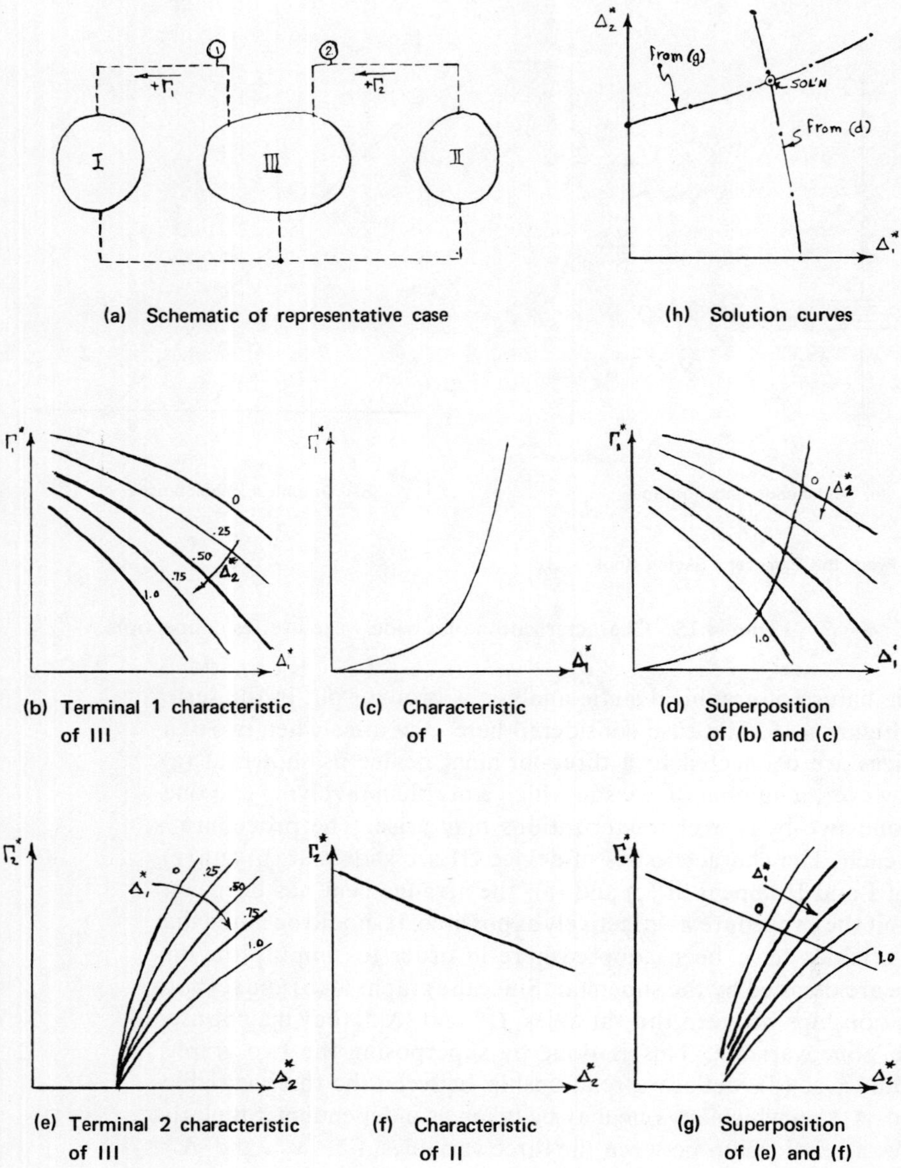

From the Engineer's Sketch Book

Figure 4.16 Static analysis when one 2 × 2 and two 1 × 1 characterizations appear.

order; this will be done throughout the following series of figures.) The values of Λ_1^* and Λ_2^* determined in (h) may be carried back into (d) and (g) in order to show the exact operating point of each component.

In preparation for the procedure given by Figure 4.16 the characterizations of the individual components must be formulated in terms of some selected system variables. This is normally just a matter of notation, but should be done carefully and explicitely. The example of Figure 4.16 represents the fortunate case where the characteristics of the three-terminal device are set up to facilitate such analysis; i.e., the terminal 1 and terminal 2 charateristics are given directly. If this were not the case, some tedious preliminary manipulations would have been required. (The characteristics of III, *or* of the network formed by I and II, could be reformulated; an example is given later.) The crux of the matter, which must always be kept in mind while setting up a problem, is that superposition is the graphical equivalent of algebraic elimination of variables. The analyst must plan to develop graphs which are superposable.

A specific example of the foregoing procedure is indicated in Figure 4.17 and involves a positioner using a solenoid. (Normalized variables are used here and in the remainder of the chapter examples.) The system, shown schematically at (a), is intended to provide for the remote positioning of some element (perhaps a valve handle) upon the rotation of a potentiometer knob. A transducer is attached to the solenoid push rod and is arranged so as to generate a dc voltage proportional to the displacement X; we assume that the transducer develops no force under static conditions. The potentiometer is set up so that a voltage proportional to its rotation is developed; it is calibrated so that its output is equal to that of the transducer when the setting Y is equal to the displacement X. The amplifier must be arranged so that the tendency is to correct an error between X and Y; if X should be moved to the left the solenoid current should increase. The mechanical load on the solenoid is assumed to be a spring and dashpot; since we are interested in the static state here, the dashpot characteristic does not enter the problem. In order to get the best "tracking" between X and Y in this system, it is useful to bias the amplifier so that X^* is equal to Y^* when Y^* is $\frac{1}{2}$; this can be done with the battery labeled V_B. The bias at the input provides the output required to give the current necessary to stretch the spring at 50% extension.

The various graphs of Figure 4.17 are arranged as in Figure 4.16. The special aspects of this case are that the electrical characteristics of the solenoid are degenerate (current is independent of displacement in a dc electromagnet) and the amplifier is assumed to draw no input current. The amplifier output curves can be labeled only after it is determined from (d) that a current of about 66% is required at 50% displacement; when entered at (g), this information provides the location of the X^* equal Y^* line for the amplifier. These lines

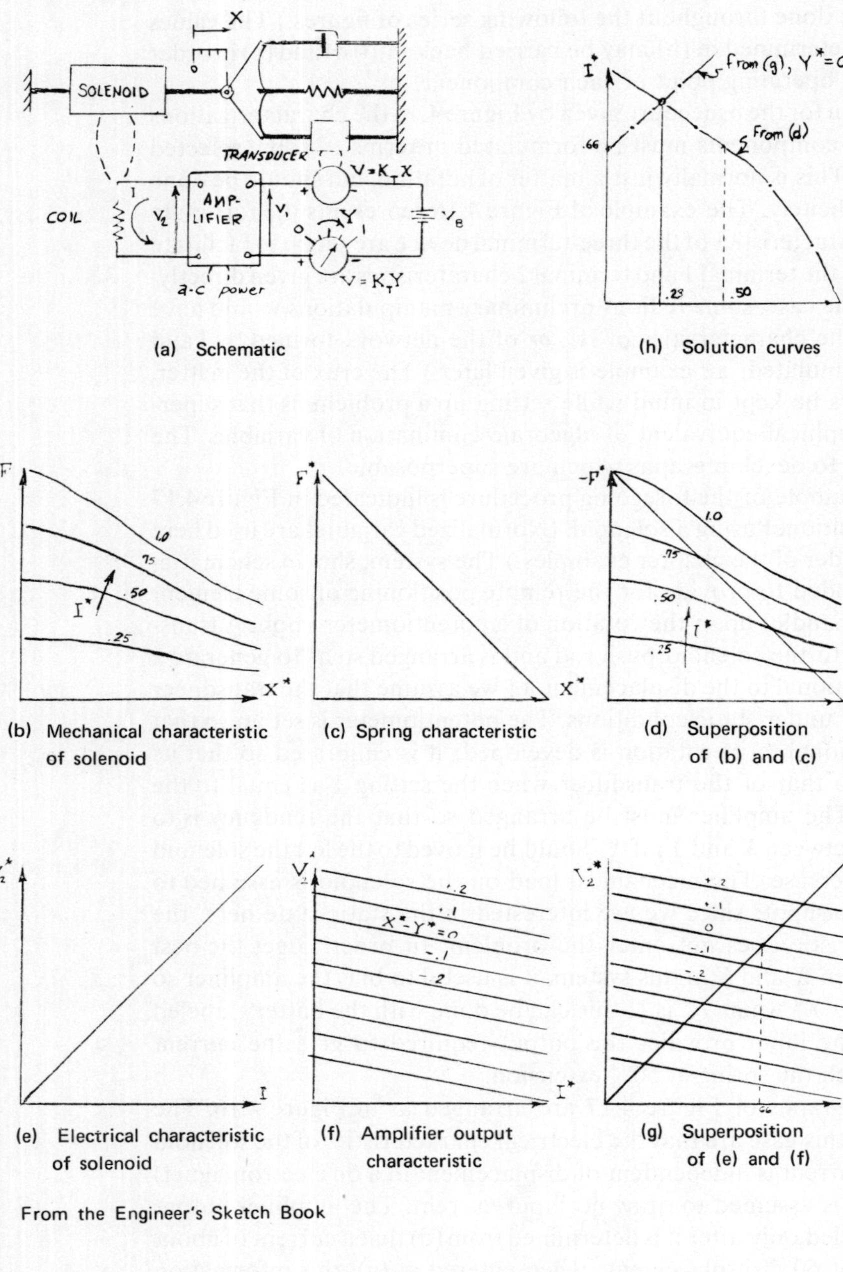

From the Engineer's Sketch Book

Figure 4.17 Static analysis of an electromagnetic positioner.

slope as an indication that the amplifier output will drop off as more and more current is drawn; this poor regulation of the amplifier output could be removed by electronic refinements and expense, but is not uncommon. The gain of the amplifier, which dictates the distance between the lines for different inputs, has been arbitrarily selected here (at a rather low value).

Now let us try to get something useful out of the analysis. Specifically, let us find out what happens when the Y knob is set to zero. From (g) we can readily find the value of X^* for each I^*, Y^* being zero. This relationship is shown entered at (h) and yields the information that X^* is actually about 23% when it should be zero! The analysis shows that the system is a very poor positioner. It must be admitted, though, that a tenfold increase in the amplifier gain would do considerable good (if the system remained stable after the change).

A more general case of static analysis is indicated in Figure 4.18; this is about the most involved kind of problem that can be solved graphically without undue effort. The case can arise in a number of ways, but is represented by the parallel connection of two general three-terminal devices. The various characterizations and superpositions are indicated as usual. Note that the consistent formulations of the two characterizations are essential to the efficiency of the process. Since the superpositions yield relationships between $\mathbf{\Gamma}_1$ and $\mathbf{\Delta}_1$, and $\mathbf{\Gamma}_2$ and $\mathbf{\Delta}_2$, the resulting lines may be labeled as "operating lines" for these junctions; this would not be essential in the procedure. For example, the graphs may yield relationships between two $\mathbf{\Gamma}$ variables at different terminals and no such physical interpretation would be possible. The usual solution curves at (h) provide the final solution to the problem. Although few physical systems have such a simple form, many may be analyzed by this method through recognition of some dividing line between two parts of the system such that each part is represented by a two-by-two characterization. The following examples show this.

A system in which a gasoline engine drives a dissipative load and is regulated by a flyball governor is considered at Figure 4.19, the schematic being shown at (a). The vertical dashed line divides the system into two parts, each having four terminals and two power transfer media (rotary and translational mechanical). Therefore, the characterizations of either half will be in the two-by-two form and we can use the technique of Figure 4.18. The graphical procedure goes as usual, the result appearing at (h). As with the positioner, some adjustment of this system would be carried out in order to optimize the performance of the controller. This has not been considered specifically in the example. The throttle characteristic of the engine has been exaggerated here, in order to emphasize the procedure, in that the actual influence of engine speed on throttle reaction (due mainly to the change in air flow past the butterfly valve) would normally be small compared to the actuating forces available from the governor. A practical need for this analysis would be in the case where the

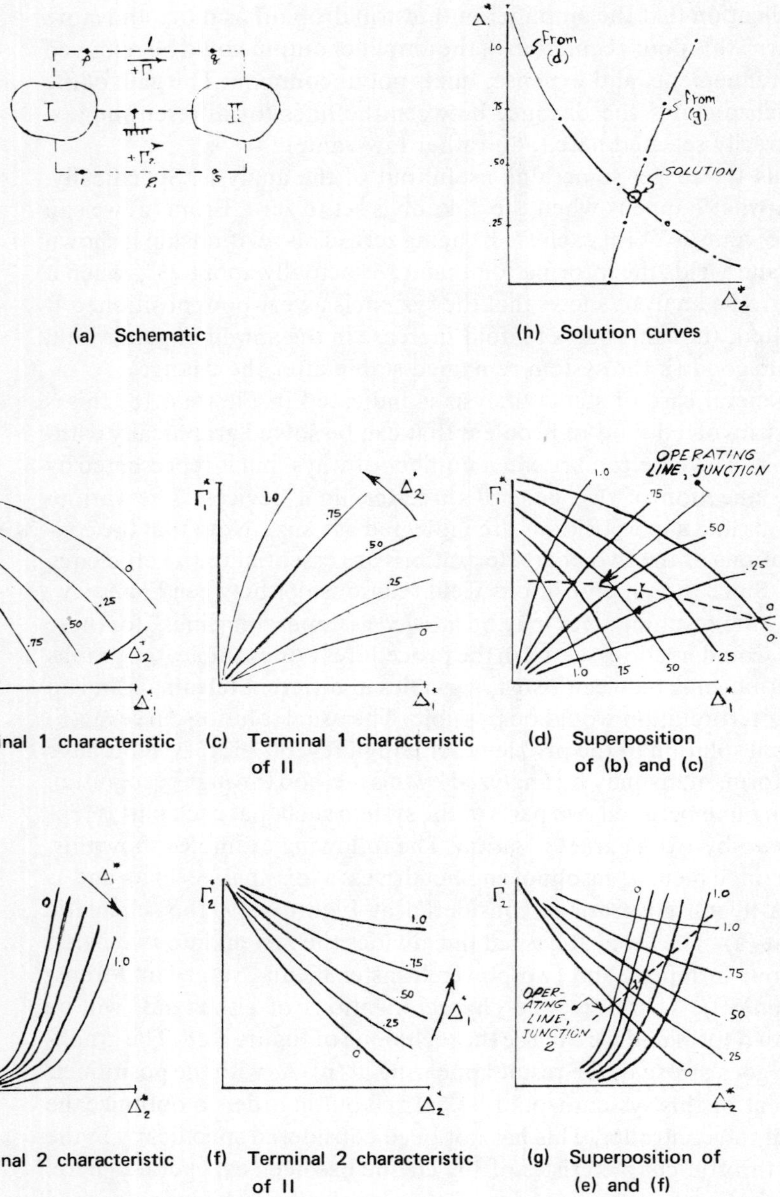

(a) Schematic

(h) Solution curves

(b) Terminal 1 characteristic of I

(c) Terminal 1 characteristic of II

(d) Superposition of (b) and (c)

(e) Terminal 2 characteristic of I

(f) Terminal 2 characteristic of II

(g) Superposition of (e) and (f)

From the Engineer's Sketch Book

Figure 4.18 Static analysis of parallel connection of two 3-terminal devices.

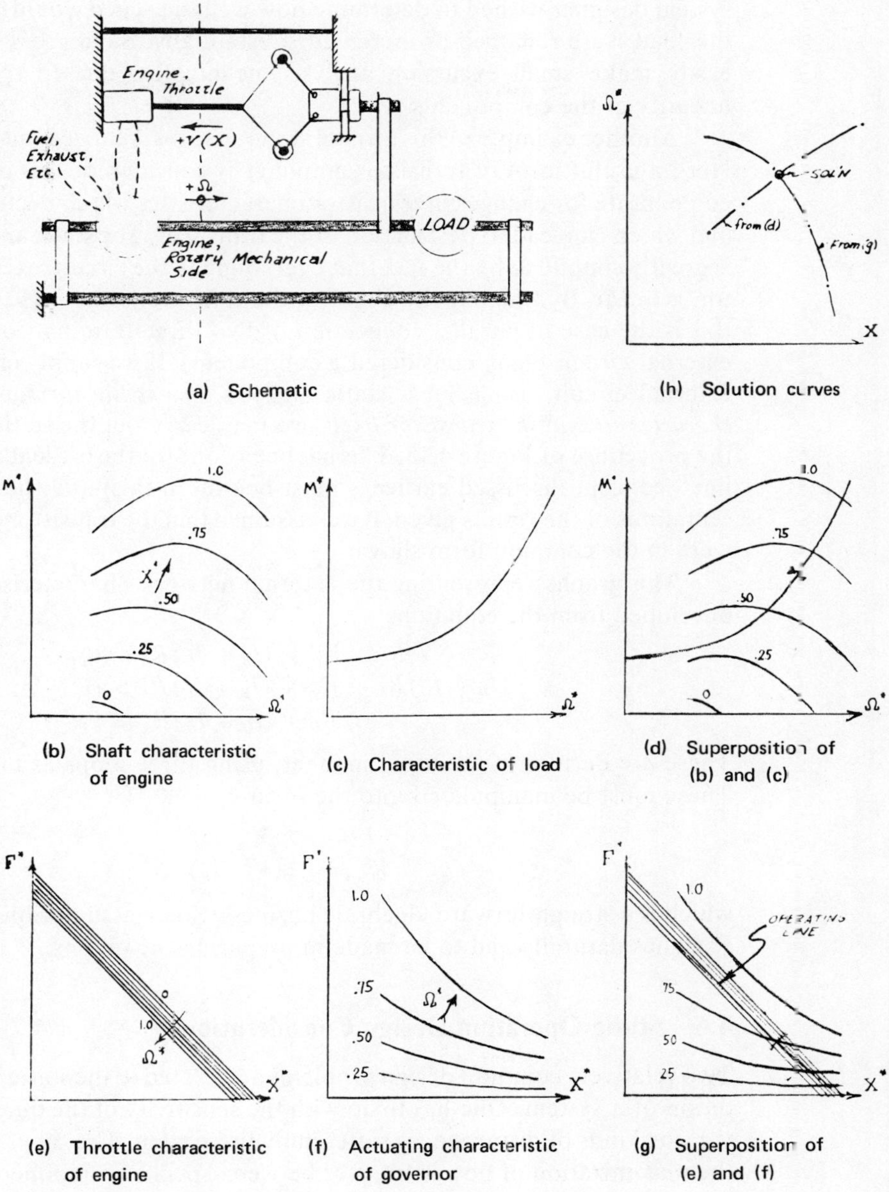

From the Engineer's Sketch Book

Figure 4.19 Static analysis of engine controlled by flyball governor.

system designer wished to determine how well the speed would be controlled if the load were removed or increased substantially. Such gross changes could easily make small excursion analysis meaningless due to the actual nonlinearity of the components.

Another example of this form of static analysis is carried out in Figure 4.20. Here a useful form of transistor amplifier is considered; this circuit tends to compensate for changes in transistor characteristics which occur upon heating and which can lead to destruction of the transistor. The static analysis problem is greatly simplified by the fact that everything in the circuit, except the transistor, is linear. By isolating the circuit without the transistor, as in (b), we see that this is the case of parallel connection of two three-terminal components (the external circuit being considered a component). If we simply characterize the external circuit, using linear static analysis, *in exactly the same form as the characteristics of the transistor itself*, we may carry out the static analysis as in the procedure of Figure 4.18. This has been done for the present case (the "load line" concept discussed earlier is most helpful in this) and the graphical presentations of the results given. It was assumed that the transistor characteristics were in the common form shown.

The graphs representing the external network characteristics have been developed from the equations

$$I_C R_C + V_{31} \times (I_B + I_C) R_E = V_{CC}$$
$$(I_B - I_D) R_A + V_{21} + (I_B + I_C) R_E = O$$
$$I_D R_D + (I_D - I_B) R_A = V_{CC}$$

These are derived in the usual manner, using three loops as indicated at (b). These must be manipulated into the form

$$I_C = \mathscr{F}(V_{31}, I_B)$$
$$V_{21} = \mathscr{F}_2(V_{31}, I_B)$$

which is a straightforward algebraic business. Some assumptions regarding the R values naturally had to be made in preparing the graphs.

4.6 Static Operation Design Considerations

Two relatively common design problems are related to the static operating condition of a system. One has to do with the sensitivity of the quiescent point to various kinds of disturbances suffered by the system. The other has to do with the maximization of power transfer between specific components.

The sensitivity of a quiescent point to disturbances can be a very serious matter. For example, the circuit considered at Figure 4.20 had to be developed when it was found that transistor amplifiers could be made inoperative by relatively small temperature changes, through changes in the transistor character-

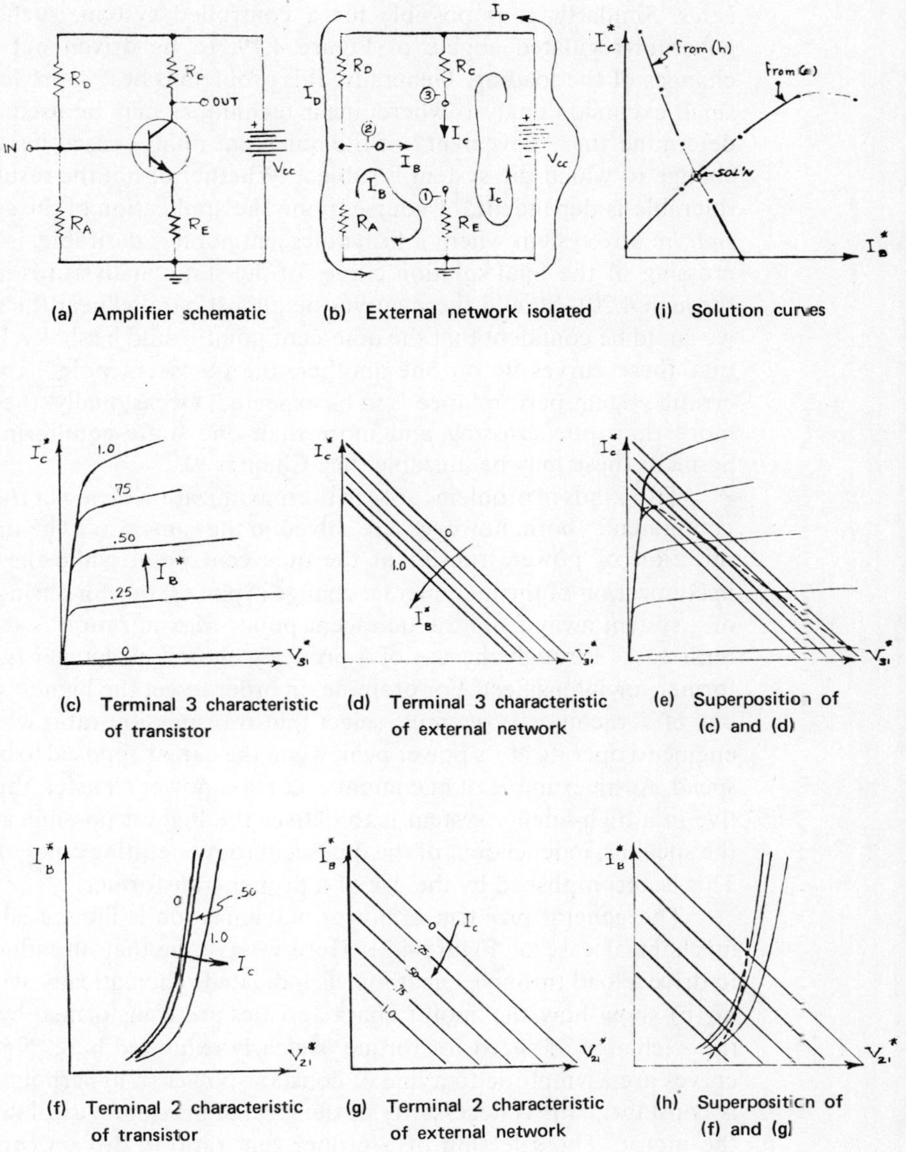

From the Engineer's Sketch Book

Figure 4.20 Static analysis of transistor amplifier.

istics. Similarly, it is possible for a controlled system, such as the flyball governor-regulated engine of Figure 4.19, to be driven out of control by changes of the loading. Generally, this problem is best approached through a small excursion analysis where linear techniques may be used. We may then determine the "movement" of the quiescent point caused by any worrisome change to which the system is subject. Whether or not the resulting change is tolerable is dependent, of course, upon the application of the system. A good sign, in any system where a firm quiescent point is desirable, is a well-defined crossing of the final solution curves of the static analysis (as in Figures 4.16 through 4.20). Should these curves be almost parallel near the solution point, we could be confident that the quiescent point would be shaky. In the rare case that these curves lie on one another, the quiescent point is ill-defined and erratic system performance is to be expected. Occasionally, these curves have more than one crossing and more than one static equilibrium is possible. Some of these may be unstable (See Chapter 9).

Two kinds of problems arise in the maximization of power transfer between components; both, however, are solved in the same way. The first is the maximization of power transfer at the quiescent point while the second is the maximization of the incremental change in power transfer during the excursion of a system away from the quiescent point. Maximization is accomplished in both cases through the use of a properly chosen exchanger (gearbox, transformer, or intensifier). For example, in order to get the highest possible speed out of a racing car, we must select that transmission ratio which allows the engine to operate at its power peak when the car is supposed to be at maximum speed. As an example of maximum excursion power transfer, the design objective in a high-fidelity system is to deliver the highest possible audio power to the speaker, independent of the quiescent (dc) operating point of the amplifier. This is accomplished by the use of a proper transformer.

The general problem of power maximization is illustrated by the rotary mechanical case of Figure 4.21. Here we assume that an induction motor is to drive a load through a gearbox, as indicated schematically at (a). The curves of (b) show how the motor characteristics are transformed by the gearbox; the exchange of speed for torque is clearly exhibited here. Note that all the curves are asymptotic to a line of constant power (a hyperbola so that $M \times \Omega$ is constant); this is necessarily at the level of maximum available power from the motor. The selection of a proper gear ratio to drive a certain load with maximum power is illustrated at (c). Here the operating point of the load is readily found as the intersection of the load characteristic with the hyperbola of maximum power. The gear ratio which causes the motor and load to operate at this power level is the ratio of motor speed to load speed when each is operated at the maximum available power level. The maximization of excursion

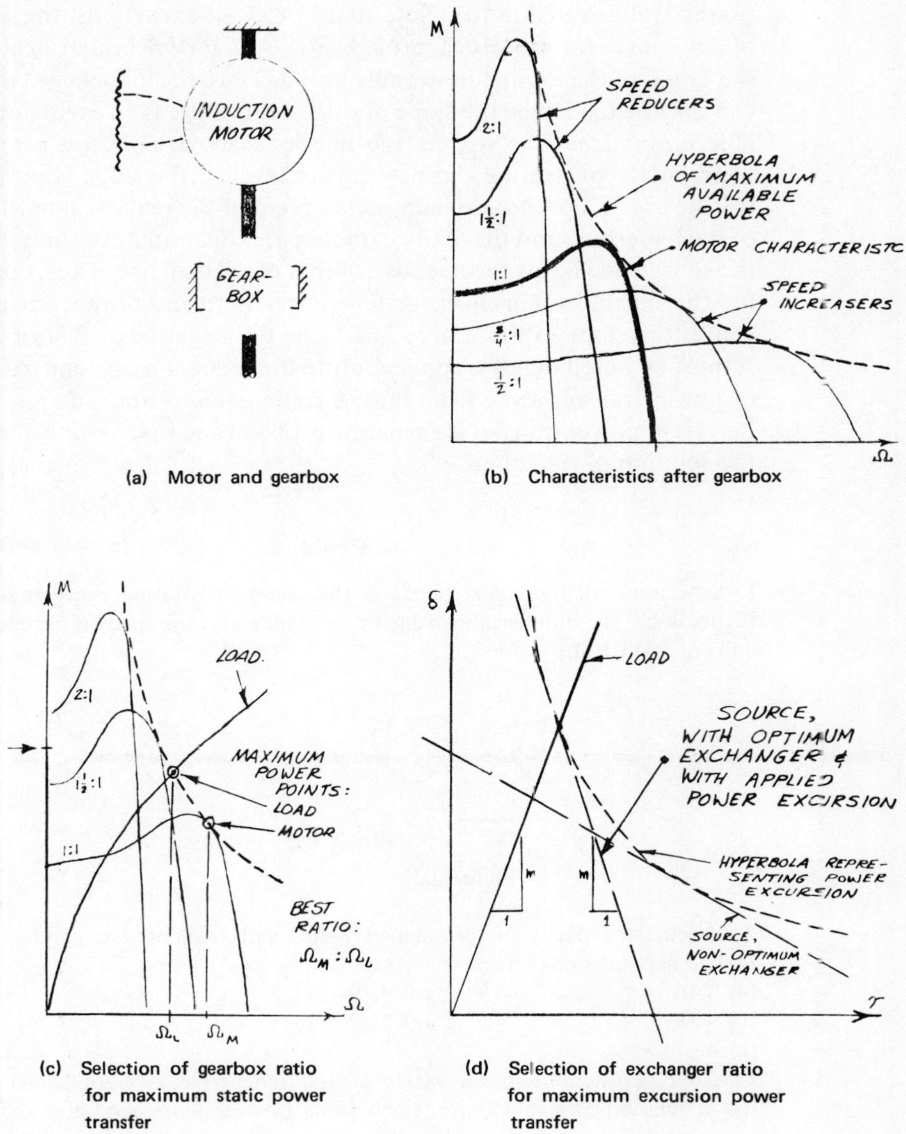

(a) Motor and gearbox

(b) Characteristics after gearbox

(c) Selection of gearbox ratio for maximum static power transfer

(d) Selection of exchanger ratio for maximum excursion power transfer

From the Engineer's Sketch Book

Figure 4.21 Maximization of power transfer.

power is illustrated at (d). Note that, for small excursions, the characteristic of any source (or load) seen through an exchanger will be a straight line through the origin of the excursion variable graph. Thus, the influence of the exchanger can only be to change the slope of this line. If there is an excursion in the available power from the source, the line must move so that it is tangent to the hyperbola representing the new power level. If the slope is properly chosen (by the use of a proper exchanger) the point of tangency will be at the crossing of the hyperbola and the load characteristic; this can occur only if the slope of the source line is just the negative of that of the load line. Thus, the general rule for the maximization of excursion power is to install that exchanger which causes the slope of the source line to be the negative of that of the load line. It must be noted that this applies only to the excursion case and to the excursion of power transfer away from that at static equilibrium. The maximization of quiescent power transfer is a separate problem and the solution of one is seldom the solution of the other.

Problems

1. Determine all the shaft speeds in the chain of rotating speed sources shown in Figure 4.22. The number above each source indicates the speed difference across it, in units of 100 RPM.

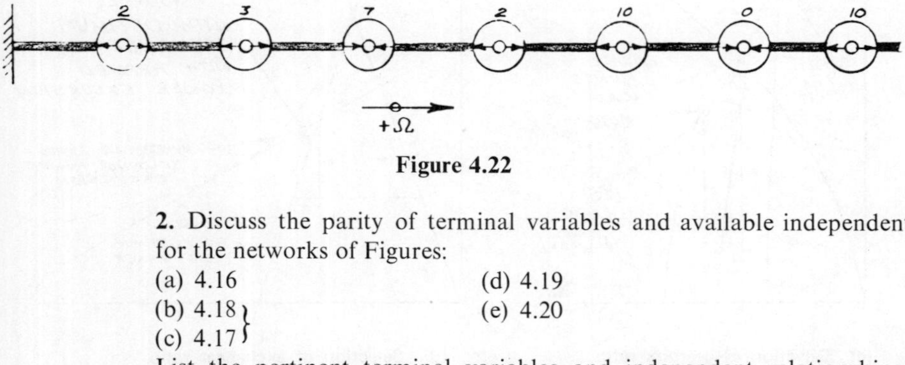

Figure 4.22

2. Discuss the parity of terminal variables and available independent relationships for the networks of Figures:

(a) 4.16
(b) 4.18 }
(c) 4.17 }
(d) 4.19
(e) 4.20

List the pertinent terminal variables and independent relationships. Divide your list of relationships into those which come from network geometry and conservation laws, and those which depend on the nature of the components.

3. Which of the following groups of components can be hooked up end to end in order to form a closed network?

(a) torque source and rotary dissipator.
(b) torque source and speed source.
(c) torque source, speed source, and rotary dissipator.
(d) two torque sources.

(e) current source and resistor.
(f) current source and voltage source.
(g) current source, voltage source, and resistor.

4. Determine the overall characteristic of a two-terminal device constructed by combining components I and II of Figure 4.14 in parallel.

5. Suppose the characteristics of devices I and II, Figure 4.14, are both symmetric about the origin. Does the orientation of either device, in the series connection, affect the overall characteristic? What if either characteristic was not symmetric about the origin?

6. Is the circuit of Figure 4.15 dissipative or generative in the second quadrant of the characterization shown at (b)? Can the circuit be operated in either the second or the fourth quadrants?

7. Suppose, in Figure 4.15, that the resistor was in parallel with the battery. Determine the graphical characterization of the combination.

8. In Figure 4.16, revise the characteristic of device I [shown at (c)] so that the value of $\Delta_1{}^*$ is 50% greater than on the original for any value of $\Gamma_1{}^*$, and of device II [shown at (f)] so that $\Gamma_2{}^* = 0.75$ independent of $\Delta_2{}^*$. Determine the new operating point of the system and show it on the revised versions of graphs (d) and (g).

Note: In the previous and next several problems, the normalized variables in the figures have a value of 1.0 at the tips of the arrowheads on the axes.

9. In Figure 4.16, change the direction of increasing $\Delta_2{}^*$ in (b) and of $\Delta_1{}^*$ in (e) so that 0 becomes 1, and vice versa. Determine the quiescent point for the revised system.

10. In Figure 4.16, show that the result of superposition (g) establishes an "operating line" for terminal 1 that may be entered on graph (b). Similarly, show that superposition (d) establishes an operating line at terminal 2 that may be entered on graph (e). Now suppose it is desired to use device III as a Γ regulator so that any device inserted in in place of I will operate at $\Delta_1{}^* = 0.6$. Determine, if possible, the characteristic of device II which would be necessary to accomplish this.

11. For the positioner of Figure 4.17(a), determine X^* for $Y^* = 0, 0.25, 0.50, 0.75$, and 1.0. Graph X^* versus Y^*. Repeat the above, but with the amplifier gain increased by a factor of 10. The lines in part (f) of the Figure will be separated by increments of 0.01 unit of $X^* - Y^*$ in this case.

12. For the positioner of Figure 4.17, graph X^* versus Y^* for the case of no bias voltage V_B. What is the error $(X^* - Y^*)$ for $Y^* = 0$?

13. For the positioner of Figure 4.17, determine the shape of a spring characteristic which would make the X^* versus Y^* graph a straight line through the point (0.5, 0.5) (due to biasing) and (0.1, 0.0).

14. Show how the technique developed at Figure 4.18 could be used to analyze networks of the form shown in Figure 4.23.

15. Carry out the static analysis of a system such as in Figure 4.18 where the curves (but not the notation) of graphs (c) and (e) are interchanged. Retain the labeling of the curves in each family (0, 0.25, 0.50, 0.75, 1.0) so that the terminal 1 characteristic of II increases clockwise, etc.

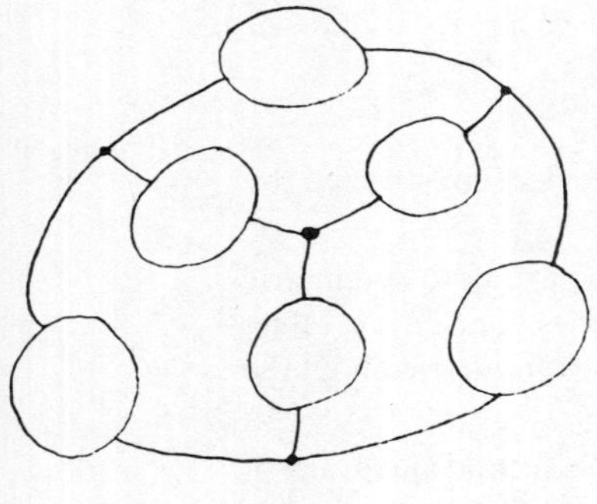

Figure 4.23

16. Suppose that the analysis of Figure 4.18 indicates that the system does not operate at the desired quiescent point and that the designer decides to evaluate several modifications to the network, as follows. For each change, determine and list the revised terminal variables for components I and II. The original reference values for normalization are to be retained.

(1) Insert a dissipator, with characteristic $\mathbf{\Delta}_{pq}{}^* = 0.1\ \mathbf{\Gamma}_1{}^*$, between p and q on the schematic.

(2) Insert a $\mathbf{\Delta}$ generator between r and s so that $\mathbf{\Delta}^*_{rs} = +0.3$.

17. The system shown in Figure 4.19 has a serious shortcoming in that the actuating characteristics of the governor and the throttle characteristics of the engine do not interact effectively. As may be seen at (g), the F^* versus X^* curves are parallel in the vicinity of $X^* = 0.25$, $\Omega^* = 0.85$. This causes $\partial\Omega^*/\partial X^*$ to be zero in this region; i.e., there is no control action. Suppose the governor is redesigned, in order to avoid this difficulty, so that the F^* versus X^* curves of (f) become horizontal ines at the F^* values originally developed at $X^* = 0.5$.

(1) Determine the the quiescent point of the revised system.

(2) Determine the region in which the load characteristic can vary and the system still keep the speed Ω^* between 0.8 and 1.0.

18. The transistor circuit shown in Figure 4.20 tends to "turn off" a transistor which might otherwise overheat. The design problem originates with transistor characteristics, which are very temperature sensitive. Basically, as temperature increases, the curves of (c) move upward so that added collector current (I_C) flows even with no change in base current (I_B); this added current tends to heat the transistor further and the effect can become accumulative. The effects may be seen as follows:

(1) Suppose the network is operating at some elevated temperature so that the family of curves in (c) has moved up one "notch"; i.e., the line for $I_B{}^* = 0.25$ becomes the $I_B{}^* = 0$ line. Determine the new solution curve "from (e)" for this case.

(2) It may be seen by inspection that $I_C{}^*$, in the above case, did not increase as much—due to temperature—as if $I_B{}^*$ had a fixed value. Suppose the increase was still determined to be too large. Should the ratio of R_E to R_C be increased or decreased,

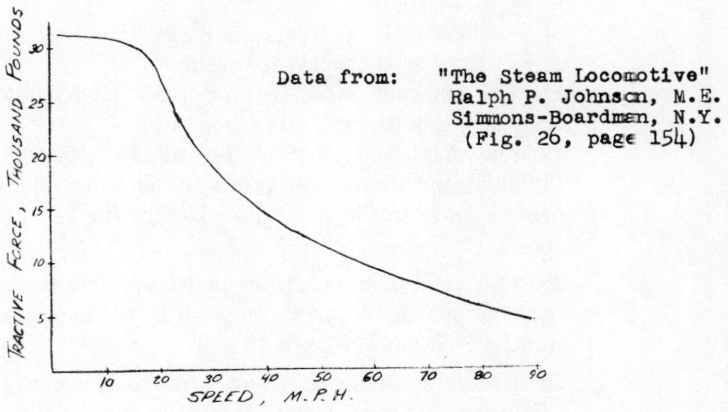

From the Engineer's Sketch Book

Figure 4.24

assuming the sum remains the unchanged, in order to further limit the increases in I_C*?

19. The drawbar characteristic of a steam locomative (taken from an old book) is shown in Figure 4.24. Explain why no transmission was required on these engines.

20. Redraw the shaft characterisitics of the engine of Figure 4.19, presuming a 2:1 reduction gearbox was attached. How does this affect the power available from the engine?

21. What gearbox ratio should be used with the motor of Figure 4.21(c) to deliver the maximum static power to a load with a constant torque characteristic, at the level indicated by the arrowhead?

22. Use a maximization technique from calculus to prove the argument of Figure 4.21(d).

23. What value resistor can dissipate the most power when connected across a 12-volt battery with 2 ohms internal resistance?

24. (Problem courtesy of Prof. D. R. Vaughan.) A Texas rancher is dissatisfied with his car because its maximum speed is only 81.6 mph. The rancher decides to increase the maximum speed by doubling the rear axle speed ratio (changing the ratio so that the engine turns only half as fast for any given velocity of the car). The rancher finds that the maximum speed is less with the new rear axle ratio than with the old. Assuming that:

(1) The supply characteristic with the original rear axle speed ratio was

$$F_S = 600 - \frac{1}{5}(V_S - 50)^2 \text{ for } 0 \leq V_S \leq 100 \text{ where } V_S \text{ is the velocity of the car, mph,}$$

F_S is the total propulsion force exerted on the road by the rear wheels, lbs.

(2) The load characteristic is

$F_D = 0.06\ V_S^2$
where F_D is the total resistive or drag force
required to propel the car, lbs.

(3) The efficiency of power transmission from engine to rear wheels is not changed by changes in the rear axle speed ratio.

Find the maximum speed that the car can achieve if the rancher installs the "best" (for obtaining top speed) rear axle ratio, and the value n_2 of the new rear axle ratio if the original ratio was n_1 and n is defined by the equation $V_S = n\Omega$, where Ω is the engine speed.

25. Use static small excursion analysis to determine an expression for the current gain (ratio of change in current through R_C to change in drive current through R_B) for the transistor circuit of Figure 4.13.

26. Use static small excursion analysis to determine the speed regulation (ratio of speed change to load torque change) of the system of Figure 4.19. Draw the small excursion equivalent network for the system, determine the various proportionality constants from the graphs of the figure (at the quiescent point), and state your answer in percent speed change per percent load torque change.

27. Draw the static small excursion equivalent network for the positioner of Figure 4.17. Determine the proportionality constants, at the equilibrium, from the graphs of the figure. Determine the "stiffness" (change in position per change in external force applied to the controlled body) of the system.

5 Small excursion dynamic characteristics of components and systems

The dynamics of systems is a complex, intriguing subject. As we have shown in Chapter 2, some systems oscillate, some vibrate, and some lock onto a predetermined static state. Some unfortunate systems destroy themselves as soon as they are turned on. Just how the engineer goes about predicting and enforcing the various kinds of system performance is the general subject of the remainder of this book. In the next three chapters, which are devoted to the dynamics of systems that are operating relatively close to some static equilibrium point (either stable or unstable), we utilize the methods of linear, ordinary, differential equations with constant coefficients. These methods are applicable only when all the equations which describe the performance of system components are linear. As was shown in Chapter 3, such representations are usually valid so long as the excursion variables do not become excessive. This technique, then, is just a beginning in the study of system dynamics. It is approximate for most systems of interest, it shows only a small aspect of the total performance, and it has no capacity for dealing with the interesting vibrator networks. It turns out, however, that a very large number of engineering decisions can be based on this form of analysis. We would probably not be exaggerating to say that it is the most widely used discipline in all of engineering. Familiarity and facility with it are essential for effective system-design work; fortunately, it is not a terribly difficult subject.

In this chapter, we shall consider the first step in the conduct of a small excursion analysis. That is the writing of the differential equations which describe the dynamic relationships between system variables. Although the student has undoubtedly studied this subject in prior courses, it is one which seems to merit considerable practice. The solving of this kind of equation can be done by rote and learned by absolutely anyone. The writing of them appears to be another matter—perhaps because it requires knowledge of mechanics, electricity, and fluids as well as an ability to generalize the details of these subjects. Since all the remainder of the material in this book is without value if the equations cannot be derived, we go to considerable bother to clarify the principles and techniques which are involved.

Previously in the text, we have considered components and systems separately. It turns out, however, that the methods used to determine the differential equations which describe a component are identical to those which are used in deriving system equations. In fact, in this matter, the physical entities which we have been calling "components" lose their distinction from those we have called "systems." Here we analyze a component by considering the action of all its parts; we look at it as a small system. Similarly, we can imagine small systems to be components of a kind; instead of observing the overall action of such a system, we investigate only the relationship between a small number of the terminal variables. Therefore, we shall consider component and system description simultaneously.

5.1 Small Excursion Analysis

Basically, there are two objectives to small excursion analysis. The first is to determine whether or not a particular static equilibrium condition for a system is stable. For example, it is quite simple to build a boat which will float, but which will not sit upright. There is an equilibrium of forces in the upright position, but a small disturbance destroys the balance and the designer's reputation. Similarly, and with similar results, it is quite easy to build a control system which *should* hold a system in some particular condition according to static analysis, but which actually contributes to the inability of the system to maintain its equilibrium. It is claimed that the steering systems of some cars used to be this way.

The second major objective of small excursion analysis is to determine how a system will respond when it is acted upon by an external agent. Some systems are subject to resonance and potential destruction because of it. Others must respond in a given amount of time when particular actions are applied. In both cases, and many others, the methods of small excursion analysis are applied in order to determine the kind of response to be expected and the usefullness of various adjustments which are proposed.

Central to the conduct of a small excursion analysis is the derivation of the equations relating the system variables. In doing this, it is convenient to draw a network schematic in which all the components have been replaced by their linear equivalents, as developed in Chapter 3. This approach has the advantage that the network variables are the excursion variables in which we are interested and the form of the linear equivalent representation may be well known because of frequent application. This approach is particularly effective when some of the components include energy storage elements which do not belong. For example, in the equivalent representation of a clutch, it is a simple matter to indicate the location and amount of the various inertias of the unit. Normally, these would not appear on the schematic used for static analysis or network explanation.

Once the dynamic equations have been developed, the analyst has his choice of several methods for their solution. The basic methods are given in the next chapter; most rely heavily on the fact that there are relatively few kinds of solutions to these equations. Given a catalog of these solutions, we can always find that which suits our conditions.

In order to clarify exactly why we are so interested in small excursion analysis and what we hope to accomplish through its use, let us investigate

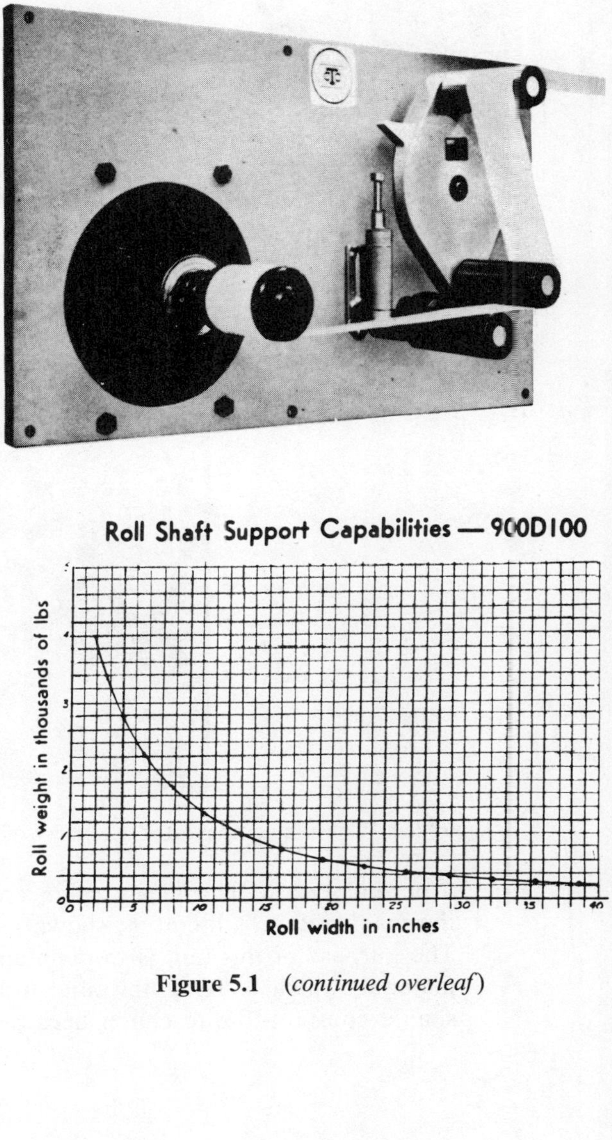

Figure 5.1 (*continued overleaf*)

CTC

HEAVY DUTY UNWIND TENSION COMPENSATOR MODEL No. 900D100

Rear view showing feed-back cable connecting sensing and braking elements.

Rear view with covers removed to show 10" dia torsion spring and 3 H. P. brake.

Figure 5.1 Tension compensator. Courtesy of Compensating Tension Controls, Inc.

a pertinent case. The nicely engineered tension compensator described by the manufacturer's literature shown at Figure 5.1 is an admirable example. The purpose of this unit is to maintain constant tension in paper, or similar material, which is being unwound from a roll. This cannot be done with a simple constant-torque clutch because of the diameter change of the feed roll. The unit utilizes a controllable brake in which the drum torque is controlled by the rotary displacement of the control element (which we have called the runner for similar units described in Chapter 1). The paper passes through a roller arrangement which serves as a tension-to-torque converter. The torque developed by this unit is utilized to control the brake; the cable shown clearly at (b) in the figure provides the torque transfer. The gear arrangement allows the initial spring torque in the tension sensor to be changed, thereby providing a means for setting the desired tension. An all-mechanical feedback system is relatively rare; the designer should be quite proud of the straight-forwardness and elegance of this system.

In a device of this nature, having the design concept is not enough, however. For one thing, it is necessary to determine the accuracy with which the unit can control tension; although this can be done by experiment, it is far more useful to have an analytical method. Small excursion analysis is one way in which this can be done; however, this is only a statics problem. More important is a determination of the stability of the system. Since it is a feedback arrangement, there is some chance that oscillations would occur. If they do, the unit would undoubtedly break the paper; this would not make a very marketable product. Another matter of concern is whether or not the unit can control tension during the time that the paper is undergoing a speed change; this requires transient analysis and is effectively handled by small excursion methods. Since this system illustrates so many aspects of our interest here, we will return to it for several examples later in the chapter.

In the following section a very useful technique for isolating the small excursion aspects of a system is introduced; this technique provides the starting point for the derivation of the small excursion equations.

5.2 Dynamic Linear Equivalent Components

In order to write the differential equations which describe the small excursion performance of a component or system, it is convenient (and a significant aid to the accomplishment of accuracy) to separate all the various physical phenomena which occur. For example, in a clutch there is a flywheel effect as well as the desired controllable dissipation. If we try to write the equations directly from the system schematic we must imagine the way in which the inertia enters; a safer way of going about the analysis is to show the inertia as a separate component of the system and then consider the clutch to be massless. Another helpful step is to eliminate the static components of the variables in the problem; this allows us to concentrate on the excursion variables without being distracted by the aspects of the system which have an influence only on the static operation. The way in which these changes may be carried out is to replace the original system with a system of linear equivalent components. In the linear equivalent system which results, and which may be thought of as an entity in itself, the terminal variables are the excursion variables of the original system and the components are basic linear components which reflect the excursion performance of the originals. The equivalent system has the advantage that it shows, explicitly, all the effects which enter the dynamic performance of the system. Also, since it is made up of two-terminal components exclusively, the process of writing its equations follows a very straightforward pattern. All this is most readily shown by example, which we shall proceed to do. We begin by determining the excursion variable equivalent for the basic energy storing elements; although the results are not particularly surprising they are of basic importance. Also, the development is a good illustration of the use of small excursion variables.

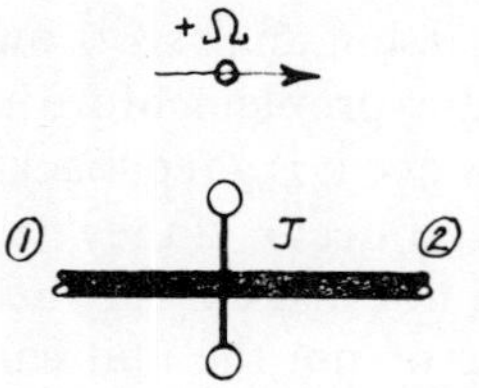

$M = M_1 - M_2$ (+M tightening)

$$M = J\frac{d\Omega}{d\tau} \quad \longleftrightarrow \quad \Omega = \frac{1}{J}\int_{\tau_0}^{\tau} M\,d\tau + \Omega(\tau_0)$$

at static equilibrium, $M = 0$ & $\frac{d\Omega}{d\tau} = 0$; $\Omega \neq 0$ necessarily

$$m = M - M_Q \; (M_Q = 0) \therefore m = M \; ; \quad \omega = \Omega - \Omega_Q \left(\frac{d\Omega_Q}{d\tau} = 0\right) \therefore \frac{d\omega}{d\tau} = \frac{d\Omega}{d\tau}$$

if static equilibrium exists at τ_0,

$$m = J\frac{d\omega}{d\tau} \qquad \omega = \frac{1}{J}\int_{\tau_0}^{\tau} m\,d\tau$$

$$\boxed{m = J[D\omega] \; ; \quad \omega = \frac{1}{J}\left[\frac{m}{D}\right]}$$

From the Engineer's Sketch Book

Figure 5.2 Excursion variable relationships for a flywheel.

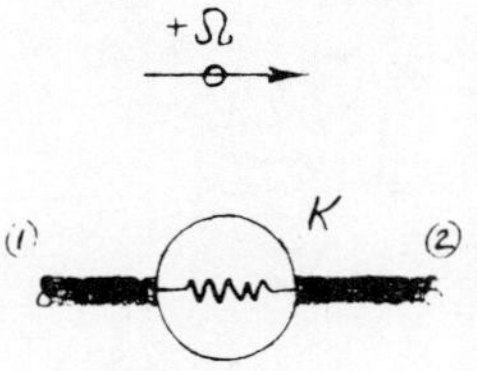

$\Omega = \Omega_1 - \Omega_2$ (+M tightening)

$$M = K\int_{\tau_0}^{\tau} \Omega\, d\tau + M(\tau_0) \quad \longleftrightarrow \quad \Omega = \frac{1}{K}\frac{dM}{d\tau}$$

at static equilibrium, $\Omega = 0$ & $\frac{dM}{d\tau} = 0$; $M \neq 0$ necessarily

$m = M - M_Q$ $\left(\frac{dM_Q}{d\tau} = 0\right) \therefore \frac{dm}{d\tau} = \frac{dM}{d\tau}$; $\omega = \Omega - \Omega_Q$ $(\Omega_Q = 0) \therefore \omega = \Omega$

if static equilibrium exists at τ_0,

$$m = K\int_{\tau_0}^{\tau} \omega\, d\tau \qquad \omega = \frac{1}{K}\frac{dm}{d\tau}$$

$$\boxed{m = K\left[\frac{\omega}{D}\right] \quad ; \quad \omega = \frac{1}{K}[Dm]}$$

From the Engineer's Sketch Book

Figure 5.3 Excursion variable relationships for a rotary spring.

The small excursion variable relationships for rotary mechanical, electrical, and one form of hydraulic energy storing elements are developed in Figures 5.2 through 5.6. These are essentially self explanatory; we will discuss 5.2 and leave the study of the others to the reader. For the flywheel of Figure 5.2, the usual mathematical description is that given just to the left of the double headed arrow below the sketch. Here, as usual, M is torque and Ω is rotational speed; J is the polar moment of inertia of the flywheel. The relationship at the right of the arrow is an equivalent description of the unit; it may be derived from the other, and vice versa. The form used in analysis would depend on the case at hand, as will be seen. Since the static equilibrium condition of any rotary mechanical system is that at which all the rotational speeds are unchanging, the given conditions for the static state apply. This

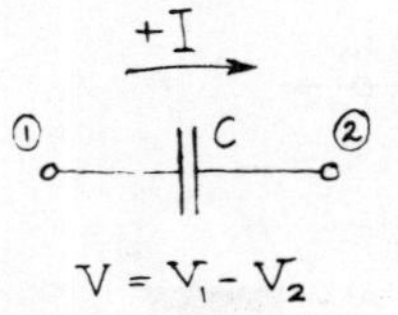

$$V = \frac{1}{C}\int_{\tau_0}^{\tau} I\,d\tau + V(\tau_0) \longleftrightarrow I = C\frac{dV}{d\tau}$$

at static equilibrium, $I = 0$ & $\frac{dV}{d\tau} = 0$; $V \neq 0$ necessarily

$$v = V - V_Q \quad \left(\frac{dV_Q}{d\tau} = 0\right) \therefore \frac{dv}{d\tau} = \frac{dV}{d\tau}; \quad i = I - I_Q \quad (I_Q = 0) \therefore i = I$$

if static equilibrium exists at τ_0,

$$v = \frac{1}{C}\int_{\tau_0}^{\tau} i\,d\tau \qquad\qquad i = C\frac{dv}{d\tau}$$

$$\boxed{v = \frac{1}{C}\left[\frac{i}{D}\right] \quad ; \qquad i = C\left[Dv\right]}$$

From the Engineer's Sketch Book

Figure 5.4 Excursion variable relationships for a capacitor.

leads to the two relationships just above the box at the bottom of the figure. The first of these states that the excursion torque is equal to the moment of inertia times the rate of change of the excursion speed. The second indicates that the excursion speed is equal to the reciprocal of the moment of inertia times the integral, with respect to time, of the excursion torque if the integration commences during static equilibrium. In summary, the only conclusion from this is that the linear equivalent of an already linear flywheel is an identical flywheel.

Inside the box, at the bottom of Figure 5.2, are shown alternative—and somewhat more convenient—ways of writing the same relationships. Since we must indicate time derivatives and integrals so frequently in this work, it is useful to have as compact a symbolism as possible. Here the D is used in the conventional way to indicate differentiation (which we agree shall, in this book, be with respect to time exclusively). Since no well-known, shorthand symbol for integration with respect to time (commencing at the static equilibrium of a system) exists, it seems reasonable to select one which suits our needs exactly.

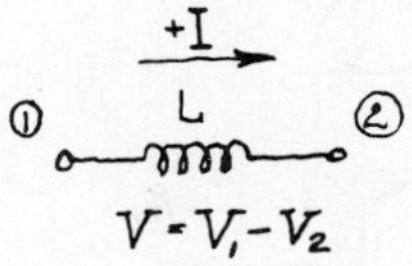

$$V = L\,\frac{dI}{d\tau} \quad \longleftrightarrow \quad I = \frac{1}{L}\int_{\tau_0}^{\tau} V\,d\tau + I(\tau_0)$$

at static equilibrium, $V = 0$ & $\frac{dI}{d\tau} = 0$; $I \neq 0$ necessarily

$v = V - V_Q \quad (V_Q = 0) \therefore\ v = V$; $i = I - I_Q \quad \left(\frac{dI_Q}{d\tau} = 0\right) \therefore\ \frac{di}{d\tau} = \frac{dI}{d\tau}$

if static equilibrium exists at τ_0,

$$v = L\,\frac{di}{d\tau} \qquad\qquad i = \frac{1}{L}\int_{\tau_0}^{\tau} v\,d\tau$$

$$\boxed{v = L\,[Di] \quad ; \quad i = \frac{1}{L}\left[\frac{v}{D}\right]}$$

From the Engineer's Sketch Book

Figure 5.5 Excursion variable relationships for an inductor.

As will be seen in the next chapter, $1/D$ is convenient in this respect. For the time being, this should be interpreted literally as a symbol for integration; i.e., m/D should be read "the integral of excursion torque" (with the specified limits of integration); it may be noted that D^2 may be used to represent the second time derivative of a variable, and that $D(m/D)$ is identical to m. Manipulations with D are discussed in more detail in the next chapter.

Inspection of the results obtained in Figures 5.2 through 5.6 shows that all energy-storing elements are their own linear equivalents; i.e., their accomodation in small excursion analysis is a trivial matter. The only case in which the matter becomes even slightly interesting is that in which the original component is not linear. One example of this situation, that of a non-linear spring, is considered in Figure 5.7; the result is that the linear equivalent, for small excursion analysis, is another spring. The constant for this is the slope of the force versus displacement curve at the static equilibrium point. Similar results, of course, are obtained for other nonlinear storage elements. Another aspect that is important in some cases is that most energy-storing elements are

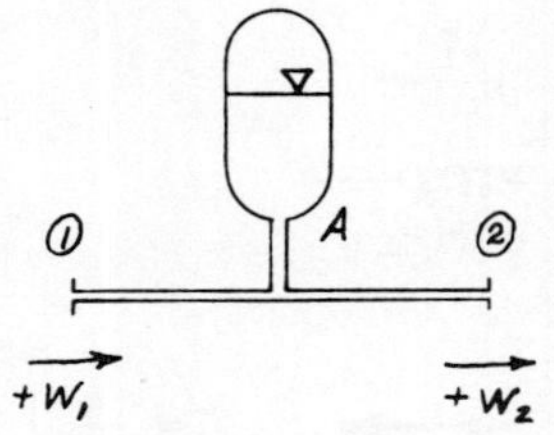

$$W = W_1 - W_2 \; ; \; P = P_1 - P_2$$

$$P = A \int_{\tau_0}^{\tau} W \, d\tau + P(\tau_0) \quad \overset{*}{\longleftrightarrow} \quad W = \frac{1}{A} \frac{dP}{d\tau}$$

at static equilibrium, $W = 0$ & $\frac{dP}{d\tau} = 0$; $P \neq 0$ necessarily

$$p = P - P_Q \left(\frac{dP_Q}{d\tau} = 0\right) \therefore \frac{dp}{d\tau} = \frac{dP}{d\tau} \; ; \quad w = W - W_Q \; (W_Q = 0) \therefore w = W$$

if static equilibrium exists at τ_o,

$$p = A \int_{\tau_0}^{\tau} w \, d\tau \qquad\qquad w = \frac{1}{A} \frac{dp}{d\tau}$$

$$\boxed{p = A\left[\frac{w}{D}\right] \; ; \qquad w = \frac{1}{A}[Dp]}$$

From the Engineer's Sketch Book

Figure 5.6 Excursion variable relationships for a hydraulic accumulator.

not perfectly efficient. For example, there would usually be some windage type of friction with a flywheel; in order to take this into account, a dissipator returned to ground would be indicated on the equivalent network schematic.

The linear equivalent representation for a dissipator, for small excrusion analysis, has been discussed in Chapter 3. We should not have to consider it further here. Also discussed in Chapter 3 were the static linear equivalent representations, for small excursions, of the controllable dissipator elements. We may utilize that material as a background for the dynamic case here; the dynamic aspects appear as appendages on the static equivalent networks, as in the next example.

The brake utilized in the paper tension regulator described earlier rep-

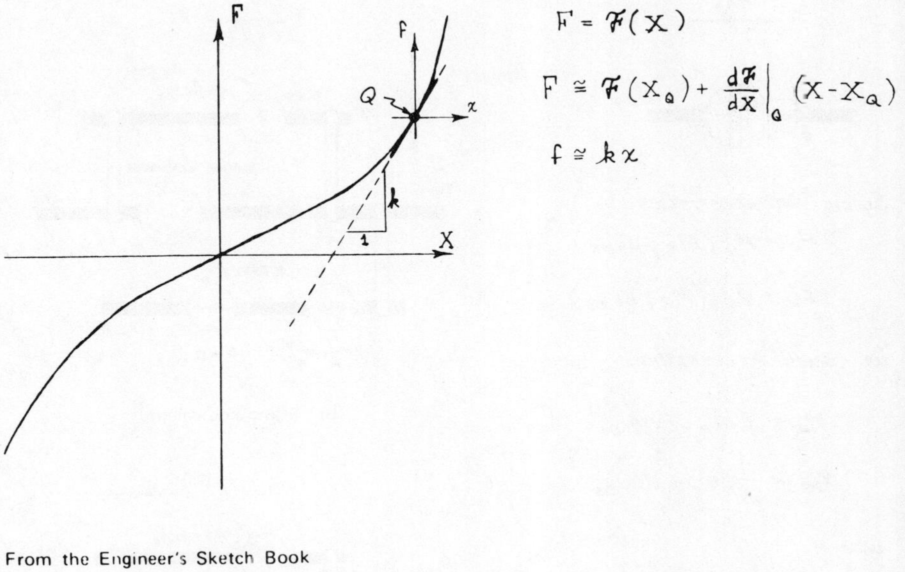

Figure 5.7 Small excursion equivalent for non-linear spring.

resents a good example for the determination of a linear equivalent to a controllable dissipator. A little extra can be gained from the analysis by considering the more general case represented by a controllable clutch; this we shall do. The device symbol and notation is shown in Figure 5.8(a) along with the determination of the parameters for the static equivalent device. The static equivalent is shown at (b) in the figure. Note that m_D is the sum of two torques; therefore, it must be developed by parallel elements, as is shown. Since a torque control (instead of speed-sensitive control, or "non-conductive" control) is assumed, the runner position relative to the spider, instead of runner speed, enters the analysis. The relative rotation of the runner is given by a sum in the characterization form chosen here; therefore, it must be generated by series elements, which have been indicated. As usual, the resulting network has one device for each of the terms in the linear approximation to the characteristics; also, each of the devices are basic two-terminal components. Finally, we may introduce the dynamic aspects of the component. This is shown at (c) of the figure. In this case, the energy storage mechanisms of most significance are the inertias associated with the moving parts of the clutch. To specify an example, we may consider the flywheel J_D. This represents the inertia of all the parts which move at a speed, relative to an inertially fixed reference, which is proportional to Ω_D. For the construction shown in Chapter 1 (Figure 1.22),

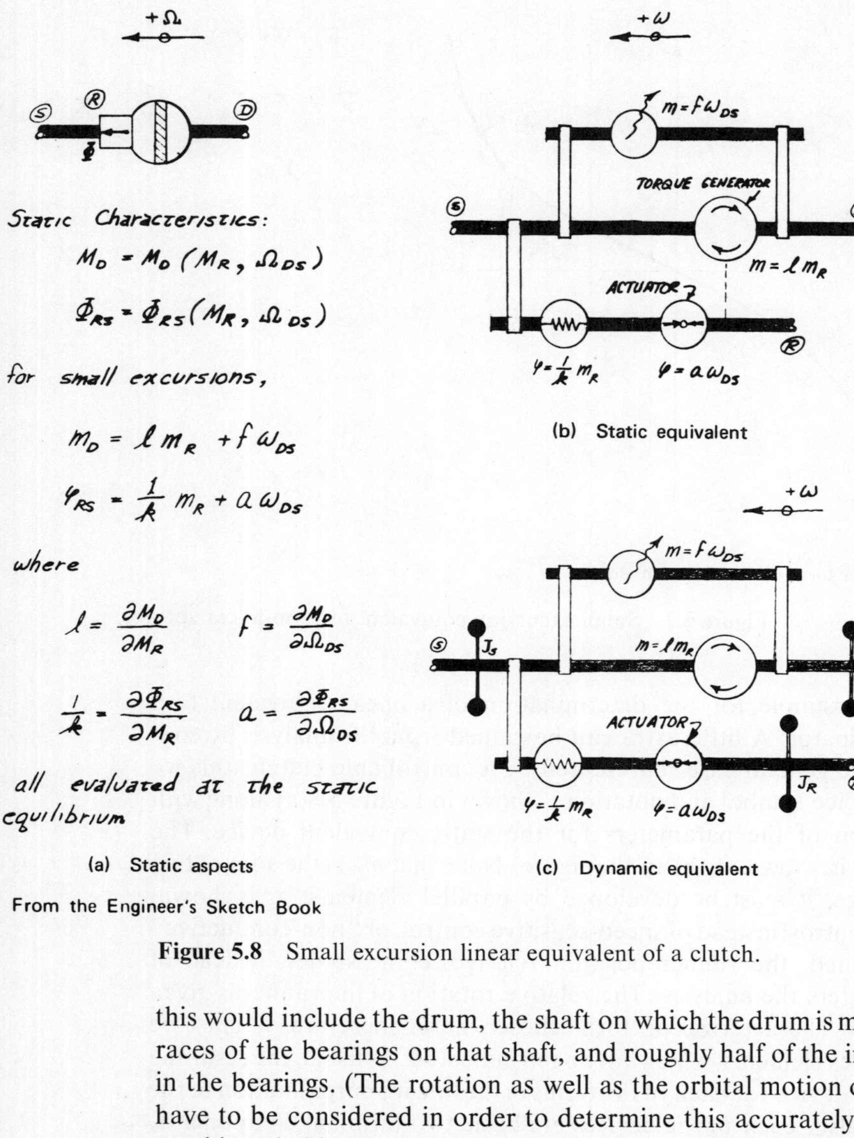

From the Engineer's Sketch Book

Figure 5.8 Small excursion linear equivalent of a clutch.

this would include the drum, the shaft on which the drum is mounted, the inner races of the bearings on that shaft, and roughly half of the inertia of the balls in the bearings. (The rotation as well as the orbital motion of the balls would have to be considered in order to determine this accurately; energy methods would probably be used to determine the inertia of an equivalent flywheel.) It should be noted that there are several possible locations for J_D on the equivalent the equivalent network; it may be placed on any shaft which is rotating at the speed ω_D. Similar reasoning leads to the values and locations for the other inertias. It is not inconceivable that a controllable dissipator might have some springiness between the drum and spider shafts; this would naturally have to be included in the equivalent network. The clutch construction here, however, is very stiff and it is hard to imagine any significant effect arising through this means.

In order to determine the differential equations describing the clutch, we must apply the methods described later in the chapter. We shall continue the example at that point.

Many examples of analogous dynamic effects in other forms of controllable dissipators may be found in specialized texts and in manufacturers' literature. For example, vacuum tubes have dynamic linear equivalent networks which include capacitors between each pair of terminals. In a transistor, the main dynamic effect is represented by a capacitor between the base and collector equivalent terminals. In a pilot-operated spool valve, the inertia of the spool results in a hydraulic inertia between the pilot and outlet terminals of the equivalent device.

Other kinds of components—i.e., actuators, exchangers, and converters—have linear equivalents which are determined by the same process as used here for the clutch. In each case, the dynamic equivalent is fabricated from the static representation; the dynamic effects are brought in by locating and evaluating the energy storage which may occur in the device. Several more examples of the process appear in later discussions in this chapter.

5.3 Dynamic Linear Equivalent Networks

Once the equivalent devices for each of a system's components have been determined, the equivalent network may be drawn almost automatically. There are a few slightly tricky aspects to the business, however; we point out a few of them in this section.

Referring back to Figure 5.8(c), it should be noticed that the clutch, a three-terminal device under static conditions, becomes a four-terminal in its dynamic equivalent. This is because of the inertial reference of the flywheels, an effect that does not have an exact parallel in the other media of power transfer. Of particular importance is that we *cannot* write an expression such as

$$m_D \pm m_R \pm m_S = 0$$

for the dynamic case. Only by including the virtual torque (as represented by D'Alembert type of inertia forces and torques) can we write such an expression; in that case, four, not three, torques would be involved. In contrast to this is the case of a transistor; the summation of current at the three terminals holds for dynamic as well as static operation. No current "escapes" by a mechanism analogous to the effect of inertia.

Sometimes confusing in the developent of equivalent networks is the role of a well-regulated power source, such as a synchronous motor in rotary mechanical systems or batteries in electrical systems. For perfect regulation the excursion variable remains identically zero; this requires the use of a perfect conductor or non-conductor as the equivalent of the regulated device. Although the replacement of a battery by a short circuit may at first seem ridiculous, it is in fact the proper excursion equivalent. The battery forces the excursion variable to be zero; the short circuit does the same in the excursion network. For a synchronous motor, the speed does not change as the amount of torque varies but the rotor does displace with respect to its no-load position. The linear equivalent representation of this effect is made by a rotary spring having one end grounded. The motor, literally, contains the magnetic equivalent of a spring and this is what is shown by the equivalent device.

In order to illustrate the process of equivalent network development, let us return to the tensioner problem. The basic system schematic is shown in Figure 5.9(a); the diagram helps to explain the operation of the system. Note that rotation of the tension sensor is restrained by the spring with value k_s; the torque of the sensor, less that of the spring, is the amount applied to the control of the brake. We must know the stiffness of the brake control in order to determine the rotation of the sensor; this is why the particular brake (clutch) characterization used in section 5.2 was adopted. The small excursion dynamic equivalent of the tensioner is shown at (b) in Figure 5.9. Since the spider of the clutch is grounded in this braking application, there is no need to indicate the spider inertia. The inertia J_1 includes both the drum inertia and that of the roll of paper (this is changing constantly, adding a little extra interest to the problem; we assume the change is slow enough to neglect in any practical design of the system). The inertia J_2 includes that of the runner as well as the tension sensor. The torques applied to the shafts result from tension excursions, as indicated by the notation. It might be noted that the motor which pulls the paper out of the tensioner always operates at the same torque (assuming the tensioner is operating); the paper speed may be just about any reasonable value and is determined by the motor speed. In the equivalent network, the actuator shown in Figure 5.8(c) has been dropped from the brake equivalent representation; this effect would be very small in most brake designs. The final equivalent network at Figure 5.9(b) leads directly to the equations for the system. These will be derived in section 5.7.

5.4 Variable Systems for Dynamic Analysis

The equations describing the dynamic performance of a system may be formulated in several ways. If loop methods of analysis are used exclusively, the only variables entering the representation will be of the **Γ** type. Similarly, junction

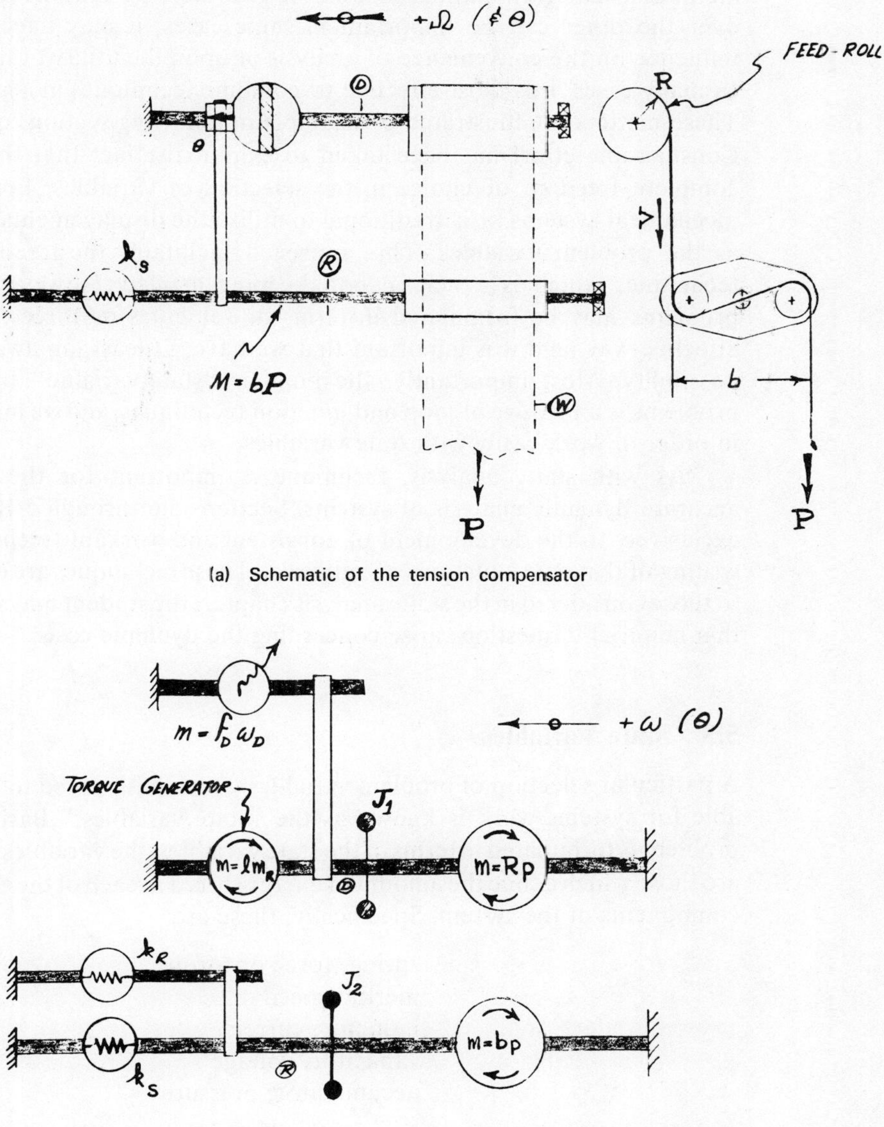

(a) Schematic of the tension compensator

(b) Small excursion dynamic equivalent network

From the Engineer's Sketch Book

Figure 5.9 Preparations for analysis of the tension compensator.

methods lead to formulations in terms of Δ variables. The choice of one method over the other can be important in some cases; it may have considerable influence on the convenience of analysis or upon the utility of the final result. In many cases, it is most effective to combine techniques in a given problem. These matters are illustrated by example in following sections of the chapter. Considerable effort has been taken to exhibit the fact that the analyst has complete freedom of choice in the selection of variables. For example, in mechanical systems, it is traditional to utilize the displacements of the masses as the problem variables. This approach facilitates the free-body-diagram technique, which is a helpful bookkeeping artifice. However, mechanical problems may be formulated in terms of velocities or forces in an equally effective way, and it is important that we have, at least, an awareness of this possibility. Most importantly, the popular "state-variable" formulation of problems is a mixture of loop and junction techniques, and we must know both in order to work easily with state variables.

As with static analysis, technique is important for the efficient and accurate dynamic analysis of systems. Sections 5.6 through 5.10 are devoted exclusively to the development of consistent and workable techniques for the writing of dynamic equations of networks. These techniques are closely related to those considered in the static analysis chapter; the student may wish to review that material if questions arise concerning the dynamic case.

5.5 State Variables

A particular selection of problem variables which has proved to be very desirable for systems work is known as the "state variables." Basically, when a problem is formulated in terms of the state variables, the variables which appear are those which define the amount of energy stored in each of the energy-storing components of the system. Specifically, these are:

spring; force or torque
inertia; speed
inductor; current
capacitor; voltage
accumulator; pressure

In terms of Γ and Δ variables, the state variables are defined as follows:

> For energy-storing elements which are static conductors, the Γ variable is the state variable.
> For energy-storing components which are non-conductors in static conditions, the Δ variable is the state variable.

As may be seen in Figures 5.2 through 5.6, the state variables appear as dependent variables in the mathematical representation of component behavior which includes the integral operator $1/D$.

Another way of defining the state variables arises from the manner in which the amount of energy stored in a device may be expressed. Specifically, the state variable for an energy-storing component is that $\mathbf{\Gamma}$ or $\mathbf{\Delta}$ variable which appears in the expression giving stored energy as a static function of one variable. That either the $\mathbf{\Gamma}$ or $\mathbf{\Delta}$ variable, but not both, can be used in this way is shown as follows: For a dissipator, the power is given by $\mathbf{\Gamma} \times \mathbf{\Delta}$. The same expression yields the power transfer to an energy storing component. Assume the stored energy E is a function of $\mathbf{\Gamma}$ as follows:

$$E = \mathscr{E}\,(\mathbf{\Gamma})$$

Then the rate of change of stored energy is the delivered power P, and

$$\frac{dE}{dt} = \frac{d\mathscr{E}(\mathbf{\Gamma})}{d\mathbf{\Gamma}}\frac{d\mathbf{\Gamma}}{dt} = \mathscr{E}'(\mathbf{\Gamma})\frac{d\mathbf{\Gamma}}{dt} = \mathbf{\Gamma}\mathbf{\Delta}$$

Thus,
$$\mathbf{\Delta} = \left[\frac{\mathscr{E}'(\mathbf{\Gamma})}{\mathbf{\Gamma}}\right]\frac{d\mathbf{\Gamma}}{dt}.$$

This is the form of characteristic representing an energy storing device which is a conductor in static conditions. All these devices, then, have energy contents which may be expressed algebraically in terms of the $\mathbf{\Gamma}$ variable.

Assume the stored energy E is a function of $\mathbf{\Delta}$ as follows:

$$E = \mathscr{E}(\mathbf{\Delta})$$

Then,
$$\frac{d\mathscr{E}(\mathbf{\Delta})}{d\mathbf{\Delta}}\frac{d\mathbf{\Delta}}{d\tau} = \mathscr{E}'(\mathbf{\Delta})\frac{d\mathbf{\Delta}}{d\tau} = \mathbf{\Gamma}\mathbf{\Delta}$$

Thus,
$$\mathbf{\Gamma} = \left[\frac{\mathscr{E}'(\mathbf{\Delta})}{\mathbf{\Delta}}\right]\frac{d\mathbf{\Delta}}{d\tau}$$

This represents a static non-conductor.

Since we have considered the general case where the device is non-linear, the result appears somewhat different from the well-known linear cases where the stored energy is given by expressions such as $\frac{1}{2}mv^2$, $\frac{1}{2}kx^2$, $\frac{1}{2}Li^2$, etc. The expressions

$$\left[\frac{\mathscr{E}'(\mathbf{\Gamma})}{\mathbf{\Gamma}}\right] \quad \text{and} \quad \left[\frac{\mathscr{E}'(\mathbf{\Delta})}{\mathbf{\Delta}}\right],$$

which represent the dynamic characteristics of energy-storing devices in the relationships above, degenerate into constants in the linear case.

One of the effects of using state variables is that dynamics problems become, from a practical point-of-view, statics problems. This occurs as follows: In order to determine the behavior of the system when state variables are used, we must determine the $\mathbf{\Gamma}$ variables in those energy-storing components were the state variable is of the $\mathbf{\Delta}$ form, and vice versa. This can be done by replacing each statically conductive storage element with a $\mathbf{\Gamma}$ generator (set to generate the appropriate state variable), and each non-conductive storage element with a $\mathbf{\Delta}$ generator (also set to produce the state variable of the original component). Since all the dynamic components have now been removed from the system, only static techniques need be used for analysis. Examples of this effect appear in the following sections and, with considerable influence, in the examples of Chapter 9.

As may be seen by inspection of the mathematical representations of energy storing devices,

for static conductors, $D\mathbf{\Gamma} = \mathscr{F}_1(\mathbf{\Gamma})\mathbf{\Delta} : \mathbf{\Gamma}$ state variable

for static non-conductors, $D\mathbf{\Delta} = \mathscr{F}_2(\mathbf{\Delta})\mathbf{\Gamma} : \mathbf{\Delta}$ state variable

Since the $\mathbf{\Gamma}$ variable in any statically non-conductive component, or the $\mathbf{\Delta}$ variable in any statically conductive component, will be a static function of the state variables of the system and the various forcings (y), the rate of change of any state variable is a static function of the state variables and forcing exclusively. This means that dynamics problems, when formulated in terms of the state variables, may always be written in this form:

$$
\begin{aligned}
Dx_1 &= \mathscr{F}_1(x_1, x_2, \ldots x_n, y_1) \\
Dx_2 &= \mathscr{F}_2(x_1, x_2, \ldots x_n, y_2) \\
&\;\;\vdots \\
Dx_n &= \mathscr{F}_n(x_1, x_2, \ldots x_n, y_n)
\end{aligned}
$$

where the *y*'s represent independent functions of time ("forcing functions") and the functions $\mathscr{F}_1$, $\mathscr{F}_2$, etc., are all static functions.

For simplicity in the remaining discussion, let us consider the case where there are three state variables X_1, X_2, and X_3.

$$
\begin{aligned}
DX_1 &= \mathscr{F}_1(X_1, X_2, X_3, Y_1) \\
DX_2 &= \mathscr{F}_2(X_1, X_2, X_3, Y_1) \\
DX_3 &= \mathscr{F}_3(X_1, X_2, X_3, Y_1)
\end{aligned}
$$

If only the small excursion performance of the system is under consideration, these equations may be written as:

$$
\begin{aligned}
Dx_1 &= a_{11}x_1 + a_{12}x_2 + a_{13}x_3 + b_1 y_1 \\
Dx_2 &= a_{21}x_1 + a_{22}x_2 + a_{23}x_3 + b_2 y_2 \\
Dx_3 &= a_{31}x_1 + a_{32}x_2 + a_{33}x_3 + b_3 y_3
\end{aligned}
$$

The matrix of coefficients a completely determine the nature of the system; an experienced engineer can predict system performance quite rapidly from little more than an examination of the signs and relative magnitudes of these coefficients. Primarily, the effect of the selection of state variables to the paper-and-pencil analyst is to provide a standard form for problem formulations. In such formulations, all analogous systems have obviously related mathematical descriptions. Thus, our experience may be applied to a broad class of problems with the aid of state variables.

The most significant, practical advantage of state variables is in the use of computers for system simulation. Both analog and digital computers are readily adapted to carrying out integration, but perform differentiation poorly and with considerable inconvenience to the operator. Examination of the state variable formulation of a problem shows that solutions may be obtained simply by integrating, simultaneously, the static functions which appear in the formulation. This is readily done by a computer. Thus, the state variable formulation of a problem is the most efficient possible means of preparation for a computer analysis.

5.6 Translational Mechanical Systems

In this and the following sections we examine the various specific kinds of systems and show how the equations of dynamics are derived. In all this work, considerable care must be taken to avoid mistakes in sign; this is probably the most difficult aspect of the process, simply because there are so many opportunities for error. In order that the examples may be as understandable as possible, the mathematical manipulations have been carried out directly on the figures. Comment in the text will be restricted to the special aspects of the case at hand. Generally, there is one figure giving simple hypothetical cases and another presenting a realistic problem carried all the way from assembly schematic to the equations of dynamics. It is hoped that all the important aspects of the procedures will be clarified in this way.

In Figure 5.10 are shown several formulations of the equations for the "old standby," a spring-mass-dashpot system. The traditional procedure, using a free-body-diagram is shown at (a). Here markers on the moving body and the

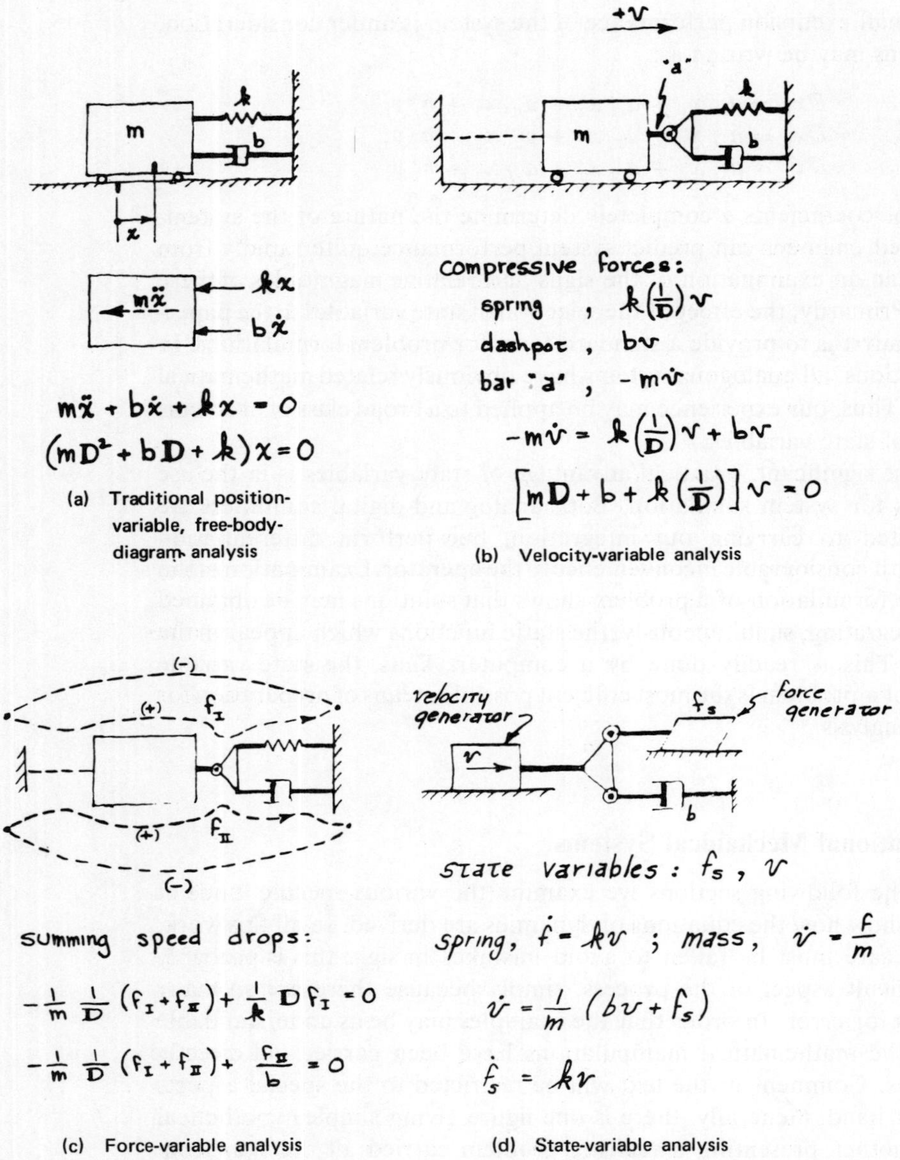

From the Engineer's Sketch Book

Figure 5.10 Determining the differential equations of translational mechanical networks.

reference frame define the problem variable, the displacement x. It is presumed that the markers are aligned in static equilibrium so that x is an excursion variable. The arrowhead here indicates the direction of a positive displacement and is absolutely essential to a successful analysis. The forces acting upon the mass are shown on the free-body-diagram and indicate the directions of the forces, for the case of positive displacement, velocity, and acceleration. It is important to note that once a direction for positive displacement is chosen, the directions for positive velocity and acceleration are fixed (they must be the same as that for positive displacement). The spring force shown is the result of a positive displacement, which in this case creates compression in the spring. The dashpot tends to resist the velocity of the mass; hence, its force is also to the left. With both of these, it is assumed that the constants (k and b) are positive, which is conventional in engineering. The handling of the $F = ma$ aspect of the problem is done by the use of D'Alembert's principle, which the author prefers since it seems to help avoid sign mistakes. In this method, an imaginary force (sometimes called the "inertia force") is indicated by a dashed vector. This vector always is directed opposite to positive acceleration (displacement) and has a magnitude equal to the mass times the acceleration. The dot over x is common way of indicating differentiation with respect to time; the interpretation is velocity. Two dots represent the second time derivative, or acceleration. A summation of the forces acting on the mass yields the equation which the system must obey; this may be written in terms of the D operator, as shown if desired.

A junction-type analysis, using Γ and Δ variables explicitly, is shown at (b). Through the mechanism of a prior selection for positive force (compression or tension), appropriately formulated descriptions of the component characteristics, specification of a direction for positive velocity, and a rule for the forces at a junction, the need for vector representations of the forces is eliminated. We simply say that the sum of the compressive force to the left of the junction must equal the sum of them on the right. The value of this approach is greater in rotary mechanical systems, where the vectors are often difficult to draw and interpret effectively.

A loop-type analysis is shown at (c) and is done in the same way as a loop analysis for a static case. Here the problem variables are the forces. Note that the force loops go from the mass to the reference frame, thereby accounting for the influence of inertia. From a network analysis point of view, the connection between the mass and the reference frame is very real. Typical of the process here is determination of the speed drop in the spring. The force f_l, which threads the spring, is defined to be positive on its pass through the system; therefore, the spring must be compressed if f_l is positive. If f_l has a positive derivative, the spring must be getting shorter and the speed on the left must be more than that on the right; thus we have a positive "speed drop" term in the equations.

The state-variable analysis is shown at (d). The rate of change of v is determined from a force summation at the left of the junction while the rate of change of f_S comes from a speed-drop type of procedure. It is quite common that a state-variable formulation requires both kinds of analysis. It may be noted that the state-variable formulation consists of two equations involving first derivatives while the others may all be written as one equation with a second derivative. Naturally, all the formulations are equivalent. The diagram used for state-variable analysis indicates the procedure whereby the dynamic components are replaced by appropriate generators in order to create an imaginary static network producing the desired static functions of the state variables. Analysis consists of determining the force in the velocity generator and the velocity in the force generator. Analogous procedures characterize state-variable analysis.

A relatively common problem which involves translational mechanical devices is the analysis of a vehicle suspension. A simplified case is considered at Figure 5.11. Part (a) shows the car body as a mass with some rotational inertia about the center of gravity; the suspension springs and "shock absorbers" operate in parallel at each end. ("Shock absorbers" are actually shock transmitters; if they "absorb" anything, it is vibration through the mechanism of energy dissipation.) In the simple case considered here, the axles (at the bottoms of the springs and dashpots) do not move vertically; motion at these points would have to be allowed if we wanted to determine the action of the vehicle as it moves over a rough road. Tire springiness and dissipation would also have to be included to get a meaningful solution. The free-body diagram is shown at (b). The inertia torque, opposite in direction to positive rotations, has the magnitude $I\ddot{\theta}$. The spring deflections are due to both the vertical motion of the center of gravity and the rotation of the body; this is accounted for the spring force expressions. The approximation $\sin\theta \cong \theta$ for small θ has been used here; thus the equations are indeed meaningful only for small excursions. The state-variable formulation of the problem is carried out at (c) and (d); other formulations would, of course, be possible.

5.7 Rotary Mechanical Systems

The virtual connection of a mass to the inertial reference system in translational mechanical systems has its counterpart in rotary systems. This peculiarty may be made more tractable by the imaginary device shown at Figure 5.12(a). In this, a flywheel is shown encased by a shell. If we imagine that the inertia torque of the flywheel is the result of some kind of interaction

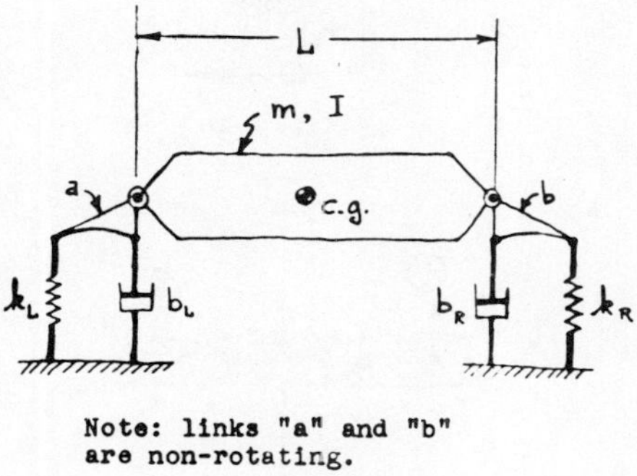

(a) Simplified schematic of automobile suspension

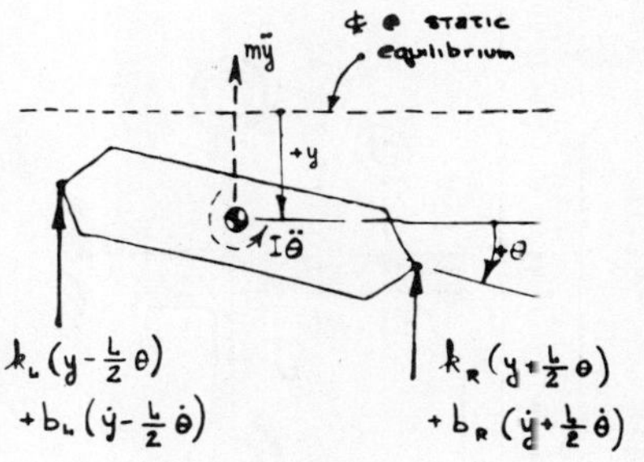

(b) Free-body-diagram for small excursions

State Variables:

$v\ (=\dot{y})$

$\omega\ (=\dot{\theta})$

f_L (= force in spring L)

f_R (= force in spring R)

$$f_L + b_L\left(v - \frac{L}{2}\omega\right) + f_R + b_R\left(v + \frac{L}{2}\omega\right) + m\dot{v} = 0$$

$$\frac{L}{2}\left[f_L + b_L\left(v - \frac{L}{2}\omega\right)\right] - \frac{L}{2}\left[f_R + b_R\left(v + \frac{L}{2}\omega\right)\right] - I\dot{\omega} = 0$$

$$\dot{f}_L = k_L\left(v - \frac{L}{2}\omega\right)\ ; \quad \dot{f}_R = k_R\left(v + \frac{L}{2}\omega\right)$$

(c) Development of equations using state variables

$$\dot{v} = -\frac{1}{m}\left[(b_L + b_R)v + \left(\frac{b_R L}{2} - \frac{b_L L}{2}\right)\omega + f_L + f_R\right]$$

$$\dot{\omega} = \frac{L}{2I}\left[(b_L - b_R)v + \left(\frac{b_R L}{2} + \frac{b_L L}{2}\right)\omega + f_L - f_R\right]$$

$$\dot{f}_L = k_L\left(v - \frac{L}{2}\omega\right)$$

$$\dot{f}_R = k_R\left(v + \frac{L}{2}\omega\right)$$

(d) Standard form for the state-variable equations

From the Engineer's Sketch Book

Figure 5.11 State-variable equations for a suspension system.

between the shell and the flywheel, the unit appears to conform to the pattern of two-terminal devices. The inertia torque would be transmitted to the reference frame by the shaft on which the shell is mounted. The analogy between a capacitor and a flywheel is emphasized by this way of making the drawing. Also, loop analysis for rotary mechanical systems is made somewhat more rational since the torques acting on a flywheel do not just disappear into the ether. The same effect on a network as produced by the arrangement at (a) comes from the more conventional layout of (b), and this is normally what is shown on the schematics of rotary mechanical systems. The fabrication at

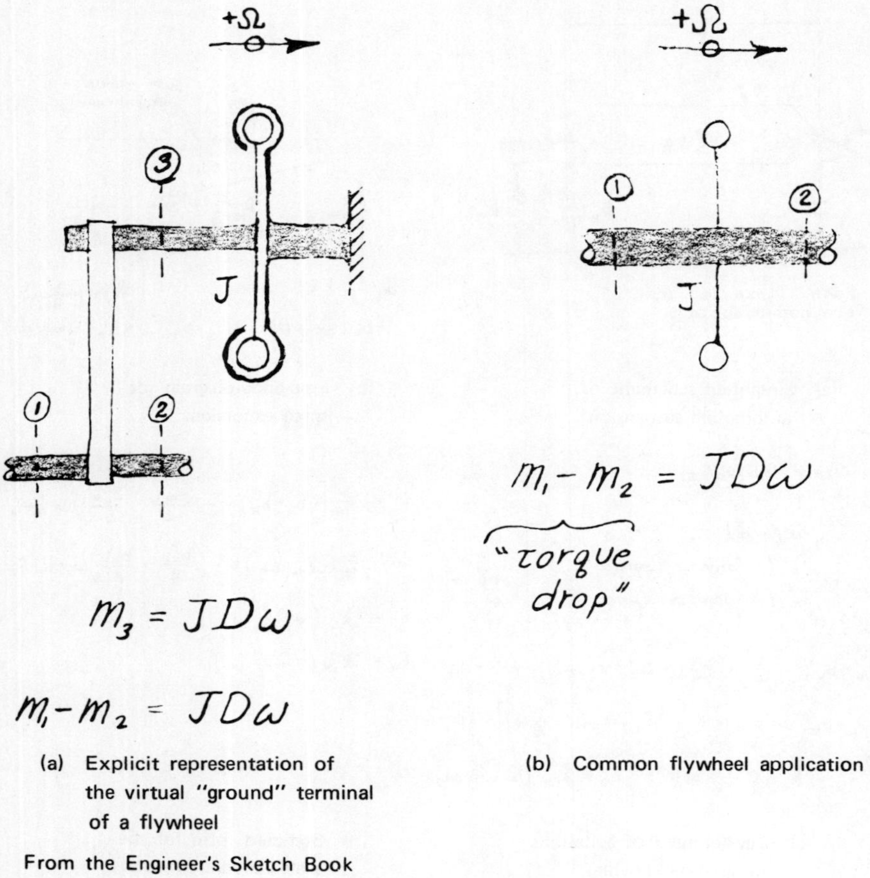

Figure 5.12 A means for resolving the peculiarity of a flywheel.

(a), however, may be helpful in understanding the loop analysis portions of the following examples.

A typical small excursion rotary mechanical system, resulting from the linearization of a real rotary system, is considered in detail at Figure 5.13. Junction, loop, and state-variable analyses are carried out in the various portions of the figure. At (a) and (b) the equations are arranged, finally, in a standard matrix form with the dependent variables and their associated operator combinations at the left, the independent variables and operators at the right. As will be seen in the next chapter, the determinant of the operator expressions on the left is an important indicator of system performance; it is

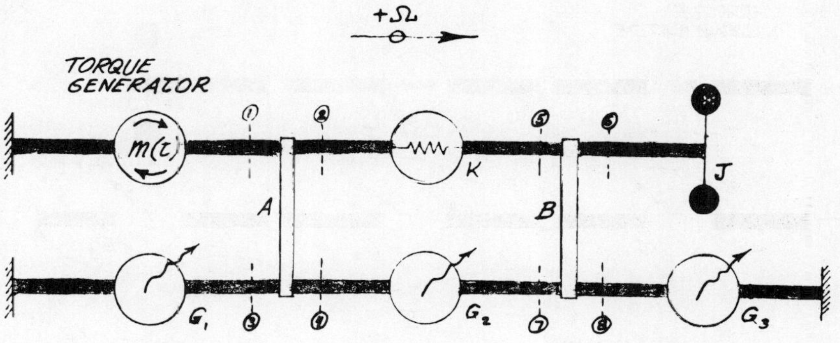

AT A,

$$m_1 + m_3 = m_2 + m_4$$

(tightening torque positive)

$$m(t) + G_1(0 - \omega_A) = K\,\frac{\omega_A - \omega_B}{D} + G_2(\omega_A - \omega_B)$$

AT B,

$$m_5 + m_7 = m_6 + m_8$$

$$K\,\frac{\omega_A - \omega_B}{D} + G_2(\omega_A - \omega_B) = JD\omega_B + G_3\omega_B$$

THESE MAY BE REWRITTEN:

$$[(G_1 + G_2)D + K]\,\omega_A - [G_2 D + K]\,\omega_B = D\,m(t)$$

$$[G_2 D + K]\,\omega_A - [JD^2 + (G_2 + G_3)D + K]\,\omega_B = 0$$

(a) Junction, or velocity-variable, analysis

From the Engineer's Sketch Book

Figure 5.13 (*continued overleaf*)

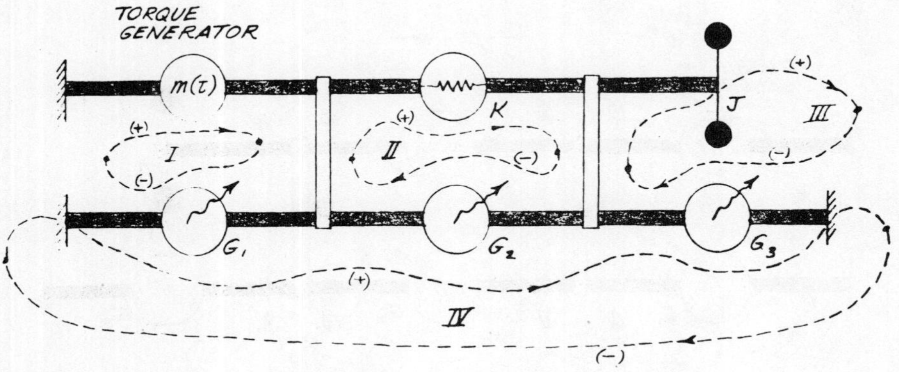

On I,

$$m_I = m(\tau)$$

On II,

$$\frac{Dm_{II}}{K} = \frac{m_{IV} - m_{II}}{G_2}$$

On III,

$$\frac{m_{III}}{JD} = \frac{m_{IV} - m_{III}}{G_3}$$

On IV

$$\frac{m_{IV} - m_I}{G_1} + \frac{m_{IV} - m_{II}}{G_2} + \frac{m_{IV} - m_{III}}{G_3} = 0$$

THESE MAY BE REWRITTEN:

$$(G_2 D + K)\, m_{II} + \quad 0\, m_{III} + \quad (-K)\, m_{IV} = 0$$

$$(0)\, m_{II} + (JD + G_3)\, m_{III} + \quad (-JD)\, m_{IV} = 0$$

$$(-G_1 G_3)\, m_{II} + \quad (-G_1 G_2)\, m_{III} + (G^2)\, m_{IV} = G_2 G_3\, m(\tau)$$

$$\text{where} \quad G^2 = G_1 G_2 + G_1 G_3 + G_2 G_3$$

(b) Loop, or torque-variable, analysis

From the Engineer's Sketch Book

Figure 5.13 *(continued)*

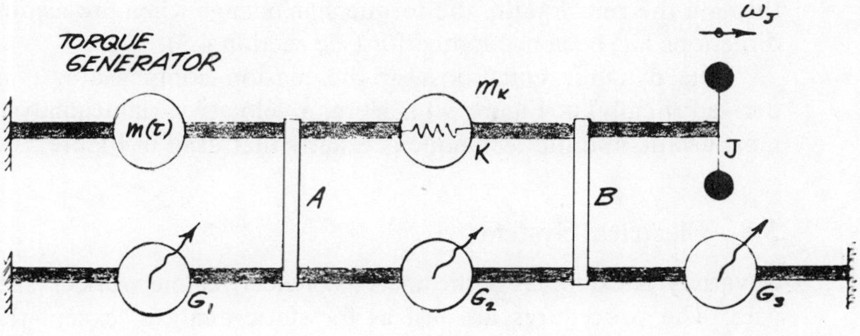

Desired form: $Dm_K = F_1[m_K, \omega_J, m(t)]$

$D\omega_J = F_2[m_K, \omega_J, m(t)]$

From static analysis:

$$\omega_A = \frac{m(t) + G_2\omega_J - m_K}{G_1 + G_2}$$

Torque balance at B:

$$m_K + G_2(\omega_A - \omega_J) = JD\omega_J + G_3\omega_J$$

$$D\omega_J = \frac{G_1}{J(G_1+G_2)} m_K - \frac{G_1G_2 + G_1G_3 + G_2G_3}{J(G_1+G_2)} \omega_J + \frac{G_2}{J(G_1+G_2)} m(t)$$

Speed change A to B:

$$\omega_A - \omega_B = \frac{KDm_K}{K}$$

$$Dm_K = -\frac{K}{G_1+G_2} m_K - \frac{KG_1}{G_1+G_2}\omega_J + \frac{K}{G_1+G_2} m(t)$$

(c) State-variable analysis

From the Engineer's Sketch Book

Figure 5.13 Small excursion dynamic equations of rotary mechanical networks.

the same for each formulation. Note that four torque variables are required at (b) because four areas are enclosed by the schematic when the return path and flywheel virtual terminal connection are shown explicitly. Since it is clumsy in this network to run the negative portion of the torque loops all through the return path, the torque sign change when proceeding in opposite directions has been accounted for (see section 4.3).

The dynamic equations for the tension compensator (Figure 5.1) are derived (finally) at Figure 5.14. Here, a velocity variable analysis appears the most useful, and the technique is exactly that used in Figure 5.13(a).

5.8 Electrical Systems

A variety of examples of the analysis of electrical networks is shown in Figure 5.15. The procedures are just as for static analysis, except the D operator appears for each inductor or capacitor. The case at (a) is a simple loop; it should be noted that the voltage generator at the left produces a voltage rise and, thus, a negative drop in the equations. The case shown at (b), which arises when a transistor is replaced by its h parameter small excursion linear equivalent, is most effectively handled by mixed analysis since junction methods would be clumsy on the left and loop methods would be clumsy at the right. The state-variable analysis at (d) is complicated because the state variables do not, directly, represent an efficient approach to the static analysis. Here four variables are introduced and four equations written; the two variables which are not state variables are then eliminated algebraically.

A typical real problem in electrical network analysis is shown in Figure 5.16. The circuit shown at (a) includes a number of useful features (most of which complicate the analysis procedure). The capacitor at the input (C_A) isolates the static behavior of the amplifier from that of its input; this prevents the chain of amplifiers from being overly sensitive to temperature or dc input noise. The input capacitor must be large enough to act like a short circuit in the frequency range for which the amplifier is intended. (Obviously, this cannot be done with dc amplifiers.) The resistor R_E is necessary to provide thermal stability of the amplifier itself, but tends to reduce the amplification factor. This loss may be overcome with another capacitor, C_E here, which also acts as a short ciruit as far as ac signals are concerned. It is assumed that the amplifier is driving a load which is partially inductive, such as a speaker coil or solenoid; this is the origin of the inductor L. At (b) is shown the static equivalent circuit which facilitates static analysis. Note that the capacitors become open circuits while the inductor becomes a short circuit. A simplified static analysis is shown at (c); this must be carried out in order to determine the coefficients in the linear representation of the transistor. The complete small excursion linear equivalent circuit is shown at (d); the h parameter euqivalent for the transistor has been indicated. Several assumptions regarding components and the frequency (f) of operation lead to the simplified circuit shown at (e). Basically, the assumptions are that the excursion currents through

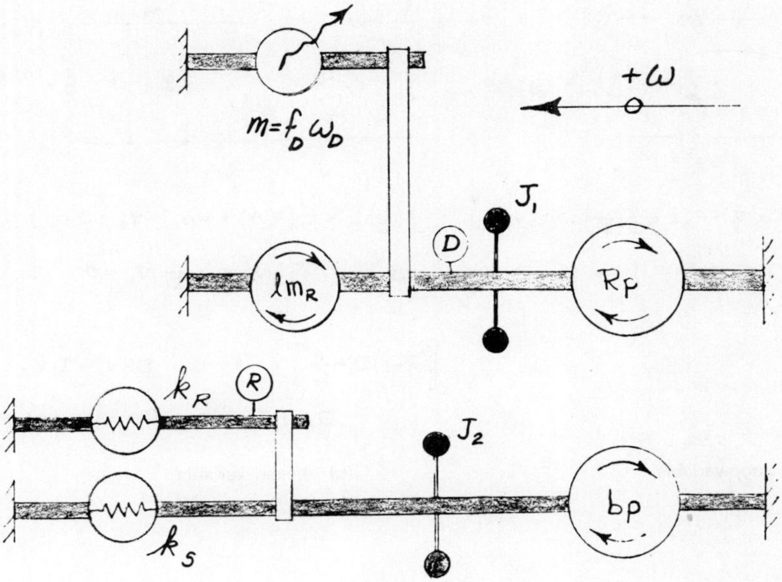

(a) Small excursion equivalent network

$$Rp - J_1 D\omega_D = \left(k_R \frac{1}{D}\omega_R\right)l + f_D\,\omega_D$$

$$bp - J_2 D\omega_R = (k_R + k_S)\frac{1}{D}\,\omega_R$$

$$DRp \qquad - k_R l\,\omega_R = (J_1 D^2 + f_D D)\,\omega_D$$

$$Dbp - (J_2 D^2 + k_R + k_S)\,\omega_R = 0$$

(b) Junction analysis

From the Engineer's Sketch Book

Figure 5.14 Small excursion dynamic equations for the tension compensator.

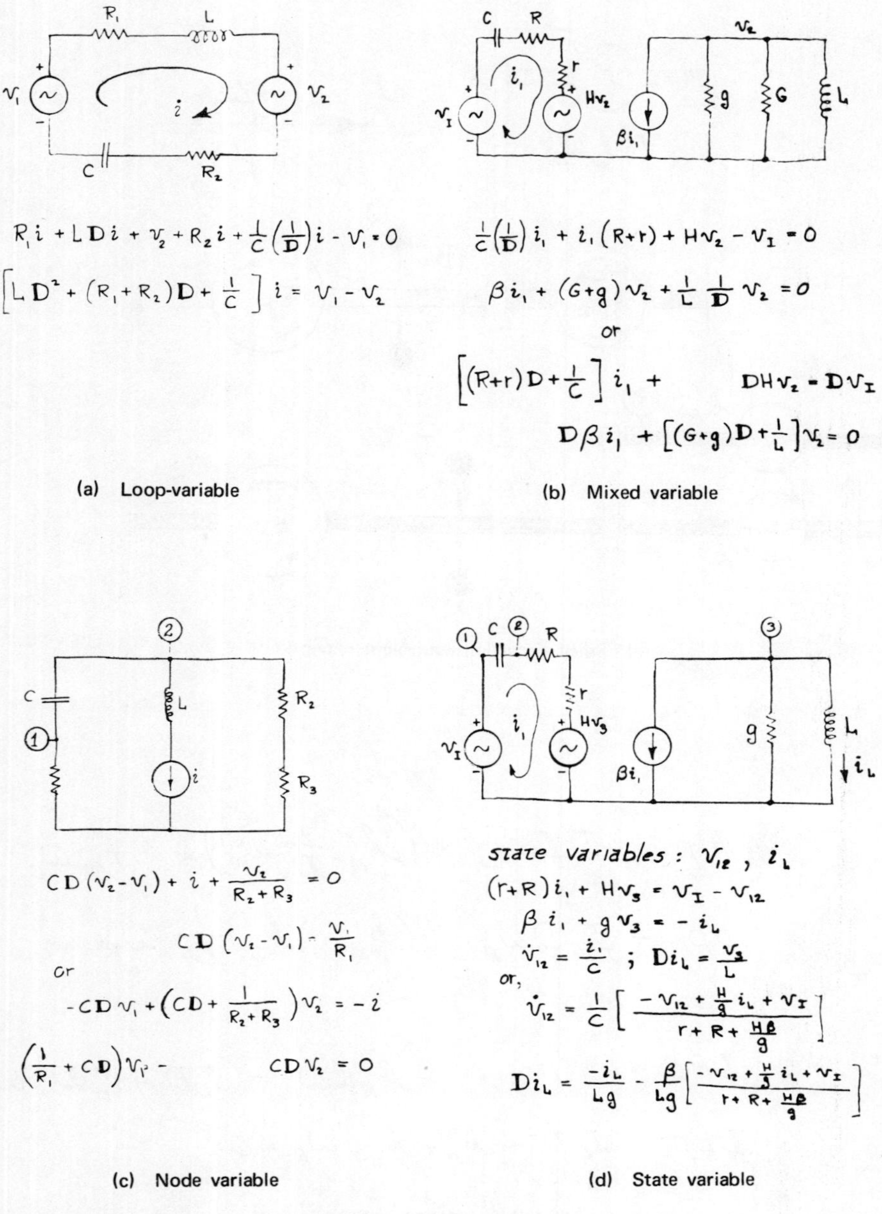

From the Engineer's Sketch Book

Figure 5.15 Small excursion dynamic equations for electrical networks.

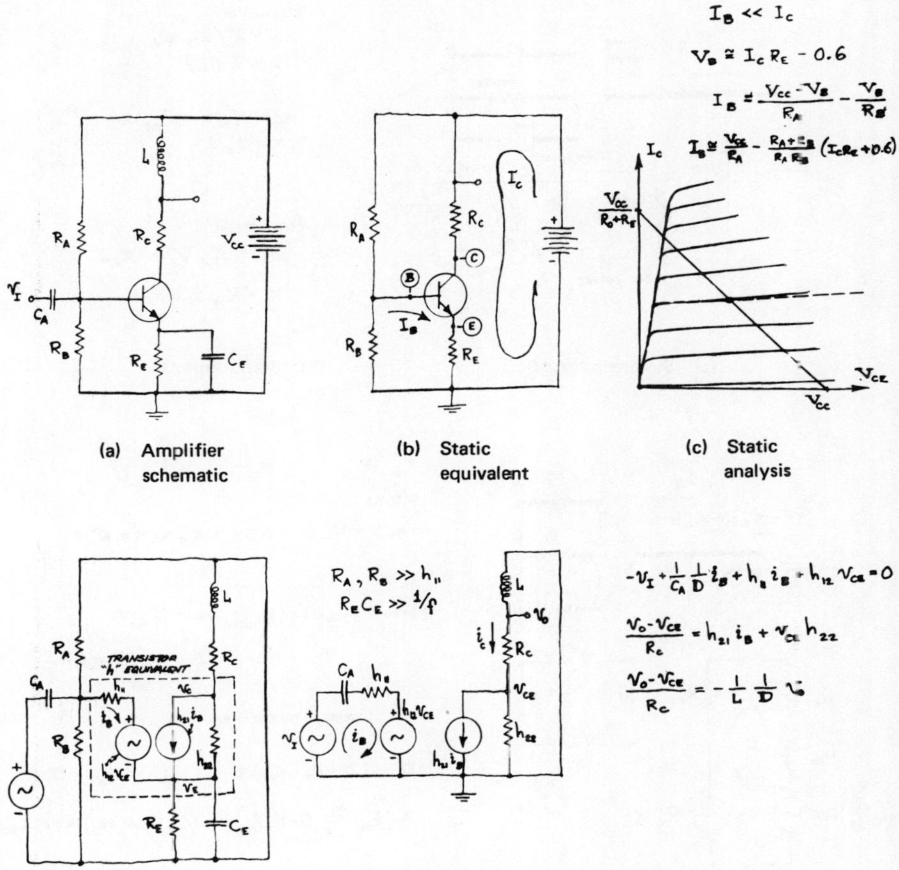

(a) Amplifier schematic

(b) Static equivalent

(c) Static analysis

(d) Complete small-excursion equivalent

(e) Simplified small-excursion equivalent

(f) equations of the simplified network

From the Engineer's Sketch Book

Figure 5.16 Dynamic equations for a transistor amplifier.

R_A and R_B are very small with respect to that through h_{11} and that C_E does, in fact, act as a short circuit in the frequency range of interest. The same might have been done with C_A; let us assume that the purpose of the analysis is to determine the effect of C_A on circuit performance. A mixed form of analysis is shown at (f); three equations in the variables i_B, v_{CE}, and v_O describe the circuit performance.

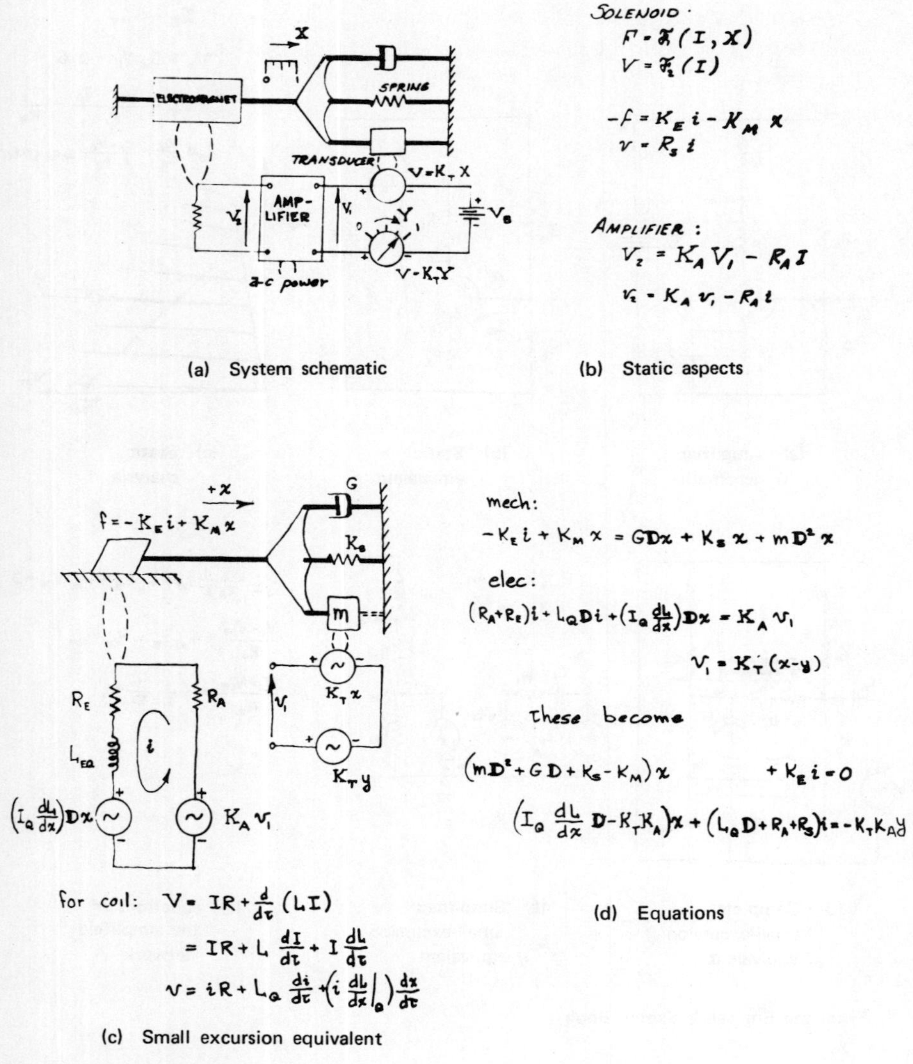

From the Engineer's Sketch Book

Figure 5.17 Dynamic equations for an electromagnetic system.

5.9 Mixed Systems

No new principles are involved in determining the equations for a mixed system, such as frequently occurs in system engineering. A typical case is shown in

Figure 5.17; the system considered is the electromagnetic positioner for which a static analysis was carried out in Chapter 4. One unusual aspect of this system is associated with the behavior of a coil containing a moving plunger; a voltage is generated through the change in inductance, as shown at the bottom of (c). For small excursions, the coil is represented by an equation having three terms. Thus, the equivalent network has three devices, one of which is a voltage generator producing a signal proportional to the speed of the plunger in the coil. The solenoid can only create tension in the bar which it operates; hence, the force generator equivalent to the mechanical side of the solenoid is a tension generator. The mass shown on the X transducer arm is that of all the moving parts of the system, not just that of the transducer itself. Once the equivalent network is determined, the equations may be written by the usual methods, as shown at (d).

5.10 Hydraulic Systems

In most hydraulic systems, the fundamental dynamic aspects of the system arise from the associated mechanical devices. Of course, there are notable exceptions where the fluids themselves create the dynamics. For example, in a hydroelectric plant, the inertia and compressibility of the water in the penstocks, the surge tanks at the top of the penstocks, and other fluid effects are very significant. Figure 5.18 shows a more typical case, however. The network shown [by equivalent schematics at (a)] is the basic part of a hydraulic positioner. As the valve spool moves up and down, the ram forces change and positioning may be carried out. The variable orifice form of the schematic is the most helpful for analysis and understanding. Note that the spool motion operates each orifice simultaneously, two opening and two closing with any motion. For good control, no orifice may close completely; thus, an underlapped form of valve must be used. The characteristics of one orifice are shown at (b); this represents an orifice which opens as X increases. The static analysis of the system is shown at (c). The ram takes no flow when it is stationary; therefore, the flow through any pair of orifices (such as a and b) must be equal and opposite. The total pressure drop must be the supply pressure. Therefore, the superposed characteristics of (c), in which one set of curves is actually from the third quadrant of the "closing" orifice characteristic and in which the two flows must be identical in magnitude, represents the static equilibrium. Although the characteristics are very non-linear, a small excursion linearization near the central equilibrium would be justified, as shown. This leads to the equivalent network of (d) and to the system equations shown there. Note that the ram has different areas on each side of the piston due to the piston rod; this has significant effects on the nature of the analysis and requires that a "make-

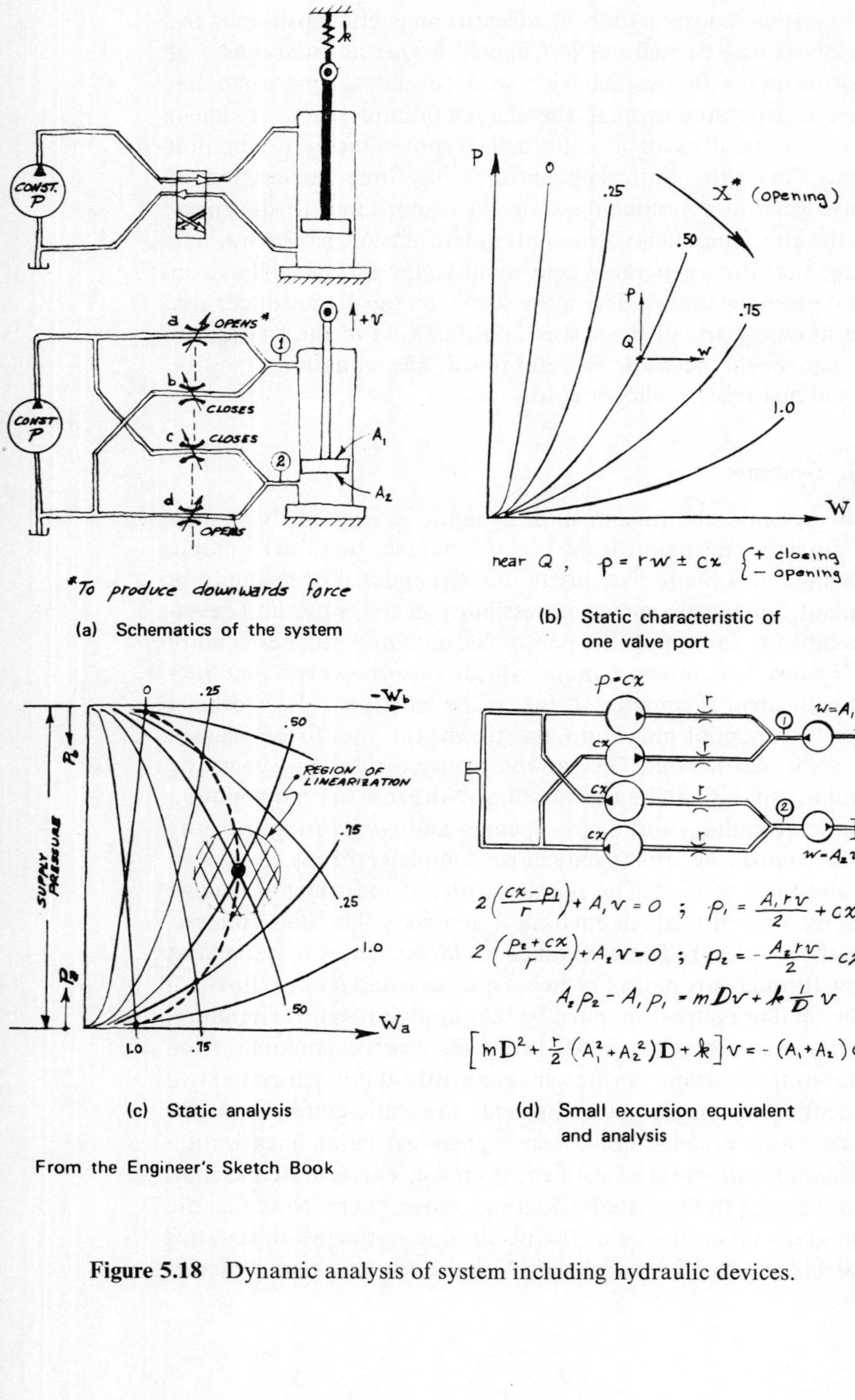

Figure 5.18 Dynamic analysis of system including hydraulic devices.

up" connection to the sump be provided. The m in the final equation is intended to represent the mass of all the moving parts of the system which move at a speed v.

Problems

1. Draw the small excursion equivalent networks for the systems of the following figures. (Be sure to use two-terminal devices exclusively.)
(1) Figure 2.9(a); include the effect of the mass of the moving parts of the relay and the spurious (parallel) capacitance of the diode.
(2) Figure 2.9(b); include the spurious capacitance of both diodes in the transistor.
(3) Figure 2.9(c); include the effect of the mass of the spools in the valves.
(4) Figure 2.9(d); include the masses associated with all the dissipators and clutches.
(5) Figure 2.24; include both transistor and diode capacitances.
(6) Figure 4.19(a); include the engine, governor, and load inertias, the springiness of the driveshaft between the engine and load, and the mass of the throttle linkage.

2. Many networks including exchangers may be simplified, for purposes of dynamic analysis, by removing the exchanger from the equivalent network and changing all the "downstream" conponents to suit. Supposing that the upstream terminals of an exchanger are A and the downstream terminals B and that the exchanger provides a Δ variable increase of ratio n from A to B, what component of what value would have to be attached to the network at A in order to replace the effect of the exchanger and the following device connected at B:
(1) Flywheel of inertia J?
(2) Rotary dissipator of value G?
(3) A resistor R?
(4) A capacitor C?
(5) An inductor L?

2. What is the single device equivalent to the following combinations:
(1) Series springs, k_1 and k_2?
(2) Parallel springs, k_1 and k_2?
(3) Series rotary dissipators, g_1 and g_2?
(4) Parallel rotary dissipators, g_1 and g_2?
(5) Parallel rotary inertias, J_1 and J_2?

4. Can any string of resistors, capacitors, and inductors be replaced, for purposes of small excursion dynamic analysis, by a string of one resistor, one capacitor, and one inductor?

5. What is the equivalent of two flywheels (j_1 and J_2) mounted end to end on the same shaft? What is the equivalent of two differential flywheels [as in Figure 1.15(b); call these J_3 and J_4] similarly mounted?

6. Suppose a spur gear with n_1 teeth and inertia J_1 drives another gear with n_2 teeth and inertia J_2. What is the inertia of the combination, as seen from the #1 end?

7. Do position, velocity, torque, and state-variable analyses of the system shown in Figure 5.19.

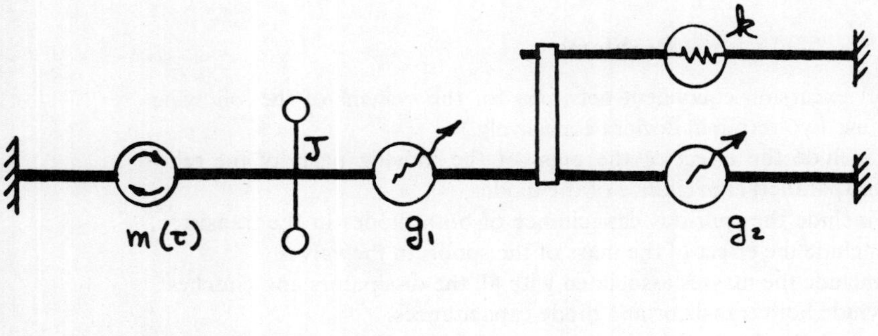

Figure 5.19

8. Do current, voltage, and state-variable analyses of the circuit shown in Figure 5.20.

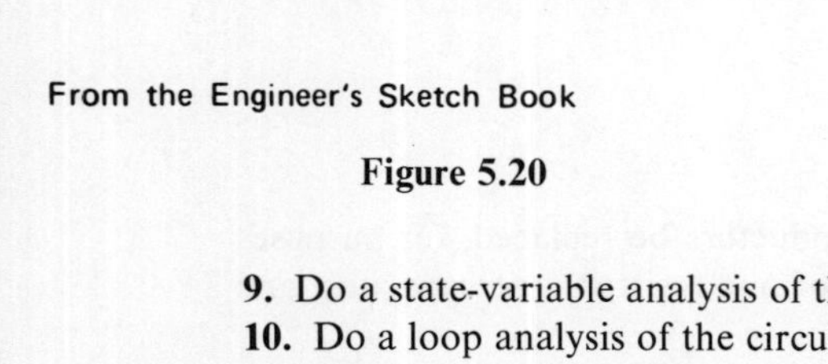

Figure 5.20

9. Do a state-variable analysis of the network of Figure 5.14(a).

10. Do a loop analysis of the circuit of Figure 5.15(c).

11. Write the state-variable equations for the equivalent system of Figure 5.17(c).

Note: In the following problem, the instructor should select specific networks from those shown in the figures.

12. Write the state-variable equations for the equivalent systems shown in Figures 5.21 and 5.22.

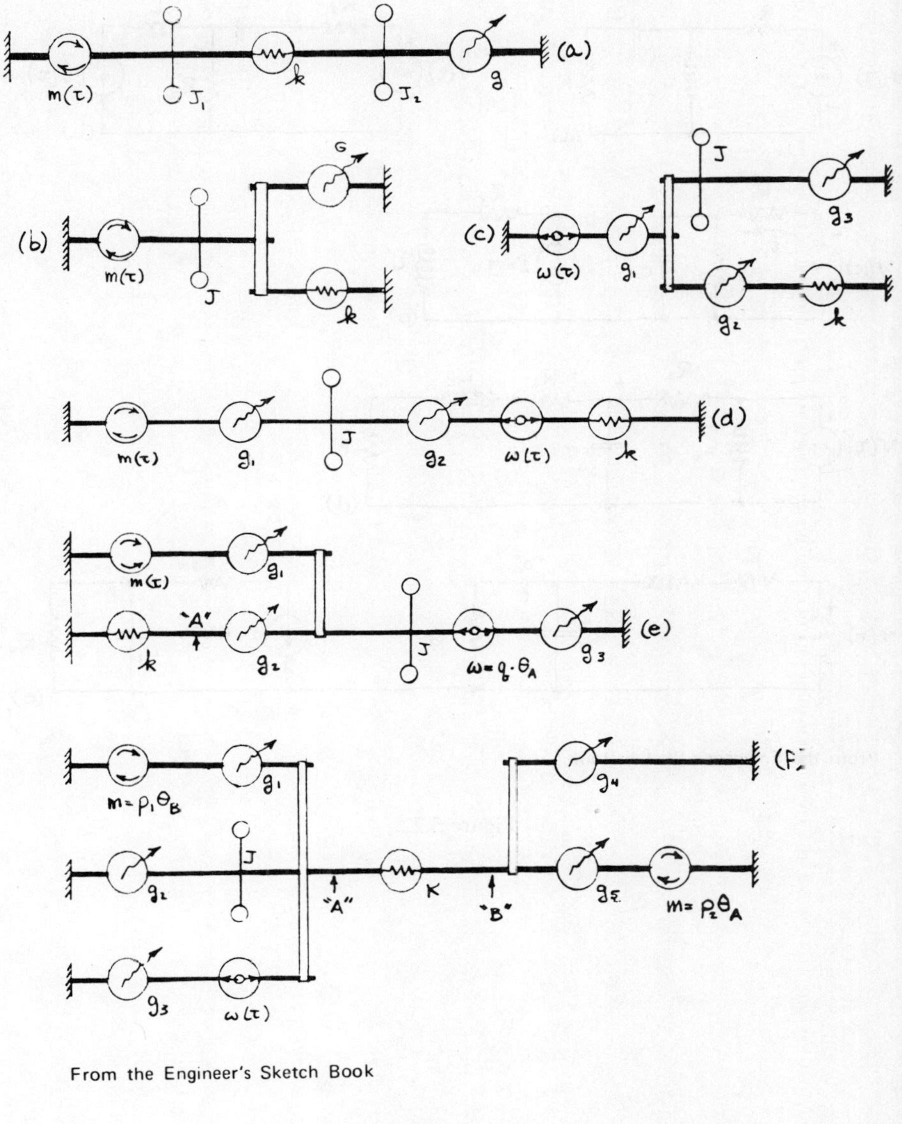
(a)
m(τ)
J₁
k
J₂
g
(b)
m(τ)
J
G
k
(c)
ω(τ)
g₁
J
g₃
g₂
k
(d)
m(τ)
g₁
J
g₂
ω(τ)
k
m(τ)
g₁
"A"
k
g₂
J
ω = q·θ_A
g₃
(e)
m = ρ₁θ_B
g₁
g₂
J
"A"
K
"B"
g₄
(F)
g₅
m = ρ₂θ_A
g₃
ω(τ)
From the Engineer's Sketch Book

Figure 5.21

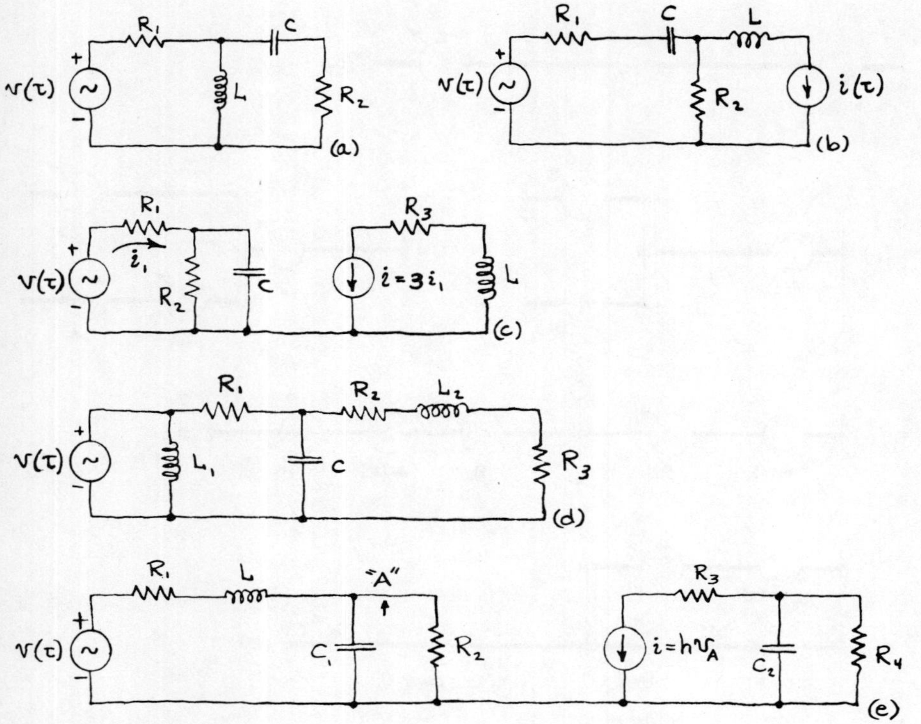

From the Engineer's Sketch Book

Figure 5.22

6 Methods of small excursion dynamic analysis

The small excursion dynamic behavior of a network is invariably described by a set of linear, ordinary differential equations with constant coefficients if analysis has been carried out by the methods of the preceding chapter. The techniques by which such equations may be solved are the subjects of this chapter. Since there is quite a bit of strictly mathematical work to be accomplished here, we will temporarily deviate from the pattern of the earlier portions of the book wherein considerable attention was given to physical devices and specific examples concerning them. When we have finished introducing the necessary mathematics, we can return to the physical world in which our real interest lies.

Chapter 5 included many examples of the kinds of equation which we must solve in dynamic analysis. A simple example is

$$3\mathbf{D}^2x + 5\mathbf{D}x + 10x = 0.$$

This is a specific case of the general form

$$a_n\mathbf{D}^n x + a_{n-1}\mathbf{D}^{n-1}x + \cdots a_1\mathbf{D}x + a_0 x = 0.$$

Such an equation represents a system in the absence of any externally applied disturbance. We refer to the solution of such an equation as the "natural" response of the system. As mentioned in Chapter 2, this might be quite dramatic since it could be oscillation, vibration, or self-destruction. The determination of natural response is one of the very fundamental steps in all system design. Also, it is of even greater importance in the design of feedback control systems, since these systems are notoriously subject to instability or self-generated oscillation.

The general form of equation above includes a variety of cases. The fundamental distinction between these cases is associated with the number of derivatives which appear. These cases are referred to by the "order" of the equation, which is the greatest exponent of **D** which appears. Thus, the specific example given first is of the second order while the general form represents an *n*th order equation.

In order to conserve space and make solution procedures as efficient as possible, the general form above is frequently written

$$A(\mathbf{D})x = 0$$

where the expression $A(\mathbf{D})$ obviously is a shorthand way of writing the polynomial

$$a_n \mathbf{D}^n + a_{n-1}\mathbf{D}^{n-1} + \cdots a_1\mathbf{D} + a_0 .$$

We shall use this convenience frequently in the following work.

Sometimes the equation to be solved has the form of

$$\mathbf{D}^2 x + 2\mathbf{D}x + 10x = y(t)$$

where $y(t)$ is some specified function of time. This function represents some action which is being applied to the system and thereby causing it to react; we refer to the function $y(t)$ as the "forcing function" for obvious reasons. The general form of such an equation could be

$$A(\mathbf{D})\, x = y(t)$$

and might also have the appearance

$$A(\mathbf{D})x = B(\mathbf{D})y$$

if the network arrangement and selection of variables did not eliminate the operator $B(\mathbf{D})$.

In most interesting system analysis problems, more than one variable is involved and it is necessary that more than one equation be written. A specific case might be:

$$\begin{aligned} (\mathbf{D}^2 + k)\, x + \quad 9v &= 0 \\ 13x + (\mathbf{D} + 3)v &= 0. \end{aligned}$$

These equations would have to be solved simultaneously. This is an example of the general two-variable case:

$$\begin{aligned} A_{11}(\mathbf{D})x_1 + A_{12}(\mathbf{D})x_2 &= 0 \\ A_{21}(\mathbf{D})x_1 + A_{22}(\mathbf{D})x_2 &= 0. \end{aligned}$$

The extension of the general form to greater numbers of variables is obvious. Although we do not have the space in this text to introduce the subject, the use of matrix techniques for representing and manipulating such equations is very effective and to be recommended.

If a two-variable system is subject to forcing, the equations become

$$A_{11}(\mathbf{D})x_1 + A_{12}(\mathbf{D})x_2 = B_1(\mathbf{D})y_1$$
$$A_{21}(\mathbf{D})x_1 + A_{22}(\mathbf{D})x_2 = B_2(\mathbf{D})y_2$$

where the extension to higher numbers of variables is again obvious.

If the system equations have been formulated by using state variables, the general form of the equations to be solved will be (using a two-variable case):

$$\mathbf{D}x_1 = a_{11}x_1 + a_{12}x_2 + b_1y_1$$
$$\mathbf{D}x_2 = a_{21}x_2 + a_{22}x_2 + b_2y_2 \qquad a\text{'s, } b\text{'s constants}$$

Here, the x's are the state variables and the y's the forcing functions. The order of such a set of equations is equal to the number of state variables. Again, matrix techniques can be very useful, especially in complex problems, but are not essential to our work here. The form of higher order state-variable equations is a simple extension of the above.

Fortunately, the techniques for solving all the kinds of system equations shown above are essentially the same. We will start with a consideration of the first general form given, but will then show that the techniques applicable to it are appropriate in the other cases.

The general kinds of question that are answered by small excursion analysis of this nature are represented by the following: Is a certain static equilibrium stable? How fast will the system return to equilibrium after a disturbance? How far away from equilibrium will a system get in response to a given disturbance? Will a network that is supposed to be an oscillator actually operate? Will the amplitude of oscillation of such a network be constant? The answers to such questions are often the central points of system design; the techniques we develop here are of the foremost importance in all system engineering.

In developing these methods, the author has taken as much care as was practically possible to show the intuitive background as well as the operational details of each approach. Also, an attempt has been made to show that no one method has magical powers; specifically, it is shown that the classical **D** operator approach can be just as powerful as the more commonly favored Laplace transform. It is the author's opinion that the student should understand both techniques in order to be prepared to practice system engineering, and that an understanding of both will increase his facility with either. This chapter presents both methods in detail and concludes with an interpretive comparison. As preparation for these developments, we open with a discussion of the simple assumed solution (Ae^{rt} forms) method which may be used in many cases.

6.1 Assumed Solutions

One of the simplest and most effective methods for solving sets of linear ordinary differential equations with constant coefficients depends on the remarkable versatility of the function e^{rt} when r is allowed to be some real or imaginary constant. As will be shown, e^{rt} is capable of representing exponential decay or rise, sinusoids and decaying or increasing sinusoids. These turn out to be the components of all the natural responses of linear systems and are frequently the forced responses as well. Since the time derivative of e^{rt} is simply re^{rt}, the manipulations required in solving linear system equations turn out to be extremely simple. In this section the techniques of using e^{rt} type solutions are developed.

6.1(a) Properties of the Function e^{rt}

The function e^{rt}, when r is real and negative, is the familiar exponential decay shown in Figure 6.1(a). If r is positive, an exponential rise, as shown in Figure 6.1(b), is represented. The less familiar situation where r is a complex number will be encountered very frequently in the work to come.

The basic relationship leading to an interpretation of e^{rt} when r is a complex number* is

$$e^{j\theta} = \cos\theta + j\sin\theta \qquad j = \sqrt{-1}. \tag{6.1}$$

This is derived from the series expansion definitions of the functions e^x, cos x, and sin x, and may be readily verified from examination of those definitions.

Available directly from equation (6.1) is the following form for a negative value of the exponent:

$$e^{-j\theta} = \cos\theta - j\sin\theta. \tag{6.2}$$

This is a result of the fact that $\cos\theta = \cos(-\theta)$ and $\sin\theta = -\sin(-\theta)$. Equations (6.1) and (6.2) may be manipulated to determine the following relationships:

$$\cos\theta = \frac{e^{j\theta} + e^{-j\theta}}{2} \tag{6.3}$$

*Complex numbers arise in system analysis as the solutions of polynomials in the variable r. For example, the solution of $r^2 + 1 = 0$ is $r = \sqrt{-1}$. Although the $\sqrt{-1}$ is an imaginary number, it is a very useful mathematical device (as we shall see). It is commonly, and here, denoted by the shorthand symbol j. Sometimes the solution of an equation for r gives a result which has both a real and an imaginary portion. For example, the solution of $r^2 - 2r + 5 = 0$ is $r = 1 + 2j$ and $r = 1 - 2j$; both of these are complex numbers. In the addition and subtraction of complex numbers, which we must do in this work, the parts must be kept separate.

$$\sin\theta = \frac{je^{j\theta} - je^{-j\theta}}{2} = -\frac{e^{j\theta} - e^{-j\theta}}{2j}. \tag{6.4}$$

If this material is new to the reader, he may be concerned about the appearance of an imaginary quantity (as $j \sin \theta$) in an expression which is supposed to represent the behavior of some real system. For example, if we

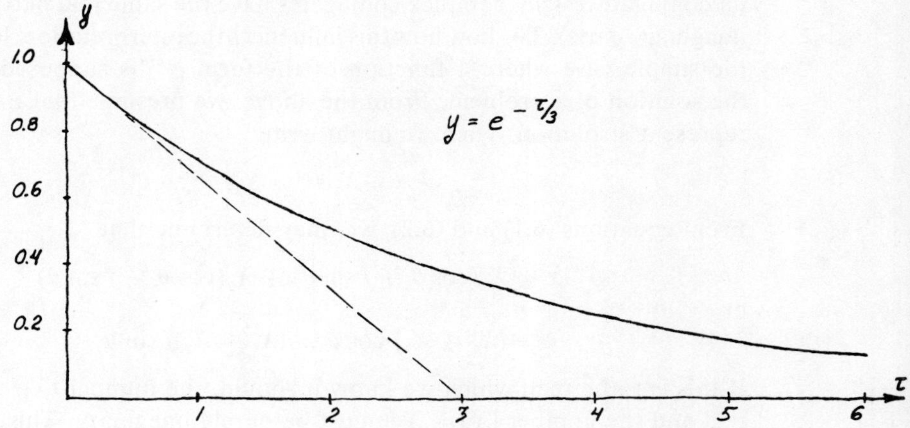

(a) Exponential decay

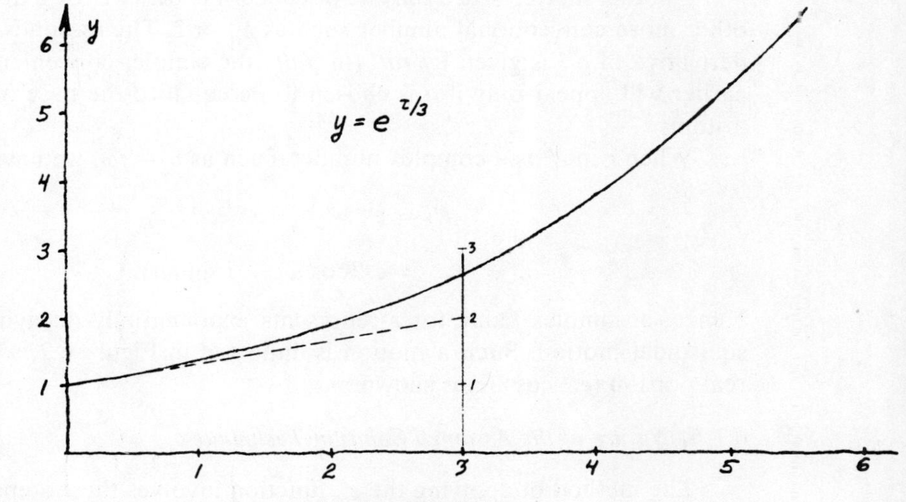

(b) Exponential rise

From the Engineer's Sketch Book

Figure 6.1 The function e^{rt} when r is real.

determined from theoretical analysis that a variable x was equal to $Xe^{j\omega t}$, where X and ω are constants, we would be led to believe that $x = X(\sin \omega t + j \sin \omega t)$; i.e., that x was partially imaginary. This would be both confusing and incorrect. The reason this is so is that complex numbers always arise, in problems of the nature we are considering, in pairs which are "complex conjugates." This means that if the complex number $a + jb$ arises, then so must its conjugate $a - jb$ (complex conjugates have the same real part and opposite imaginary parts). To show how this influences the interpretation, let us consider the simple case where a function of the form e^{jwt} is supposed to represent the solution of a problem. From the above, we presume that $e^{-j\omega t}$ must also represent a solution. Then we might write

$$x = X_1 e^{j\omega t} + X_2 e^{-j\omega t}.$$

From equations (6.1) and (6.2), we may determine that

$$x = X_1(\cos \theta + j \sin \theta) + X_2(\cos \theta - j \sin \theta)$$

or

$$x = (X_1 + X_2) \cos \theta + (X_1 - X_2)j \sin \theta.$$

If this is to be real, which we know it should, the number $(X_1 + X_2)$ must be real and the number $(X_1 - X_2)$ must be purely imaginary. This is the case if, and only if, the numbers X_1 and X_2 are themselves complex conjugates.

Another matter which may be of concern is why we use e instead of some other more conventional number such as 10 or 2. The reason is that the time derivative of a^{rt} is given by $ra^{rt}(ln\ a)dt$; the simple, convenient form given earlier will appear only if a is chosen to be equal to the base of natural logarithms.

When r in e^{rt} is a complex number, such as $\sigma + j\omega$, we have

$$e^{rt} = e^{(\sigma + j\omega)t} = e^{\sigma t}\, e^{j\omega t}$$

$$= e^{\sigma t}(\cos \omega t + j \sin \omega t).$$

Hence, a complex value for r represents exponentially decaying (or rising) sinusoidal motion. Such a motion is indicated in Figure 6.2, where only the real portion ($e^{\sigma\tau} \cos \omega\tau$) is shown.

6.1(b) Nature of the Assumed Solution Technique

The method of applying the e^{rt} function involves these steps:

(a) We guess that the solution has the form $A\, e^{rt}$, where A is some constant (which may be a complex number).

(b) We insert the guess in the differential equations. After some manipulations, an algebraic expression for r appears.

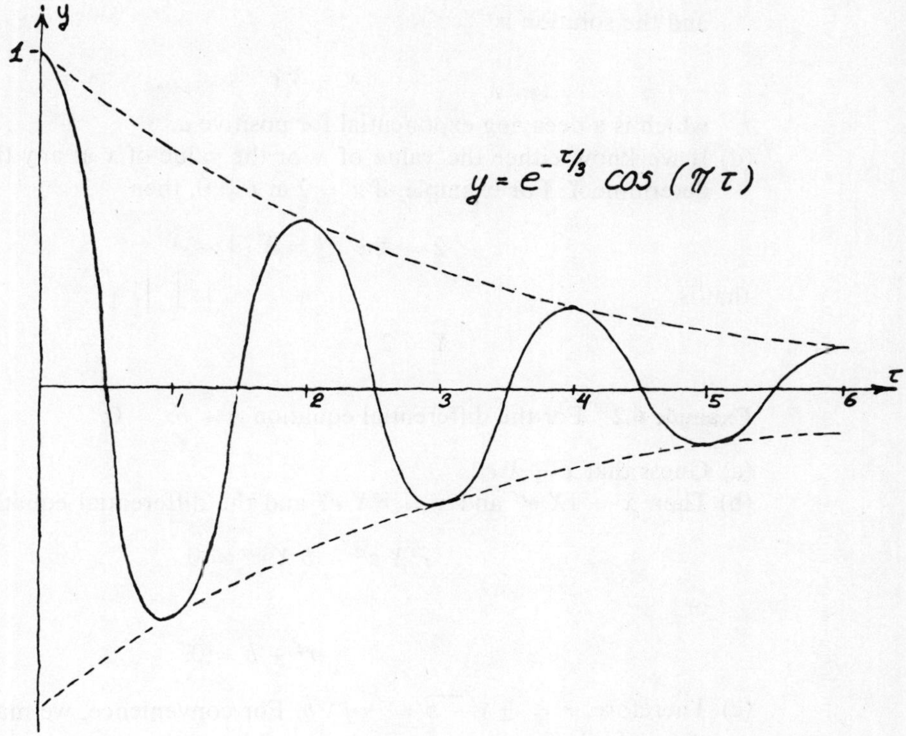

From the Engineer's Sketch Book

Figure 6.2 Exponentially decaying sinusoidal motion.

(c) The equation for r is solved. The form of the motion is then known.
(d) The value of A is determined from the starting conditions, or in some cases, from ending or intermediate conditions.

Example 6.1 For the differential equation $\dot{x} + ax = 0$:

(a) Guess that $x = X\,e^{rt}$ (X is a constant).
(b) Then $\dot{x} = Xr\,e^{rt}$ and the differential equation may be transformed into

$$Xr\,e^{rt} + aX\,e^{rt} = 0$$

or

$$r + a = 0.$$

(c) Therefore,

$$r = -a,$$

and the solution is

$$x = X\,e^{-at}$$

which is a decaying exponential for positive a.

(d) If we know either the value of x or the value of $\dot{x}$ at any time, we may determine X. For example, if $x = 2$ at $t = 0$, then

$$2 = X\,e^{-a0} = X \cdot 1 = X$$

that is,

$$X = 2.$$

Example 6.2 For the differential equation $\ddot{x} + bx = 0$:

(a) Guess that $x = X\,e^{rt}$.

(b) Then $\dot{x} = rX\,e^{rt}$ and $\ddot{x} = r^2X\,e^{rt}$ and the differential equation becomes

$$r^2X\,e^{rt} + b\,X\,e^{rt} = 0$$

or

$$r^2 + b = 0.$$

(c) Therefore, $r = \pm\sqrt{-b} = \pm j\sqrt{b}$. For convenience, we may write ω in place of $\sqrt{b}$. Then, there are two possible solutions:

$$x = X_1\,e^{+j\omega t}$$

and

$$x = X_2\,e^{-j\omega t}.$$

Since either one will satisfy the differential equation, a sum of the two will also. (This may be seen from inspection of the equation.) In order to be complete, we must write the sum, which we may do as follows:

$$x = X_1\,e^{+j\omega t} + X_2\,e^{-j\omega t}.$$

This may be rewritten in terms of sinusoids,

$$\begin{aligned} x &= X_1\,(\cos\omega t + j\sin\omega t) + X_2[\cos(-\omega t) + j\sin(-\omega t)] \\ &= X_1\,(\cos\omega t + j\sin\omega t) + X_2(\cos\omega t - j\sin\omega t) \\ &= (X_1 + X_2)\cos\omega t + j(X_1 - X_2)\sin\omega t. \end{aligned}$$

This represents pure sinusoidal motion, but is not usable as is because of the imaginary quantities which appear. Let us go on to determine X_1 and X_2 in order to see if any clarification might appear.

(d) In order to determine X_1 and X_2, let us assume that at $t = 0$, $x = 3$ in. and $\dot{x} = 10$ in./sec. From

$$x = (X_1 + X_2)\cos\omega t + j(X_1 - X_2)\sin\omega t$$

we have

$$x = -\omega(X_1 + X_2)\sin\omega t + j\omega(X_1 - X_1)\cos\omega t.$$

Then

$$3 = X_1 + X_2$$
$$10 = j\omega(X_1 - X_2)$$

so that

$$2X_1 = 3 + \frac{10}{j\omega}, \qquad X_1 = 1.5 - j\frac{5}{\omega}$$

$$2X_2 = 3 - \frac{10}{j\omega}, \qquad X_2 = 1.5 + j\frac{5}{\omega}$$

and

$$x = 3\cos\omega t + \frac{10}{\omega}\sin\omega t.$$

This solution contains no imaginary quantities and is therefore acceptable as an answer to a physical problem.

It is instructive to review this portion of the analysis, noting that the final solution would have contained no imaginary terms *no matter what initial conditions were chosen*. This may be shown simply by writing the general quantities $x(0)$ and $\dot{x}(0)$ in place of "3" and "10" in part (d). Even though the equations at (c) above appear to include imaginary terms, they do not since $(X_1 + X_2)$ is real, $(X_1 - X_2)$ is imaginary.

The student may feel that a better guess to the solution of this problem would have been $x = X_3 \cos \Omega t + X_4 \sin \Omega t$ since no imaginary numbers would have to be used. Analysis would yield $\Omega = \omega$ and $X_3 = 3$, $X_4 = 10/\omega$. However, this guess would be appropriate for a limited kind of linear system equations and is therefore not a good approach for general use.

Example 6.3 The equation of motion for a spring-mass-dashpot system is $m\ddot{x} + b\dot{x} + kx = 0$. In order to solve the equation, we assume that $x = Xe^{rt}$. Taking numerical values for m, b, and k as follows,

$$m = 1 \text{ lb sec}^2/\text{in.}$$
$$k = 100 \text{ lb/in.}$$
$$b = 3 \text{ lb sec/in.}$$

The equation becomes, using $\dot{x} = Xr\,e^{rt}$, $\ddot{x} = Xr^2\,e^{rt}$,

$$Xr^2\,e^{rt} + 3Xr\,e^{rt} + 100X\,e^{rt} = 0$$

$$r^2 + 3r + 100 = 0$$

then

$$r = \frac{-3 \pm \sqrt{9 - 400}}{2} = -1.5 \pm \sqrt{\frac{-391}{4}}$$
$$= -1.5 \pm j\,9.9.$$

The solution is

$$x = e^{-1.5t}\,(X_1\,e^{j9.9t} + X_2\,e^{-j9.9t}).$$

Taking the initial conditions to be $x(10) = 1$ and $\dot{x}(0) = 0$,

$$1 = X_1 + X_2$$

from

$$\dot{x} = rX\,e^{rt}$$

$$\dot{x} = (-1.5 + j9.9)e^{-1.5t}\,X_1\,e^{j9.9t} + (-1.5 - j9.9)e^{-1.5t}\,X_2\,e^{-j9.9t}$$

$$0 = (-1.5 + j9.9)X_1 + (-1.5 - j9.9)X_2$$
$$= -3(X_1 + X_2) + j9.9(X_1 - X_2)$$
$$= -3 + j9.9(X_1 - X_2)$$
$$-j\frac{3}{9.9} = X_1 - X_2$$

$$x = e^{-1.5t}\,[(X_1 + X_2)\cos 9.9t + j(X_1 - X_2)\sin 9.9t]$$
$$x = e^{-1.5t}\,[\,\cos 9.9t + \frac{3}{9.9}\sin 9.9t\,].$$

Note that it has been more convenient to work directly in terms of the exponential functions than to change over to the trigonometric form at the beginning.

The analysis indicates that the motion of the system is a decaying sinusoid having a frequency of 9.9 rad/sec.

In general, in the e^{rt} type of analysis, the equation from which r is determined is a polynomial in r. It is obvious from the fact that $(d^n/dt^n)e^{rt} = r^n e^{rt}$ that this polynomial may be written directly from the differential equation as indicated by the following:

$$a_n \frac{d^n x}{dt^n} + a_{n-1} \frac{d^{n-1} x}{dt^{n-1}} + \ldots + a_2 \ddot{x} + a_1 \dot{x} + a_0 x = 0$$

$$a_n r^n + a_{n-1} r^{n-1} + \ldots + a_2 r^2 + a_1 r + a_0 = 0.$$

Here we are temporarily concerned only with those cases where the right-hand side of the equation is zero; i.e., where the forcing function is absent.

It is well known that an nth degree polynomial has n solutions; therefore, an nth order differential equation leads to a solution having n terms; i.e., the above equation would have a solution of the form

$$x = X_1 e^{r_1 t} + X_2 e^{r_2 t} + \ldots + X_{n-1} e^{r_{n-1} t} + X_n e^{r_n t}.$$

In order to determine all the X's, we will need n conditions on the solution. For example, $d^{n-1}x/dt^n$, d^{n-2}/dt^{n-2}, . . . , $\ddot{x}$, $\dot{x}$, and x might be specified at $t = 0$. We would then have n linear simultaneous algebraic equations in n unknowns (X_1, X_2, etc.) and, except for special cases, would be able to determine the solution. Notice that we should never try to specify more than n initial conditions on the solution because the differential equation must be satisfied by the conditions. Since there are $n + 1$ terms in the differential equation, the specification of only n conditions allows one term to be free; this term may then take whatever value is necessary to satisfy the equation. In those cases where the conditions set on the solution occur at different times, the proper number of conditions is still n. Any other choice will make a solution for the X's impossible to achieve in general.

The polynomial in r, obtained by substituting a guess of the form $A e^{rt}$ in a differential equation containing no forcing function, is known as a "characteristic equation." This equation is used frequently and will be discussed further in later sections.

There are several aspects of the e^{rt} form of analysis which must still be considered. Specifically, we have not considered the case where a forcing function appears, where there is more than one variable and equation, and also some special situations which arise. These matters are taken up in the following paragraphs.

6.1(c) Assumed Solutions When Forcing Is Present

When the forcing function in a linear system equation is not zero, the solution to the equation is made up of two parts as follows:

(1) The solution to the equation with the forcing function set equal to zero (the "complementary" solution). This represents the "natural" behavior of the system; i.e., the behavior when there is no external influence.

(2) A solution which is related to the forcing function and is called the "particular" solution. This is the reaction to the external influence.

The complete solution of the equation is the sum of these parts. That the complementary solution is always a part of such a solution is clear from the following.

Take the equation

$$a_2\ddot{x} + a_1\dot{x} + a_0x = f(\tau)$$

If x_0 is a solution of

$$a_2\ddot{x} + a_1\dot{x} + a_0x = 0$$

then

$$a_2\ddot{x}_0 + a_1\dot{x}_0 + a_0x_0 = 0$$

If x_1 is a solution of the original equation ($x_1 \neq x_0$),

$$a_2\ddot{x}_1 + a_1\dot{x}_1 + a_0x_1 = f(\tau)$$

Then

$$a_2(\ddot{x}_0 + \ddot{x}_1) + a_1(\dot{x}_0 + \dot{x}_1) + a_0(x_0 + x_1) = f(t)$$

If $x = x_0 + x_1$, the original equation is satisfied.

If the forcing function can be represented by an expression of the form e^{rt} (or $t^m\, e^{rt}$), the e^{rt} approach may be applied. Otherwise, the method becomes relatively inconvenient. This is the reason that we go to other techniques in order to be able to solve *all* forced response problems.

In order to determine the particular solution of the differential equation, let us again guess a solution of the form $x = A\, e^{rt}$. We assume that the forcing is a term $Y\, e^{qt}$, where Y and q are constants, as follows:

$$a_n \frac{d^n x}{dt^n} + a_{n-1} \frac{d^{n-1}x}{dt^{n-1}} + \ldots + a_2\ddot{x} + a_1\dot{x} + a_0x = Y\, e^{q\tau}.$$

Assuming a particular solution $x = X\, e^{rt}$,

$$(a_n r^n + a_{n-1} r^{n-1} + \ldots + a_2 r^2 + a_1 r + a_0)Xe^{r\tau} = Ye^{q\tau}$$

We note that the a's, r, X, Y, and q are all constants. The only factors which change with time are e^{rt} and e^{qt}. If the equation is to be satisfied for all time, these factors must change by the same ratio in any time interval. The only way this can be true is if $r = q$. Therefore, *the particular solution associated with a forcing function e^{qt} must have a factor e^{qt}.* The equation then becomes

$$(a_n q_n^n + a_{n-1} q^{n-1} + \ldots + a_2 q^2 + a_1 q + a_0)X = Y$$

and the particular solution is completely determined.

Example 6.4 Consider the equation

$$\ddot{x} + 2\dot{x} + 9x = 7e^{-4t}.$$

The complementary solution:

$r^2 + 2r + 9 = 0$

$$r = \frac{-2 \pm \sqrt{4^2 - 3^6}}{2} = -1 \pm j4.48$$

$$x_c = X_1 e^{(-1 + j4.48)t} + X_2 e^{(-1 - j4.48)t}.$$

The particular solution:

– Guess $x_p = X_3 e^{-4\tau}$

$$(16 - 8 + 9)X_3 = 7$$

$$X_3 = \frac{7}{17}$$

The values for X_1 and X_2 would be determined from initial conditions as before.

If the forcing function has the form $t^m e^{qt}$, the appropriate form for the particular solution is

$$x_p = X_m t^m e^{qt} + X_{m-1} t^{m-1} e^{qt} + \ldots + X_1 t e^{qt} + X_0 e^{qt}$$

This is a summation of a term with the same time dependent factor as the forcing function and all the terms which may be derived from that factor by differentiation.

A number of special situations arise in application of the e^{rt} method. These are discussed in the following section. Proofs are not presented; the fact that the methods given work is assumed to be adequate justification for their use.

6.1(d) Special Cases in the Assumed Solution Technique

In some instances the characteristic equation leads to repeated values for a root. For example, the characteristic equation $(r + 1)^3(r + 1) = 0$ has $r = -1$ as root three times. In these cases there are as many distinct solutions associated with the root as there are repetitions of the root, and these are of the form

$$X_0 e^{rt} + X_1\tau e^{rt} + X_2\tau^2 e^{rt} + \ldots + X_{q-1}t^{q-1} e^{rt}$$

where r is the root which is repeated q times.

In some cases the values of the exponent q in a forcing function coincide with the values of roots r determined by the characteristic equation. If a guess for the particular solution of $x_p = X e^{qt}$ is inserted in the differential equation, it disappears. The guess in this case must be modified to be

$$x_p = t^p e^{qt}$$

where p is the number of times the number q is a root of the characteristic equation.

6.1(e) Multi-Variable Problems

As mentioned in the introduction, many system analysis problems involve several variables. No essential change in solution technique is required, however, as may be seen from consideration of the general two-variable case as follows:
For

$$A_{11}(\mathbf{D})x_1 + A_{12}(\mathbf{D})x_2 = B_1(\mathbf{D})y_1$$
$$A_{21}(\mathbf{D})x_1 + A_{22}(\mathbf{D})x_2 = B_2(\mathbf{D})y_2$$

the characteristic equation is developed, as usual, by assuming $y_1 = y_2 = 0$, $x_1 = X_1e^{rt}$, $x_2 = X_2e^{rt}$ so that

$$A_{11}(r)X_1 + A_{12}(r)X_2 = 0 \quad (6.5)$$
$$A_{21}(r)X_1 + A_{22}(r)X_2 = 0. \quad (6.6)$$

These equations may be combined into one by normal algebraic manipulations. We get

$$A_{11}(r)\,A_{22}(r)\,X_1\,X_2 - A_{21}(r)\,A_{12}(r)\,X_1\,X_2 = 0$$

so that the characteristic equation is

$$A_{11}(r)\,A_{22}(r) - A_{21}(r)\,A_{12}(r) = 0.$$

Note: The use of determinants is quite convenient in determining the characteristic equation in this, and other multi-variable, cases. Writing an expression, in determinants, for X_1 in the case that there is no forcing, we have

$$X_1 = \frac{\begin{vmatrix} O & A_{12}(r) \\ O & A_{22}(r) \end{vmatrix}}{\begin{vmatrix} A_{11}(r) & A_{12}(r) \\ A_{21}(r) & A_{22}(r) \end{vmatrix}}.$$

The numerator of the fraction must be zero since it includes a column in which all the entries are zero. The only way that X_1 can have a finite solution is for the denominator to be zero also. Thus,

$$\begin{vmatrix} A_{11}(r) & A_{12}(r) \\ A_{21}(r) & A_{22}(r) \end{vmatrix} = 0$$

must be true when there is no forcing. This represents a polynomial in r. The order of the polynomial is the greater of that of (A_{11} plus that of A_{22}) or (A_{21} plus that of A_{12}); this is easily shown to be the same as the order of the polynomial obtained by substitution. Thus, the determinant must produce the characteristic equation when evaluated.

From equations (6.5) and (6.6) it may be seen that

$$\frac{X_1}{X_2} = -\frac{A_{12}(r)}{A_{11}(r)} = -\frac{A_{22}(r)}{A_{21}(r)}.$$

Thus, for each value of the root r there is a specified ratio between X_1 and X_2, and the two constants are not independent. Assuming that the order of the characteristic equation is n, there will be n roots of the characteristic equation and n values of X_1 (and thus, n values of X_2) to be determined. These will come, as usual, from n specifications of initial (or other) conditions on the solution.

The procedure for determining the particular solution also goes in the usual way. For $y_1 = Y_1 e^{q_1 t}$ and $y_2 = Y_2 e^{q_2 t}$ the usual guesses for the x's may be made. Considering only the y_1 portion, we have

$$A_{11}(q_1)X_1 + A_{12}(q_1)X_2 = B_1(q_1)Y_1$$

$$A_{21}(q_1)X_1 + A_{22}(q_1)X_2 = 0.$$

Thus, there are two independent equations with two unknowns, X_1 and X_2. The determination of these is quite straightforward. (Again, determinants are useful.)

A matter which disturbs some students is the use of the same exponent in the guesses for the variables—in this case, x_1 and x_2. How do we know that the motions will be of the same form? We may look at the situation in this way: Suppose we assume that $x_1 = X_1 e^{rt}$, where r is some particular constant. We make no initial assumption about x_2. If we insert our guess for x_1 in the differential equations, the terms including e^{rt} may be thought of as forcing functions. These forcing functions act on x_2. It has already been shown that the response in such a situation *must have* the same exponent. Therefore, the guess for x_2 should also have the factor e^{rt}. This is just one way of saying that if x_1 contains some e^{rt} in its response, then x_2 must also contain some. Since the characteristic equation yields several values of r, there is a mathematical mechanism through which the various natural motions of the individual components may be considered.

Example 6.5 The system shown in Figure 6.3 is described by the pair of differential equations

$$(m_1 + m_2)\ddot{x} + kx + \frac{m_2 L}{2}\ddot{\theta} = 0$$

$$\frac{m_2 L}{2}\ddot{x} + \left(I + \frac{m_2 L}{2}\right)\ddot{\theta} + k_2 L^2 \theta = 0$$

The guesses

$$x = Xe^{rt}$$

$$\theta = \Theta e^{rt}$$

may be made. When these are substituted into the differential equations, the following algebraic equations result:

$$(m_1 + m_2)r^2 X + kX + \frac{m_2 L}{2} r^2 \Theta = 0$$

$$\frac{m_2 L}{2} r^2 X + \left(I + \frac{m_2 L}{2}\right) r^2 \Theta + k_2 L^2 \Theta = 0$$

$$[(m_1 + m_2)r^2 + k]X + \left[\frac{m_2 L r^2}{2}\right]\Theta = 0$$

or

$$\left[\frac{m_2 L r^2}{2}\right]X + \left[\frac{I r^2 + m_2 L r^2}{2 + k_2 L^2}\right]\Theta = 0$$

$$\frac{X}{\Theta} = \frac{-m_2 L r^2/2}{(m_1 + m_2)r^2 + k} = -\frac{I r^2 + m_2 L r^2/2 + k_2 L^2}{m_2 L r^2/2};$$

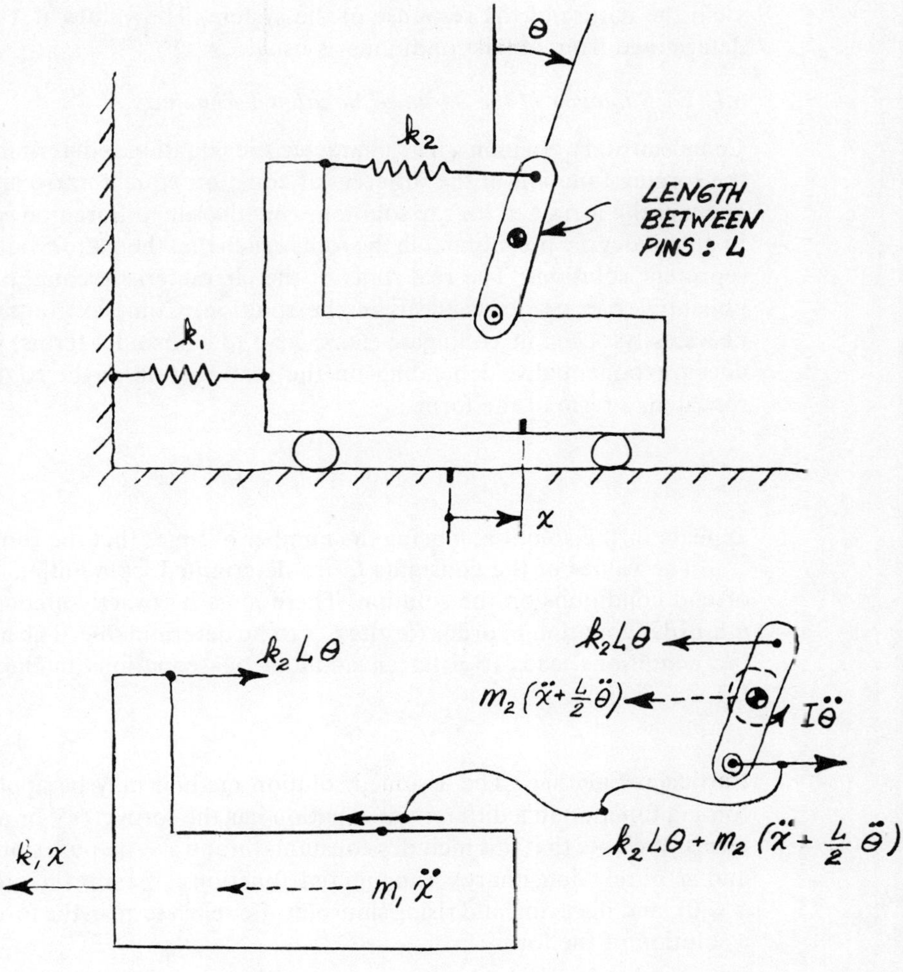

From the Engineer's Sketch Book

Figure 6.3 A system leading to two simulatneous equations in two variables.

therefore

$$[(m_1 + m_2)r^2 + k]\left[\frac{Ir^2 + m_2 Lr^2}{2 + k_2 L^2}\right] - \left(\frac{m_2 Lr^2}{2}\right)\left(\frac{m_2 Lr^2}{2}\right) = 0.$$

This equation, of fourth degree in r, is the characteristic equation of the system. The four roots (which will be two pair of imaginary roots in this case) will yield the nature of the response of the system. The values of X and Θ will be determined from initial conditions as usual.

6.1 (F) Summary of the Assumed Solution Technique

Complementary Solution The characteristic equation is determined by setting the forcing function in the differential equation equal to zero and inserting a guess of the form Ce^{rt} for the solution. An nth order differential equation leads to an nth degree polynomial in the roots r such that there are n values of r which represent solutions. The real roots of the characteristic equation lead to exponential rise or decay terms in the solution. Complex roots, which must necessarily occur in conjugate pairs, lead to sinusoidal terms, which rise or decay exponentially, depending on the real portion of the root. If a root is repeated, a term of the form

$$(C_1 + C_2 t + \ldots + C_h t^{h-1})e^{rt}$$

appears in the solution, h being the number of times that the root is repeated.

The values of the constants C are determined from initial, intermediate, or end conditions on the solution. There must be exactly n conditions on an nth order equation in order for the C's to be determinable. The application of the conditions leads to n linear simultaneous equations in the n unknowns $C_1, C_2, \ldots, C_n$.

Particular Solution The assumed solution method may be applied when the forcing function in a differential equation has the form $ct^m e^{qt}$ or a sum of such terms. We note that this includes constants (m and $q = 0$), pure sinusoids ($m = 0$ and q purely imaginary), exponential functions of time (m = any number, $q = 0$), and decaying and rising sinusoids. For each term in the forcing function a solution of the form

$$(C_0 + C_1\tau + C_2 t^2 + \ldots + C_m t^m)e^{qt}$$

is assumed. Insertion of this expression into the differential equation and cancellation of the e^{qt} factors leads to an algebraic equation for the C's. The coefficient of any particular power of t must be the same on each side of the equation, hence all the C's may be determined.

In the case that a term in the forcing function is the same as some term in the complementary solution multiplied by a constant, the assumption for the particular solution is multiplied through by t^k, where k is the lowest integer which will make the terms differ by a factor of t.

6.2 Manipulations of Differential Equations

In many situations it is convenient to be able to manipulate differential equations, or sets of differential equations, before a solution is attempted. In other cases, appropriate manipulation is all that is required in performing engineering analysis. The **D** operator represents a means for facilitating such manipulation; it also has an interesting relationship to the Laplace transform and its study represents further preparation for that method. It is possible to formulate all the commonly used aspects of the Laplace transform method in terms of the **D** operator; because of its simplicity and straightforwardness, the **D** operator is preferred over the Laplace transform by some engineers.

The **D** operator is defined, as before, so that

$$\mathbf{D}x = \frac{dx}{dt}.$$

Then

$$\mathbf{D}[\mathbf{D}x] = \mathbf{D}\left[\frac{dx}{dt}\right] = \frac{d^2x}{dt^2}$$

and we write

$$\mathbf{D}^2x = \mathbf{D}[\mathbf{D}x] = \frac{d^2x}{dt^2}.$$

A linear system equation may then be rewritten as

$$a_n\frac{d^nx}{dt^n} + a_{n-1}\frac{d^{n-1}x}{dt^n} + \cdots + a_2\frac{d^2x}{dt^2} + a_1\frac{dx}{dt} + a_0x = F(t)$$

$$(a_n\mathbf{D}^n + a_{n-1}\mathbf{D}^{n-1} + \cdots + a_2\mathbf{D}^2 + a_1\mathbf{D} + a_0)x = F(t).$$

The expression in parentheses is a linear operator and is of a kind that is commutative and distributive. That is, if

$$L_1 = a_n\mathbf{D}^n + a_{n-1}\mathbf{D}^{n-1} + \ldots + a_2\mathbf{D}^2 + a_1\mathbf{D} + a_0$$

$$L_2 = b_n\mathbf{D}^m + b_{m-1}\mathbf{D}^{m-1} + \ldots + b_2\mathbf{D}^2 + b_1\mathbf{D} + b_0,$$

then

$$L_1[L_2(x)] = L_2[L_1(x)]$$

and

$$(a_n\mathbf{D}^n + a_{n-1}\mathbf{D}^{n-1} + \ldots + a_2\mathbf{D}^2 + a_1\mathbf{D} + a_0)x$$

$$= a_n\mathbf{D}^n x + a_{n-1}\mathbf{D}^{n-1}x + \ldots + a_2\mathbf{D}^2 x + a_1\mathbf{D}x + a_0 x$$

as has been utilized already. Although these properties are not of much direct interest to the engineer, they lead to the following extremely important result:

The **D** *operator, or linear operators such as the L's above, may be manipulated as if they were algebraic quantities; i.e., all the rules of algebra, such as factorization, apply.* Note that it is not actually necessary to introduce **D** in place of d/dt; the writing is simply facilitated thereby.

To illustrate the use of this kind of manipulation of a differential equation, consider the equation

$$(\mathbf{D}^2 + 3\mathbf{D} + 2)x = 0.$$

The operator may be factored and the equation rewritten

$$\left[(\mathbf{D}+1)(\mathbf{D}+2)\right]x = 0$$

or simply

$$(\mathbf{D}+1)(\mathbf{D}+2)x = 0.$$

The order in which the operators are applied is immaterial. The equation will be satisfied if $(\mathbf{D}+1)x = 0$ or $(\mathbf{D}+2)x = 0$; hence, the second-order equation has been reduced to first order. It will be shown later that any first-order equation, irrespective of the nature of the right-hand side, may be solved; therefore, there is some potential here for a very powerful method.

As a second illustration, suppose that two variables are related by operators as follows:

$$A(\mathbf{D})x = B(\mathbf{D})y.$$

Then we should be able to write

$$x = \frac{B(\mathbf{D})}{A(\mathbf{D})}y \quad \text{or} \quad y = \frac{A(\mathbf{D})}{B(\mathbf{D})}x.$$

These expressions require interpretation of an expression having the form

$$\frac{b_n\mathbf{D}^n + b_{n-1}\mathbf{D}^{n-1} + \ldots + b_2\mathbf{D}^2 + b_1\mathbf{D} + b_0}{a_m\mathbf{D}^m + a_{m-1}\mathbf{D}^{m-1} + \ldots + a_2\mathbf{D}^2 + a_1\mathbf{D} + a_0}y. \tag{6.7}$$

This will be carried out in the next section.

6.3 Tabulated Solutions to Differential Equations

When the assumed solution technique of solving differential equations may not be applied because the forcing function is not of the e^{qt} form, another approach must be taken. One effective scheme is to use a Fourier expansion representation of the offending function. Fourier expansions are normally written in terms of sine or cosine functions, but for linear systems work an exponential form ($e^{i\omega t}$) is most convenient. The resulting expansion consists of a summation of terms which we may handle by the assumed solution technique. Therefore, the approach does accomplish our objective. The subject of Fourier expansions, however, is really best investigated in a more mathematically inclined location than this text. Therefore, the approach to be considered here is of a very practical nature wherein the solutions of ordinary differential equations with constant coefficients are simply tabulated. This provides an extremely efficient means for the solution of engineering problems, without getting very involved with theoretical considerations. Surprisingly, though, the technique by which the tabulations are formulated does offer some extra insight into linear system analysis.

6.3(a) Interpretation of D *Operator Expressions*

The tabulation of interpretations for all possible specific forms of expression (6.7) above would be a most formidable task were it not for some fortunate simplifications which can be made. These can be best explained in terms of the **D** operator manipulations of section 6.2. Let us first note that expansion of the **D** operator expression (6.8) results in a general form of equation as (6.9):

$$(\mathbf{D} - r_1)(\mathbf{D} - r_2) \ldots (\mathbf{D} - r_n)x = y \tag{6.8}$$

$$(\mathbf{D}^n - A_{n-1}\mathbf{D}^{n-1} - \ldots A_1\mathbf{D} - A_0)x = y \tag{6.9}$$

if the quantities r_1, r_2, etc., are the roots of the characteristic equation

$$(r^n - A_{n-1}r^{n-1} - \ldots A_1 r - A_0) = 0.$$

This means that an operator expression such as (6.7) may always be written in the form

$$\frac{(\mathbf{D} - z_1)(\mathbf{D} - z_2) \ldots (\mathbf{D} - z_n)}{(\mathbf{D} - p_1)(\mathbf{D} - p_2) \ldots (\mathbf{D} - p_m)} y \tag{6.10}$$

where the quantities z are the roots of the equation $B(\mathbf{D}) = 0$ and the p's are the roots of $A(\mathbf{D}) = 0$.

Determining the significance of the numerator of (6.10) is straightforward since it is simply a multiple differentiation and summation on the function y. Therefore, let us concentrate our attention on the denominator, which we do by assuming a numerator of unity:

$$\frac{1}{(\mathbf{D} - p_1)(\mathbf{D} - p_2)\ldots(\mathbf{D} - p_3)}y. \tag{6.11}$$

As a start on the interpretation of this expression, let us attempt to find the meaning of

$$\frac{1}{\mathbf{D} - p_1}y. \tag{6.12}$$

The result should give us considerable insight into the general case represented by (6.10). First, however, let us see what possibilities exist as the nature of p changes:

For $p = 0$, we must interpret $\frac{1}{\mathbf{D}}y$.

For p a pure real number, we have $\frac{1}{\mathbf{D} - p}y$.

For p a pure imaginary number (such as $j\omega$), there must be another p which is just the negative (*i.e.*, $-j\omega$) since roots always appear as complex conjugates. If we choose, we may write

$$\frac{1}{(\mathbf{D} - j\omega)(\mathbf{D} + j\omega)}y = \frac{1}{\mathbf{D}^2 - \omega^2}y.$$

In this way, we avoid the appearance of any imaginary quantities in the expression (which may, or may not, be desirable from the point of view of the analyst).

For p a complex number $(a - jb)$, there must be another root p which is $(a + jb)$. Again, if we choose, we may write

$$\frac{1}{(\mathbf{D} - a - jb)(D - a + jb)}y = \frac{1}{\mathbf{D}^2 - 2a\mathbf{D} + a^2 + b^2}y$$

which also avoids the appearance of complex numbers in the expression to be interpreted.

The group of expressions above must be the basic kinds of expressions to be interpreted since we have considered all of the possible forms of the root p. The interpretation of these basic operator expressions follows.

If $x = (1/\mathbf{D})y$, then x is the solution of the differential equation

$$\mathbf{D}x = y$$

or

$$x = \int_0^t y\,dt + x_0.$$

Then

$$\frac{1}{\mathbf{D}}y = \int_0^t y\,dt + x_0 \tag{6.13}$$

and we may think of $1/\mathbf{D}$ as an "integrating operator"; this represents a natural extension of the method.*

We would then expect that $[1/\mathbf{D}^2]y$ would have something to do with the second integral of y. Starting from

$$\mathbf{D}^2x = y \longrightarrow$$

$$\mathbf{D}x = \int_0^t y\,dt + \dot{x}_0$$

we get

$$x = \int_0^t \left(\int_0^t y\,dt + \dot{x}_0\right) dt + x_0$$

and

$$= \int_0^t \int_0^t y\,dt\,dt + \int_0^t \dot{x}_0\,dt + x_0$$

$$= \int_0^t \int_0^t y\,dt\,dt + \dot{x}_0 t + x_0.$$

Therefore,

$$\frac{1}{\mathbf{D}^2}y = \int_0^t \int_0^t y\,dt\,dt + c_1 t + c_2 \tag{6.14}$$

The extension to higher orders is obvious.

A situation which is not quite so straightforward arises when the original differential equation is

$$(1 + T\mathbf{D})x = y$$

* The integrating operator was used previously in connection with the writing of network dynamic equations. There (in Chapter 5) it was specified that the initial conditions were to be zero; i.e., that the integration was starting from an equilibrium condition of the system wherein all the excursion variables were zero. This condition is not actually necessary, since normal forms of the differential equations are always obtained by successive differentiation (multiplication by the **D** operator) so that no integrating operators or constants of integration remain. By avoiding the specification of initial conditions at that stage of analysis a considerable simplification of procedures is obtained.

Note: We write $(1 + T\mathbf{D})$ instead of $(\mathbf{D} - p)$ for later convenience. The coefficient of $\mathbf{D}$ must have units of time; therefore, T is called a "time constant."

We are interested in the meaning of

$$\left(\frac{1}{1 + T\mathbf{D}}\right)y.$$

A completely general solution to the differential equation may be achieved through the use of an integrating factor. We note that if we had some function $p(t)$ such that

$$(1 + T\mathbf{D})x = y \text{ or } \left(\frac{1}{T} + \mathbf{D}\right)x = \frac{y}{T}$$

could be written

$$\mathbf{D}(px) = py/T \quad \text{(which is integrable),}$$

we would be able to solve the equation for any $y(t)$. That this can be done is shown by the following:

$$\mathbf{D}(px) = \frac{py}{T}$$

$$x\mathbf{D}p + p\mathbf{D}x = \frac{py}{T}$$

$$\left(\frac{\mathbf{D}p}{p} + \mathbf{D}\right)x = \frac{y}{T}$$

Therefore, if $\mathbf{D}p/p$ can be made equal to $1/T$, the equation can be solved. If

$$\frac{\mathbf{D}p}{p} = \frac{1}{T} \qquad p - T\mathbf{D}p = 0,$$

we have an equation which may be readily solved by the methods of the last section. Assume

$$p = P\,e^{rt}.$$

Then

$$1 - Tr = 0$$

$$r = \frac{1}{T}$$

and the integrating factor $p(t)$ is given by $Pe^{t/T}$. Since any function of this form is satisfactory, P may be chosen to be unity. Then, from (6.16),

$$\mathbf{D}(e^{t/T}x) = e^{t/T}y/T$$

$$e^{t/T}x = \frac{1}{T}\int_0^t e^{t/T}y\,dt + x_0$$

$$x = e^{-t/T}\left[\frac{1}{T}\int_0^t e^{t/T}y\,dt + x_0\right].$$

The interpretation desired is then clearly

$$\left(\frac{1}{1+T\mathbf{D}}\right)y = e^{-t/T}\left[\frac{1}{T}\int_0^t e^{t/T}y\,dt + c\right] \tag{6.18}$$

where the constant c is the initial value of the dependent function.

6.3(b) Extension to the General Case

We may go to the general case quite readily from this point. Consider the equation

$$x = \left(\frac{1}{1+T_1\mathbf{D}}\right)\left(\frac{1}{1+T_2\mathbf{D}}\right)y = \frac{T_1}{T_1-T_2}\left(\frac{1}{1+T_1\mathbf{D}}\right)y + \frac{T_2}{T_2-T_1}\left(\frac{1}{1+T_2\mathbf{D}}\right)y.$$

From (6.18), and using superposition,

$$x = \left(\frac{T_1}{T_1-T_2}\right)e^{-t/T_1}\left[\frac{1}{T_1}\int_0^t e^{t/T_1}y\,dt + c_1\right] + \frac{T_2}{T_2-T_1}\,e^{-t/T_2}\left[\frac{1}{T_2}\int_0^t e^{t/T_2}y\,dt + c_2\right]$$

Because this is a second-order equation, two initial conditions are required in order to determine the constants c_1 and c_2. These would most often be conditions on initial value and initial rate of change of x.

Extending this approach, we recognize that

$$\left(\frac{1}{1+T_1\mathbf{D}}\right)\left(\frac{1}{1+T_2\mathbf{D}}\right)\cdots\left(\frac{1}{1+T_m\mathbf{D}}\right)B(\mathbf{D}) = \frac{B(\mathbf{D})}{A(\mathbf{D})}$$

if $(1 + T_1\mathbf{D})(1 + T_2\mathbf{D})\ldots(1 + T_m\mathbf{D}) = A(\mathbf{D})$. Then

$$\left(\frac{K_1}{1+T_1\mathbf{D}} + \frac{K_2}{1+T_2\mathbf{D}} + \cdots \frac{K_m}{1+T_m\mathbf{D}}\right) = \frac{1}{A(\mathbf{D})}$$

if the K's are chosen properly.* If

* The K's must be chosen so that, when the indicated addition is carried out, using a common denominator, the numerator will be unity.

$$x = \frac{1}{A(\mathbf{D})}y,$$

then

$$x = K_1 e^{-t/T_1}\left(\frac{1}{T_1}\int_0^t e^{t/T_1} y\, dt + c_1\right) + K_2 e^{-t/T_2}\left(\frac{1}{T_2}\int_0^t e^{t/T_2} y\, dt + c_2\right) + \text{etc.}$$

In the general case where

$$x = \frac{B(\mathbf{D})}{A(\mathbf{D})}y$$

the solution may be written as

$$x = K_1 e^{-t/T_1}\left(\frac{1}{T_1}\int_0^t e^{t/T_1} y^*\, dt + c_1\right) + \text{etc.}$$

where $y^* = B(\mathbf{D})y$ and is readily computable.

The crucial step in this analysis, changing

$$\frac{1}{A(\mathbf{D})} = \left(\frac{1}{1 + T_1\mathbf{D}}\right)\left(\frac{1}{1 + T_2\mathbf{D}}\right)\cdots\left(\frac{1}{1 + T_m\mathbf{D}}\right)$$

to

$$\frac{1}{A(\mathbf{D})} = \left(\frac{K_1}{1 + T_1\mathbf{D}}\right) + \left(\frac{K_2}{1 + T_2\mathbf{D}}\right) + \cdots + \left(\frac{K_m}{1 + T_m\mathbf{D}}\right) \tag{6.19}$$

is known as a "partial fraction expansion" and is probably more familiar to the student in the reversed form, that of addition using a common denominator. This technique is considered in detail in the next section.

6.4 Partial Fraction Expansions

As seen in section 6.3, the result from

$$\frac{1}{1 + T\mathbf{D}}y$$

may be applied to the general case

$$\frac{1}{A(\mathbf{D})}y$$

if the denominator is factored as

$$\frac{1}{(\mathbf{D} - p_1)(\mathbf{D} - p_2)\cdots(\mathbf{D} - p_n)}y$$

and the resulting fraction broken up into its component parts to give

$$\frac{K_1}{\mathbf{D} - p_1}y + \frac{K_2}{\mathbf{D} - p_2}y + \cdots + \frac{K_n}{\mathbf{D} - p_n}y.$$

Techniques for determining the K's are the subject of this section.

6.4(a) Basic Principles of Partial Fraction Expansion

In order to determine the constants K in the partial fraction expansion, the addition of (6.19) may be written out:

1st term $$\frac{K_1(1 + T_2\mathbf{D})(1 + T_3\mathbf{D})\ldots(1 + T_m\mathbf{D})}{\Pi}$$

2nd term $$\frac{K_2(1 + T_1\mathbf{D})(1 + T_3\mathbf{D})\ldots(1 + T_m\mathbf{D})}{\Pi}$$

*m*th term $$\frac{K_m(1 + T_1\mathbf{D})(1 + T_2\mathbf{D})\ldots(1 + T_{m-1}\mathbf{D})}{\Pi}$$

where Π is the product of all the $(1 + T\mathbf{D})$'s.

The sum of all numerators must equal unity. In order for this to happen, the sum of all the terms in the numerators having like powers of **D** must be zero. Therefore, m expressions involving the T's and K's can be found and the K's determined. Although considerable algebra is involved, the method is completely general and is easy to recall.

Example 6.6 Suppose it is desired to determine the partial fraction expansion of

$$\frac{1}{A(\mathbf{D})} = \frac{1}{30\mathbf{D}^3 + 31\mathbf{D}^2 + 10\mathbf{D} + 1}.$$

The denominator has been chosen to factor readily as follows:

$$\frac{1}{A(\mathbf{D})} = \frac{1}{(1 + 2\mathbf{D})(1 + 3\mathbf{D})(1 + 5\mathbf{D})}.$$

The partial fraction expansion is then

$$\frac{1}{A(\mathbf{D})} = \frac{K_1}{1 + 2\mathbf{D}} + \frac{K_2}{1 + 3\mathbf{D}} + \frac{K_3}{1 + 5\mathbf{D}}$$

$$\frac{1}{A(\mathbf{D})} = $$
$$= \frac{K_1(1 + 3\mathbf{D})(1 + 5\mathbf{D}) + K_2(1 + 2\mathbf{D})(1 + 5\mathbf{D}) + K_3(1 + 2\mathbf{D})(1 + 3\mathbf{D})}{(1 + 2\mathbf{D})(1 + 3\mathbf{D})(1 + 5\mathbf{D})}$$

$$\frac{1}{A(\mathbf{D})} = $$
$$= \frac{K_1(1 + 8\mathbf{D} + 15\mathbf{D}^2) + K_2(1 + 7\mathbf{D} + 10\mathbf{D}^2) + K_3(1 + 5\mathbf{D} + 6\mathbf{D}^2)}{(1 + 2\mathbf{D})(1 + 3\mathbf{D})(1 + 5\mathbf{D})}.$$

Then

$$\begin{aligned} K_1 + K_2 + K_3 &= 1 \\ K_1 8\mathbf{D} + K_2 7\mathbf{D} + K_3 5\mathbf{D} &= 0 \\ K_1 15\mathbf{D}^2 + K_2 10\mathbf{D}^2 + K_3 6\mathbf{D}^2 &= 0 \end{aligned} \qquad \begin{aligned} K_1 + K_2 + K_3 &= 1 \\ 8K_1 + 7K_2 + 5K_3 &= 0 \\ 15K_1 + 10K_2 + 6K_3 &= 0 \end{aligned}$$

$$K_1 = \frac{\begin{vmatrix} 1 & 1 & 1 \\ 0 & 7 & 5 \\ 0 & 10 & 6 \end{vmatrix}}{\begin{vmatrix} 1 & 1 & 1 \\ 8 & 7 & 5 \\ 15 & 10 & 6 \end{vmatrix}} = \frac{42 - 50}{-8 + 27 - 25} = \frac{-8}{-6} = \frac{4}{3}$$

$$K_2 = \frac{\begin{vmatrix} 1 & 1 & 1 \\ 8 & 0 & 5 \\ 15 & 0 & 6 \end{vmatrix}}{-6} = \frac{+27}{-6} = -\frac{9}{2}$$

$$K_3 = \frac{\begin{vmatrix} 1 & 1 & 1 \\ 8 & 7 & 0 \\ 15 & 10 & 0 \end{vmatrix}}{-6} = \frac{5}{-6} = -\frac{5}{6}.$$

Then

$$\frac{1}{B(\mathbf{D})} = \frac{4/3}{1 + 2\mathbf{D}} + \frac{-9/2}{1 + 3\mathbf{D}} + \frac{-5/6}{1 + 5\mathbf{D}}.$$

The solution to

$$x = \frac{1}{B(\mathbf{D})} y$$

could then be written out directly with the aid of equation (6.18).

In many cases the denominator of the fraction which is to be expanded contains factors which are imaginary, or has repeated factors. In the first case, a factor of the form

$$(\mathbf{D}^2 + \omega^2) = (\mathbf{D} + j\omega)(\mathbf{D} - j\omega)$$

might appear. The partial fraction expansion would then contain the terms

$$\frac{K_1}{\mathbf{D} + j\omega} \quad \text{and} \quad \frac{K_2}{\mathbf{D} - j\omega}.$$

This would lead to imaginary values for the K's and an increase in the amount of algebra required. It is more effective to leave $(\mathbf{D}^2 + \omega^2)$ unfactored in the expansion. In this case it turns out that the terms

$$\frac{K_1}{\mathbf{D}^2 + \omega^2} + \frac{K_2\mathbf{D}}{\mathbf{D}^2 + \omega^2}$$

appear. Although this requires interpretation of the meaning of terms like

$$x = \frac{K}{\mathbf{D}^2 + \omega^2}\,y \quad \text{and} \quad \frac{K\mathbf{D}}{\mathbf{D}^2 + \omega^2}\,y$$

the interpretation is possible and the constants K are all real.

A primary rule of partial fraction expansion is that the number of terms in the expansion is equal to the degree of the polynomial in D which is the denominator of the fraction being expanded. This is illustrated by the fact that when we choose to avoid an expansion of the form

$$\frac{K_1}{\mathbf{D} + j\omega} + \frac{K_2}{\mathbf{D} - j\omega}$$

we must use two terms:

$$\frac{K_1}{\mathbf{D}^2 + \omega^2} + \frac{K_2\mathbf{D}}{\mathbf{D}^2 + \omega^2}.$$

The need for the **D** factor in the second term is to make the second term distinct from the first.

In the case of a quadratic factor in the denominator we wish to avoid using

$$\frac{K_1}{\mathbf{D} + \sigma + j\omega_n} + \frac{K_2}{\mathbf{D} + \sigma - j\omega_n}$$

since the K's would be complex. The alternative is

$$\frac{K_1}{\mathbf{D}^2 + 2\zeta\omega_n\mathbf{D} + \omega_n^2} + \frac{K_2\mathbf{D}}{\mathbf{D}^2 + 2\zeta\omega_n\mathbf{D} + \omega_n^2}.$$

6.4(b) Special Cases in Partial Fraction Expansion

When a factor of the denominator is repeated, the proper number of terms is included as follows:

If $(1 + T\mathbf{D})$ is a factor 3 times—i.e., $1/(1 + T\mathbf{D})^3$ is a factor of $1/A(\mathbf{D})$—the expansion has the form

$$\frac{K_1}{(1 + T\mathbf{D})} + \frac{K_2}{(1 + T\mathbf{D})^2} + \frac{K_3}{(1 + T\mathbf{D})^3}.$$

In general, if $(1 + T\mathbf{D})$ is a factor q times, then the expansion has the form

$$\frac{K_1}{1 + T\mathbf{D}} + \frac{K_2}{(1 + T\mathbf{D})^2} + \cdots + \frac{K_q}{(1 + T\mathbf{D})^q}.$$

If a quadratic factor is repeated q times, the expansion will have the form

$$\frac{K_{11} + K_{12}\mathbf{D}}{\mathbf{D}^2 + 2\delta\omega_n^2\,\mathbf{D} + \omega_n^2} + \frac{K_{21} + K_{22}\mathbf{D}}{(\mathbf{D}^2 + 2\delta\omega_n\mathbf{D} + \omega_n^2)^2} + \cdots + \frac{K_{q1} + K_{q2}\mathbf{D}}{(\mathbf{D}^2 + 2\delta\omega_n\mathbf{D} + \omega_n^2)^q}$$

which has the proper number of terms. Since any polynomial with real coefficients can be factored with factors of the the form $(1 + T\mathbf{D})$, $(\mathbf{D}^2 + \omega^2)$, $(\mathbf{D}^2 + \gamma\mathbf{D} + \omega^2)$ and repeated factors of these forms, all the necessary possibilities have been considered.

6.4(c) A Convenient Technique for Partial Fraction Expansion

Here is a neat trick for doing partial fraction expansion:
Suppose we are trying to expand $1/A(\mathbf{D})$, and we have

$$\frac{1}{A(\mathbf{D})} = \frac{K_1}{1 + T_1\mathbf{D}} + \frac{K_2}{1 + T_2\mathbf{D}} + \cdots + \frac{K_m}{1 + T_m\mathbf{D}}.$$

The algebra involved in finding the K's is rather tedious; let us try to avoid some of it. This can be done if we temporarily think of the **D**'s as being algebraic quantities (dummy variables) instead of differential operators. We can then assign values to the **D**'s; if we pick some value for **D** which facilitates the algebra, we will accomplish our goal. Since the **D** operators obey the rules of algebra, any result we get by using the dummy variables and algebraic manipulation will represent a valid relationship between the **D** operator expressions.

Let us try to get K_1. Multiply both sides of (6.20) by $1 + T_1\mathbf{D}$. We then have

$$\frac{1 + T_1\mathbf{D}}{A(\mathbf{D})} = \frac{1 + T_1\mathbf{D}}{(1 + T_1\mathbf{D})(1 + T_2\mathbf{D}) \cdots (1 + T_m\mathbf{D})} = K_1 + \frac{K_2(1 + T_1\mathbf{D})}{1 + T_2\mathbf{D}} + \cdots + \frac{K_m(1 + T_1\mathbf{D})}{1 + T_m\mathbf{D}}.$$

Now select (the dummy variable) $\mathbf{D} = -1/T_1$ so that $1 + T_1\mathbf{D} = 0$. We then have

$$\frac{1}{(1 + T_2\mathbf{D})(1 + T_3\mathbf{D}) \cdots (1 + T_m\mathbf{D})} = K_1$$

or

$$K_1 = \frac{1}{(1 - T_2/T_1)(1 - T_3/T_1) \cdots (1 - T_m/T_1)}.$$

The same device will work with $(\mathbf{D}^2 + \omega^2)$ and $(\mathbf{D}^2 + 2\zeta\omega_n\mathbf{D} + \omega_n^2)$ factors as well; in the first case $\mathbf{D}^2$ must be taken to be $-\omega^2$, in the second the quadratic must be factored in order to determine the value of $\mathbf{D}$ which will make the quantity zero.

Example 6.7 Let us work the earlier example (6.6) by utilizing this approach. We had

$$\frac{1}{A(\mathbf{D})} = \frac{1}{(1 + 2\mathbf{D})(1 + 3\mathbf{D})(1 + 5\mathbf{D})}$$

$$\frac{1}{A(\mathbf{D})} = \frac{K_1}{1 + 2\mathbf{D}} + \frac{K_2}{1 + 3\mathbf{D}} + \frac{K_3}{1 + 5\mathbf{D}}.$$

According to the above,

$$K_1 = \frac{1}{(1 + 3\mathbf{D})(1 + 5\mathbf{D})} \quad \text{if } \mathbf{D} \text{ is imagined equal to } -1/2$$

then

$$K_1 = \frac{1}{(1 - 3/2)(1 - 5/2)}$$

$$= \frac{1}{(-1/2)(-3/2)} = \frac{4}{3}$$

which agrees with the earlier analysis and is much simpler.

Example 6.8 To illustrate the case where the fraction contains imaginary factors, consider

$$\frac{1}{B(\mathbf{D})} = \frac{1}{\mathbf{D}^3 + 3\mathbf{D}^2 + 16\mathbf{D} + 48}$$

$$= \frac{1}{(\mathbf{D}^2 + 16)(\mathbf{D} + 3)}.$$

We take

$$\frac{1}{\mathrm{B(D)}} = \frac{\mathrm{K}_1}{\mathrm{D}^2 + 16} + \frac{\mathrm{K}_2\mathrm{D}}{\mathrm{D}^2 + 16} + \frac{\mathrm{K}_3}{\mathrm{D} + 3}$$

$$\frac{\mathrm{D}^2 + 16}{\mathrm{B(D)}} = \mathrm{K}_1 + \mathrm{K}_2\mathrm{D} + \frac{\mathrm{D}^2 + 16}{\mathrm{D}+3}$$

at

$$\mathrm{D}^2 = -16$$

$$\frac{1}{\mathrm{D}+3} = \mathrm{K}_1 + \mathrm{K}_2\mathrm{D} \qquad \mathrm{K}_1 = \frac{3}{25}$$

$$\frac{\mathrm{D}-3}{\mathrm{D}^2 - 9} = \frac{\mathrm{D}-3}{-25} = \mathrm{K}_1 + \mathrm{K}_2\mathrm{D} \qquad \mathrm{K}_2 = -\frac{1}{25}$$

The analysis shows that the $(\mathbf{D}^2 + 16)$ factor does require two terms in the expansion, one having a **D** factor in the numerator. To complete the expansion,

$$K_3 = \frac{\mathbf{D}+3}{B(\mathbf{D})} \qquad \text{for } \mathbf{D} = -3$$

$$K_3 = \frac{1}{\mathbf{D}^2 + 16} = \frac{1}{9 + 16} = \frac{1}{25}.$$

The expansion is then

$$\frac{1}{\mathbf{D}^3 + 3\mathbf{D}^2 + 16\mathbf{D} + 48} = \frac{3/25}{\mathbf{D}^2 + 16} - \frac{\mathbf{D}/25}{\mathbf{D}^2 + 16} + \frac{1/25}{\mathbf{D}+3}.$$

Example 6.9 An expansion in which repeated roots occur is

$$\frac{1}{A(\mathbf{D})} = \frac{K_1}{(1 + T_1\mathbf{D})^2} + \frac{K_2}{1 + T_1\mathbf{D}} + \frac{K_3}{1 + T_2\mathbf{D}}$$

$$K_1 = \frac{(1 + T_1\mathbf{D})^2}{B(\mathbf{D})} = \frac{1}{1 + T_2\mathbf{D}} \qquad \mathbf{D} = -\frac{1}{T_1}$$

$$K_1 = \frac{1}{1 - \dfrac{T_2}{T_1}}$$

$$K_3 = \frac{1}{(1+T_1\mathbf{D})^2} \qquad \mathbf{D} = -\frac{1}{T_2}$$

$$= \frac{1}{(1 - \dfrac{T_1}{T_2})^2}$$

The evaluation of K_2 is troublesome since if we use the usual method; i.e.,

$$\frac{(1+T_1\mathbf{D})}{A(\mathbf{D})} = \frac{1}{(1+T_1\mathbf{D})(1+T_2\mathbf{D})} = \frac{K_1(1+T_1\mathbf{D})}{(1+T_1\mathbf{D})^2} + K_2,$$

we find that the term with K_1 does not go to zero and both sides become infinite if $\mathbf{D}$ is assigned the value $-1/T_1$. In order to resolve this difficulty, we may either substitute the known values of K_1 and K_3 in the equation and determine K_2 algebraically, or the following approach may be used:

$$\frac{(1+T_1\mathbf{D})^2}{A(\mathbf{D})} = K_1 + K_2(1+T_1\mathbf{D}) + K_3\frac{(1+T_1\mathbf{D})^2}{1+T_2\mathbf{D}}$$

$$\frac{1}{1+T_2\mathbf{D}} = K_1 + K_2(1+T_1\mathbf{D}) + K_3\frac{(1+T_1\mathbf{D})^2}{(1+T_2\mathbf{D})}.$$

Differentiating both sides with respect to $\mathbf{D}$ gives

$$-\frac{T_2}{(1+T_2\mathbf{D})^2} = K_2T_1 + \frac{(1+T_2\mathbf{D})^2(1+T_1\mathbf{D})T_1 - (1+T_1\mathbf{D})^2T_2}{1+T_2\mathbf{D}} \qquad \mathbf{D} = -\frac{1}{T_1}$$

$$-\frac{T_2}{(1 - \dfrac{T_2}{T_1})^2} = K_2T_1$$

$$K_2 = -\frac{T_2/T_1}{[1-(T_2/T_1)]^2}.$$

The general procedure for repeated roots is illustrated by the following:

$$A(\mathbf{D}) = (1+T_1\mathbf{D})^q(1+T_2\mathbf{D}).$$

The expansion has the form

$$\frac{1}{A(\mathbf{D})} = \frac{K_q}{(1+T_1\mathbf{D})^q} + \frac{K_{q-1}}{(1+T_1\mathbf{D})^{q-1}} + \dots + \frac{K_1}{(1+T_1\mathbf{D})} + \frac{L}{(1+T_2\mathbf{D})}$$

multiplying both sides by $(1 + T_1\mathbf{D})^q$,

$$\frac{(1 + T_1\mathbf{D})^q}{B(\mathbf{D})} = K_q + (1 + T_1\mathbf{D})K_{q-1} + (1 + T_1\mathbf{D})^2K_{q-2} + (1 + T_1\mathbf{D})^3K_{q-3}$$

$$+ \cdots \text{ evaluate } K_q \text{ at } \mathbf{D} = -\frac{1}{T_1}$$

$$\frac{d}{d\mathbf{D}}\left[\frac{(1 + T_1\mathbf{D})^q}{B(\mathbf{D})}\right] = 0 + T_1K_{q-1} + 2(1 + T_1\mathbf{D})T_1K_{q-2} + 3(1 + T_1\mathbf{D})^2T_1K_{q-3}$$

$$+ \cdots \text{ evaluate } K_{q-1} \text{ at } \mathbf{D} = -\frac{1}{T_1}$$

$$\frac{d^2}{d\mathbf{D}^2}\left[\frac{(1 + T_1\mathbf{D})^q}{B(\mathbf{D})}\right] = 2T_1{}^2K_{q-2} + 6(1 + T_1\mathbf{D})T_1{}^2K_{q-3}$$

$$+ \text{ other terms with a } (1 + T_1\mathbf{D}) \text{ factor}$$

$$\frac{d^3}{d\mathbf{D}^3}\left[\frac{(1 + T_1\mathbf{D})^q}{B(\mathbf{D})}\right] = 6T_1{}^3K_{q-3} + \text{ terms which go to zero}$$

$$\frac{d^n}{d\mathbf{D}^n}\left[\frac{(1 + T_1\mathbf{D})^q}{B(\mathbf{D})}\right] = n!\,T_1{}^nK_{q-n} + \text{ terms which go to zero}$$

$$K_{q-n} = \frac{1}{n!\,T_1{}^n}\frac{d^n}{d\mathbf{D}^n}\left[\frac{(1 + T_1\mathbf{D})^q}{B(\mathbf{D})}\right] \text{ at } \mathbf{D} = -\frac{1}{T_1}.$$

6.4(d) *Summary of Partial Fraction Expansion*

The objective of partial fraction expansion is to separate a fraction such as

$$\frac{1}{A(\mathbf{D})}$$

into components of the form

$$\frac{K}{\mathbf{D}},\quad \frac{K}{1 + T\mathbf{D}},\quad \frac{K}{\mathbf{D}^2 + \omega_n^2},\quad \frac{K}{\mathbf{D}^2 + 2\zeta\omega_n\mathbf{D} + \omega_n^2}$$

so that

$$\frac{1}{A(\mathbf{D})} = \frac{K_1}{A_1(\mathbf{D})} + \frac{K_2}{A_2(\mathbf{D})} + \ldots + \frac{K_n}{A_n(\mathbf{D})}$$

and thus facilitate the use of tabular presentations of the solutions of linear ordinary differential equations with constant coefficients. The technique is also applied in conjunction with the Laplace transform method of solution.

(1) For a single, real factor of $A(\mathbf{D})$ of $(1 + T_k\mathbf{D})$,

$$K_k = \frac{1 + T_k\mathbf{D}}{A(\mathbf{D})} \quad \text{for} \quad \mathbf{D} = -\frac{1}{T_k}.$$

(2) For a repeated, real factor of $A(\mathbf{D})$ of $(1 + T_k)^q$, the expansion has the form

$$\frac{K_q}{(1 + T_k\mathbf{D})^q} + \frac{K_{q-1}}{(1 + T_k\mathbf{D})^{q-1}} + \cdots + \frac{K_1}{(1 + T_k\mathbf{D})}$$

$$K_q = \frac{(1 + T_k\mathbf{D})^q}{B(\mathbf{D})} \text{ for } \quad \mathbf{D} = -\frac{1}{T_k}$$

$$K_{q-1} = \frac{1}{T_k}\frac{d}{d\mathbf{D}}\left[\frac{(1 + T_k\mathbf{D})^q}{B(\mathbf{D})}\right] \quad \text{for} \quad \mathbf{D} = -\frac{1}{T_k}$$

$$K_{q-2} = \frac{1}{2T_k^2}\frac{d^2}{d\mathbf{D}^2}\left[\frac{(1 + T_k\mathbf{D})^q}{B(\mathbf{D})}\right] \quad \text{for} \quad \mathbf{D} = -\frac{1}{T_k}$$

$$K_{q-n} = \frac{1}{n!\,T_k^n}\frac{d^n}{d\mathbf{D}^n}\left[\frac{(1 + T_k\mathbf{D})^q}{B(\mathbf{D})}\right] \quad \text{for} \quad \mathbf{D} = -\frac{1}{T_k}.$$

(3) For a quadratic factor $(\mathbf{D}^2 + 2\zeta\,\omega_n\mathbf{D} + \omega_n^2)$, $\zeta < 1$, the expansion has the form

$$\frac{K_1}{\mathbf{D}^2 + 2\zeta\omega_n\mathbf{D} + \omega_n^2} + \frac{K_2\mathbf{D}}{\mathbf{D}^2 + 2\zeta\omega_n\mathbf{D} + \omega_n^2}$$

$$\frac{\mathbf{D}^2 + 2\zeta\omega_n\mathbf{D} + \omega_n^2}{A(\mathbf{D})} = K_1 + K_2\mathbf{D} \quad \text{for} \quad \mathbf{D} = -(\zeta - \omega_n\sqrt{\zeta^2 - 1})$$

$$[\text{or} \quad \mathbf{D} = -(\zeta + \omega_n\sqrt{\zeta^2 - 1})].$$

In order to determine both K_1 and K_2, the real and imaginary parts of the above expression are equated separately.

(4) For repeated quadratic factors, $(\mathbf{D}^2 + 2\zeta\,\omega_n\mathbf{D} + \omega_n^2)^q$, $\zeta < 1$, the expansion has the form

$$\frac{K_{1q} + K_{2q}\mathbf{D}}{(\mathbf{D}^2 + 2\zeta\omega_n\mathbf{D} + \omega_n^2)^q} + \frac{K_{1,q-1} + K_{2,q-1}\mathbf{D}}{(\mathbf{D}^2 + 2\zeta\omega_n\mathbf{D} + \omega_n^2)^{q-1}} + \cdots \tag{6.21}$$

$$K_{1q} + K_{2q}\mathbf{D} = \frac{(\mathbf{D}^2 + 2\zeta\omega_n\mathbf{D} + \omega_n{}^2)^q}{A(\mathbf{D})} \qquad \mathbf{D} = -(\zeta + \omega_n\sqrt{\zeta^2 - 1})$$

$$K_{1,q-1}(2\mathbf{D} + 2\zeta\omega_n) + K_{2,q-1}(3\mathbf{D}^2 + 4\zeta\omega_n\mathbf{D} + \omega_n^2)$$

$$= \frac{d}{d\mathbf{D}}\left[\frac{(\mathbf{D}^2 + 2\zeta\omega_n\mathbf{D} + \omega_n{}^2)^q}{A(\mathbf{D})}\right] - K_{2q} \qquad \mathbf{D} = -(\zeta + \omega_n\sqrt{\zeta^2 - 1})$$

Further K values may be determined by repeated differentiation of (6.21), however, such cases arise very infrequently. A more efficient scheme would be to factor $\mathbf{D}^2 + 2\zeta\,\omega_n\mathbf{D} + \omega_n^2$ and handle as for repeated simple roots.

6.5 Solutions Directly from D Operator Expressions

The partial fraction expansion method indicates that if we had interpretation of all the following expressions,

$$\frac{y(t)}{\mathbf{D}} \qquad \frac{1}{\mathbf{D}^2 + \omega^2}y(t)$$

$$\frac{y(t)}{1 + T\mathbf{D}} \qquad \frac{\mathbf{D}}{\mathbf{D}^2 + \omega^2}y(t)$$

$$\frac{y(t)}{(1 + T\mathbf{D})^q} \qquad \frac{1}{\mathbf{D}^2 + 2\zeta\,\omega_n\mathbf{D} + \omega_n^2}y(t)$$

$$\frac{\mathbf{D}}{\mathbf{D}^2 + 2\zeta\,\omega_n\mathbf{D} + \omega_n^2}y(t),$$

we would be able to solve any linear system equation irrespective of the nature of the Forcing Function; i.e., the constraint imposed by the assumed solution approach would be removed. The first two expressions in the list have already been interpreted. Let us investigate the repeated $(1 + T\mathbf{D})$ factor: For

$$x(t) = \frac{y(t)}{(1 + T\mathbf{D})^q}$$

we may write

$$x(t) = \cdots \frac{1}{1 + T\mathbf{D}}\left\{\frac{1}{1 + T\mathbf{D}}\left[\left(\frac{1}{1 + T\mathbf{D}}\right)y(t)\right]\right\}$$

that is, we may investigate this case as simply repeated application of the $1/(1 + T\mathbf{D})$ operator. From

$$\frac{1}{1+T\mathbf{D}}y(\tau) = e^{-t/T}\left(\frac{1}{T}\int_0^t e^{t/T} y\,dt + c\right),$$

$$x(t) = \cdots \frac{1}{1+T\mathbf{D}}\left\{\frac{1}{1+T\mathbf{D}}\left[e^{-t/T}\left(\frac{1}{T}\int_0^t e^{t/T} y\,dt + c_1\right)\right]\right\}$$

$$= \cdots \frac{1}{1+T\mathbf{D}}\left\{e^{-t/T}\left[\frac{1}{T}\int_0^t e^{t/T} y\,dt + c_2\right]\right\}$$

so that

$$= \cdots e^{-t/T}\left\{\frac{1}{T}\int_0^t\left[\frac{1}{T}\int_0^t\left(\frac{1}{T}\int_0^t e^{t/T} y\,dt + c_1\right)dt + c_2\right]dt + c_3\right\}$$

$$\frac{1}{(1+T\mathbf{D})^q}y = e^{-t/T}\frac{1}{T^q}\underbrace{\int_0^t\int_0^t\cdots\int_0^t}_{q\text{ times}} e^{t/T} y \underbrace{dt\,dt\cdots dt}_{q\text{ times}} + c_1 t^{q-1} + c_2 t^{q-2} + \cdots c_{q-1}.$$

For the case

$$x(t) = \frac{1}{(\mathbf{D}^2+\omega^2)}y(t)$$

$$\frac{1}{(\mathbf{D}^2+\omega^2)} = \frac{K_1}{\mathbf{D}+j\omega} + \frac{K_2}{\mathbf{D}-j\omega}$$

where $K_1 = -(1/2j\omega)$ and $K_2 = (1/2j\omega)$.
Taking

$$x_1 = \frac{K_1}{\mathbf{D}+j\omega}y \qquad \text{and} \qquad x_2 = \frac{K_2}{\mathbf{D}-j\omega}y,$$

we attempt to find an integrating factor p so that $\mathbf{D}px = Kpy$. For x_1,

$$p_1\mathbf{D}x_1 + x_1\mathbf{D}p_1 = K_1 p_1 y$$

$$\mathbf{D}x_1 + x_1\frac{\mathbf{D}p_1}{p_1} = K_1 y = (\mathbf{D}+j\omega)x_1$$

$$x_1\frac{\mathbf{D}p_1}{p_1} = j\omega x_1$$

$$p_1 = p_1 e^{r_1 t}, \qquad r_1 = j\omega.$$

Similarly, for x_2,

$$p_2 = p_2 e^{r_2 t}, \qquad r_2 = -j\omega.$$

Then

$$x_1 = -\frac{e^{-j\omega t}}{2j\omega}\int e^{j\omega t} y\,dt, \qquad x_2 = \frac{e^{j\omega t}}{2j\omega}\int e^{-j\omega t} y\,dt$$

or

$$x_1 = -\frac{e^{-j\omega t}}{2j\omega}\int (\cos\omega t + j\sin\omega t) y\,dt,$$

$$x_2 = \frac{e^{j\omega t}}{2j\omega}\int (\cos\omega t - j\sin\omega t) y\,dt.$$

Finally,

$$x = x_1 + x_2 = \left[\frac{e^{j\omega t} - e^{-j\omega t}}{2j\omega}\right]\int y\cos\omega t\,dt - \left[\frac{e^{j\omega t} + e^{-j\omega t}}{2\omega}\right]\int y\sin\omega t\,dt$$

or

$$x = \frac{\sin\omega t}{\omega}\int y\cos\omega t\,dt - \frac{\cos\omega t}{\omega}\int y\sin\omega t\,dt.$$

Identical procedures lead to the results for $\mathbf{D}/(\mathbf{D}^2 + \omega^2)$ and $1/(\mathbf{D}^2 + 2\zeta\omega_n\mathbf{D} + \omega_n^2)$. These are displayed in Table 6.1 which may be used, in conjunction with a table of integrals, to determine in a relatively direct way the forced response of systems described by all the basic transfer functions. In combination with the partial fraction expansion technique, this table allows the determination of the total response of many systems. However, this approach is not as general or as well known as the Laplace Transform technique. It is exhibited here more as an instructive intermediate between the e^{rt} substitution and Laplace transform techniques than as an end in itself.

Table 6.1 Tabulated Interpretations of the Basic Operator Expressions.

$$x = \frac{1}{\mathbf{D}} y = \int_0^t y\,dt + x_0$$

$$x = \frac{1}{\mathbf{D}^2} y = \int_0^t \int_0^t y\,dt\,dt + \dot{x}_0 t + x_0$$

$$x = \frac{1}{\mathbf{D}^n} y = \underbrace{\int_0^t \int_0^t}_{n\text{ times}} y\,dt^n + \frac{1}{q}\frac{d^q x}{dt^q}\bigg|_0 t^q + \cdots + \frac{1}{2}\ddot{x}_0 t^2 + \dot{x}_0 t + x_0 \qquad q = n - 1$$

$$x = \frac{1}{1 + T\mathbf{D}} y = e^{-t/T}\left[\frac{1}{T}\int_0^t e^{t/T} y\,dt + x_0\right]$$

$$x = \frac{1}{\mathbf{D}^2 + \omega^2} y = \frac{\sin \omega t}{\omega} \int_0^t y \cos \omega t \, dt - \frac{\cos \omega t}{\omega} \int_0^t y \sin \omega t \, dt$$
$$+ \frac{\dot{x}_0}{\omega} \sin \omega t + x_0 \cos \omega t.$$

$$x = \frac{\mathbf{D}}{\mathbf{D}^2 + \omega^2} y = \cos \omega t \int_0^t y \cos \omega t \, dt + \sin \omega t \int_0^t y \sin \omega t \, dt$$
$$+ \frac{\dot{x}_0}{\omega} \sin \omega t + x_0 \cos \omega t.$$

$$x = \frac{1}{\mathbf{D}^2 + 2\zeta\omega_n \mathbf{D} + \omega_n{}^2} y = \frac{e^{-at}}{b} \sin bt \int_0^t e^{at} \cos bty \, dt$$
$$- \frac{e^{-at}}{b} \cos bt \int_0^t e^{at} \sin bty \, dt$$
$$+ e^{-at} \left[\frac{\dot{x}_0 + ax_0}{\omega} \sin \omega t + x_0 \cos \omega t \right]$$

where $a = \zeta\omega_n$, $b = \omega_n\sqrt{1 - \zeta^2}$, $\omega = \sqrt{1 - \zeta^2}\, \omega_n$

6.5(a) The Unit Impulse

A hypothetical forcing function of considerable practical value is known as the "unit impulse." This function is shown in Figure 6.4 and defined as follows:

$$\delta = \lim_{a \to 0} \left\{ \begin{array}{ll} x = \dfrac{1}{a} & \text{for } 0 \le t \le a \\ x = 0 & \text{otherwise} \end{array} \right\}.$$

It is clear from the definition that the integral of the unit impulse, over any time beginning at or before zero and lasting beyond zero, is unity. Besides a number of applications to be discussed later, the unit impulse allows us to utilize Table 6.1 in a meaningful way. First, note that the integrals in the table can be evaluated quite readily if the forcing function is the unit impulse δ. This is true because the other functions within the integral may be assumed to be constants (having the value of the function at time zero) during the short interval when the impulse is non-zero. The effect of the unit impulse is to start the natural response of a system in a well-defined way; this may be seen from examination of Table 6.1.

It is interesting to note that the most common forcing functions to which

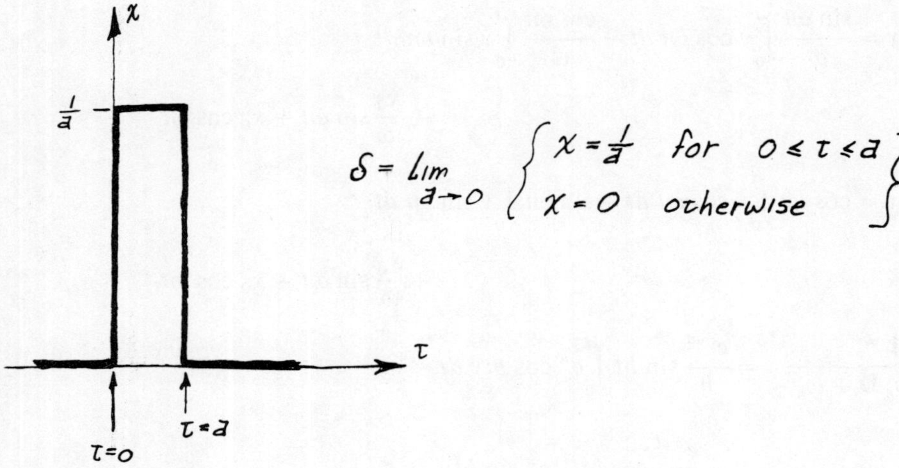

From the Engineer's Sketch Book

Figure 6.4 The δ function.

systems are subjected are the impulse responses associated with the basic operator expressions, as in Table 6.1. For example, a unit step function $u(t)$—in which y changes from 0 to 1 at time zero—is the impulse response of a system described by the $1/\mathbf{D}$ operator expression. It is then possible to approach all problems in forced response, where the forcing is suitable, by using only unit impulse response determinations and partial fraction expansion. The procedure is to suppose that the forcing function is the impulse response of some system described by the operator $I(\mathbf{D})$, and the system under study is described by $G(\mathbf{D})$. Then the forced response of $G(\mathbf{D})$ is given by

$$G(\mathbf{D})I(\mathbf{D}) \cdot (\delta)$$

and may be determined with partial fraction expansion. For example, if we wish to find the solution to $(1 + T\mathbf{D})x = u(T)$ we use the relationship

$$x = \frac{1}{1 + T\mathbf{D}} \cdot \frac{1}{\mathbf{D}} \cdot (\delta)$$

$$\frac{1}{1 + T\mathbf{D}} \cdot \frac{1}{\mathbf{D}} = \frac{K_1}{1 + T\mathbf{D}} + \frac{K_2}{\mathbf{D}}$$

$$K_1 = \frac{1}{\mathbf{D}} \qquad \text{for } \mathbf{D} = -\frac{1}{T}, K_1 = -T$$

$$K_2 = \frac{1}{1 + T\mathbf{D}} \qquad \text{for } \mathbf{D} = 0, K_2 = 1$$

$$x = \frac{-T}{1 + T\mathbf{D}}(\delta) + \frac{1}{\mathbf{D}}(\delta)$$

$$= -Te^{-t/T} + 1.$$

6.6 The Laplace Transform

A technique of even greater power than the **D** operator method utilizes the Laplace transform. The most important extension available through the transform technique is the inclusion of "noise" type excitations. By this is meant signals which cannot be described as particular functions of time, but which can be given only through representations of the statistical probability of the signal containing certain components. A great amount of research in this area has been done, especially in the area of communications. In order to utilize the results of this research, the engineer must be acquainted with the Laplace transform.

Another reason for studying the Laplace transform is that much of the original and present fundamental automatic control literature is written by utilizing transform notation. In a sense, we must study the technique simply in order to develop our technical vocabulary.

Many of the mathematical manipulations and forms associated with the Laplace transform method are identical to those associated with the e^{rt} or **D** operator methods. As might be expected, the algebra involved in solving any particular problem is identical for all three methods and often involves factoring a polynomial, partial fraction expansion, and the solution of some simultaneous linear algebraic equations. It is truly unrealistic then to consider the Laplace transform method to have any magical power—such as was often implied when the method was taught in the past. Because of the extremely close relationship between the three methods, the author has included a summary section, following this, in which the similarities of the methods are explored. The student will notice these similarities developing as the Laplace transform method is presented.

6.6(a) Nature of the Laplace Transform Technique

The objective of the Laplace transform is to change problems in linear system dynamics from the time-dependent variable system (or "time domain") into some other system where differentiation and integration become algebraic operations. The manipulations required for the problem may be carried out in this new variable system and the result, when achieved, transformed back into the time domain. The process is somewhat analogous to the transformation of certain geometry problems from Cartesian to polar coordinates in order to facilitate solution.

The variable system which is utilized in the Laplace transform is called the "s domain." The variable s may be thought of as a dummy variable—one which is artificially introduced into a problem and is eliminated before the problem is completed. It turns out that the variable s must be complex; i.e., it must have real and imaginary parts. Denoting the Laplace transform of a function $f(t)$ by $\mathscr{L}\{f(t)\}$, the general form of the transformation becomes

$$\mathscr{L}\{f(t)\} = F(s).$$

$F(s)$ is some function in the s domain which is derived from the original function $f(t)$ in the time domain. Since s is a complex quantity, the function $F(s)$ may be complex; hence, two functions actually arise, one giving $\text{Re}[F(s)]$ and another giving $\text{Im}[F(s)]$. That the s domain is a more complicated place than the time domain is fairly obvious; the special attributes of the s domain which make the method practical will be shown as we continue.

The question of what kind of transformation will achieve the desired goal was the one which was brilliantly answered by Laplace. Here is his result:

$$F(s) = \int_0^\infty e^{-st} f(t)\, dt$$

Let us look at this result briefly. We notice that:

(1) The integration process serves to eliminate the variable t; i.e., if we take the definite integral of any function, we eliminate its independent variable.
(2) The use of the limits of integration of zero and infinity insure that every aspect of $f(t)$ of interest will have some influence of $F(s)$, so long as we properly account for conditions at $t = 0$.
(3) It is obvious that a simple algebraic transformation, such as replacing t with s, s^2, e^s or some such expression in s, will not change the nature of the differentiation and integration processes. Hence, such a transformation would not be of interest.
(4) Because of the nature of the transformation, there is no point in the s plane which corresponds to some particular time; i.e., we do not have a point-by-point mapping from the time domain to the s domain, nor vice versa.
(5) It is possible that the integral cannot be evaluated for some values of s. For example, suppose $f(t) = 1$. The integral will be unbounded and $F(s)$ is not defined for $\text{Re}(s) < 0$. This *does not* limit the usefulness of the transform, as may be seen informally from the following line of reasoning.
   (a) Suppose we have two transforms, $F_1(s)$ and $F_2(s)$, of two different functions, $f_1(t)$ and $f_2(t)$. Although $F_1(s)$ and $F_2(s)$ will coincide in some places in the s domain, it will be most unusual for them to coincide along a finite line in the s plane. Such coincidence may be proved to require extremely strong mathematical interdependence of $f_1(t)$ and $f_2(t)$.

(b) If this is the case, any infinite line in the s plane (such as from $s = 100 - j\infty$ to $s = 100 + j\infty$) will most certainly be adequate for the description of functions of interest to us. In other words, every function with which we are concerned will have a transform with a different cross section at this line; hence, although the dummy variable s might take on any value whatsoever in theory, in our work we need let it take on only the values along some particular line in the s plane. In a sense, "the values of s" is a meaningless phrase since s is a dummy variable; as long as we confine our attention to a portion of the s plane where the transform exists, we may achieve meaningful results.

The most important special attribute of the Laplace transformation, from our point of view, is the way in which differentiation is transformed. This is

$$\left[\frac{df(t)}{dt}\right] = s\,\{f(t)\} - f(0+)$$

$$= s\,F(s) - f(0+) \tag{6.22}$$

Equation (6.22) indicates that the Laplace transform of the derivative of a of a function is the product of s and the transform of the function itself, adjusted by the initial value of the function (as approached from positive values of time). Hence, differentiation in the time domain is replaced by multiplication by s in the s domain.

Since this result is so important, let us prove it immediately:

$$\int_0^\infty e^{-st}\frac{df(t)}{dt}\,dt = e^{-st}f(t)\Big|_0^\infty + s\int_0^\infty e^{-st}f(t)\,dt$$

using integration by parts as follows:

$$\int_a^b u\,dv = uv\Big|_a^b - \int_a^b v\,du \qquad \begin{aligned} u &= e^{-st} \\ v &= f(t) \end{aligned}$$

then, assuming $e^{-ts}f(t) \to 0$ as $t \to \infty$ for sufficiently large s,*

$$\int_0^\infty e^{-st}\frac{df(t)}{dt}\,dt = s\int_0^\infty e^{-st}f(t)\,dt - f(0+)$$

$$\mathscr{L}\left\{\frac{df(t)}{dt}\right\} = s\mathscr{L}\{f(t)\} - f(0+)$$

By a repeated application of the same technique it may be shown that

* This formula is valid in only a portion of the s domain. However, as has been noted previously, we do not need the complete s domain in order to apply the Laplace transform technique.

$$\mathscr{L}\left\{\frac{d^n f(t)}{dt^n}\right\} = s^n \mathscr{L}\{f(t)\} - s^{n-1} f(0) - s^{n-2}\frac{df(0)}{dt} \cdots - \frac{d^{n-1}f(0)}{dt^{n-1}}.$$

Furthermore, it is easy to show that

$$\mathscr{L}\{af(t)\} = a\,\mathscr{L}\{f(t)\}$$

and $$\mathscr{L}\{f_1(t)\} + \mathscr{L}\{f_2(t)\} = \mathscr{L}\{f_1(t) + f_2(t)\} = F_1(s) + F_2(s).$$

Example 6.10 The application of the Laplace transform method may be previewed by considering the usual spring-mass-dashpot system problem:

$$m\ddot{x} + b\dot{x} + kx = F(t)$$

$$\mathscr{L}\{m\ddot{x} + b\dot{x} + kx\} = \mathscr{L}F(t)$$

$$m\mathscr{L}\{\ddot{x}\} + b\mathscr{L}\{\dot{x}\} + k\mathscr{L}\{x\} = \mathscr{L}\{F(t)\}$$

$$m[s^2\mathscr{L}\{x\} - sx(0) - \dot{x}(0)] + b[s\mathscr{L}\{x\} - x(0)] + k\mathscr{L}\{x\} = \mathscr{L}\{F(t)\}$$

$$(ms^2 + bs + k)\mathscr{L}\{x\} - (ms + b)x(0) - m\dot{x}(0) = \mathscr{L}\{F(t)\}$$

$$\mathscr{L}\{x\} = \frac{\mathscr{L}\{F(t)\}}{ms^2 + bs + k} + \frac{(ms + b)x(0) + m\dot{x}(0)}{ms^2 + bs + k} \tag{6.23}$$

In order to determine $x(t)$, we must determine the function having the right-hand side of (6.23) as its transform. Denoting the function of time which has $F(s)$ as its transform by $\mathscr{L}^{-1}\,[F(s)]$ (which we read as the "inverse Laplace transform of $F(s)$"), we have that

$$x(t) = \mathscr{L}^{-1}\left[\frac{\mathscr{L}\{F(t)\} + (ms + b)x(0) + m\dot{x}(0)}{ms^2 + bs + k}\right] \tag{6.24}$$

6.6 (b) Derivations of Some Laplace Transforms

Although analytical methods may be applied to determine the inverse transform, the value of Laplace transform technique depends on the fact that transform pairs may be readily tabulated. In order to determine the solution of a specific problem, we would simply refer to such a table, find the functions making up the transform of the solution, and write out the answer.

It is instructive to consider the derivation of several transform pairs such as we find tabulated. The following are offered for this purpose:

1. Unit impulse, $f(\tau) = \delta$

$$\mathscr{L}\{\delta\} = \int_0^\infty e^{-st}\delta\,dt = \lim_{a\to 0}\left[\int_0^a e^{-st}\cdot\frac{1}{a}\,dt + \int_a^\infty s^{-st}\cdot 0\cdot dt\right]$$

$$= \lim_{a\to 0}\left[-\frac{1}{as}\,e^{-st}\Big|_0^a - 0\right]$$

$$= \lim_{a\to 0}\left[-\frac{1}{as}e^{-as} + \frac{1}{as}\right]$$

$$= \lim_{a\to 0}\left[\frac{1 - e^{-as}}{as}\right]$$

$$= \lim_{a\to 0}\left[\frac{0 + se^{-as}}{s}\right] \quad \text{(by L'Hospital's rule)}$$

$$= 1$$

$$\mathscr{L}\{\delta\} = 1. \tag{6.25}$$

2. Step function, $u(t) = \begin{cases} 0 \text{ for } t < 0 \\ 1 \text{ for } t \geq 0 \end{cases}$

$$\mathscr{L}\{u(t)\} = \int_0^\infty e^{-st}\,dt = -\frac{1}{s}e^{-st}\Big|_0^\infty = 0 + \frac{1}{s}$$

$$\mathscr{L}\{u(t)\} = \frac{1}{s} \tag{6.26}$$

This transform exists only if the real part of s is positive.

3. Exponential decay, e^{-at}

$$\mathscr{L}\{e^{-at}\} = \int_0^\infty e^{-st}e^{-at}\,dt = \int_0^\infty e^{-(a+s)t}\,dt$$

$$= -\frac{1}{a+s}e^{-(a+s)t}\Big|_0^\infty = 0 + \frac{1}{a+s}$$

$$\mathscr{L}\{e^{-at}\} = \frac{1}{a+s} \tag{6.27}$$

This transform exists only if $\text{Re}(a + s) > 0$.

4. Sinusoid, $\sin \omega t$

$$\mathscr{L}\{\sin \omega t\} = \int_0^\infty e^{-st}\sin \omega t\,dt$$

$$= \int_0^\infty e^{-st}\left(\frac{e^{j\omega t} - e^{-j\omega t}}{2j}\right)dt$$

$$= \frac{1}{2j}\int_0^\infty [e^{(j\omega - s)t} - e^{-(j\omega + s)t}]\,dt$$

$$= \frac{1}{2j}\left[\frac{1}{j\omega + s}e^{(j\omega - s)t} + \frac{1}{j\omega + s}e^{-(j\omega + s)t}\right]_0^\infty$$

$$= \frac{1}{2j}\left[-\frac{1}{j\omega - s} - \frac{1}{j\omega + s}\right]$$

$$= \frac{1}{2j}\left[-\frac{j\omega + s}{-\omega^2 - s^2} - \frac{j\omega - s}{-\omega^2 - s^2}\right]$$

$$= \frac{1}{2j}\left[\frac{2j\omega}{s^2 + \omega^2}\right] = \frac{\omega}{s^2 + \omega^2}$$

$$\mathscr{L}\{\sin \omega t\} = \frac{\omega}{s^2 + \omega^2} \tag{6.28}$$

This transform exists for Re(s) > 0.

A table of transform pairs is included at the end of this section.

The most useful property of Laplace transforms, the way in which time differentiation is transformed, has already been presented. There are other useful properties of the transform as follows:

1. *Transform of integration with respect to time*

If
$$g(t) = \int_0^t f(t)\,dt + g(0) \tag{6.29}$$

$$F(s) = \mathscr{L}\{f(t)\}$$

and
$$\mathscr{L}\{g(t)\} = \frac{F(s)}{s} + \frac{g(0)}{s} \tag{6.30}$$

where $g(0)$ is the value of $g(t)$ as time zero is approached from positive values of time.

If
$$h(t) = \underbrace{\int_0^t \int_0^t \cdots \int_0^t}_{n\text{ times}} f(t)\,\underbrace{dt\,dt \cdots dt}_{n\text{ times}} \tag{6.31}$$

Proof: If
$$\mathscr{L}\{h(t)\} = \frac{F(s)}{s^n} + \frac{\mathbf{D}^{n-1}h(0)}{s^n} + \frac{\mathbf{D}^{n-2}h(0)}{s^{n-1}} + \frac{\mathbf{D}^{n-3}h(0)}{s^{n-2}} + \cdots + \frac{h(0)}{s}. \tag{6.32}$$

$$\mathscr{L}\{g(t)\} = \frac{F(s)}{s} + \frac{g(0)}{s}$$

$$s\mathscr{L}\{g(t)\} = F(s) + g(0)$$

$$\mathscr{L}\{\mathbf{D}g(t)\} + g(0) = F(s) + g(0) \qquad \text{from (6.22)}$$

$$\mathscr{L}\{\mathbf{D}g(t)\} = F(s)$$

$$dg(t) = f(t)$$

$$g(t) = \int_0^t f(t)\,dt + g(0).$$

2. *Final value theorem*

$$\lim_{t\to\infty} f(t) = \lim_{s\to 0} sF(s) \tag{6.33}$$

if $F(s) = \mathscr{L}\{f(t)\}$ and $\mathscr{L}\{\mathbf{D}f(t)\}$ exists.

IMPORTANT: There must be no value of s with a zero or negative real part which can make the denominator of $sF(s)$ equal to zero. Otherwise the equation *does not hold.*

This is an extremely useful theorem which allows us to determine the condition of a system after the natural response has disappeared by the simple artifice of letting s go to zero in $sF(s)$, where $f(t)$ is a variable in the system. Of course, we may not expect this theorem to give a correct result if the natural motion of a system never dies out, or if a system is being forced by a sinusoid, a rising exponential, etc.

Proof:

$$\int_0^\infty \mathbf{D}f(t)e^{-st}\,dt = sF(s) - f(0+)$$

$$\lim_{s\to 0}\int_0^\infty \mathbf{D}f(t)e^{-st}\,d\tau = \lim_{s\to 0}[sF(s) - f(0+)]$$

$$\int_0^\infty \mathbf{D}f(t)\,dt = \lim_{s\to 0}[sF(s)] - f(0+)$$

$$\lim_{t\to\infty} f(t) - f(0+) = \lim_{s\to 0}[sF(s)] - f(0+)$$

$$\lim_{t\to\infty} f(t) = \lim_{s\to 0} sF(s).$$

3. *Initial value theorem*

$$\lim_{t\to 0} f(t) = \lim_{s\to\infty} sF(s) \tag{6.34}$$

if $F(s) = \mathscr{L}\{f(t)\}$ and $\mathscr{L}\{\mathbf{D}f(t)\}$ exists.

This is a companion theorem to the final value theorem and is also quite useful. The proof is similar in nature to the final value theorem.

4. *Translation in time*

If $$\mathscr{L}\{f(t)\} = F(s)$$

and

$$f(t) = 0 \qquad \text{for } t < a,\ a > 0$$

$$\mathscr{L}\{f(t-a)\} = e^{-as}F(s). \tag{6.35}$$

This relationship allows one to account for time delays which occur in some systems.

5. *Relationships of minor usefulness*

(a) $$\frac{d}{ds}F(s) = -\mathscr{L}\{tf(t)\} \tag{6.36}$$

(b) $$\int_s^\infty F(s)\,ds = \mathscr{L}\left\{\frac{f(t)}{t}\right\} \tag{6.37}$$

(c) $$F(s-a) = \mathscr{L}\{e^{at}f(t)\} \tag{6.38}$$

These relationships are sometimes helpful in evaluating transforms.

6.6(c) Existence of the Laplace Transform

Although the existence of a Laplace transform for a particular function of time usually becomes clear when one attempts to find the transform, it is advisable to know the mathematical conditions which must be satisfied in order for a transform to exist. These are as follows:

If $\mathscr{L}\{f(t)\}$ is to exist,

(1) $f(t)$ must be continuous or piecewise continuous for $0 < t < \infty$.
(2) $\lim t^m|f(t)|$ must be finite for some number m, where $m < 1$, $t \to 0$
(3) $\lim e^{-n\tau}|f(t)|$ must be finite for some number n.

If these conditions are not met, the integral $\int_0^\infty e^{-st} f(t)dt$ does not exist.

Condition 1, requiring that $f(t)$ be at least piecewise continuous, allows finite discontinuities but indicates that functions such as $1/(t-a)$ and $\tan \omega t$ cannot have Laplace transforms. Condition 2, which concerns the behavior of the function near $t = 0$, indicates that the function $1/e^{t^2}$ cannot have a Laplace transform. Finally, condition 3, concerning the behavior of $f(t)$ for large t, indicates that e^{t^2} cannot have a Laplace transform. These conditions should be checked whenever functions not found in tables of Laplace transforms are considered.

6.6 (d) Tabulation of Laplace Transforms

Following is a table of Laplace transforms which will be found adequate for all problems encountered in this text and most found in practice.

Table 6.2 Dimensionless $\mathscr{L}$ Transform Pairs

No.	$F(s)$	$f(t) \quad 0 \le t$	Comment
1	$1/s$	1 or $u(t)$	Unit function or step function
2	$\dfrac{1}{\tau s + 1}$	$\dfrac{1}{\tau} e^{(-t/\tau)}$	Exponential lag
3	$\dfrac{\omega}{s^2 + \omega^2}$	$\sin \omega t$	
4	$\dfrac{s}{s^2 + \omega^2}$	$\cos \omega t$	
5	$\dfrac{1}{s^2 + 2\zeta\omega_n s + \omega_n{}^2}$	where $\zeta < 1$: $\dfrac{1}{\omega_n\sqrt{1-\zeta^2}} e^{-\xi\omega_n t} \sin \omega_n\sqrt{1-\zeta^2}\, t$	Underdamped quadratic factor
		where $\zeta = 1$: $te^{-\omega_n t}$	Critically damped quadratic factor
		where $\zeta > 1$: $\dfrac{1}{\omega_n\sqrt{\zeta^2 - 1}} e^{-\xi\omega_n t} \sinh \omega_n\sqrt{\zeta^2 - 1}\, t$	Overdamped quadratic factor
6	$1/s^2$	t	
7	$1/s^n$	$\dfrac{1}{(n-1)!} t^{(n-1)}$	
8	$\dfrac{1}{(\tau s - 1)^n}$	$\dfrac{1}{(n-1)!} \dfrac{t^{(n-1)}}{\tau^n} e^{(-t/\tau)}$	nth order exponential lag
9	$\dfrac{1}{s} e^{-as}$	$u(t-a)$	Delayed unit function, step function lag
10	$\dfrac{1}{s}(1 - e^{-bs})$	$u(t) - u(t-b) \quad b > 0$	Impulse of b time units in duration beginning at time $t = 0$
11	$\dfrac{1}{s}(e^{-as} - e^{-bs})$	$u(t-a) - u(t-b) \quad b > a$	Impulse of $b - a$ time units in duration delayed by the time

From: Principles of Servomechanisms Gordon S. Brown and Donald L. Campbell John Wiley & Sons, Inc. New York 1948

6.7 Comparison of Solution Techniques

Three approaches to the solution of linear system equations have been presented. Although the reasoning associated with each method is different, the results are necessarily identical and many of the manipulations and forms which appear are similar. Although the practical differences between the methods are small, they are significant enough that one method can handle a specific situation quite gracefully, while another might be rather clumsy. In particular, notation and the way in which initial conditions enter can strongly influence a given situation.

A question that might be considered is: "Are there any *real* differences between the methods?" In this section we will explore some of the striking similarities a bit further and try to determine whether or not the differences are worth our further concern.

Consider the differential equation

$$a_n\frac{d^nx}{dt^n} + a_{n-1}\frac{d^{n-1}x}{dt^{n-1}} + \cdots + a_1\frac{dx}{dt} + a_0 = b_m\frac{d^my}{dt^m} + \cdots.$$

Assume a solution of the form Ae^{rt}

$$(a_n r^n + a_{n-1}r^{n-1} + \cdots + a_1 r + a_0)X = (b_m r^m + \cdots)Y;$$

then

$$\frac{X}{Y} = \frac{b_m r^m + b_{m-1}r^{m-1} + \cdots + b_1 r + b_0}{a_n r^n + a_{n-1}r^{n-1} + \cdots + a_1 r + a_0}. \tag{6.39}$$

Rewrite the equation, using **D** operators:

$$\frac{x(t)}{y(t)} = \frac{b_m \mathbf{D}^m + b_{m-1}\mathbf{D}^{m-1} + \cdots + b_1\mathbf{D} + b_0}{a_n \mathbf{D}^n + a_{n-1}\mathbf{D}^{n-1} + \cdots + a_1\mathbf{D} + a_0}\, u(t). \tag{6.40}$$

Laplace transform the equation (assuming all initial conditions equal to zero for simplicity):

$$\frac{\mathscr{L}\{x(t)\}}{\mathscr{L}\{y(t)\}} = \frac{b_m s^m + b_{m-1}s^{m-1} + \cdots + b_1 s + b_0}{a_n s^n + a_{n-1}s^{n-1} + \cdots + a_1 s + a_0}. \tag{6.41}$$

The functions on the right-hand side of (6.39), (6.40), and (6.41) have very similar appearances but distinctly different meanings. The first is simply an algebraic quantity giving the ratio of two complex constants. The second is a highly codified way of writing a function of time. The last is a code, using a dummy variable s, for a change which might be applied to some other function of s. The last is traditionally called a "transfer function," the name arising from the fact that it allows the transform of $y(t)$ to be transferred into the transform of $x(t)$. There is no good reason why the right-hand sides of (6.39) and (6.40) should not be called "assumed solution" or "**D** operator" transfer functions. So let us call them such in the remainder of this text in order to accomodate those individuals who prefer those methods. We need remember only *which* approach we are using, and not be tempted to shift carelessly from one to the other in the middle of a problem. The transfer function then is understood to mean some way of indicating the relationship between the response and forcing functions in a linear system.

Suppose that we have two differential equations,

$$A(\mathbf{D})x(t) = B(\mathbf{D})y(t)$$
$$G(\mathbf{D})y(t) = H(\mathbf{D})z(t)$$

where

$$A(\mathbf{D}) = a_n\mathbf{D}^n + a_{n-1}\mathbf{D}^{n-1} + \ldots + a_1\mathbf{D} + a_0, \quad \text{etc.}$$

Then

$$\frac{y(t)}{z(t)} = \frac{H(\mathbf{D})}{G(\mathbf{D})}u(t)$$

so that

$$\frac{x(t)}{z(t)} = \frac{x(t)}{y(t)} \cdot \frac{y(t)}{z(t)} = \frac{B(\mathbf{D})}{A(\mathbf{D})} \cdot \frac{H(\mathbf{D})}{G(\mathbf{D})}u(t)$$

and also, of course,

$$\frac{X}{Z} = \frac{X}{Y} \cdot \frac{Y}{Z} = \frac{B(r)}{A(r)} \cdot \frac{H(r)}{G(r)}$$

if the e^{rt} approach is used.

Finally,

$$\frac{\mathscr{L}\{x(t)\}}{\mathscr{L}\{y(t)\}} = \frac{\mathscr{L}\{x(t)\}}{\mathscr{L}\{y(t)\}} \frac{\mathscr{L}\{y(t)\}}{\mathscr{L}\{z(t)\}} = \frac{B(s)}{A(s)} \cdot \frac{H(s)}{G(s)}.$$

Another widely used and extremely important relationship is the characteristic equation. This is traditionally the equation which results when using the e^{rt} approach and the forcing function $y(t)$ is taken equal to zero. The equation has the form

$$a_n r^n + a_{n-1} r^{n-1} + \ldots + a_1 r + a_0 = 0 \tag{6.42}$$

If we are working with **D** operators, the assumption that the forcing is zero leads to

$$(a_n\mathbf{D}^n + a_{n-1}\mathbf{D}^{n-1} + \ldots + a_1\mathbf{D} + a_0)x(t) = 0 \qquad (6.43)$$

Setting the first quantity in parentheses equal to zero would be meaningless, since it has significance only when operating on some function of time. It is known, however, that such an equation has a solution of the form Xe^{rt}. Equation (6.43) then leads to an equation which is identical to (6.42), and this is properly known as the characteristic equation when the D operator is applied.

In the case of the Laplace transform, if $y(t) = 0$, then $\mathscr{L}\{y(t)\} = 0$ and

$$(a_n s^n + a_{n-1}s^{n-1} + \ldots + a_1 s + a_0)\,\mathscr{L}\{x(t)\} = 0$$

which indicates that $x(t) = 0$, as it should. Obviously, the natural response of the system has not been exposed by this approach. This can be resolved by realizing that if we make $y(t) = \delta$, once the impulse is over the differential equation controlling the system has a right-hand side of zero and the conditions for the characteristic equation are satisfied. Then

$$(a_n s^n + a_{n-1}s^{n-1} + \cdots + a_1 s + a_0)\mathscr{L}\{x(t)\} = 1$$

$$\mathscr{L}\{x(t)\} = \frac{1}{a_n s^n + a_{n-1}s^{n-1} + \cdots + a_1 s + a_0}$$

$$= \frac{1}{(s - r_1)(s - r_2)\cdots(s - r_n)}$$

$$= \frac{K_1}{s - r_1} + \frac{K_2}{s - r_2} + \cdots + \frac{K_n}{s - r_n}.$$

The natural response is then a summation of terms of the form $K\,e^{rt}$, where the r's are determined from the equation

$$a_n s^n + a_{n-1}s^{n-1} + \ldots + a_1 s + a_0 = (s - r_1)(s - r_2)\ldots(s - r_n);$$

that is, the values of r are the values of s which make the original polynomial equal to zero. Hence,

$$a_n s^n + a_{n-1}s^{n-1} + \ldots + a_1 s + a_0 = 0 \qquad (6.44)$$

has a very specific significance, in fact identical to that of (6.42), and may rightly be called the characteristic equation.

Another interesting observation on the similarities of the methods arises from the final value theorem of the Laplace transform theory:

$$\lim_{t \to \infty} f(t) = \lim_{s \to 0} sF(s) \qquad (6.45)$$

Suppose that $f(t)$ is generated as the cascade response of a system described by $G(\mathbf{D})$ to the response of a system $I(\mathbf{D})$ to an impulse, i.e.,

$$f(t) = G(\mathbf{D})I(\mathbf{D})\delta.$$

Then from (6.45),

$$\lim_{t \to \infty} f(t) = \lim_{\mathbf{D} \to 0} \mathbf{D}\, G(\mathbf{D})I(\mathbf{D})$$

and we see that the final value theorem certainly exists for use in the D operator method. Its proof, without resort to Laplace transforms, is an interesting exercise.

6.8 Conclusion

An extensive amount of material has been included in this chapter. As the student must realize, a thorough understanding will require some work to achieve. However, a thorough understanding is an absolute necessity; much of the material which follows is based on these methods. Also, the techniques presented in this chapter are perhaps the most extensively used mathematical methods in engineering today. Their understanding can be a significant contribution to professional capacity.

Problems

In the following problems $A(r)$, $A(\mathbf{D})$, $A(s)$, $B(r)$, $B(\mathbf{D})$, and $B(s)$ indicate general polynomials. For example,

$$A(s) = a_n s^n + a_{n-1} s^{n-1} + \ldots + a_1 s + a_0.$$

1. Suppose x and y are related by the differential equation $A(\mathbf{D})x = B(\mathbf{D})y$. Show that if x_1 is the forced solution of the equation when $y = \cos \omega t$ and x_2 is the solution when $y = e^{j\omega t}$, then $x_1 = Re(x_2)$.

2. Consider the equation $A(\mathbf{D})x = 0$ where $A(\mathbf{D})$ is of the nth degree. Prove that, when $A(\mathbf{D})$ has complex factors, the imaginary portion of the solution must vanish if n real conditions are imposed on the solution.

3. Prove that a polynomial $A(r)$ is identical in magnitude when evaluated at $r = +j\omega$ and $r = -j\omega$. Similarly, for $\sigma + j\omega$ and $\sigma - j\omega$.

4. Develop a formula for $\mathbf{D}^m (t^n e^{rt})$. Use this formula to determine the conditions on an equation $A(\mathbf{D})x = 0$ which lead to a solution of the form $t^n e^{rt}$.

5. Show from $e^{j\theta} = \cos A + j \sin \theta$ that

$$\cos \theta = \tfrac{1}{2}(e^{j\theta} + e^{-j\theta})$$
$$\sin \theta = \tfrac{1}{2}j(e^{j\theta} - e^{-j\theta}):$$

6. Prove that, except for special cases, n conditions on x allow determination of the constants in the solution to $A(\mathbf{D})x = 0$. Consider conditions other than initial conditions; i.e., such as $x(1)$, $x(2)$, $x(3)$, and state the special conditions under which n conditions are inadequate.

7. Solve the following differential equations, using the e^{rt} method, for y = 0.

(1) $(1 + 8\mathbf{D})x = y$	$x(0) = 1$
(2) $(\mathbf{D}^2 + 36)x = y$	$x(0) = 1, \dot{x}(0) = 0$
(3) $(\mathbf{D}^2 + 36)x = y$	$x(0) = 0, \dot{x}(0) = 0$
(4) $(\mathbf{D}^2 + 36)x = y$	$x(0) = 0, \dot{x}(0) = 0$
(5) $(\mathbf{D}^2 + 2\mathbf{D} + 36)x = y$	$x(0) = 1, \dot{x}(0) = 0$
(7) $(1 + 7\mathbf{D})(\mathbf{D}^2 + 25)x = y$	$x(0) = 1, \dot{x}(0) =, \ddot{x}(0) = 0$
(7) $(1 + 7\mathbf{D})(\mathbf{D}^2 + \mathbf{D} + 16)x = y$	same
(8) $(1 + 7\mathbf{D})(\mathbf{D}^2 + 8\mathbf{D} + 16)x = y$	same
(9) $(1 + 7\mathbf{D})(\mathbf{D}^2 + 10\mathbf{D} + 16)x = y$	same
(10) $(\mathbf{D}^2 - \mathbf{D} + 9)x = y$	$x(0) = 1, \dot{x}(0) = 0$
(11) $(\mathbf{D} + 3)(\mathbf{D} - 1)x = y$	same
(12) $(\mathbf{D} + 3)(\mathbf{D} - 1)x = y$	$x(0) = 10^{-6}, \dot{x}(0) = 0$

How long will it take before $x = 1$?

(13) $(\mathbf{D} + 4)^3 x = y$	$x(0) = 1, \dot{x}(0) = 0, \ddot{x}(0) = 0$
(14) $(\mathbf{D} + 4)^2(\mathbf{D} - 3)x = y$	same
(15) $\mathrm{D}(\mathbf{D}^2 - 2\mathbf{D} + 9)x = y$	same
(16) $\mathbf{D}^2(\mathbf{D} + 3)x = y$	same
(17) $(\mathbf{D}^2 + \mathbf{D} + 49)^2 x = y$	same, plus $\dddot{x}(0) = 0$
(18) $(\mathbf{D}^3 + 2\mathbf{D}^2 + \mathbf{D} + 16)x = y$	same
(19) $(\mathbf{D}^3 + 8)x = y$	same
(20) $(\mathbf{D}^3 + 3\mathbf{D}^2 + 3\mathbf{D} + 1)x = y$	same.

8. Problem 7, $y = 1$ for $t > 0$, 0 for $t < 0$.

9. Problem 7, $y = t$.

10. Problem 7, $y = e^{-3t}$.

11. Problem 7, $y = e^{-7t}$.

12. Problem 7, $y = \cos 2t$.

13. Problem 7, $y = \cos 6t$.

14. Problem 7, $y = \tau^2 e^{-t} \cos 5t$.

15. Solve the following differential equations, using the e^{rt} method with $y_1 = 0$ and $y_2 = 0$.

(1) $(\mathbf{D}+1)x_1 + (\mathbf{D}+3)x_2 = y_1$	$x_1(0) = 1, \dot{x}_1(0) = 0$
$(\mathbf{D}+4)x_1 + (\mathbf{D}+1)x_2 = y_2$	$x_2(0) = 0, \dot{x}_2(0) = 0$
(2) $\mathbf{D}x_1 + (\mathbf{D}+1)x_2 = y_1$	same, plus $\ddot{x}_1(0) = 0$
$(\mathbf{D}+4)x_1 + \mathbf{D}^2 x_2 = y_2$	
(3) $(\mathbf{D}^2+16)x_1 + x_2 = y_1$	same, plus $\ddot{x}_1(0) = 0$
$\mathbf{D}^2 x_1 + (\mathbf{D} + 3)x_2 = y_2$	
(4) $(\mathbf{D}^2+2\mathbf{D}+9)x_1 + (\mathbf{D}+3)x_2 = y_1$	same, plus $\ddot{x}_1(0) = 0$
$(\mathbf{D}+4)x_1 + (\mathbf{D}^2 + \mathbf{D} + 8)x_2 = y_1$	$\ddot{x}_2(0) = 0.$

16. Solve 15(a) with $y_1 = e^{-5t}$, $y_2 = 0$.

17. Solve 15(b) with $y_1 = \sin 5t, y_2 = \sin 3t$.

18. Solve 15(c) with $y_1 = \cos 4t$, $y_2 = 0$.

19. Determine the interpretation of

$$y = \frac{1}{\mathbf{D}^2 + 2\zeta\,\omega_n\mathbf{D} + \omega_n^2}x.$$

20. Determine the partial fraction expansions of the following (check your solutions by addition of the partial fractions):

(1) $\dfrac{1}{(1+3\mathbf{D})(1+4\mathbf{D})(1+7\mathbf{D})}$

(2) $\dfrac{1}{(1+2\mathbf{D})(1+6\mathbf{D})(1+8\mathbf{D})}$

(3) $\dfrac{1}{(1+\mathbf{D})(1+2\mathbf{D})(1+3\mathbf{D})}$

(4) $\dfrac{1}{\mathbf{D}(1+\mathbf{D})(1+2\mathbf{D})}$

(5) $\dfrac{1}{(1+\mathbf{D})(\mathbf{D}^2+16)}$

(6) $\dfrac{1}{\mathbf{D}(1+\mathbf{D})(\mathbf{D}^2+49)}$

(7) $\dfrac{1}{(1+\mathbf{D})(\mathbf{D}^2+3\mathbf{D}+81)}$

(8) $\dfrac{1}{\mathbf{D}^3(1+\mathbf{D})}$

(9) $\dfrac{1}{(1+4\mathbf{D})^3(1+\mathbf{D})}$

(10) $\dfrac{1}{\mathbf{D}^2(1+\mathbf{D})^2(1+3\mathbf{D})}$

(11) $\dfrac{1}{(\mathbf{D}^2+25)^2(1+\mathbf{D})}$

(12) $\dfrac{1}{(\mathbf{D}^2+\mathbf{D}+16)(\mathbf{D}^2+2\mathbf{D}+8)}$

(13) $\dfrac{1}{(\mathbf{D}^2+8\mathbf{D}+16)(\mathbf{D}^2+4\mathbf{D}+4)}$

21. Tabulate, for future reference, the solutions to the following operator equations:

(1) $y = \dfrac{1}{\mathbf{D}} \cdot 1^*$ $\quad 1^* = 0$ for $t \leq 0$; 1 for $t > 0$

(2) $y = \dfrac{1}{\mathbf{D}^m} \cdot 1^*$

(3) $y = \dfrac{1}{1 + T\mathbf{D}} \cdot 1^*$

(4) $y = \dfrac{1}{\mathbf{D}^2 + \omega^2} 1^*$

(5) $y = \dfrac{1}{\mathbf{D}^2 + 2\zeta\,\omega_n \mathbf{D} + \omega_n^2} 1^*.$

22. Determine the response of a system described by the operator equation

$$y = \frac{1}{(1 + \mathbf{D})(\mathbf{D}^2 + \mathbf{D} + 16)} x$$

when

$$x = \begin{cases} 0 \text{ for } t < 0 \\ 10 \text{ for } 0 \leq t \leq 0.1 \\ 0 \text{ for } t > 0.1 \end{cases}$$

and

$$x(0) = 0, \dot{x}(0) = 0, \ddot{x}(0) = 0.$$

Compare this with the impulse response of the system.

Under what conditions is the impulse response of a system a good approximation to the response of the system for a constant input of finite duration? State your answer in terms of a transfer function $A(\mathbf{D})/B(\mathbf{D})$ and an input duration of ϵ. Consider the cases where $B(\mathbf{D})$ has factors $\mathbf{D}$, $(1 + T\mathbf{D})$, $(\mathbf{D}^2 + \omega^2)$ and $(\mathbf{D}^2 + 2\zeta\,\omega_n \mathbf{D} + \omega_n^2)$.

23. Use Table 6.1 and the "cascade impulse response" method to solve the following differential equations:

(1) $\dfrac{1}{1 + \mathbf{D}} x = e^{-5t}$ $\qquad x(0) = 0$

(2) $\dfrac{1}{(1 + \mathbf{D})(\mathbf{D}^2 + 16)} x = t^3$ $\qquad x(0) = 0,\ \dot{x}(0) = 0,\ \ddot{x}(0) = 0$

(3) $\dfrac{1}{\mathbf{D}^2 + \mathbf{D} + 10} x = \cos 3t$ $\qquad x(0) = 0,\ \dot{x}(0) = 0.$

24. Show that the integrating factor solution of first order differential equations may be used when the coefficient T in $(1 + T\mathbf{D})x = y$ is not a constant, but is some predetermined function of time $T(t)$. Discuss a general method for solving linear ordinary differential equations with *nonconstant* coefficients.

25. Use the **D** operator method to determine the forced portion of the response of the equations in Problem 7 for $y = \cos^2 3t$.

26. Problem 25, but with $y = \text{sgn}(\cos t)$,

$$\text{sgn}(x) = +1 \quad \text{if } x \geq 0$$
$$\text{sgn}(x) = -1 \quad \text{if } x < 0.$$

27. Derive the Laplace transform of the following functions:

(1) t^2
(2) t^m
(3) $\cos \omega t$
(4) $e^{-at} \cos \omega t$
(5) $t \sin \omega t$

28. Prove the initial value theorem.

29. Make neat and accurate sketch contour maps of both the real and imaginary portions of the functions:

(1) $\dfrac{1}{1 + s}$

(2) $\dfrac{1}{s^2 + \omega^2}$

(3) $\dfrac{1}{s^2 + 2\zeta \omega_n s + \omega_n^2}$

(4) $\dfrac{1}{s^3}$.

6.30. Show that if two functions

$$\frac{A_1(s)}{B_1(s)} \text{ and } \frac{A_2(s)}{B_2(s)}$$

where the A's are of the mth degree and the B's are of the nth degree, coincide at $m + n$ values of s, they must be identical. Use this result in preparing an informal argument showing that a line in the s domain contains more than an adequate range of values of s than are needed for usual applications of the Laplace transform.

31. Prove the final value theorem for **D** operators without resorting to the Laplace transform. It will be helpful to utilize the fact that the theorem is not valid unless the final value is a constant.

32. Write a set of rules which would allow a Laplace transform table to be utilized in the **D** operator method. Be precise regarding initial conditions.

33. Suppose

$$e(t) = r(t) - d(t)$$

so that

$$\mathscr{L}\{e(t)\} = \mathscr{L}\{r(t)\} - \mathscr{L}\{d(t)\}$$

$$\mathscr{L}\{c(t)\} = G(s)\mathscr{L}\{e(t)\}$$

and

$$\mathscr{L}\{d(t)\} = H(s)\mathscr{L}\{c(t)\}.$$

Find

(1) The characteristic equation of the system. [The forcing function is $r(t)$.]
(2) The transfer function relating $\mathscr{L}\{c(t)\}$ and $\mathscr{L}\{r(t)\}$.
(3) The transfer function relating $\mathscr{L}\{e(t)\}$ and $\mathscr{L}\{r(t)\}$.

7 Small excursion response of systems

The purpose of this chapter is to interpret, with emphasis on physical aspects, the mathematical methods and results of the preceding chapter. Although any linear system problem can be solved purely mathematically, it is a convenience for the engineer to be so familiar with the kinds of solutions which are found that he does not always have to carry out a detailed analysis. The objective here, then, is to develop that familiarity. Where possible, graphical interpretations are offered since they seem to be easy to recall. The applications of linear system analysis methods which appear here should also be useful to the student in solidifying his understanding of the preceding material.

The general problem of linear system analysis is represented by equation (7.1), which may be used as a reference framework for a study of the subject.

$$
\begin{aligned}
A_{11}(\mathbf{D})x_1 + A_{12}(\mathbf{D})x_2 + \ldots + A_{1n}(\mathbf{D})x_n &= B_1(\mathbf{D})y_1 \\
A_{21}(\mathbf{D})x_1 + A_{22}(\mathbf{D})x_2 + \ldots + A_{2n}(\mathbf{D})x_n &= B_2(\mathbf{D})y_2 \\
&\vdots \\
A_{n1}(\mathbf{D})x_1 + A_{n2}(\mathbf{D})x_2 + \ldots + A_{nn}(\mathbf{D})x_n &= B_n(\mathbf{D})y_n
\end{aligned}
\tag{7.1}
$$

Here, as usual, the $A(\mathbf{D})$'s represent polynomials in the differential operator $\mathbf{D}$. These will be referred to as "polynomial coefficients." This form may be written compactly by using matrix notation; however, for the purposes of the chapter, the matrix notation would veil the issues of interest and will be avoided. Instead, we will consider the single-dependent variable case, represented by

$$A(\mathbf{D})x = B(\mathbf{D})y \tag{7.2}$$

and in a section at the end of the chapter we show how the general case, as represented by (7.1), may be considered in terms of the simpler case given by (7.2).

7.1 Natural Response

In the preceding chapter it was shown that if $x_1(t)$ was a solution to the differential equation

$$A(\mathbf{D})x = 0, \tag{7.3}$$

then $x_1(t)$ would be a component of the solution to the equation

$$A(\mathbf{D})x = B(\mathbf{D})y.$$

The equation

$$A(r) = 0$$

was identified as the characteristic equation of the system, while $x_1(t)$, the solution to the characteristic equation, was called the complementary solution. This solution describes the "natural response" of the system.

In terms of transfer functions, the characteristic equation may be derived by setting the denominator of the operator equal to zero and making the proper notational changes; i.e., if we have

$$\frac{\mathscr{L}\{x\}}{\mathscr{L}\{y\}} = \frac{B(s)}{A(s)} \quad \text{or} \quad \frac{x}{y} = \frac{B(\mathbf{D})}{A(\mathbf{D})}$$

the characteristic equation may be written directly $A(s) = 0$ or $A(r) = 0$, depending on whether or not Laplace transforms are being applied.

7.1(a) Natural Response by Pole Location

Since natural response is always of the form e^{rt} (or in special cases $t^m e^{rt}$), let us continue the discussion in that terminology. Writing the transfer function (assuming the e^{rt} method) in its factored form, we have

$$\frac{X}{Y} = \frac{(r - z_1)(r - z_2) \ldots (r - z_m)}{(r - p_1)(r - p_2) \ldots (r - p_n)}. \tag{7.4}$$

The p's in this expression have a particularly important role in linear system analysis since they indicate the natural response directly. We call them "poles" of the transfer function.

> A pole of a transfer function is a value of r which makes the denominator become zero; i.e., which causes the ratio X/Y to become infinite. Associated with each pole of the transfer function is a term in the natural response of the system; hence, the natural response is given by a sum
>
> $$c_1 e^{p_1 t} + c_2 e^{p_2 t} + \ldots + c_n e^{p_n t}$$

when the denominator is of the nth degree.

A more meaningful form of the transfer function for this discussion is

$$\frac{X}{Y} = \frac{(r - z_1)(r - z_2)\cdots(r - z_m)}{r^{q_1}(r - p)^{q_2}(r^2 + \omega^2)^{q_3}(r^2 + 2\zeta\omega_n r + \omega_n{}^2)^{q_4}\cdots}.$$

That is, we recognize that when we factor $A(r)$, we get in general some zero factors, some real factors, some purely imaginary factors, some quadratic factors, and these may be repeated. If these poles are marked on a graph of the complex plane, they appear as follows:

r factors → at the origin
$(r - p)$ factors → on the real axis
$(r^2 + \omega^2)$ factors → on the imaginary axis, one at $+j\omega$ and one at $-j\omega$
$(r^2 + 2\zeta\,\omega_n r + \omega_n^2)$ factors ($\zeta < 1$) → one at $\omega_n(-\zeta + j\sqrt{1 - \zeta^2})$
one at $\omega_n(-\zeta - j\sqrt{1 - \zeta^2})$

On a graph of the poles of a transfer function, as in Figure 7.1, we may immediately identify the nature of the component of the natural response associated with each pole.

When factors of $A(r)$ are repeated, terms of the form $t^m\, e^{pt}$ appear. It is important to note that if $\text{Re}(p) < 0$, so that e^{pt} represents a component which dies out with increasing time, then $t^m\, e^{pt}$ also dies out, for any m. This is true because $\lim_{t\to\infty} t^m e^{pt} = 0$ for any m; i.e., e^{pt} gets small faster than t^m gets large. Hence, any pole in the left half-plane, repeated or not, represents a natural motion which dies out with time.

As further interpretation we note that any poles on the imaginary axis, including the origin, represent a natural response which never dies out. If such a pole is repeated, the response will increase in magnitude in direct proportion to time. Any pole in the right half-plane represents a motion which increases in magnitude with time. Since the objective of system design is normally to make the natural motion disappear as quickly as possible, we may understand why "poles in the right half-plane" are normally considered undesirable.

It should be reassuring to the student that we have identified *all* of the possible components of the natural motion of a system. We have noticed that the denominator of a transfer functions is always a polynomial and that there is a limited number of forms which the factors of a polynomial may take. Corresponding to each form of factor is a kind of natural response, the details of that response depending on the numerical values associated with the factor.

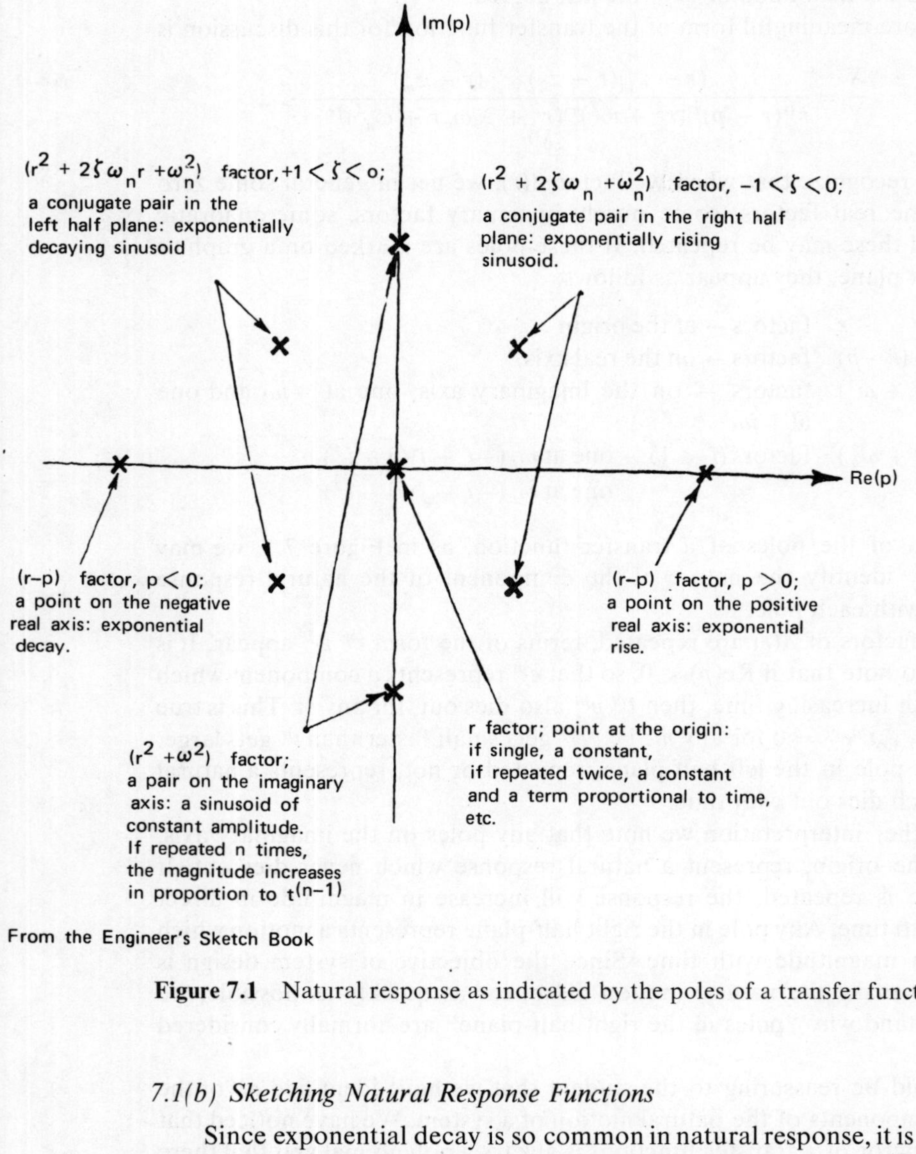

Figure 7.1 Natural response as indicated by the poles of a transfer function.

7.1(b) Sketching Natural Response Functions

Since exponential decay is so common in natural response, it is convenient to be able to sketch it rapidly. Here are a few hints for $x = e^{-t/T}$:

(1) x decreases to approximately 37% of its original value during any period T.

(2) The slope of x is given by a line through $x(t)$ to the abscissa at $(t + T)$.

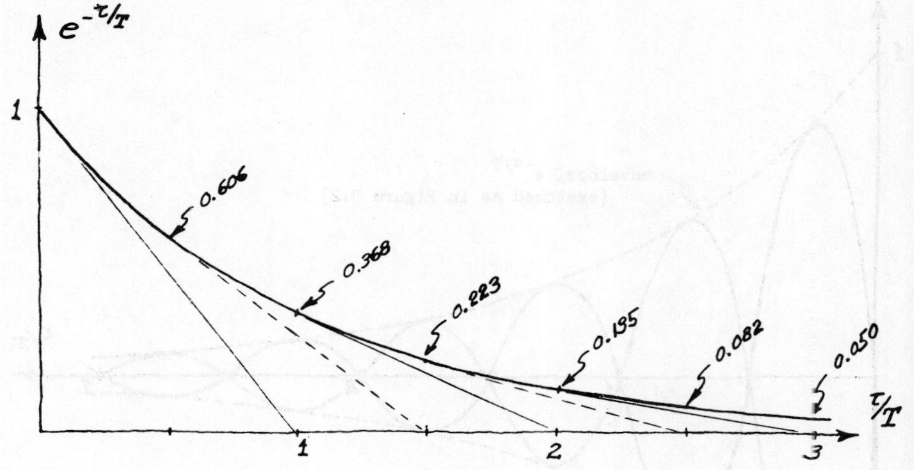

From the Engineer's Sketch Book

Figure 7.2 Sketching $e^{-t/T}$.

Application of these hints is indicated in Figure 7.2. Sketching a decaying sinusoid is also facilitated by these rules. Since $|\sin \omega t| \leq 1$, $e^{-t/T} \sin \omega t$ can never be larger than $e^{-t/T}$; therefore, $e^{-t/T}$ must be an envelope of the response and when $\sin \omega t = \pm 1$, the response must be equal and tangent to $e^{-t/T}$. Of course, when $\sin \omega t = 0$, the response is at zero. A sketch of such a situation is indicated in Figure 7.3, where the period of the sinusoid is taken to be $T/2$.

Similar techniques may be used when a response includes a t factor. For example, if an undamped spring mass system is excited at its natural frequency, the forced response has a component $t \sin \omega_n t$. The envelope of this motion is the pair of lines t and $-t$. It is left as a simple exercise for the student to develope rules for sketching exponentially rising responses.

7.2 Impulse Response

Although the major aspects of the performance of a system are divulged by the natural response, as discussed in section 7.1, the impulse response is actually a somewhat more useful device for determining the same information. For one thing, in order to determine the relative magnitude of the various components of the natural response of a system, we must solve the initial condition equations. However, in the comparison of the behavior of different systems it is difficult to find a set of meaningful standard initial conditions. The impulse response approach avoids these difficulties since we may apply the

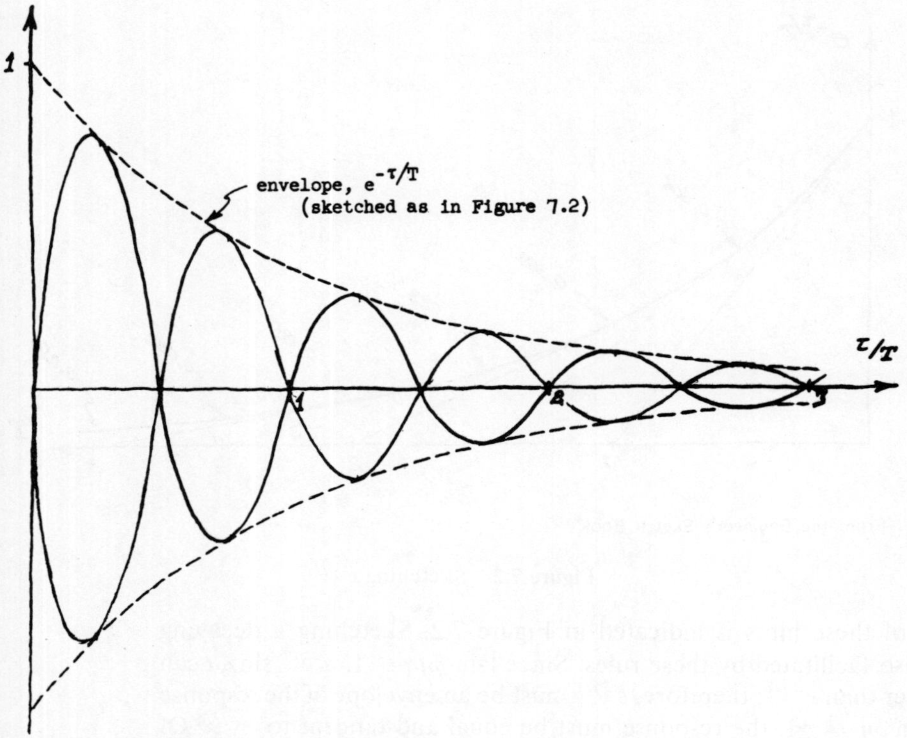

From the Engineer's Sketch Book

Figure 7.3 Sketch of $e^{-t/T}\,\mathrm{Sin}\,(\frac{2\pi t}{T})$.

partial fraction expansion rules in determining magnitudes, and we may meaningfully and conveniently always start from an initial condition of all variables and their derivatives equal to zero. A further attractive aspect of the impulse response is that impulse tests on systems with unknown transfer functions may often be carried out quite simply. For example, in determining the "yaw" transfer function of an aircraft, a single shell might be fired from a gun mounted on the wing. Recordings of the response of the plane allow the transfer function to be determined.

7.2(a) Determination of Impulse Response

Either Laplace transforms or the **D** operator method can be used in determining impulse response. Let us illustrate the **D** operator method, since it brings out some of the basic aspects of the situation. We again start with the differential equation

$$A(\mathbf{D})x = B(\mathbf{D})y.$$

Then

$$x = \frac{B(\mathbf{D})}{A(\mathbf{D})}\,\delta.$$

If we expand the transfer function by partial fraction expansion [assuming real roots of $A(\mathbf{D})$ for simplicity here],

$$x = \left(\frac{K_1}{1 + T_1\mathbf{D}} + \frac{K_2}{1 + T_2\mathbf{D}} + \ldots + \frac{K_n}{1 + T_n\mathbf{D}}\right)\delta. \tag{7.5}$$

From the analysis of Chapter 6 we have

$$x = e^{-t/T}\left[\frac{1}{T}\int_0^t e^{t/T} y\, dt + x_0\right]$$

if

$$(1 + T\mathbf{D})x = y.$$

For the first term in (7.5) we would have

$$x_1 = K_1\, e^{-t/T_1}\left[\frac{1}{T_1}\int_0^t e^{t/T_1}\,\delta\, dt\right] \tag{7.6}$$

assuming that $x_0 = 0$ (δ is defined in Chapter 6). Then

$$x_1 = \frac{K_1}{T_1}\, e^{-t/T_1} \tag{7.7}$$

The total response is given by

$$x = \frac{K_1}{T_1} e^{-t/T_1} + \frac{K_2}{T_2} e^{-t/T_2} + \ldots + \frac{K_n}{T_n} e^{-t/T_n} \tag{7.8}$$

It is clear that this response includes all the terms of the natural response, and that the magnitude of each term is determined by the simple partial fraction expansion method.

There are four points which now arise relating to the impulse response method:

(1) Upon inspection of (7.8) it may be seen that $\dot{x}(0)$ may not necessarily be zero. In fact, on further contemplation the whole initial condition situation when an impulse is applied has suspicious aspects.

(2) What happens when $A(\mathbf{D})$ has a factor $(\mathbf{D}^2 + 2\zeta\,\omega_n\mathbf{D} + \omega_n^2)$?

(3) What happens when there are repeated factors of $A(\mathbf{D})$, so that some of the partial fractions have $\mathbf{D}$ in the numerator?

(4) How does the solution relate to the location of the poles of the transfer function?

Taking these up in order, let us look into the initial condition situation. For the simple case $(1 + T\mathbf{D})x = \delta$ we have

$$x = \frac{1}{T}e^{-t/T} \tag{7.9}$$

so that

$$x(0) = \frac{1}{T}, \quad \dot{x}(0) = -\frac{1}{T^2}.$$

Putting these values back in the differential equation at time zero, we have

$$x(0) + T\dot{x}(0) = \delta(0)$$

$$\frac{1}{T} - T \cdot \frac{1}{T^2} = \delta(0),$$

but $\delta(0)$ is supposedly infinite; and besides, the solution was made with the assumption that $x(0) = 0$. The trouble arises because we are dealing with a non-physical forcing function; anything that goes to infinity and back in zero time is certainly non-physical. Furthermore, the mathematics is all involved with limit processes—which sometimes yield non-intuitive results. In order to try to resolve the situation, let us investigate the case where the impulse is finite; i.e., where it takes on the value $1/a$ at time zero, holds it for a period a, and then returns to zero. We can analyze this by considering the impulse to be made up of two parts, a $(1/a)$ step at time zero and a $-(1/a)$ step at time a.

For the first portion,

$$(1 + T\mathbf{D})x = \frac{1}{a} \qquad x(0) = 0$$

$$x_1 = \frac{1}{aT} e^{-t/T} \int_0^t e^{t/T}\, dt = \frac{1}{a} e^{-t/T}(e^{t/T} - 1) = \frac{1}{a}(1 - e^{-t/T}). \tag{7.10}$$

For the second portion,

$$(1 + T\mathbf{D})x = -\frac{1}{a}$$

$$x_2 = -\frac{1}{aT} e^{-(t-a)/T}\left[\int_a^{t-a} e^{(t-a)/T}\, dt + x_2(a)\right] \qquad a < t$$

but
$$x_2 = -\frac{1}{a} e^{-(t-a)/T}[e^{(t-a)/T} - 1 + x_2(a)] \tag{7.11}$$

$$= -\frac{1}{a}\{1 - [1 - x_2(a)]e^{-(t-a)/T}\} \tag{7.12}$$

$$x_2(a) = x_1(a) = \frac{1}{a}(1 - e^{-a/T}). \tag{7.13}$$

The component and total responses are indicated in Figure 7.4. From the figure

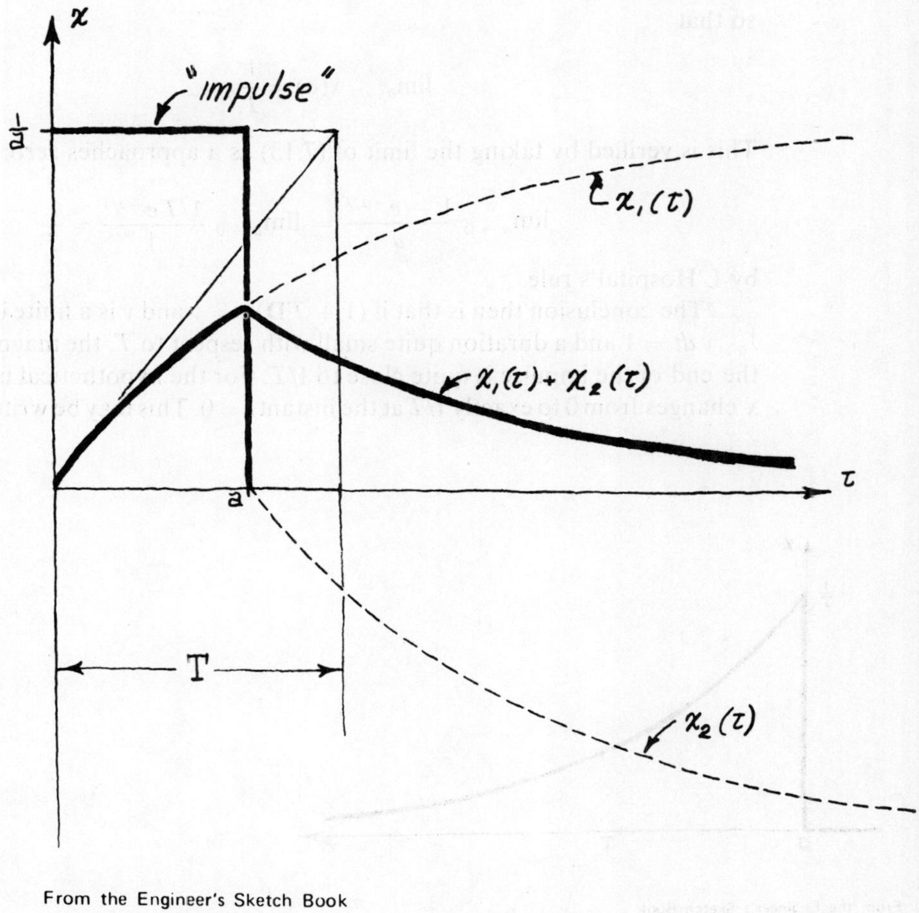

Figure 7.4 Solution of $(1 + TD)x$ to a finite impulse.

we may visualize what would happen if the period a were made shorter and shorter. The initial slope would get larger and larger, but the intersection of the solution and the end of the impulse would move closer and closer to the origin, thereby tending to keep $x(a)$ equal to a finite value. In fact, for times that are very small with respect to T,

$$x(t) \cong \frac{t}{aT}$$

$$x(a) \cong \frac{1}{T}$$

so that

$$\lim_{a \to 0} x(a) = \frac{1}{T}.$$

This is verified by taking the limit of (7.13) as a approaches zero:

$$\lim_{a \to 0} \frac{1 - e^{-a/T}}{a} = \lim_{a \to 0} \frac{1/Te^{-a/T}}{1} = \frac{1}{T}$$

by L'Hospital's rule.

The conclusion then is that if $(1 + T\mathbf{D})x = y$ and y is a finite impulse with $\int_0^\infty y\,dt = 1$ and a duration quite small with respect to T, the magnitude of x at the end of the impulse is quite close to $1/T$. For the hypothetical unit impulse, x changes from 0 to exactly $1/T$ at the instant $t = 0$. This may be written formally

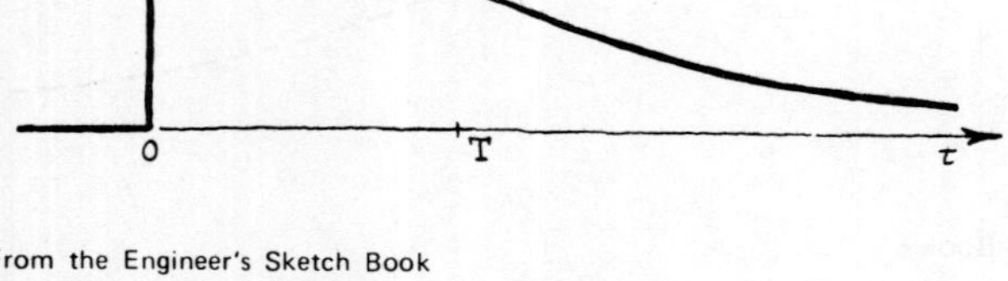

From the Engineer's Sketch Book

Figure 7.5 Unit impulse response for $(1 + T\mathbf{D})x = \delta$.

$$x(0-) = 0$$

$$x(0+) = \frac{1}{T}.$$

That is, if $t = 0$ is approached from $t < 0$, $x(0) = 0$, whereas if $t = 0$ is approached from $t > 0$, $x(0) = 1/T$.

The unit impulse response of x in $(1 + T\mathbf{D})x = \delta$ is shown in Figure 7.5. It is clear from the figure that we will never be able to satisfy an initial condition such as $\dot{x}(0) = 0$ if an impulse response is applied to a system with a real root in the denominator of the transfer function. We may, however, satisfy a condition such as $\dot{x}(0-) = 0$, and this is what is normally done, even though most discussions of the subject do not make that clear.

The matter of initial conditions deserves even more attention, since we have not considered other kinds of factors in $A(\mathbf{D})$, nor have we considered the overall initial condition situation. By arguments similar to the preceding, the impulse response associated with other forms of terms in the expansion of a transfer function may be found to be the following:

Table 7.1 Impulse Response*

$\frac{1}{\mathbf{D}}\delta = 1$
$\frac{1}{\mathbf{D}^2}\delta = t$
$\frac{1}{\mathbf{D}^n}\delta = \frac{t^{n-1}}{(n-1)!}$
$\frac{1}{1 + T\mathbf{D}}\delta = \frac{1}{T}e^{-t/T}$
$\frac{1}{(1 + T\mathbf{D})^2}\delta = \frac{1}{T^2}t\,e^{-t/T}$
$\frac{1}{(1 + T\mathbf{D})^n}\delta = \frac{1}{(n-1)!}\frac{1}{T^n}t^{n-1}\,e^{-t/T}$
$\frac{\mathbf{D}}{1 + T\mathbf{D}}\delta = -\frac{1}{T^2}e^{-t/T}$
$\frac{\mathbf{D}}{(1 + T\mathbf{D})^2}\delta = \frac{1}{T^2}(1 - \frac{t}{T})\,e^{-t/T}$
$\frac{1}{\mathbf{D}^2 + 2\zeta\omega_n\mathbf{D} + \omega_n^2}\delta = \frac{1}{\omega_n\sqrt{1-\delta^2}}e^{-\delta\omega nt}\quad \sin_{\omega n}\sqrt{1-\delta^2 t}$

* $x(0-)$, $\dot{x}(0-)$, $\ddot{x}(0-)$, etc., equal to zero

The more important impulse responses are sketched in Figure 7.6.

We note that for all these responses, the displacement, velocity, etc., are all zero up to time zero when they each take on some finite value. If then, for an nth-order system, we have the initial conditions

$$\begin{aligned} x(0-) &= 0 \\ \dot{x}(0-) &= 0 \\ \ddot{x}(0-) &= 0 \\ &\vdots \\ \mathbf{D}^{n-1}x(0-) &= 0 \end{aligned}$$

and apply an impulse to the system, a response made up of the components given in Table 7.1 will satisfy the conditions. Furthermore, if the magnitudes of the components have been determined by partial fraction expansion, the response will be a true indication of the response of the system to the impulse. On the other hand, if we try to satisfy the initial conditions

$$\begin{aligned} x(0+) &= 0 \\ \dot{x}(0+) &= 0 \\ \ddot{x}(0+) &= 0 \\ &\vdots \\ \mathbf{D}^{n-1}x(0+) &= 0 \end{aligned}$$

by assuming some initial condition on each term of the response, we will have a long algebraic exercise and the result, although having some meaning, will not indicate the impulsive response of the system. For one thing, the only solution to $A(\mathbf{D})x = 0$ (which will govern the system for $t > 0$) with these initial conditions is $x = 0$. That result is certainly not of interest.

7.2(b) Impulse Response Magnitude from Pole Locations

The preceding discussion, which was concerned with the initial condition difficulties encountered with impulse response, also—necessarily—covered the matters of quadratic and repeated roots. Therefore, we may go on to the last question, regarding the relationship between impulse response and the location of the poles of a transfer function. Of course, the form of the impulsive response associated with each pole of a transfer function is readily determined from the location of the pole in the complex plane, since it is the same as that for the natural response. An interesting and useful result, however, is that the

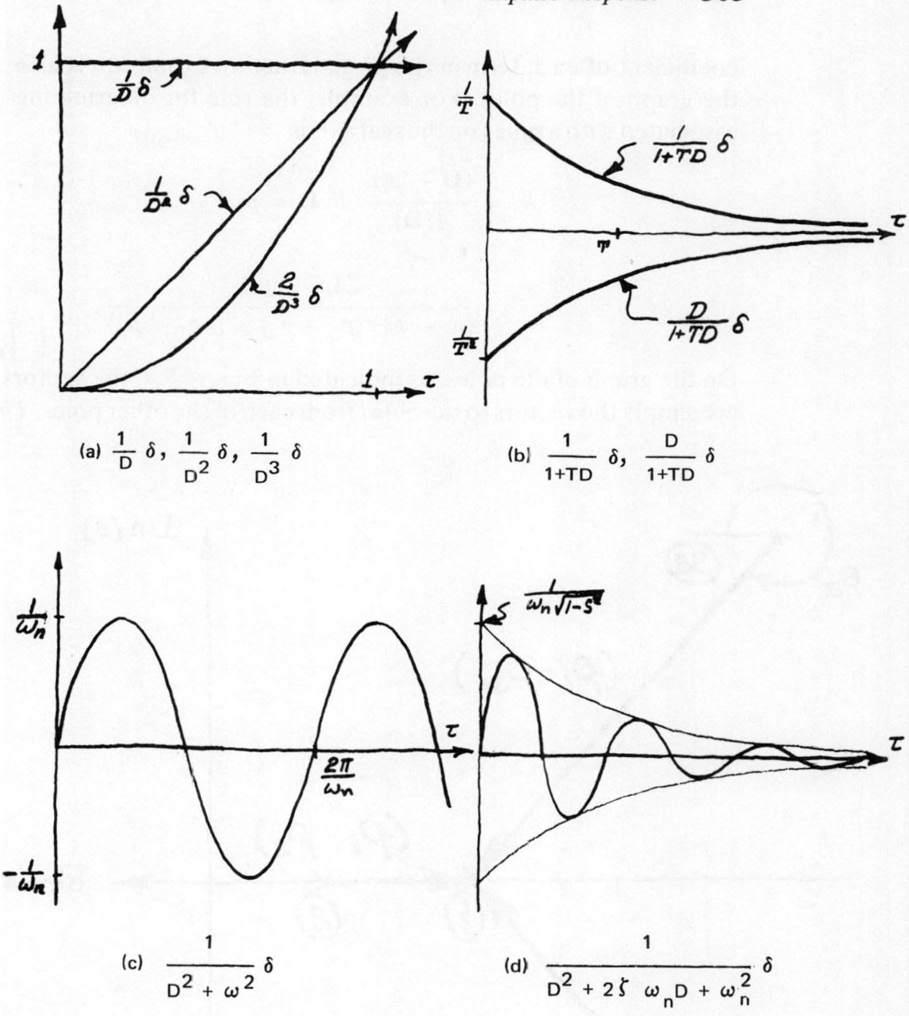

From the Engineer's Sketch Book

Figure 7.6 Some important impulse responses.

coefficient of each term in the partial fraction expansion is also available from the graph of the poles. For example, the rule for determining the coefficient associated with a pole *i* on the real axis is

$$K_i = \frac{(\mathbf{D} - p_i)}{A(\mathbf{D})} \qquad \mathbf{D} = p_i$$

$$= \frac{(p_i - p_i)}{(p_i - p_1)(p_i - p_2)\ldots(p_i - p_n)}.$$

On the graph of the poles, as indicated in Figure 7.7, the vectors $(p_i - p_2)$, etc., are simply the vectors to the pole *i* from each of the other poles. *The magnitude of*

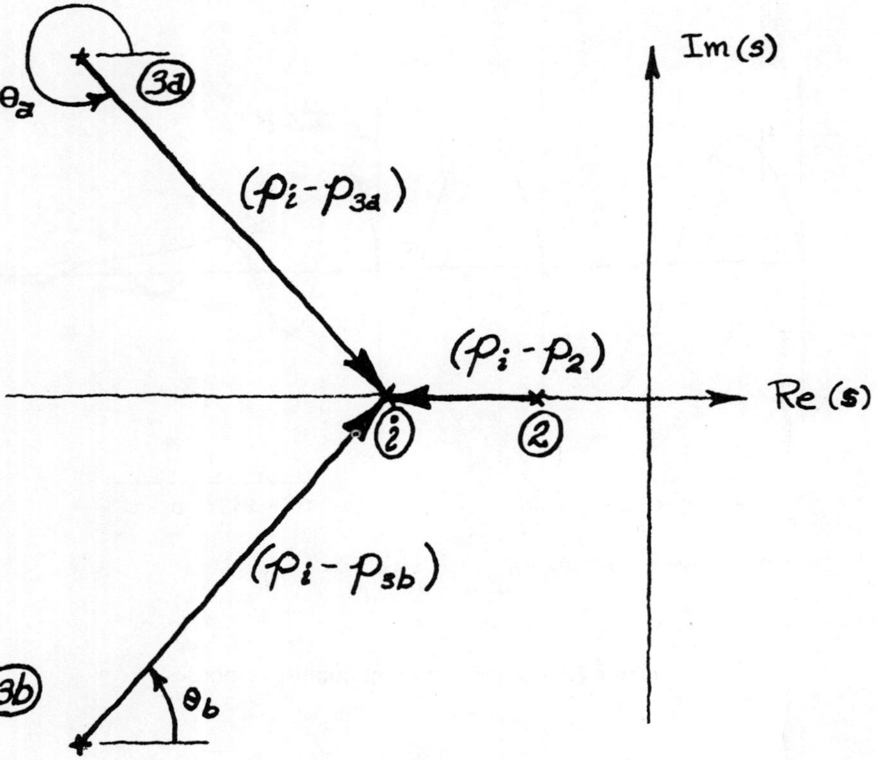

From the Engineer's Sketch Book

Figure 7.7 Determination of the coefficients in a partial fraction expansion.

K_i is then the reciprocal of the product of the lengths of these vectors, while the angle of K_i is the negative of the sum of the angles (such as θ in Figure 7.7).

For example, from Figure 7.7,

$$|K_1| = \frac{1}{1 \cdot 3 \cdot 3} = \frac{1}{9}$$

$$\angle K_1 = -\theta - (-\theta) - 180° = 180°$$

$$K_1 = -\frac{1}{9}$$

Example 7.1 Determine the partial fraction expansion of

$$\frac{1}{(\mathbf{D} + 1)(\mathbf{D} + 3)(\mathbf{D}^2 + \mathbf{D} + 4)}$$

using the graphical method.

The expansion has the form

$$\frac{K_1}{\mathbf{D} + 1} + \frac{K_2}{\mathbf{D} + 3} + \frac{K_3}{\mathbf{D}^2 + \mathbf{D} + 4} + \frac{K_4 \mathbf{D}}{\mathbf{D}^2 + \mathbf{D} + 4}.$$

Taking the value of the K's analytically, in order to substantiate the graphical results we have

$$K_1 = \frac{1}{(-1 + 3)(1 - 1 + 4)} = \frac{1}{2 \cdot 4} = \frac{1}{8}$$

$$K_2 = \frac{1}{(-3 + 1)(9 - 3 + 4)} = \frac{1}{-2 \cdot 10} = -\frac{1}{20}$$

$$K_3 + K_4\left(-\frac{1}{2} + j1{\cdot}94\right) = \frac{1}{\left(-\frac{1}{2} + j1.94 + 1\right)\left(-\frac{1}{2} + j1.94 + 3\right)}$$

$$= \frac{1}{\left(\frac{1}{2} + j1.94\right)(2.5 + j1.94)} = \frac{-2.5 - j5.81}{40}$$

$$1.94\, K_4 = -\frac{5.81}{40} \qquad K_4 = -\frac{5.81}{40 \cdot 1.94} = -0.075$$

$$K_3 + 0.037 = \frac{-2.5}{40} \qquad K_3 = -0.062 - 0.037 = -0.099$$

$$K_3 = -0.099, \qquad K_4 = -0.075.$$

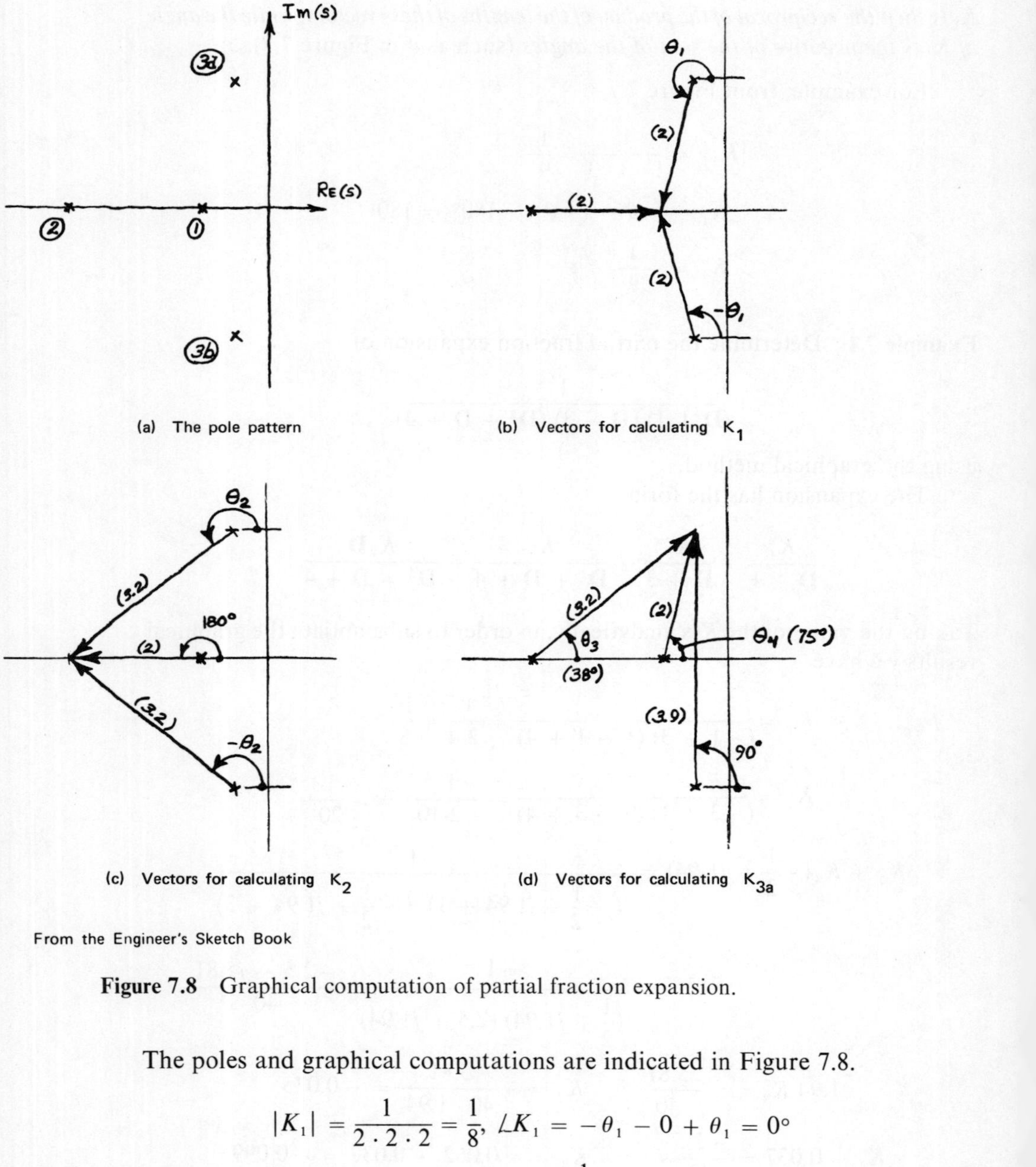

Figure 7.8 Graphical computation of partial fraction expansion.

The poles and graphical computations are indicated in Figure 7.8.

$$|K_1| = \frac{1}{2 \cdot 2 \cdot 2} = \frac{1}{8}, \ \angle K_1 = -\theta_1 - 0 + \theta_1 = 0°$$

$$K_1 = \frac{1}{8}$$

$$|K_2| = \frac{1}{2 \cdot 3.2 \cdot 3.2} = \frac{1}{20.4}, \angle K_2 = -\theta_2 - 180 + \theta_2 = -180$$

$$K_2 = -\frac{1}{20.4}$$

$$|K_{3a}| = |K_{3b}| = \frac{1}{3.2 \cdot 2 \cdot 4} = \frac{1}{25.6} = 0.039$$

$$\angle K_{3a} = -38° - 75° - 90° = -203° = 157°$$
$$\angle K_{3b} = -157°$$

$$K_{3a} = -0.036 + j0.015$$

$$K_{3b} = -0.036 - j0.015$$

The coefficients K_{3a} and K_{3b} are associated with the roots of the factored quadratic and yield a result

$$\frac{K_{3a}}{\mathbf{D} + \frac{1}{2} - j1.94} + \frac{K_{3b}}{\mathbf{D} + \frac{1}{2} + j1.94}$$

This may be rewritten as follows:

$$\frac{K_{3a}(\mathbf{D} + \frac{1}{2} + j1.94) + K_{3b}(\mathbf{D} + \frac{1}{2} - j1.94)}{\mathbf{D}^2 + \mathbf{D} + 4} =$$

$$\frac{(K_{3a} + K_{3b})(\mathbf{D} + \frac{1}{2}) + j1.94(K_{3a} - K_{3b})}{\mathbf{D}^2 + \mathbf{D} + 4} = \frac{-0.096 - 0.072\mathbf{D}}{\mathbf{D}^2 + \mathbf{D} + 4}.$$

This agrees very well with the analytically determined result of

$$\frac{-0.099 - 0.075\mathbf{D}}{\mathbf{D}^2 + \mathbf{D} + 4}.$$

If the function being expanded includes zeros, then the expression for the coefficient has the form

$$K_i = (p_i - p_i)\frac{(p_i - z_1)(p_i - z_2)\cdots(p_i - z_m)}{(p_i - p_1)(p_i - p_2)\cdots(p_i - p_n)}.$$

The magnitude of the coefficient, in this case, is the product of the distances from the zeros to the pole of interest, divided by the product of the distances from the other poles. The angle is the sum of the angles of the vectors from the zeros minus the sum of the angles from the other pole.

The most important use of the graphical method is that it allows the engineer to make subjective judgments of system performance. For example, the following interpretations may be made:

(1) If there is a zero very close to a pole, the contribution of that pole to the impulse response will be small compared to that of other poles.
(2) A pole which is widely separated from all the other poles of a transfer function will contribute relatively little to the impulse response.
(3) Two closely spaced poles tend to dominate the impulsive response.
(4) A zero which is widely separated from the poles of a transfer function tends to make the magnitude of the impulse response quite large.

The graphical method is also extremely effective when the partial fraction expansion of transfer function containing repeated poles must be found. This is true because the derivative if a rational function may be determined graphically in a simple manner as follows:

Suppose

$$\frac{B(\mathbf{D})}{A(\mathbf{D})} = \frac{(\mathbf{D} - z_1)(\mathbf{D} - z_2)\cdots(\mathbf{D} - z_m)}{(\mathbf{D} - p_1)^q(\mathbf{D} - p_2)\cdots(\mathbf{D} - p_n)}$$

$$\frac{B(\mathbf{D})}{A(\mathbf{D})}(\mathbf{D} - p_1)^q = (\mathbf{D} - z_1)(\mathbf{D} - z_2)\cdots(\mathbf{D} - z_m)(\mathbf{D} - p_2)^{-1}\ldots(\mathbf{D} - p_n)^{-1}$$

$$\frac{d}{d\mathbf{D}}\left[\frac{B(\mathbf{D})}{A(\mathbf{D})}(\mathbf{D} - p_1)^q\right]$$

$$= \left[\frac{d}{d\mathbf{D}}(\mathbf{D} - z_1)\right](\mathbf{D} - z_2)\cdots(\mathbf{D} - z_m)(\mathbf{D} - p_2)^{-1}\cdots(\mathbf{D} - p_n)^{-1}$$

$$+ \left[\frac{d}{d\mathbf{D}}(\mathbf{D} - z_2)\right](\mathbf{D} - z_1)\cdots(\mathbf{D} - z_m)(\mathbf{D} - p_2)^{-1}\cdots(\mathbf{D} - p_n)^{-1} + \cdots$$

$$+ \left[\frac{d}{d\mathbf{D}}(\mathbf{D} - p_2)^{-1}\right](\mathbf{D} - z_1)(\mathbf{D} - z_2)\cdots(\mathbf{D} - z_m)\cdots(\mathbf{D} - p_n)^{-1} + \text{etc.}$$

$$= \left[\frac{d}{d\mathbf{D}}(\mathbf{D} - z_1)\right]\left[\frac{1}{\mathbf{D} - z_1}\cdot\frac{B(\mathbf{D})}{A(\mathbf{D})}(\mathbf{D} - p_1)^q\right] +$$

$$+ \left[\frac{d}{d\mathbf{D}}(\mathbf{D} - z_2)\right]\left[\frac{1}{\mathbf{D} - z_2}\cdot\frac{B(\mathbf{D})}{A(\mathbf{D})}(\mathbf{D} - p_1)^q\right]$$

$$+ \dots$$

$$+ \left[\frac{d}{d\mathbf{D}} (\mathbf{D} - p_2)^{-1}\right]\left[(\mathbf{D} - p_2)\frac{B(\mathbf{D})}{A(\mathbf{D})}(\mathbf{D} - p_1)^q\right]$$

$$+ \dots$$

$$= \left[\frac{1}{\mathbf{D} - z_1} + \frac{1}{\mathbf{D} - z_2} + \dots + \frac{1}{\mathbf{D} - z_m}\right]\frac{B(\mathbf{D})}{A(\mathbf{D})}(\mathbf{D} - p_1)^q$$

$$+ \left[-\frac{1}{(\mathbf{D} - p_2)^2}\right]\left[(\mathbf{D} - p_2)\frac{B(\mathbf{D})}{A(\mathbf{D})}(\mathbf{D} - p_1)^q\right] + \dots$$

Then

$$\frac{d}{d\mathbf{D}}\left[\frac{B(\mathbf{D})}{A(\mathbf{D})}(\mathbf{D} - p_1)^q\right]_{\mathbf{D}=p_1} = \left[\frac{B(\mathbf{D})}{A(\mathbf{D})}(\mathbf{D} - p_1)^q\right] \cdot \left[\frac{1}{\mathbf{D} - z_1} + \frac{1}{\mathbf{D} - z_2} + \dots\right.$$

$$\left. + \frac{1}{\mathbf{D} - z_m} - \frac{1}{\mathbf{D} - p_2} - \frac{1}{\mathbf{D} - p_3} - \dots - \frac{1}{\mathbf{D} - p_n}\right]$$

at $\mathbf{D} = p_1$

The first term in brackets is the coefficient K_q in $K_q/(\mathbf{D} - p_1)^q$, which may be determined by the usual analytical or graphical method. The second term is the sum of the reciprocals of the vectors from the zeros minus the sum of the reciprocals of the vectors from the poles and is readily determined graphically. This is much more convenient to apply than the direct analytical method.

In the case of higher-order poles the formula may be expanded with the use of $d(uv) = u\,dv + v\,du$. The result, though tedious to carry out, is probably preferable to the analytical approach.

7.3 The Root Locus Method

A display of the location of the roots of the characteristic equation of a system is of extreme usefulness to a designer since the root locations may be interpreted in terms of stability, impulse response and other quantities. The root locus method is simply a technique for convenient generation of such a display. It has the particularly useful attribute of showing how the roots move in the complex plane as some selected parameter is varied. If this parameter is one of the adjustments available to the designer, then the desired value of the adjustment is obvious from the diagram.

The root locus method is essentially a set of sketching rules. It is formulated to apply to a characteristic equation written in the form

$$1 + K\frac{B(r)}{A(r)} = 0 \tag{7.15}$$

or

$$1 + K\frac{B_G(r)B_H(r)}{A_G(r)B_H(r)} = 0$$

which is the form normally generated in control system analysis. The constant K is available for accomodating the factors which are to be taken out of the original polynomials in order to isolate the effect of a particular adjustment. The method may also be applied to characteristic equations in the form

$$a_n r^n + a_{n-1} r^{n-1} + \ldots + a_1 r + a_0 = 0$$

by *dividing through by all the terms except those including the factor which is of interest*. For example, if we wish to see how the roots move as a_{n-1} is varied in the above equation, we would rewrite the equation as follows:

$$1 + a_{n-1}\frac{r^{n-1}}{a_n r^n + a_{n-2} r^{n-2} + \cdots + a_1 r + a_0} = 0$$

This is then in the desired form, and we may proceed.

The sketching rules develop from the following basic facts:

Since

$$1 + K\frac{B(r)}{A(r)} = 0,$$

(1) $\dfrac{B(r)}{A(r)}$ must be a real number. (7.16)

(2) $K\dfrac{B(r)}{A(r)} = -1$ (7.17)

for r a solution to the characteristic equation. The first tells us about the angles of the vector components of $A(r)$ and $B(r)$, while the second pertains to magnitudes.

In the following paragraphs the root locus method is presented as a set of step-by-step rules:

(1) Write the characteristic equation of the system in the form

$$1 + K\frac{B(r)}{A(r)} = 0.$$

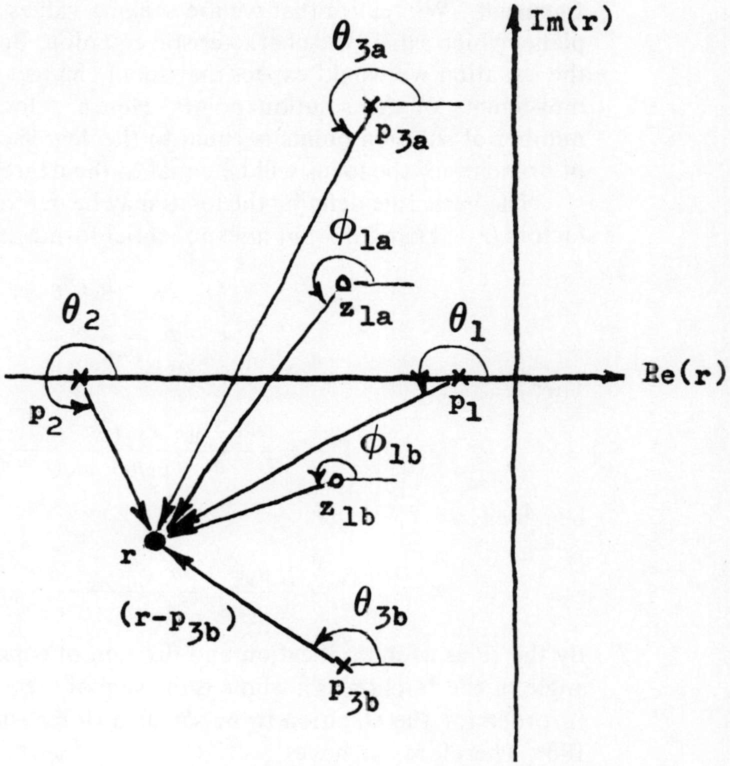

From the Engineer's Sketch Book

Figure 7.9 Pole-zero pattern and vector factors of a rational fraction in r.

(2) Factor the polynomials $A(r)$ and $B(r)$ so that the characteristic equation takes the form

$$1 + K\frac{(r-z_1)(r-z_2)\cdots(r-z_m)}{(r-p_1)(r-p_2)\cdots(r-p_n)} = 0 \tag{7.18}$$

The p's and z's are real or complex constants. Note that at $r = z_1$ the numerator of the fraction is zero, while at $r = p_1$ the denominator is zero. Because of this, z_1 is called a zero of the fraction and p_1 is a pole (the fraction goes to infinity). Similarly, for the other p's and z's.

(3) Mark the poles (with X's) and the zeros (with 0's) on the complex plane as shown in Figure 7.9. Note that the factors of $B(r)$ and $A(r)$ are the vectors from the zeros and poles to some point r in the plane.

Comment We realize that we are seeking values of r (points in the complex plane) which satisfy the characteristic equation. Because of the smoothness of the equation we would expect that small changes in K should result in small movements of the solution points. Hence a locus should exist. Since the number of solution points is equal to the degree of the equation, the number of branches of the locus will be equal to the degree of the equation.

The basic rule defining the locus may be determined by writing the vector factors $(r - z)$ and $(r - p)$ in exponential form; i.e., let

$$r - z_p = b_p e^{i\phi_p} \tag{7.19}$$

$$r - p_q = a_q e^{j\phi_q} \tag{7.20}$$

Then

$$1 + K\frac{(r - z_1)(r - z_2)\ldots(r - z_m)}{(r - p_1)(r - p_2)\ldots(r - p_n)} = 0$$

becomes

$$1 + K\frac{b_1 e^{j\phi_1}\cdot b_2 e^{j\phi_2}\cdots b_m e^{j\phi_m}}{a_1 e^{j\theta_1}\cdot a_2 e^{j\theta_2}\cdots a_n e^{j\theta_n}} = 0. \tag{7.21}$$

By the rules of multiplication and division of exponentials, we know that the angle of the fraction as a whole is the sum of the ϕ's minus the sum of the θ's. In order for the equation to be satisfied (for positive K), this angle must be 180°. Therefore, we have

$$\phi_1 + \phi_2 + \cdots + \phi_m - \theta_1 - \theta_2 - \cdots \theta_n - 180° = 0. \tag{7.22}$$

In words, this is:

At any point in the complex plane which is a solution of the characteristic equation, the sum of the angles of the vectors from the zeros minus the the sum of the angles from the poles is equal to 180°.

The angles in question are shown in Figure 7.9. The majority of the following rules are derived from this condition. Also, the condition is applied at selected points when greater accuracy than is available from the sketching rules is desired.

(4) (*a*) The loci start at the poles. The value of K at the poles is zero.
(*b*) The loci end at the zeros, where the value of K is infinity.
(*c*) If P^* is the number of finite poles and Z^* is the number of finite zeros, then $P^* - Z^*$ of the loci go to infinity. (We assume that $P^* > Z^*$.)

(5) The locus exists at any point along the real axis where an odd number of poles plus zeros is found to the right of the point. (Since complex poles, or complex zeros, occur in pairs, one need count only the poles and zeros on the real axis.) If an acceptable region lies between two poles or two zeros, the loci break away from the real axis at some intermediate point.

(6) The points at which the loci break away from the real axis are given by the expression

$$\frac{d}{dr}\left[\frac{B(r)}{A(r)}\right] = 0 \tag{7.23}$$

There is an odd number of such points between adjacent real axis poles and between adjacent real axis zeros. There is an even number of such points on the real axis between poles and zeros which are adjacent.

(7) In the immediate vicinity of a break-away point, the loci have equal angular spacing, as shown in Figure 7.10.

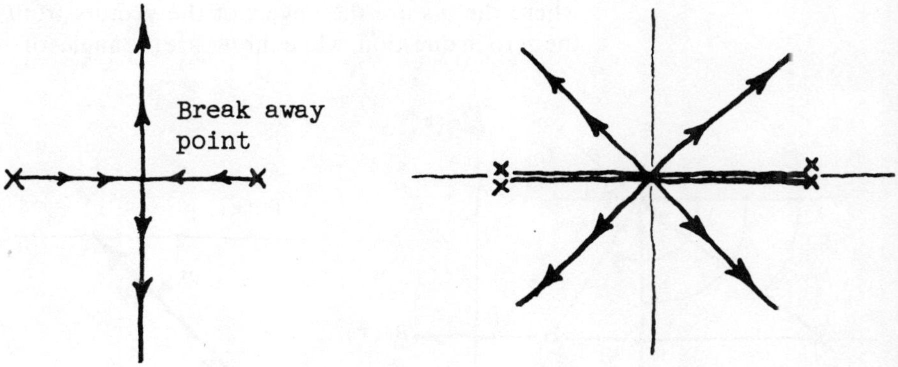

From the Engineer's Sketch Book

Figure 7.10 The equal angular spacing of loci near a break-away point.

(8) For large values of K, the loci are asymptotic to the angles

$$\frac{(2k+1)180^\circ}{P^* - Z^*} \qquad \text{for } k = 0,1,2,\ldots. \tag{7.24}$$

Angles are measured counterclockwise from the positive real axis.

(9) The asymptotes intersect the real axis at a point called the center of gravity of the roots given by

$$CG = \frac{\Sigma \text{ poles} - \Sigma \text{ zeros}}{P^* - Z^*} \tag{7.25}$$

Since complex poles (and zeros) always occur in conjugate pairs, the imaginary parts must cancel. Therefore one need add only the real parts, and CG must be on the real axis.

(10) (*a*) The angle at which a locus leaves a complex pole may be determined from the equation

$$\Psi = \phi_1 + \phi_2 + \cdots - \theta_1 - \theta_2 - \cdots 180° \tag{7.26}$$

where the angles ϕ are the angles of the vectors from the zeros to the pole in question and the angles θ are the angles of the vectors from the other poles to the pole in question. These angles, in a simple case, are illustrated in Figure 7.11.

(*b*) The angle at which a locus arrives at a complex zero is given by

$$-\Gamma = \phi_1 + \phi_2 + \cdots - \theta_1 - \theta_2 - \cdots - 180° \tag{7.27}$$

where the ϕ's are the angles of the vectors from other zeros to the zero in question, while the θ's are the angles of the vectors from

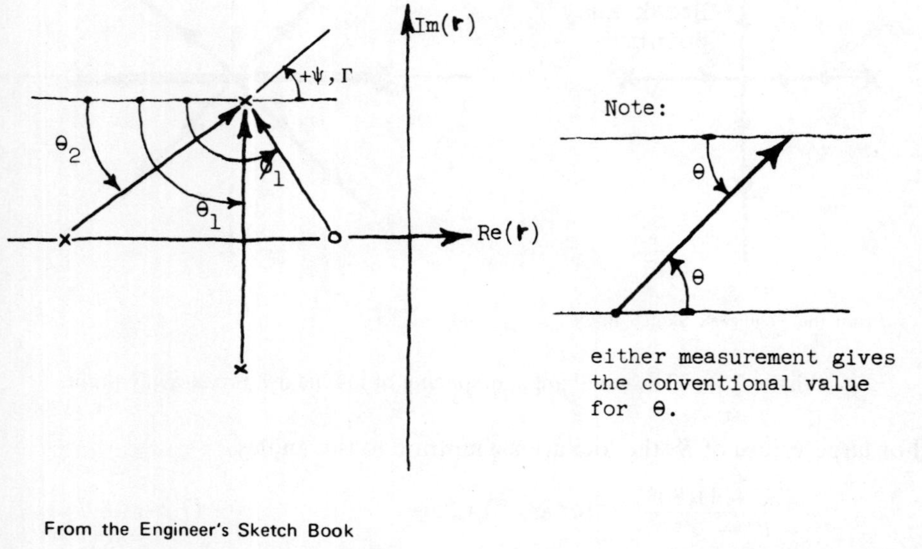

Figure 7.11 Angles for determination of tangent to locus at a pole.

the poles to the zero in question. Note that the sign of Γ is the only difference in rules 10(a) and 10(b).

(11) The values of K along the locus may be determined from the relation

$$K = \frac{\text{product of pole distances}}{\text{product of zero distances}}$$

where the distances are those between the point on the locus where K is to be determined and the various poles and zeros. The pole or zero distance associated with a pole or zero at infinity is unity; i.e., is not considered.

Example 7.2 The application of the root locus rules is very well exhibited by a detailed consideration of the familiar characteristic equation

$$mr^2 + br + k = 0$$

Let us first examine the root locus for varying spring constant k. We write the equation as

$$1 + \frac{k}{mr^2 + br} = 0 \qquad \text{(rule 1);}$$

$$1 + \frac{k}{mr^2 + br} = 0 \qquad \text{(rule 2).}$$

There is one pole at the origin and another at $r = -b/m$ (rule 3). The locus starts at these points (rule 4) and must exist between them (rule 5). A break-away point must exist between $r = 0$ and $r = -b/m$ (rule 6), and is given by

$$\frac{d}{dr}\left(\frac{k}{r(mr + b)}\right) = 0$$

$$\frac{2(2mr + b)}{mr^2 + br} = 0$$

$$r = -b/2m.$$

Therefore, the break-away point is halfway between the two poles. The locus must leave the real axis at $\pm 90°$ (rule 7). The asymptotes of the locus are at the angles (rule 8)

$$\frac{2k + 1}{P^* - Z^*} \cdot 180° = \frac{2k + 1}{2} \cdot 180°, \qquad k = 0,1,2,\ldots$$

$$= 90°, 270° \ (450°, \text{etc.}).$$

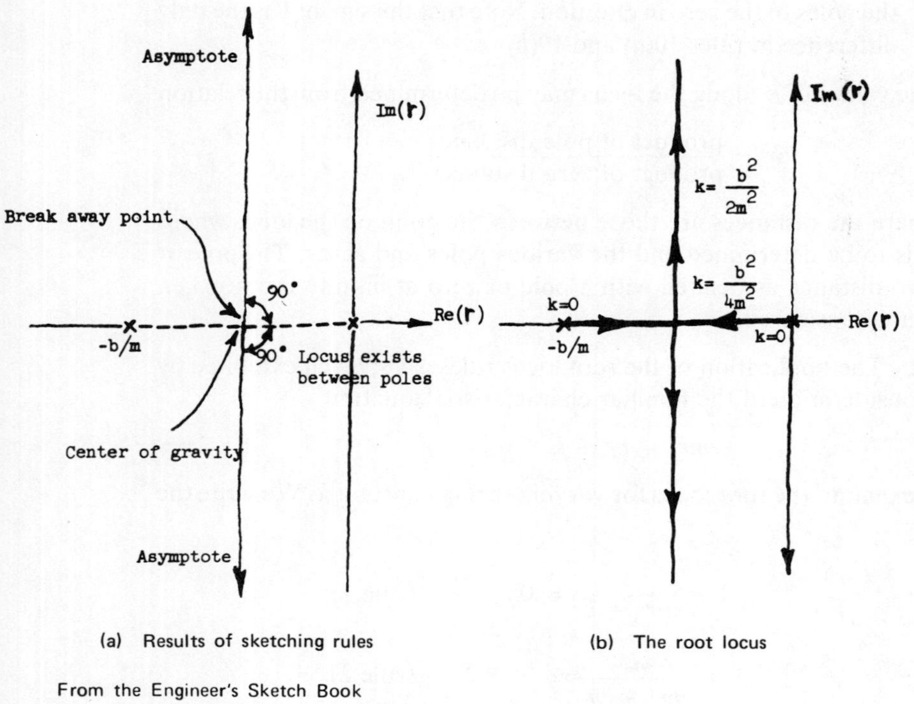

Figure 7.12 The root locus for $1 + [k/r(mr + b)] = 0$.

The asymptotes intersect the real axis at (rule 9)

$$\frac{0 + (-b/m)}{2} = -\frac{b}{2m}$$

This coincides (coincidentally) with the BAP. Rule 10 need not be applied since there are no complex poles or zeros. A sketch of these results is shown in Figure 7.12. The value of k at the BAP (where the two roots have coalesced) is given by (rule 11)

$$\frac{(b/2m)\,(b/2m)}{1} = \frac{b^2}{4m^2}$$

while the value at a distance b/m above the real axis is

$$\frac{(\sqrt{2}\,\frac{b}{2m})\,(\sqrt{2}\,\frac{b}{2m})}{1} = \frac{b^2}{2m^2}$$

The root locus provides a ready means for interpretation of the natural behavior of the system. For example, in this case we see that for low values of k there are two distinct real roots; the system is known as an "over-damped" system. The impulse response has no oscillatory response, it is purely exponential decay. As k increases, the roots coalesce and become complex conjugates. Therefore, the impulse response becomes a damped sinusoid. As k increases further, the damping ratio becomes smaller. It is obvious from the diagram that k can never become so large that the system will be unstable.

Now let us generate the root locus which exhibits the variation with damping coefficient b. We have

$$mr^2 + br + k = 0$$

$$1 + \frac{br}{mr^2 + k} = 1 + \frac{br}{(r + j\omega_n)(r - j\omega_n)} = 0 \qquad \omega_n = \sqrt{k/m},$$

There is a zero at the origin and one pole at $r = +j\omega_n$, another at $r = -j\omega_n$. The locus must exist everywhere on the negative real axis. The break-away point is given by

$$\frac{d}{dr}\left(\frac{br}{mr^2 + k}\right) = 0$$

$$-\frac{2mbr^2}{(mr^2 + k)^2} + \frac{b}{mr^2 + k} = 0$$

$$-2mbr^2 + mbr^2 + kb = 0$$

$$mr^2 = k$$

$$r^2 = \frac{k}{m}$$

$$r \pm \frac{k}{m} \qquad (= \pm\omega_n)$$

Since we are interested in the break-away point to the left of the origin, we take $r = -\omega_n$. The angles of the asymptotes are

$$\frac{(2k + 1)}{1} 180° = 180°, (540°, \ldots) \qquad k = 0, 1, 2, \ldots$$

which is no new information. Since the asymptote is the negative real axis, there is no need to compute a CG.

From the sketch shown in Figure 7.13a, the angle at which the locus leaves the complex poles may be found as follows:

$$\psi = 90° - 90° - 180° = -180°.$$

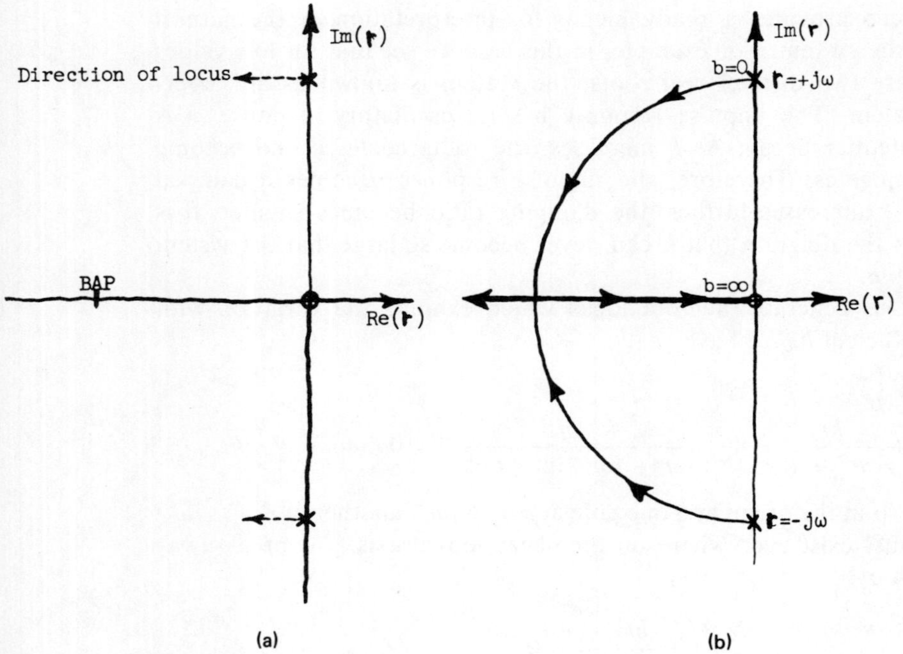

From the Engineer's Sketch Book

Figure 7.13 The root locus for $1 + br/mr^2 + k) = 0$.

Therefore, the locus leaves perpendicular to the imaginary axis. The locus is shown in Figure, 7.13b. It is circular in the region between the poles and the break-away point, but this cannot be determined simply by application of the sketching rules.

The locus for this form of the characteristic equation clearly shows the transition from an undamped system to a critically damped system (at the BAP) and finally to an overdamped system (both roots real).

A variety of root loci are shown for further clarification of the method in Figure 7.14.

In application to design, the root locus method is very effective. We may, for example, determine—at a glance—whether or not a system might be unstable, what a system's damping ratio might be, whether or not there are excess poles at the origin,* and most importantly, how proposed changes in the design would change the system response. These uses will be shown in later sections.

*Which pertains to final response; discussed in Section 7.6.

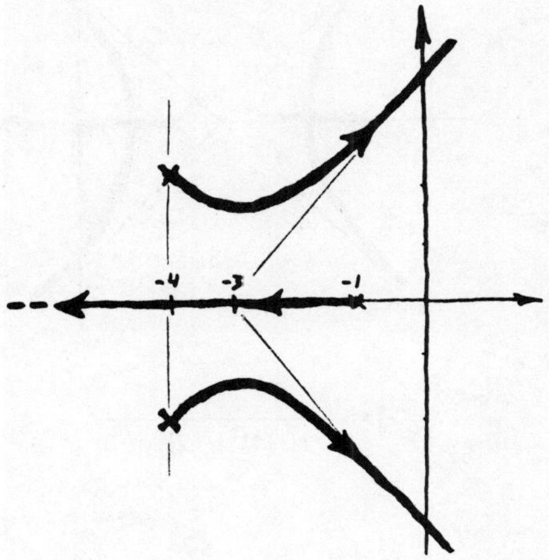

(a) $1+K \dfrac{1}{(1+r)\ (r^2+8r+20)} = 0$

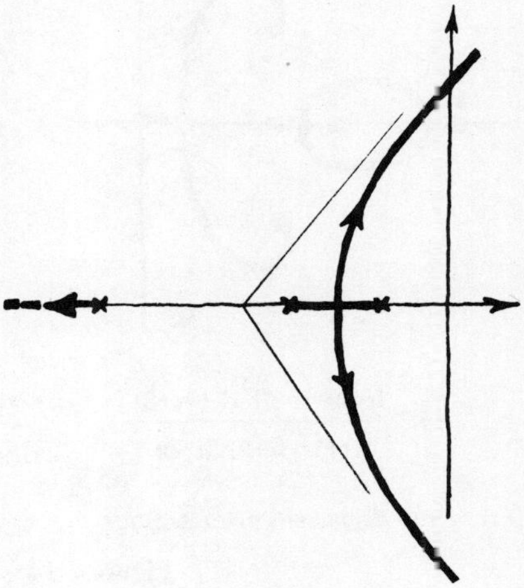

(b) $1+K \dfrac{1}{(1+3r)\ (1+7r)\ (1+15r)} = 0$

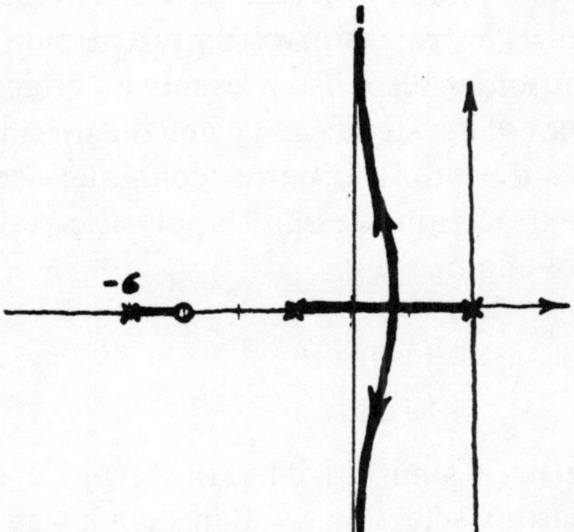

(c) $!+K \dfrac{(r+5)}{r(r+3)\ (r+6)} = 0$

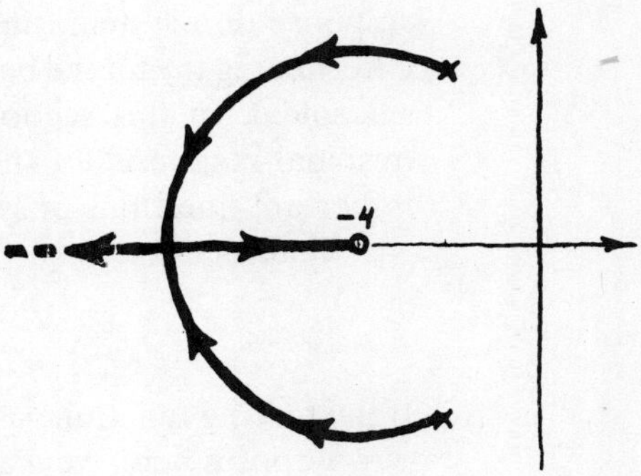

(d) $I+K \dfrac{(r+4)}{(r^2+4r+20)}$

From the Engineer's Sketch Book

Figure 7.14 (*continued overleaf*)

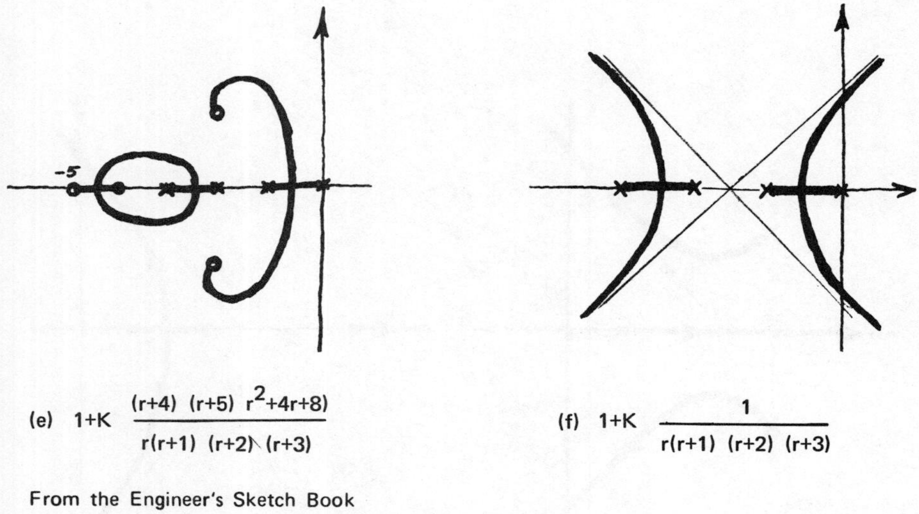

Figure 7.14 Example root loci.

7.4 Sinusoidal Response

The response of systems to sinusoidal forcing is a particularly important subject since it arises so frequently in practice and because the techniques used in its study offer considerable insight into the nature of a particular system. Just as the impulse response gives us a straightforward presentation of the natural response of a system, the sinusoidal response presents an effective means for determining the forced behavior of a system, even when the forcing is not simply sinusoidal. In this section we develop the basic technique used to study sinusoidal response. In the next section we shall apply the results to several important questions of system engineering.

Consider the usual case

$$x = \frac{B\,(\mathbf{D})}{A\,(\mathbf{D})}y.$$

If the forcing function y is a general sinusoid, it has the form $Y \cos(\omega t - \phi)$. For our purposes here, we may start by selecting $t = 0$ in such a way that ϕ is zero; this saves us some trouble and the result can be generalized when the study is otherwise complete. Then

$$y = \frac{Y}{2}(e^{j\omega t} + e^{-j\omega t}).$$

Using the e^{rt} solution method, we may handle the components of the forcing function separately as follows:

$$\text{for } y_1 = \frac{Y}{2}e^{j\omega t}, \quad \text{guess } x_1 = \frac{X_1}{2}e^{j\omega t}$$

$$\text{for } y_2 = \frac{Y}{2}e^{-j\omega t}, \quad \text{guess } x_2 = \frac{X_2}{2}e^{-j\omega t};$$

$$\text{then} \quad \frac{X_1}{Y} = \frac{B(j\omega)}{A(j\omega)} \qquad \frac{X_2}{Y} = \frac{B(-j\omega)}{A(-j\omega)}$$

At this point, we can save ourselves considerable trouble by applying a property of polynomials of complex numbers which states that

$$A(j\omega) \text{ is the complex conjugate of } A(-j\omega). \tag{7.29}$$

This is readily demonstrated as follows:

For

$$A(\mathbf{D}) = a_5\mathbf{D}^5 + a_4\mathbf{D}^4 + a_3\mathbf{D}^3 + a_2\mathbf{D}^2 + a_1\mathbf{D} + a_0$$

$$A(j\omega) = a_5 j\omega^5 + a_4\omega^4 - a_3 j\omega^3 - a_2\omega^2 + a_1 j\omega + a_0$$

$$A(-j\omega) = -a_5 j\omega^5 + a_4\omega^4 + a_3 j\omega^3 - a_2\omega^2 - a_1 j\omega + a_0$$

These expressions, and the extensions to higher degrees, may be verified from the following table:

n	$(j)^n$	$(-j)^n$
1	j	$-j$
2	-1	-1
3	$-j$	j
4	1	1
5	j	$-j$
6	-1	-1
7	$-j$	$-j$
	$-$etc.	

Upon examination of the above equations, we may see that the real parts of each are identical while the imaginary parts are equal in magnitude but opposite in sign. Thus, when the polynomials are evaluated, the results are complex conjugates. The same is true for fractions formed by polynomials. We may see this by taking.

$$B(j\omega) = b_1 + jb_2 \quad \text{so that } B(-j\omega) = b_1 - jb_2$$

$$A(-j\omega) = a_1 + ja_2 \text{ so that } A(-j\omega) = a_1 - ja_2$$

$$\frac{B(j\omega)}{A(j\omega)} = \frac{b_1 + jb_2}{a_1 + ja_2} = \frac{(a_1b_1 + a_2b_2) + j(a_1b_2 - a_2b_1)}{a_1^2 + a_2^2}$$

$$\frac{B(-j\omega)}{A(-j\omega)} = \frac{b_1 - jb_2}{a_1 - ja_2} \quad \frac{(a_1b_1 + a_2b_2) - j(a_1b_2 - a_2b_1)}{a_1^2 + a_2^2}.$$

Thus, the values of X_1 and X_2 must necessarily be complex conjugates, and we need only carry out the evaluation of one expression in order to find the value of each quantity.

It is convenient at this point to denote the polynomial fraction $B(\mathbf{D})/A(\mathbf{D})$ simply by $G(\mathbf{D})$. Thus,

$$\frac{B(j\omega)}{A(j\omega)} = G(j\omega) \quad \frac{B(-j\omega)}{A(-j\omega)} = G(-j\omega). \tag{7.30}$$

The response x may then be written

$$x = \frac{X_1}{2} e^{j\omega t} + \frac{X_2}{2} e^{-j\omega t}$$

$$= \tfrac{1}{2}G(j\omega)Y\, e^{j\omega t} + \tfrac{1}{2}G(-j\omega)Y\, e^{-j\omega t}$$

This represents a response

$$x = C_1 \cos \omega t + C_2 \sin \omega t$$

where $\quad C_1 = \text{Re } G(j\omega),\ C_2 = \text{Im } G(j\omega).$

Note:

Re $G(j\omega)$ is read "The real part of $G(j\omega)$,"
Im $G(j\omega)$ is read "The imaginary part of $G(j\omega)$."

It is a simple matter to show that the response may also be written as

$$x = X \cos(\omega t + \phi)$$

where $\quad X = Y\,|G(j\omega)| = Y\sqrt{[\text{Re } G(j\omega)]^2 + [\text{Im } G(j\omega)]^2}$

$$\phi = \arctan \frac{\text{Im } G(j\omega)}{\text{Re } G(j\omega)} = \underline{/G(j\omega)} \tag{7.32}$$

The result of the analysis may be stated in words: For $x = G(\mathbf{D})y$ where y is a sinusoid of magnitude Y, x must also be a sinusoid and have the same frequency. The amplitude ratio X/Y is equal to the magnitude of $G(j\omega)$ where ω is the frequency of both sinusoids. The phase angle by which x lags y is given by the arctan of the real portion of $G(j\omega)$ divided by the imaginary portion of $G(j\omega)$.

It is sometimes convenient to represent a sinusoid by an expression such as

$$y = \text{Re}[\vec{Y} e^{j\omega t}].$$

In this case, the response may be represented by

$$x = \text{Re}[\vec{X} e^{j\omega t}].$$

Here, the arrow over the constants is a reminder that they may be complex numbers. It may be shown directly from the results of the foregoing analysis that

$$\left|\frac{\vec{X}}{\vec{Y}}\right| = |G(j\omega)| \qquad \phi_x - \phi_y = \underline{/G(j\omega)} \qquad \left\{\frac{\vec{X}}{\vec{Y}} = G(j\omega)\right\}. \tag{7.33}$$

This notation is especially helpful where a series of transfer functions are being considered.

Example 7.3 Determine the magnitude of the response of a system described by

$$\frac{x}{y} = \frac{1 + 3\mathbf{D}}{(1 + 5\mathbf{D})(\mathbf{D}^2 + 16)} = G(\mathbf{D})$$

when

$$y = 4 \cos 3_t$$

Find $G(j\omega) = G(j3)$:

$$\frac{X}{4} = \frac{1 + j\,9}{(1 + j\,15)(-9 + 16)} = \frac{1 + j\,9}{7 + j\,15} = \frac{(7 - j\,15)(1 + j\,9)}{49 + 225}$$

$$= \frac{7 + j63 - j15 + 135}{247} = \frac{142 + j48}{274}$$

$$X = \frac{4}{274}\sqrt{142^2 + 48^2} = \frac{600}{274} = 2.19$$

Therefore, the amplitude of the response is 2.19 units.

7.5 Sinusoidal Response by Graphical Methods

Considerable good use may be made of the fact that a linear system is completely described by its sinusoidal response. That is to say, if two systems have identical sinusoidal response, then they respond identically to any and all forcing functions. Therefore, anything which may be accomplished by the root locus method may also be accomplished by the so-called "frequency response" method. The major advantage of the frequency response method over other possibilities is that when a complicated existing piece of hardware must be built into a system, it is usually rather simple to empirically determine its frequency response. Analysis and design of the complete system may then be carried out without the need for a hypothetical mathematical model for the component.

Both amplitude and phase response must be included in a frequency response analysis. The way these are handled without inordinate amounts of labor is that we determine and tabulate the amplitude and phase response of all the kinds of factors which can appear in a transfer function. Once this is

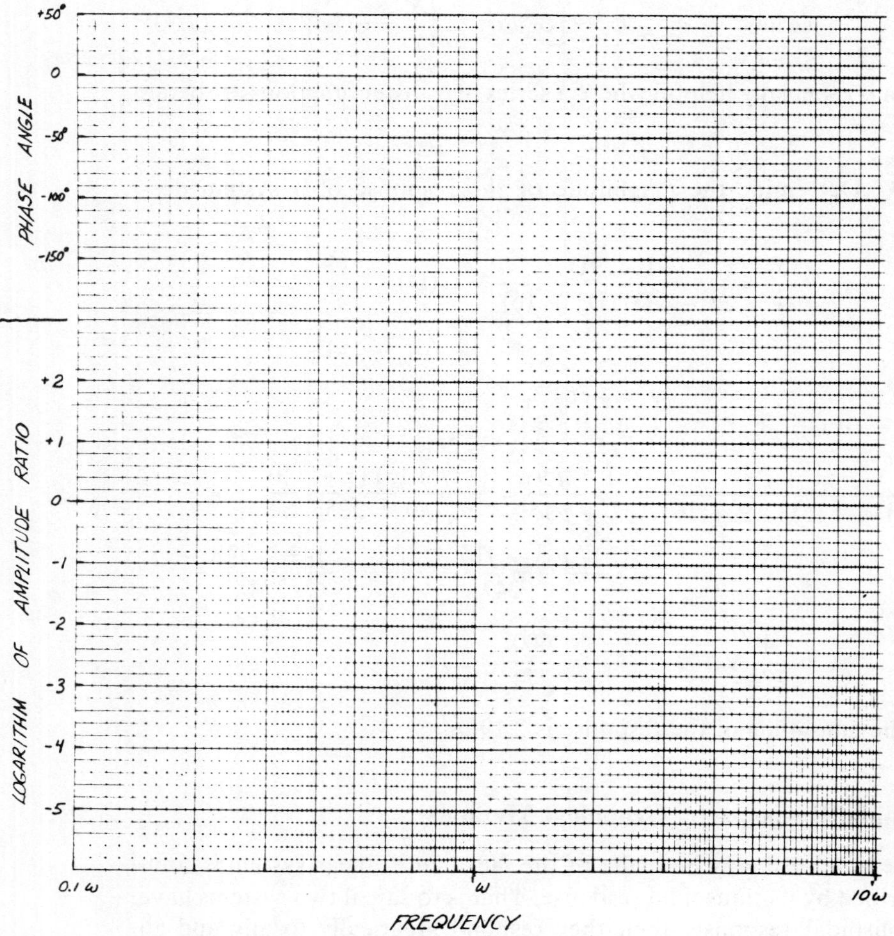

Figure 7.15 Form for frequency response graphs.

done, any transfer function may be handled as a combination of these known results.

The graphical form which is used is indicated in Figure 7.15. The logarithmic amplitude scale facilitates the development of combined response patterns from the known components, while the logarithmic frequency scale compresses the graph and also provides some simplifications in graph sketching. The "decibel" electrical engineering convention for amplitude measurement has been foregone here in favor of the simpler and more pertinent "logarithm of the amplitude ratio." Note that the phase scale is linear, not logarithmic.

The possible factors of a transfer function are:

$$\mathbf{D}, \quad (\mathbf{D})^{-1}$$

$$1 + T\mathbf{D}, \quad (1 + T\mathbf{D})^{-1}$$

$$\mathbf{D}^2 + 2\zeta\omega_n\mathbf{D} + \omega_n^2, \quad (\mathbf{D}^2 + 2\zeta\omega_n\mathbf{D} + \omega_n^2)^{-1}$$

We will develop the frequency response characteristics for the $(1 + T\mathbf{D})^{-1}$ factor. The other developments will be left to the student, although the results for all will be presented.

$$\text{For } x = \vec{X}e^{j\omega t},\ y = \vec{Y}e^{j\omega t},\ x = \frac{1}{1 + T\mathbf{D}}\, y$$

$$\frac{\vec{Y}}{\vec{X}} = \frac{1}{1 + Tj\omega}$$

$$\frac{|\vec{Y}|}{|\vec{X}|} = A = \left| \frac{1 - Tj\omega}{1 + T^2\omega^2} \right|$$

$$= \frac{\sqrt{1 + T^2\omega^2}}{1 + T^2\omega^2} = \sqrt{\frac{1}{1 + T^2\omega^2}}.$$

If $T\omega \ll 1 \quad \left(\omega \ll \frac{1}{T}\right)$

$A \cong 1$: a horizontal line on the graph.

If $T\omega \gg 1 \quad \left(\omega \gg \frac{1}{T}\right)$

$$A \cong \frac{1}{T\omega}$$

$\log A = \log 1 + \log\left(\frac{1}{T}\right) - \log \omega$: a line with slope -1 through the point $A = 1$ at $\omega = 1/T$.

These two lines represent asymptotes for the amplitude response curve. The following specific values for A are helpful in sketching the curve:

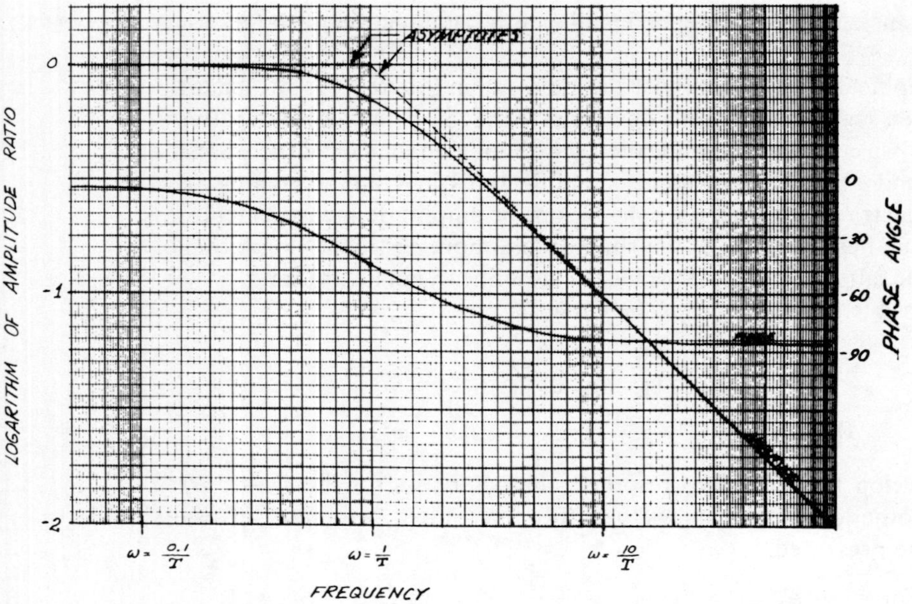

Figure 7.16 Frequency response of $(1 + T\mathbf{D})^{-1}$.

$$\omega = 1/2T, \quad A = 0.93$$
$$\omega = 1/T, \quad A = 0.707$$
$$\omega = 2/T, \quad A = 0.586$$

To compute the phase, we have

$$\angle \frac{\vec{X}}{\vec{Y}}\left[= \phi\right] = \angle \frac{1}{1 + Tj\omega} = \angle(1 - Tj\omega)$$

$\Phi = \arctan(-T\omega)$.
If $T\omega \ll 1$, $\phi \cong 0$.
If $T\omega \gg 1$, $\phi \cong -90°$.

The two curves are shown in Figure 7.16. This is the first of the reference curves for the frequency response method.

The response curves for the remaining factors are given in Figures 7.17, 7.18, and 7.19.

The procedure for determining the frequency response of a complex transfer function is now quite straightforward. For the amplitude response, the use of the logarithmic scale allows graphical addition to be used to determine

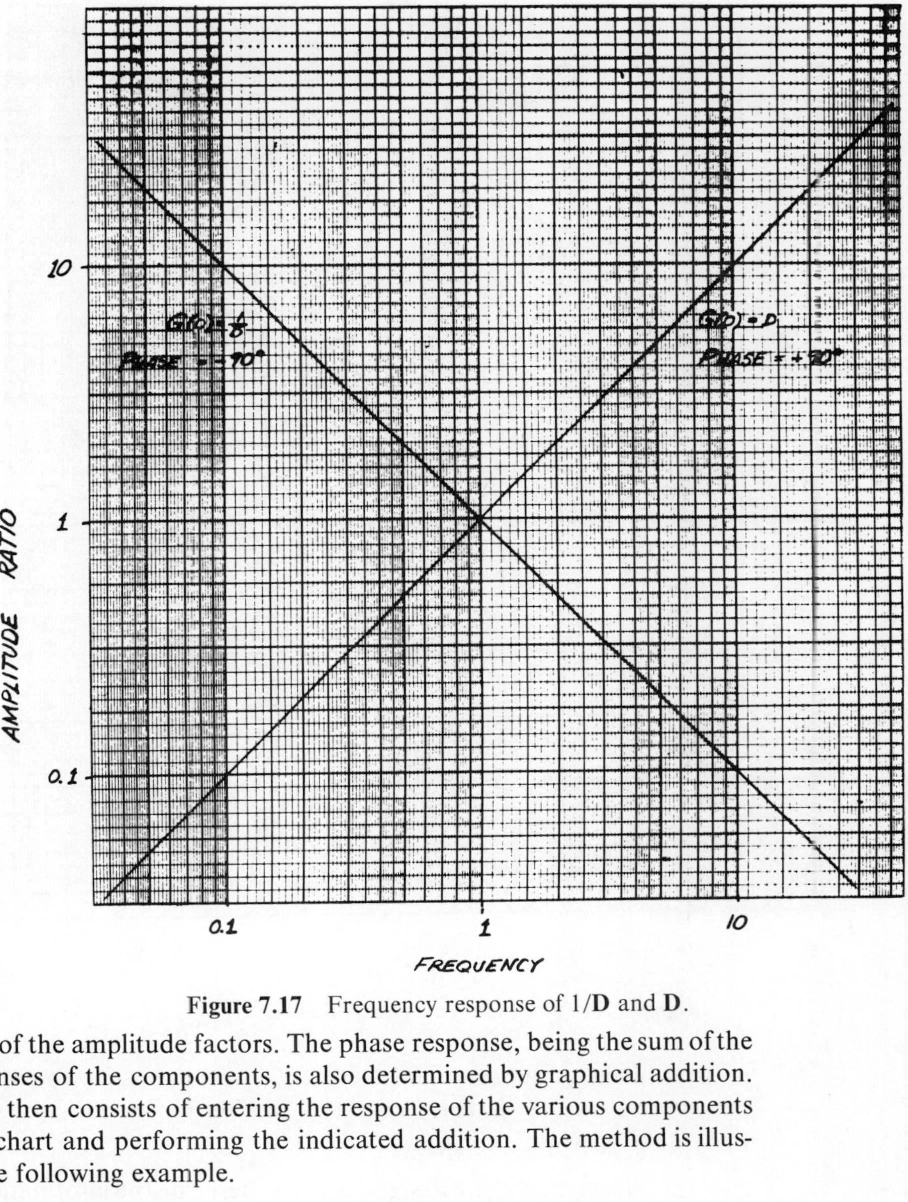

Figure 7.17 Frequency response of 1/**D** and **D**.

the product of the amplitude factors. The phase response, being the sum of the phase responses of the components, is also determined by graphical addition. The process then consists of entering the response of the various components on a single chart and performing the indicated addition. The method is illustrated by the following example.

Example 7.4 The frequency response of the transfer function

$$\frac{Y(\mathbf{D})}{X(\mathbf{D})} = \frac{(1 + 0.2\mathbf{D})}{(1 + 3\mathbf{D})(\mathbf{D}^2 + 2\mathbf{D} + 25)}$$

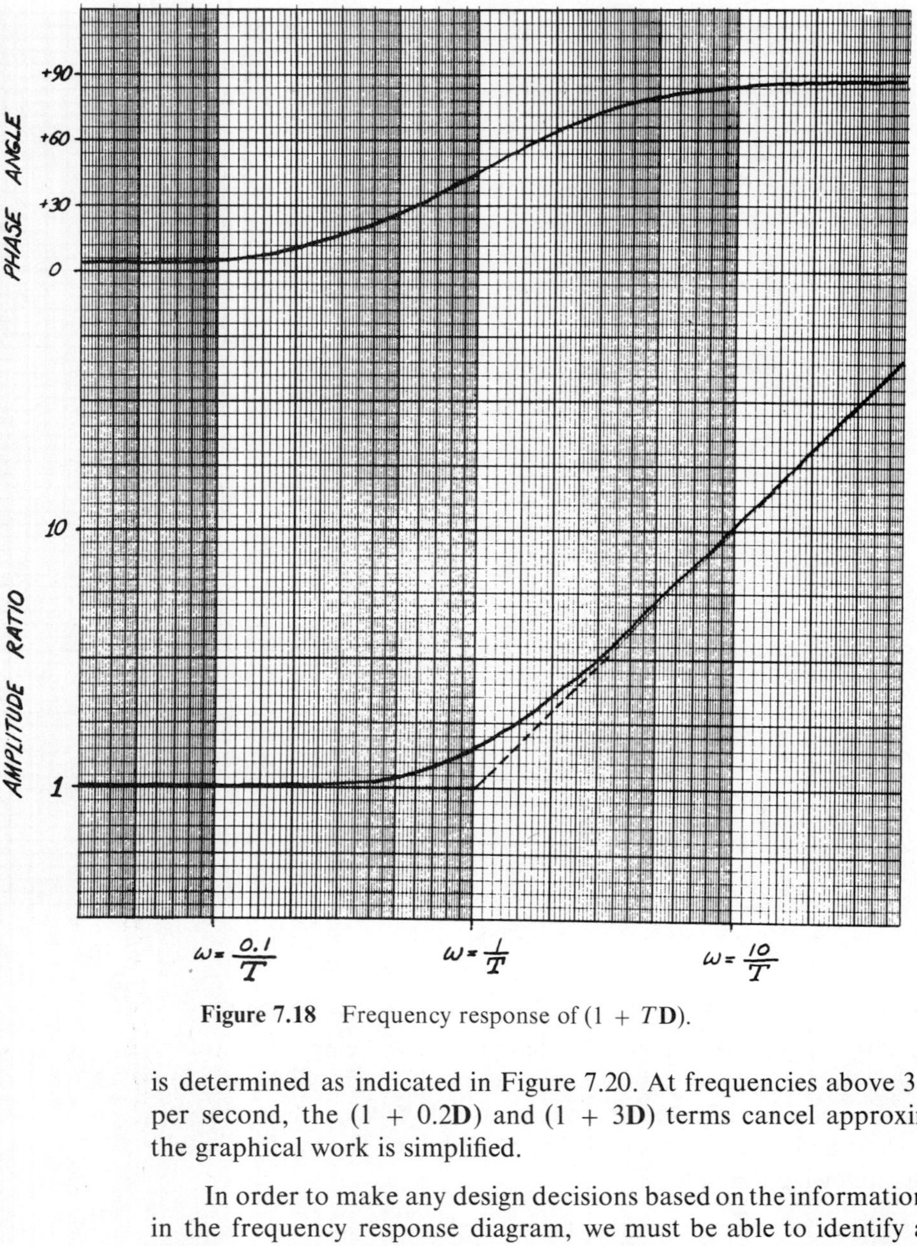

Figure 7.18 Frequency response of $(1 + T\mathbf{D})$.

is determined as indicated in Figure 7.20. At frequencies above 3 to 4 cycles per second, the $(1 + 0.2\mathbf{D})$ and $(1 + 3\mathbf{D})$ terms cancel approximately and the graphical work is simplified.

In order to make any design decisions based on the information contained in the frequency response diagram, we must be able to identify an unstable system, interpret the diagram, in terms of impulse or step response, and determine the steady state performance of the system.

With regard to stability, let us determine the frequency response condition for marginal stability by finding the nature of a system which is capable of self-sustained constant amplitude oscillations. This limit of stability will then serve the same role as does the imaginary axis in the root locus technique; it will divide systems into the two cases of interest—stable and unstable.

Consider a system in which the variables r, c, and e are related as implied by the diagram in Figure 7.21, which is that of a feedback system. If this system exhibits self-sustained oscillations—i.e., oscillations with $\vec{R} = 0$, then each signal in the system must be a sinusoid of constant amplitude. Suppose we were to sever the system at some point, say, at the dashed line in Figure 7.21. We should be able to recreate all the signals found in the self-sustained oscillation case by providing the proper signal for E_2. If the frequency of the applied signal were precisely equal to the frequency of the self-sustained oscillation, then the signal E_1 should have exactly the same amplitude and phase as the applied signal, since that had to be the case under self-sustained oscillation. Hence, we have a criterion for marginal stability in a feedback system:

> If the transfer function relating the variables on each side of an imaginary severance of a feedback system has unity amplitude ratio at the frequency where the phase change is 180°, the system can have self-sustained oscillations at that frequency.

Note that the "adder" inverts the output of the H function, thereby providing an automatic 180° of phase shift. The total shift, if the system provides 180°, is then 360° or 0°, and a sinusoidal signal undergoes precisely no alteration in being acted upon by the severed system. It is easy to see that, if the transfer function has an amplitude ratio greater than unity where it has 180° phase shift, the signal magnitudes will increase with time and the system would be unstable. Conversely, if the amplitude ratio is less than unity where the phase shift is 180°, the signals would decrease with time, self-sustained oscillation would be impossible, and the system would be stable.

The graphical stability criterion is shown in Figure 7.22. It is clear that either the 180° phase shift point or the unity amplitude ratio point may be selected for test. Since the designer is interested in the degree of stability as well as the existence, the measures known as "gain margin"* and "phase margin" have been introduced. The idea of these measures is simply that a system having a gain of one half at the 180° phase shift point is more stable than a system with 0.9 gain at that point, while a system with 90° phase shift at the unity gain point is more stable than one with 170°. The definitions of these terms are as follows:

Phase margin: $180° + \Phi^*$

$$\Phi^* = \text{phase shift of "severed system" transfer function where amplitude ratio is unity} \tag{7.34}$$

*The term "gain" is used in electrical engineering for amplitude ratio.

Gain margin: $1/A^*$

$$A^* = \text{amplitude ratio of "severed system" transfer function where phase shift is } 180°. \quad (7.35)$$

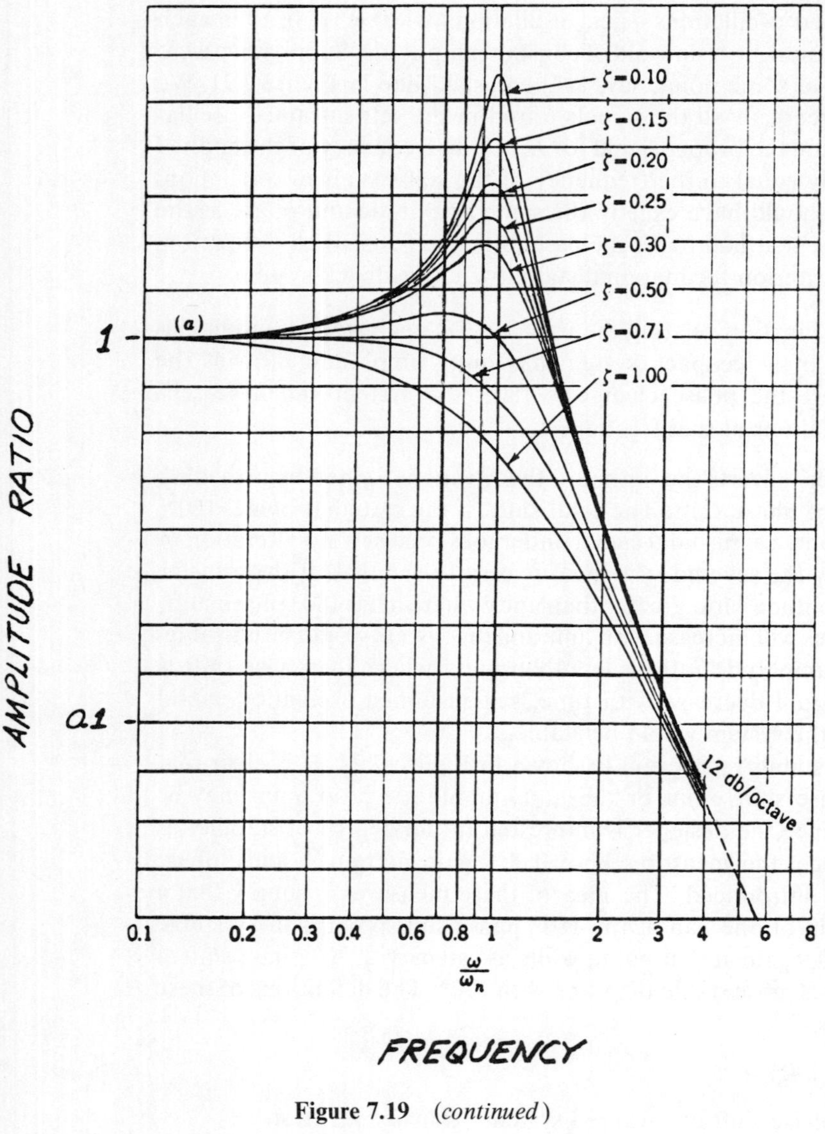

Figure 7.19 (*continued*)

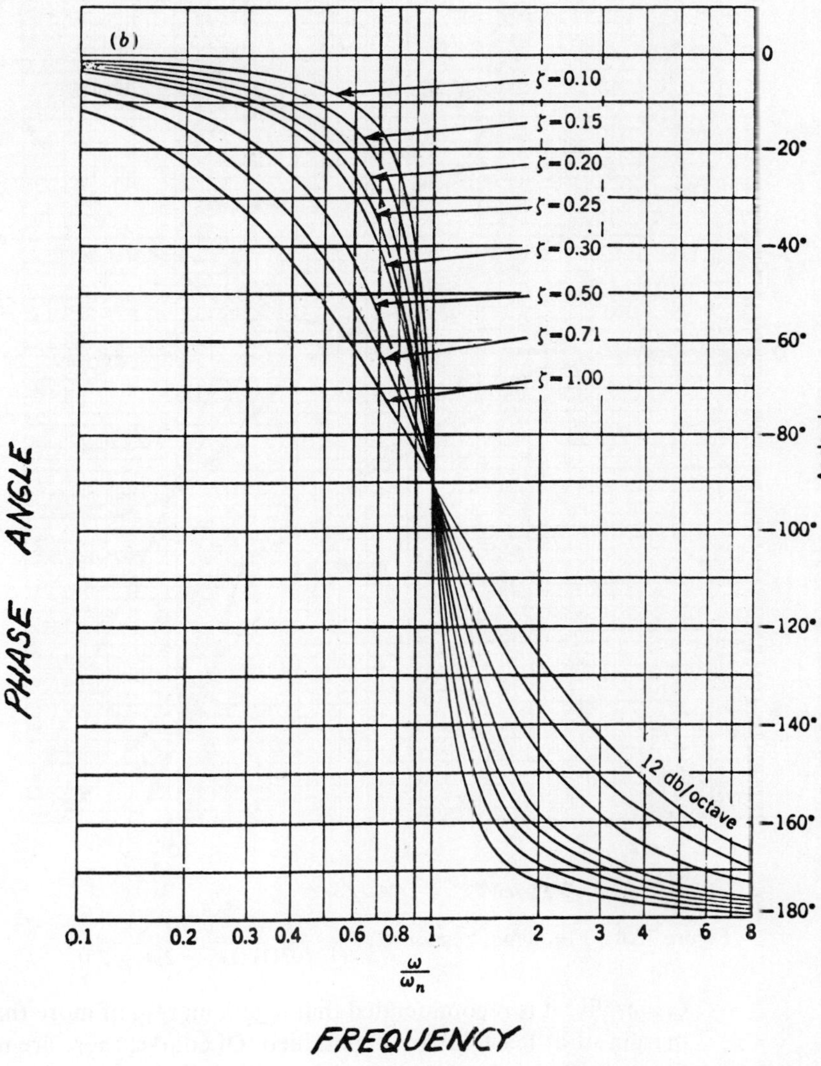

Figure 7.19 Frequency response of $(\mathbf{D}^2 + 2\zeta W_n \mathbf{D} + Wn^2)^{-1}$.

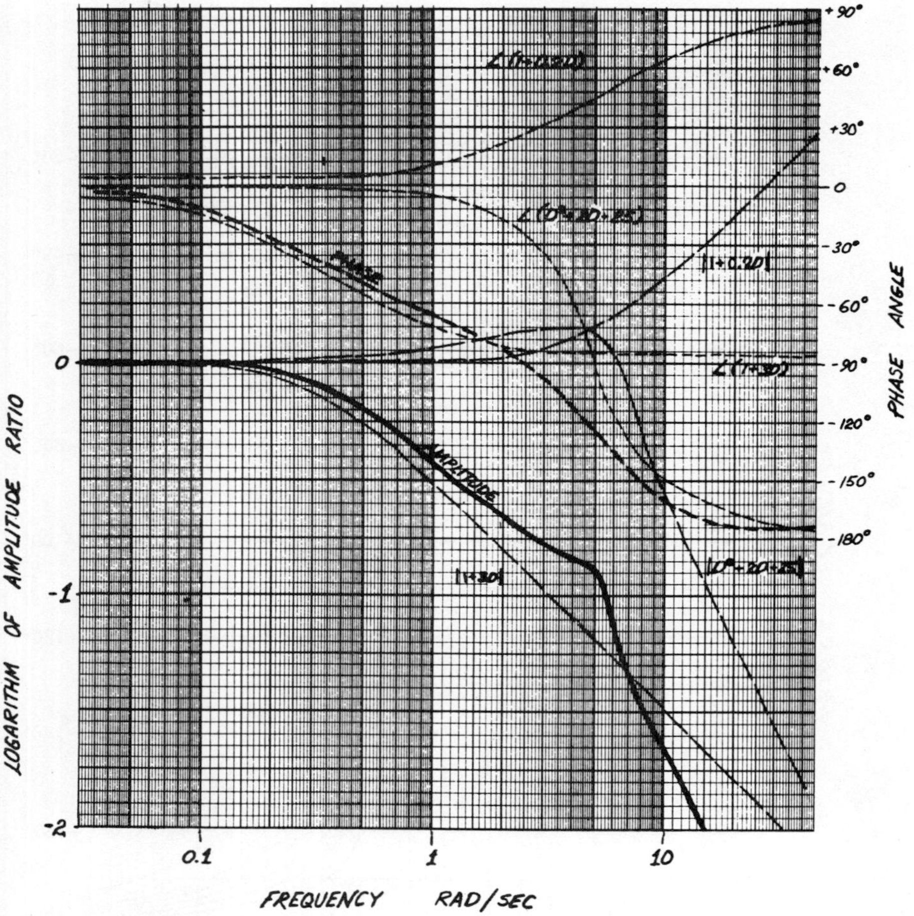

Figure 7.20 Frequency response of $\frac{1 + 0.2\mathbf{D}}{(1 + 3\mathbf{D})(\mathbf{D}^2 + 2\mathbf{D} + 25)}$.

Generally, it is recommended that a gain margin of more than 1.3 and a phase margin of at least 30° be maintained. Of course, there are many special situations which justify violation of these guidelines.

Many engineers utilize the frequency response diagram to predict system impulse response in the following informal, but useful, way. They reason that if the overall frequency response of the system is generally similar to that of a $1/(1 + Ts)$ system, then the transient responses are assuredly rather similar. Also, they might note sharp peaks on the diagram and conclude that the system would produce an oscillatory response at the frequency of the peak, if the system is impulsively excited. Low frequency response also yields to this kind of interpretation.

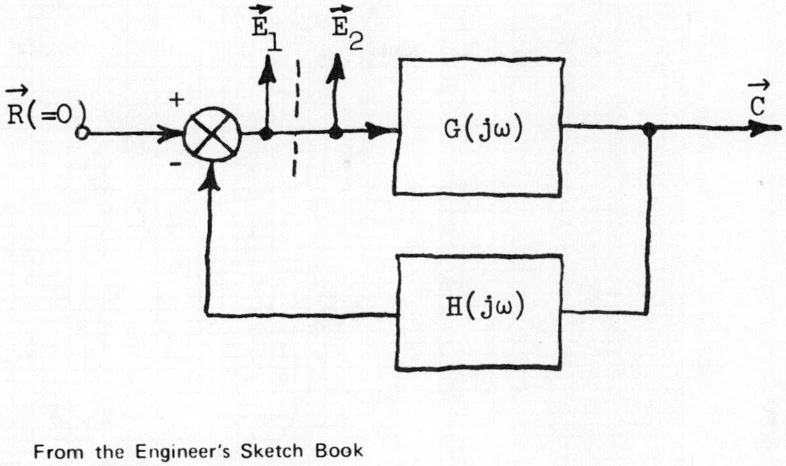

Figure 7.21 For determination of the marginal stability condition.

7.6 Final Response

In many system engineering problems, one of the goals is to achieve a specified kind of behavior after the natural response has died out. For example, in the automatic control systems which point telescopes at stars in order to allow for photographic time exposures, the system must have a final response which has a velocity exactly equal to the apparent velocity of the star.

The most convenient means for studying the final response of a system is the final value theorem of Laplace transform theory:

$$\lim_{t \to \infty} x(t) = \lim_{s \to 0} sX(s) \quad X(s) = \mathscr{L}\{x(t)\}.$$

We recall that it is necessary that all the poles of the transform lie in the left half-plane in order for the relationship to hold.

It turns out that the aspect of a transfer function which exposes the steady state behavior of a system is the "number of poles at the origin." By this is meant that, for a transfer function as in

$$X(s) = \frac{B(s)}{A(s)} Y(s)$$

which might be written

$$X(s) = \frac{b_m s^m + b_{m-1}s^{m-1} + \cdots + b_1 s + b_0}{a_n s^n + a_{n-1}s^{n-1} + \cdots + a_1 s + a_0} Y(s)$$

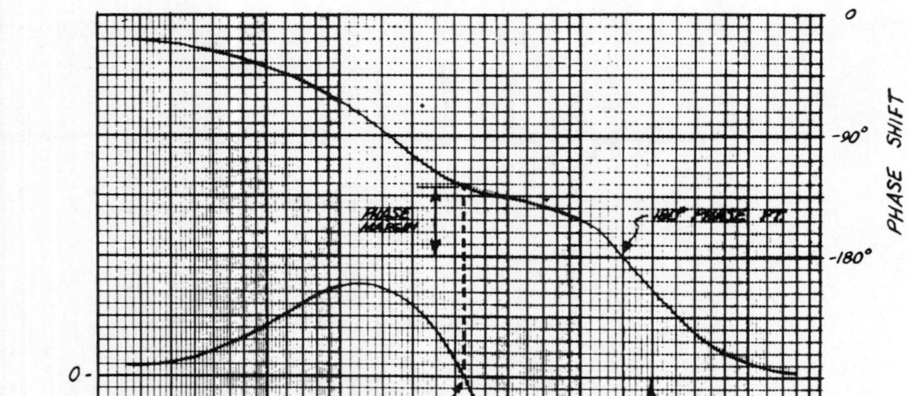

Figure 7.22 Stability criteria as shown on frequency response diagram.

we frequently find that some power of s may be factored out of the denominator (a_0 in the above might be zero) such that the alternative form

$$X(s) = \frac{b_m s^m + b_{m-1} s^{m-1} + \cdots + b_1 s + b_0}{s^p(a_n s^n + a_{n-1} s^{n-1} + \cdots + a_1 s + a_0)} Y(s) \tag{7.37}$$

may be written. The excess s's, being p in number in the above form, cause the transfer function to be infinite at $s = 0$ and therefore represent poles at the origin.

Let us apply the final value theorem, assuming that $y(t)$ is a unit step function so that $Y(s) = 1/s$. We recall, however, that the final value theorem cannot be applied if the transform $sX(s)$ has poles in the right half-plane or on the imaginary axis; we are forced to multiply through by s^p before applying the theorem. Then

$$s^pX(s) = \frac{b_m s^m + b_{m-1}s^{m-1} + \cdots + b_1 s + b_0}{a_n s^n + a_{n-1}s^{n-1} + \cdots + a_1 s + a_0} \cdot \frac{1}{s}. \tag{7.38}$$

The quantity $s^p X(s)$ is the transform of the pth derivative of $x(t)$. Therefore, we may write

$$\lim_{t\to\infty}\{\mathbf{D}^p x(t)\} = \lim_{s\to 0} \frac{b_m s^m + b_{m-1}s^{m-1} + \cdots + b_1 s + b_0}{a_n s^n + a_{n-1}s^{n-1} + \cdots + a_1 s + a_0} = \frac{b_0}{a_0}, \tag{7.39}$$

assuming that the polynomials $A(s)$ have no roots with positive and real parts. The conclusion is that the value of the pth derivative of $x(t)$ in the steady state is b_0/a_0. If we integrate

$$\mathbf{D}^p x(\infty) = \frac{b_0}{a_0}$$

p times, we find that $x(\infty)$ includes a term proportional *to* t^p. This result is identical to what would be achieved if the system were a computing machine, set up to perform p integrations on $y(t)$.

Most engineers who work in the systems area prefer to determine the final response by thinking of the system in this way; i.e., by temporarily considering the system to be a p-fold integrating computer. The nature of the response is quickly determined in this fashion.

A similar result is found if the input is proportional to time, time squared, etc.; i.e., the system acts as a p-fold integrator.

Example 7.5 Consider a spring-mass-dashpot system on which a constant load W has been applied. The transformed equation is

$$(ms^2 + bs + k)X(s) = \frac{W}{s}$$

$$\lim_{\tau\to\infty} x(\tau) = \lim_{s\to 0} \frac{\mathrm{W}}{ms^2 + bs^2 + k} = \frac{W}{k} \tag{7.39}$$

$$x(\infty) = \frac{W}{k}.$$

If the spring is cut so that k becomes zero,

$$(ms^2 + bs)X(s) = \frac{W}{s}$$

$$\lim_{t \to \infty} \{\mathbf{D}x(t)\} = \lim_{s \to 0} \frac{W}{ms + b} = \frac{W}{b}.$$

The steady state is a constant velocity of W/b. Since this system has a single pole at the origin, there is one integration of the input. The response x is the time integral of a constant, which is, of course, a constant velocity.

7.7 Response of Multi-Variable Systems; Modes

At the beginning of the chapter the equations describing a general system were given in the form

$$\begin{aligned} A_{11}(\mathbf{D})x_1 + A_{12}(\mathbf{D})x_2 + \cdots + A_{1n}(\mathbf{D})x_n &= B_1(\mathbf{D})y_1 \\ A_{21}(\mathbf{D})x_1 + A_{22}(\mathbf{D})x_2 + \cdots + A_{2n}(\mathbf{D})x_n &= B_2(\mathbf{D})y_2 \\ \vdots \qquad\qquad & \qquad \vdots \\ A_{n1}(\mathbf{D})x_1 + A_{n2}(\mathbf{D})x_2 + \cdots + A_{nn}(\mathbf{D})x_n &= B_n(\mathbf{D})y_n \end{aligned} \tag{7.40}$$

and it was stated that the response of such a system could be considered to be made up of components found in the response of simpler systems. The objective of this section is to present the techniques for handling the general case.

Equations of the form shown above arise when a system includes coupled physical phenomena and more than one variable is required to express the energy content of the system.

7.7 (a) Solution of Equations for Coupled Systems

In order to investigate the fundamental aspects of the analysis of coupled systems, let us consider the two variable case where the equations are

$$\begin{aligned} A_{11}(\mathbf{D})x_1 + A_{12}(\mathbf{D})x_2 &= B_1(\mathbf{D})y_1 \\ A_{21}(\mathbf{D})x_1 + A_{22}(\mathbf{D})x_2 &= B_2(\mathbf{D})y_2. \end{aligned} \tag{7.41}$$

Using substitution, or determinants, we may find that an equivalent pair of equations is

$$x_1 = \frac{A_{22}B_1y_1 - A_{12}B_2y_2}{A_{11}A_{22} - A_{12}A_{21}} \tag{7.42}$$

$$x_2 = \frac{A_{11}B_2y_2 - A_{21}B_1y_1}{A_{11}A_{22} - A_{12}A_{21}}. \tag{7.43}$$

In order to determine the characteristic equation of the system, y_1 and y_2 may be set equal to zero. Then

$$(A_{11}A_{22} - A_{12}A_{21})x_1 = 0$$
$$(A_{11}A_{22} - A_{12}A_{21})x_2 = 0$$

and

$$A_{11}(r)A_{22}(r) - A_{12}(r)A_{21}(r) = 0 \tag{7.44}$$

must be the characteristic equation. This will necessarily be a polynomial in r, since the A's are all polynomials, and the solutions of the characteristic equation must fall within the same group as for the single variable case. The natural motions found in coupled systems are then identical to the natural motions found in single variable systems. The only modification required in the analysis method is that the equations of motion must be manipulated so as to expose the characteristic equation. Considering the general case represented by equation 7.1 at the beginning of the chapter, we see that:

> The characteristic equation of a coupled system, as described by a set of equations of the form of (7.1), is the determinant of the polynomial coefficients of the dependent variables.

The handling of initial conditions in coupled systems is straightforward and is indicated in the following example.

Example 7.6 Consider the arrangement shown in Figure 7.23, where a wheel may roll and slide with viscous friction on the horizontal surface and is attached through a frictionless bearing to a spring connected to the vertical surface. The equations of motion, written from the free-body diagram also shown in Figure 7.23, are

$$m\ddot{x} + b\dot{x} + kx + br\omega = 0$$
$$rb\dot{x} + I\dot{\omega} + br^2\omega = 0$$

or

$$(m\mathbf{D}^2 + b\mathbf{D} + k)x + br\omega = 0$$
$$rb\mathbf{D}x + (I\mathbf{D} + br^2)\omega = 0$$

Taking the following numerical values for the constants,

$m = 0.1$ lb sec^2/in.
$b = 1$ lb/in./sec
$k = 10$ lbs/in.
$I = 0.1$ in. lb sec^2
$r = 2$ in.,

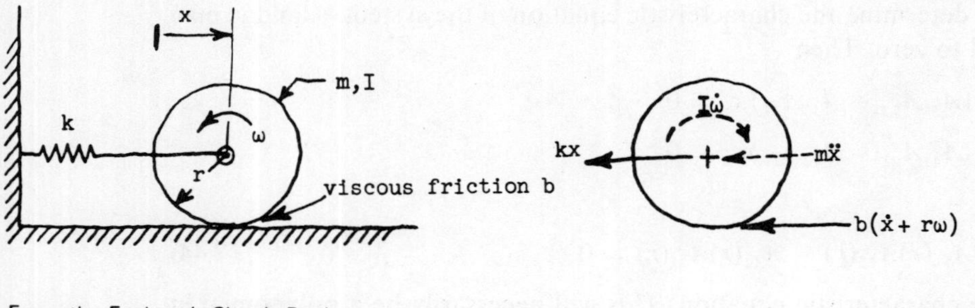

Figure 7.23 System for Example 7.6.

we have

$$(0.1\,\mathbf{D}^2 + \mathbf{D} + 10)x + 2\omega = 0$$
$$2\mathbf{D}x + (0.1\,\mathbf{D} + 4)\omega = 0. \qquad (7.45)$$

The characteristic equation is

$$(0.1\,r^2 + r + 10)(0.1\,r + 4) - 4r = 0$$
$$0.01\,r^3 + 0.5\,r^2 + r + 40 = 0.$$

This has a root near $r = -50$. Applying Newton's method,

$$\frac{dF}{dr} = 0.03r^2 + r + 1.$$

At

$$r = -50, \quad \frac{dF}{dr} = 75 - 50 + 1 = 26$$

$$26\,\Delta r = -10, \qquad \Delta = -0.385.$$

The root is very close to -49.615. Then the characteristic equation is

$$(r + 49.615)(0.01r^2 + 0.00385r + 0.806) = 0$$
$$0.01\,(r + 49.6)(r + 0.192 + j18)(r + 0.192 - j18) = 0.$$

The natural response is made up of an exponential decay with a time constant of 1/49.6 sec and a decaying sinusoid with a frequency of 18 rad/sec and a damping ratio of 0.0107.

Suppose we have the initial conditions $x(0) = 1$, $\dot{x}(0) = 0$, $\omega(0) = 0$. The response has the form

$$x = X_1 e^{-49.6t} + X_2 e^{-(0.192 + j18)t} + X_3 e^{-(0.192 - j18)t}$$
$$\omega = \Omega_1 e^{-49.6t} + \Omega_2 e^{-(0.192 + j18)t} + \Omega_3 e^{-(0.192 - j18)t}.$$

From the initial conditions we have

$$x(0) = 1 = X_1 + X_2 + X_3$$
$$\omega(0) = 0 = \Omega_1 + \Omega_2 + \Omega_3$$
$$\dot{x}(0) = 0 = -49.6X_1 - (0.192 + j18)X_2 - (0.192 - j18)X_3.$$

Three more equations must be found if the six coefficients are to be determined. These equations are available from the original equations (7.45) as follows:

$$(0.1\,\mathbf{D}^2 + \mathbf{D} + 10)x + 2\omega = 0.$$
$$2\mathbf{D}x + (0.1\mathbf{D} + 4)\omega = 0.$$

These equations are solved if

$$x = X_1 e^{-49.6t} \qquad \omega = \Omega_1 e^{-49.6t}$$

Then

$$(0.1r^2 + r + 10)X_1 + 2\Omega_1 = 0 \tag{7.46}$$

and

$$2rX_1 + (0.1r + 4)\Omega_1 = 0 \qquad \text{for } r = -49.6. \tag{7.47}$$

Therefore, X_1 and Ω_1 are related. Using either equation (6.24) or (6.25), we get the result

$$X_1 = -0.00967\,\Omega_1 \qquad \Omega_1 = -103X_1.$$

Similarly, for

$$x = X_2 e^{-(0.192 + j18)t} \qquad \omega = \Omega_2 e^{-(0.192 + j18)t}$$

$$X_2 = -\frac{0.1r + 4}{2r}\Omega_2 \qquad \text{at } r = -(0.192 + j18)$$
$$= -(0.049 + j0.112)\Omega_2$$

$$X_3 = -\frac{0.1r + 4}{2r}\Omega_3 \qquad \text{at } r = -(0.192 - j18)$$
$$= -(0.049 - j0.112)\Omega_3.$$

Then

$$1 = X_1 + X_2 + X_3$$
$$0 = -103X_1 - 67(0.049 - j0.112)X_2 - 67(0.049 + j0.112)X_3$$
$$0 = -49.6X_1 - (0.192 + j18)X_2 - (0.192 - j18)X_3$$

Solving for the X's and Ω's, and rewriting in terms of a cosine function, we have

$$x = \cdot 0.03e^{-49.6t} + 1.03e^{-0.192t} \cos (18t - \arctan 0.071)$$
$$\omega = 3.1e^{-49.6t} + 8.5e^{-0.192t} \cos (18t - \arctan 2.84)$$

The aspect of this example which bears particular mention is the fact that relationships between the coefficients in the terms of the response, determined by inserting the roots of the characteristic equation in the assumed solution, must be used in finding the coefficients. Unlike the single variable case, the initial conditions alone will not yield the solution.

7.7 (b) Natural Modes

One of the most interesting aspects of coupled motion, especially when the oscillation of coupled systems is observed physically, is that there are different "modes" of oscillation. One familiar example is the vibration of a banjo string; if the string is plucked near the center, the major portion of the response is made up of the fundamental, or lowest, tone which the string may give. The "mode shape" associated with this response is indicated in Figure 7.24 (a). If the string is plucked near an end, the response is rich in harmonics and a mode shape such as shown in Figure 7.24(b) develops. These may be seen quite easily if the instrument is held in the proper light. The banjo string is one example of a distributed system and is therefore somewhat different from the systems being considered here, where the physical phenomena are associated with a finite number of elements.

(a) Fundamental mode (b) Third harmonic

From the Engineer's Sketch Book

Figure 7.24 Mode shapes of a banjo string.

One major quality of an oscillatory mode is that it can exist alone. A second is that the various modes of which a system is capable combine to make up the natural motion of the system; i.e., any natural motion is simply a combination of the basic modes. For example, consider the bar suspended by two springs as shown in Figure 7.25. One of the natural modes of this system is a vertical vibration wherein the bar remains horizontal. Another is an angular vibration in which the center point of the bar does not move at all. It is easy to see that

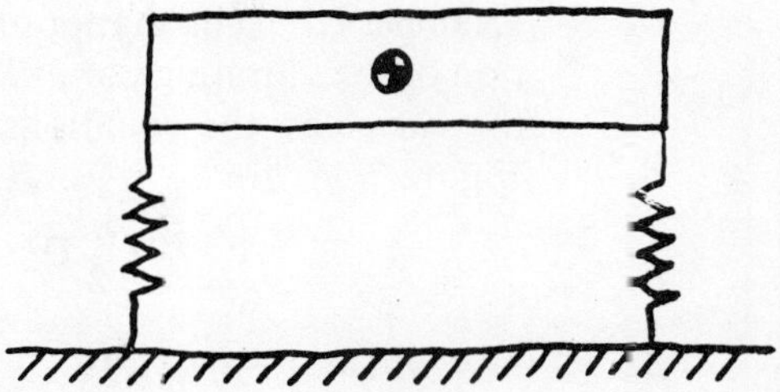

From the Engineer's Sketch Book

Figure 7.25 A system with two natural modes.

anything the bar might do could be considered a combination of these motions. This example, incidentally, can be readily observed in practice. If we remove the shock absorbers from an automobile and drive on a highway with some regular pattern of bumps—such as an old-fashioned concrete road—we will encounter the two modes of vibration in a most violent way. The amount of each which is present will depend on the speed at which the automobile is operated.

Generally speaking, oscillatory modes are associated with the natural motion of undamped systems and with the steady state forced sinusoidal vibration of systems in general. Although certain other situations are related to these cases, the undamped and forced cases are normally used as standards for comparison. For example, in the preceding problem (Example 7.6), the relationship between X_2 and Ω_2 amounts to the definition of a mode, as will be seen. To call it a mode, however, would not be conventional practice.

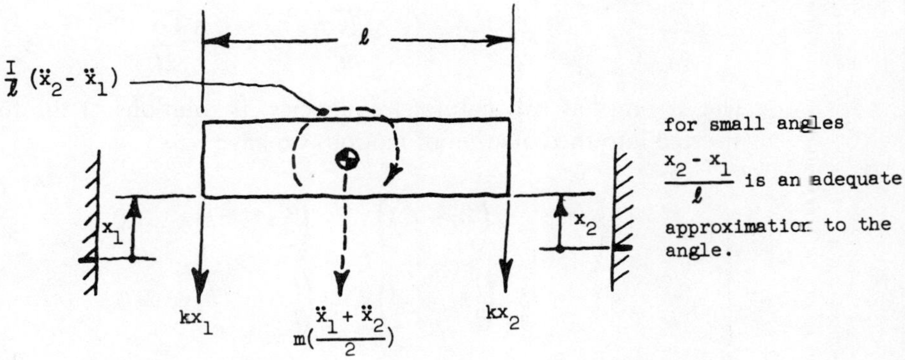

From the Engineer's Sketch Book

Figure 7.26 Free body diagram of vibrating bar.

Example 7.7 The matter of natural modes is quite well exposed by consideration of the vibrating bar of Figure 7.25. Taking the deflection of each spring as the variables, the equations of motion, written from the free body diagram of Figure 7.26, are

$$\left(\frac{m}{2}\mathbf{D}^2 + k\right)x_1 + \left(\frac{m}{2}\mathbf{D}^2 + k\right)x_2 = 0$$

$$\left(\frac{I}{l}\mathbf{D}^2 + \frac{kl}{2}\right)x_1 - \left(\frac{I}{l}\mathbf{D}^2 + \frac{kl}{2}\right)x_2 = 0.$$

Then

$$-2\left(m\mathbf{D}^2 + k\right)\left(\frac{I}{l}\mathbf{D}^2 + \frac{kl}{2}\right) = 0$$

is the determinant of the polynomial coefficients. The characteristic equation is

$$\left(\frac{m}{2}r^2 + k\right)\left(\frac{I}{l}r^2 + \frac{kl}{2}\right) = 0$$

$$\frac{mI}{2l}r^4 + \left(\frac{kI}{l} + \frac{mkl}{4}\right)r^2 + \frac{k^2l}{2} = 0$$

$$r^4 + \left(\frac{2k}{m} + \frac{kl^2}{2I}\right)r^2 + \frac{k^2l^2}{mI} = 0$$

$$r^4 + \left(\frac{2k}{m} + \frac{kl^2}{2I}\right)r^2 + \left(\frac{2k}{m}\right)\left(\frac{kl^2}{2I}\right) = 0$$

$$\left(r^2 + \frac{2k}{m}\right)\left(r^2 + \frac{kl^2}{2I}\right) = 0$$

$$r = \pm j\frac{\sqrt{2k}}{m}, \qquad r = \pm\frac{\sqrt{kl^2}}{2I}.$$

The system has two natural frequencies. If solutions of the form $X\,e^{rt}$ are inserted into the equation of motion, we have

$$\left(\frac{m}{2}r^2 + k\right)X_1 + \left(\frac{m}{2}r^2 + k\right)X_2 = 0 \tag{7.48}$$

$$\left(\frac{I}{l}r^2 + \frac{kl}{2}\right)X_1 - \left(\frac{I}{l}r^2 + \frac{kl}{2}\right)X_2 = 0. \tag{7.49}$$

As in the preceding example, each root of the characteristic equation has an associated relationship between the coefficients. Specifically, at $r = +j\sqrt{2k/m}$ [from Equation (6.27)],

$$\left[\frac{I}{l}\left(-\frac{2k}{m}\right)+\frac{kl}{2}\right]X_1-\left[\frac{I}{l}\left(-\frac{2k}{m}\right)+\frac{kl}{2}\right]X_2=0$$

$$X_1-X_2=0$$

$$X_1=X_2.$$

Here we have a clear statement that the motion at frequency $\sqrt{2k/m}$ occurs only with x_1 and x_2 exactly in phase; i.e., identical at all times. Hence, this is the motion where the bar remains horizontal. The equality of X_1 and X_2 when $r = \sqrt{2k/m}$ is the definition of the mode shape for this frequency.

At $r = \pm j\sqrt{kl^2/2I}$,

$$\left[\frac{m}{2}\left(-\frac{kl^2}{2I}\right)+k\right]X_1+\left[\frac{m}{2}\left(-\frac{kl^2}{2I}\right)+k\right]X_2=0$$

$$X_1+X_2=0$$

$$X_1=-X_2.$$

Therefore, the motion at the frequency $\sqrt{kl^2/2I}$ is an angular motion; the center of the bar remains fixed since the ends always move an equal amount in opposite directions.

Now let us look at the same problem in another way. We write the equations of motion using the deflection of the center of the bar and the angle, with the free body diagram of Figure 7.27 as a guide.

$$(m\mathbf{D}^2+2k)y=0 \tag{7.50}$$

$$\left(I\mathbf{D}^2+\frac{kl^2}{2}\right)\theta=0 \tag{7.51}$$

Using this set of coordinates, the equations of motion have a much different form than with those used earlier. Specifically, no coupling appears in the equations. The student should notice that the frequencies of the natural motions are obvious with this formulation. What has happened here is indicative of a general aspect of the determination of the natural motion of undamped systems. That is, there is always some set of coordinates in which the natural motions appear decoupled. The use of these coordinates is very desirable because of analytical simplification and simplicity of interpretation.

From Example 7.7 it may be seen that the selection of variables can be quite important. The proof that there is some set of "natural coordinates" in

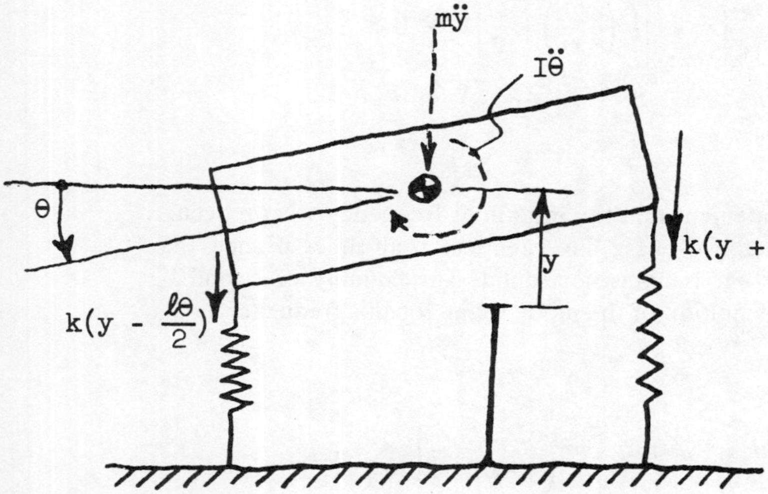

From the Engineer's Sketch Book

Figure 7.27 A new variable system for the problem of the vibrating bar.

which no coupling appears is beyond the plan of this book. The general idea of the proof is easy to comprehend, however. Suppose that a problem is originally formulated by using the coordinates x_1, x_2 and x_3 (taking a three-independent-variable case for example). The equations would have the form

$$
\begin{aligned}
A_{11}(\mathbf{D})x_1 + A_{12}(\mathbf{D})x_2 + A_{13}(\mathbf{D})x_3 &= 0\\
A_{21}(\mathbf{D})x_1 + A_{22}(\mathbf{D})x_2 + A_{23}(\mathbf{D})x_3 &= 0\\
A_{31}(\mathbf{D})x_1 + A_{32}(\mathbf{D})x_2 + A_{33}(\mathbf{D})x_3 &= 0.
\end{aligned}
\tag{7.52}
$$

If natural coordinates do exist (let us call them u_1, u_2, and u_3) the equations written in terms of them would have the form

$$
\begin{aligned}
B_{11}(\mathbf{D})u_1 + 0 + 0 &= 0\\
0 + B_{22}(\mathbf{D})u_2 + 0 &= 0\\
0 + 0 + B_{33}(\mathbf{D})u_3 &= 0.
\end{aligned}
\tag{7.53}
$$

If either the u's or the x's are adequate to completely determine the configuration of a system, they cannot be independent. In fact, with linear systems they must be related by a set of equations such as

$$
\begin{aligned}
u_1 &= c_{11}x_1 + c_{12}x_2 + c_{13}x_3\\
u_2 &= c_{21}x_1 + c_{22}x_2 + c_{23}x_3\\
u_3 &= c_{31}x_1 + c_{32}x_2 + c_{33}x_3.
\end{aligned}
\tag{7.54}
$$

In order to determine the constants c, and thereby find the natural coordinates, we must

(1) Solve for the x's in terms of the u's and c's.
(2) Rewrite the original equations in terms of the u's and c's.
(3) Collect terms in the new set of equations.
(4) Choose the c's so that the polynomial coefficients of u_2 and u_3 in the first equation, of u_1 and u_3 in the second, and u_1 and u_2 in the third go to zero.

All this, of course, amounts to considerable algebraic manipulation. The use of a matrix methods is the only practical approach in most cases. It is emphasized that this will work only when no damping appears in the system; i.e., when the polynomials in the original set of equations are of the form $(a_2\mathbf{D}^2 + a_0)$. A good reference for this material is *Handbook of Engineering Mechanics*, edited by W. Flügge and published by McGraw Hill.

7.7(c) Forced Modes

The forced motion of coupled systems has many aspects in common with the natural motion. In particular, when a coupled system is sinusoidally excited, the situation is very similar to natural motion in that modes, or fixed relationships between the phases and magnitudes, may be found. This is obvious from a set of equations such as

$$A_{11}(\mathbf{D})x_1 + A_{12}(\mathbf{D})x_2 = e^{j\omega t}$$
$$A_{21}(\mathbf{D})x_1 + A_{22}(\mathbf{D})x_2 = 0.$$

We would have

$$A_{11}(j\omega)X_1 + A_{12}(j\omega)X_2 = 1$$
$$A_{21}(j\omega)X_1 + A_{22}(j\omega)X_2 = 0$$

and it is clear that both X_1 and X_2 must be constants. Therefore, we may define a mode of vibration, and we note that the mode changes as frequency changes. In fact, one of the most spectacular aspects of the forced vibration of coupled systems is the violence with which mode changes appear as the forcing frequency is varied.

Example 7.8 Consider the case of two beads on a string under constant tension T as shown in Figure 7.28. At the right of the figure is shown the geometry of a single span of the string. The angle θ is given by $\arctan(y_n - y_{n-1}/a)$. The vertical component of the tension T is then $T\sin\theta$. For small angles, $\sin\theta \cong \tan\theta \cong 0$, therefore the vertical component is approximated by $(y_n - y_{n-1})(T/a)$. The total downward force on any bead is given by

$$(y_n - y_{n-1})(\frac{T}{a}) + (y_n - y_{n+1})(\frac{T}{a})$$

or

$$(2y_n - y_{n-1} - y_{n+1})(\frac{T}{a}).$$

The equations of motion for the two bead problem are then

$$m\ddot{y}_1 + (\frac{T}{a})(2y_1 - y_2) = 0$$

$$m\ddot{y}_2 + (\frac{T}{a})(2y_2 - y_1) = (\frac{T}{a})y_3$$

or

$$(m\mathbf{D}^2 + 2\frac{T}{a})y_1 - (\frac{T}{a})y_2 = 0$$

$$-(\frac{T}{a})y_1 + (m\mathbf{D}^2 + 2\frac{T}{a})y_2 = (\frac{T}{a})y_3$$

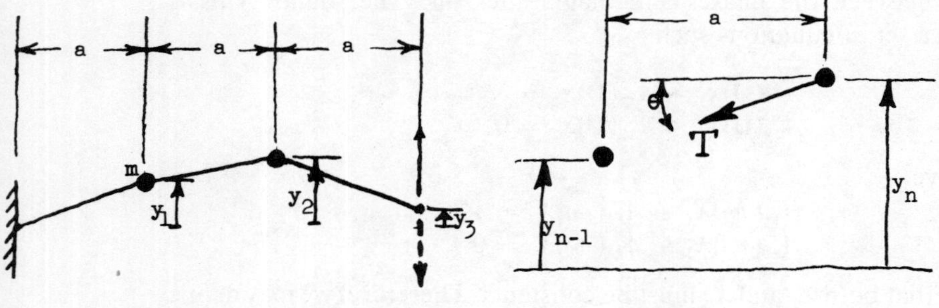

From the Engineer's Sketch Book

Figure 7.28 For Example 6.5.

Suppose y_3 is varied sinusoidally with unit magnitude. For $y_3 = e^{j\omega t}$ and making the usual assumptions, we have

$$(-m\omega^2 + 2\frac{T}{a})T_1 - (\frac{T}{a})Y_2 = 0$$

$$-(\frac{T}{a})Y_1 + (-m\omega^2 + 2\frac{Y}{a})Y_2 = \frac{T}{a}$$

$$Y_1 = \frac{(\frac{T}{a})}{(2\frac{T}{a} - m\omega^2)} Y_2, \qquad Y_2 = (\frac{T}{a}) \frac{(2\frac{T}{a} - m\omega^2)}{(2\frac{T}{a} - m\omega^2)^2 - (\frac{T}{a})^2}.$$

Here the general nature of the response is quite clear. For very low exciting frequencies Y_1 would approach $\frac{1}{2}$ of Y_2 and Y_2 would approach $\frac{2}{3}$(of Y_3). Hence the mode shape will be a straight string, as would be expected. As ω gets very large, both Y_1 and Y_2 go to zero. There are two natural frequencies in the system, given by the solutions to

$$2\frac{T}{a} - m\omega^2 = 0 \qquad (\omega = \frac{\sqrt{2T}}{am})$$

$$(2\frac{T}{a} - m\omega^2)^2 - (\frac{T}{a})^2 = 0 \qquad (\omega = \frac{\sqrt{T}}{am}).$$

We cannot meaningfully interpret the response at or close to natural frequencies because our approximate equations of motion do not apply to large amplitude motions. Otherwise the following interpretation will hold:

For $0 < \omega < \sqrt{T/am}$, Y_1 has the same sign as Y_2 and is greater than $\frac{1}{2}Y_2$. Y_2 is positive and increases without bound. A typical mode shape would be that shown in Figure 7.29(*a*).

For $\sqrt{T/am} < \omega < \sqrt{2T/am}$, Y_1 has the same sign as Y_2 and is greater than Y_2. Y_2 is negative and approaches zero as frequency increases. A representative mode shape is shown in Figure 7.29(*b*).

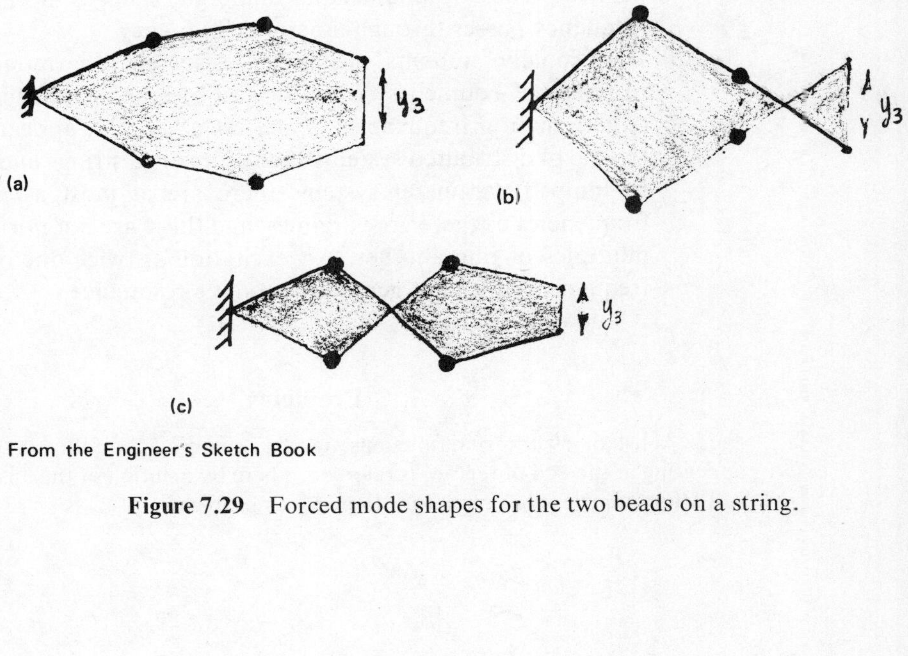

Figure 7.29 Forced mode shapes for the two beads on a string.

For $\sqrt{2T/am} < \omega < \infty$, Y_1 has the opposite sign from Y_2. Y_2 is positive and approaches zero at each end of the interval. See Figure 7.29(*c*).

This system makes an interesting physical demonstration. In particular, if we change rapidly from a frequency somewhat below to somewhat above either natural frequency, the mode shape is readily seen to make the appropriate change, usually accompanied by violent transients.

Regarding coupled systems in general, the following statements may be made:

(1) The natural motion of a coupled system is determined from the characteristic equation which is, in turn, obtained from the determinant of the polynomial coefficients [the $A(\mathbf{D})$'s] in the dynamic equations.
(2) Associated with each solution of the characteristic equation is a set of relationships between the coefficients of the response terms. In the case of an undamped system these relationships define oscillatory modes.
(3) In systems where all the polynomial coefficients in the equations of motion are of the form $(a_2\mathbf{D}^2 + a_0)$, there is always a set of "natural coordinates" in which no coupling appears. There are as many natural frequencies as there are variables. Associated with each natural frequency is a mode of vibration. If natural coordinates are used, all the variables except one will be zero in any particular mode.
(4) In the forced vibration of coupled systems, mode shapes may be defined. These mode shapes undergo changes in nature as the frequency passes through a natural frequency.
(5) Occasionally students mistakenly associate "harmonic" with the response of coupled systems. It is to be emphasized that harmonies (where natural frequencies of ω_n, $2\omega_n$, $3\omega_m$, . . . appear) are characteristic of distributed systems such as musical strings and organ pipes. In lumped parameter systems there are at most as many natural frequencies as there are variables, and these are not normally integral multiples of one another. An excitation at twice one of the natural frequencies does not generally produce resonance.

Problems

Note: A list of 39 transfer functions, used in many of the following problems and appearing at the end of the set, is referenced here by mention of the "list." Instructors will select appropriate samples for the various problems.

1. Describe the components of the natural response of the systems described by the transfer functions selected from the "list" by your instructor.

2. Use the limiting procedure of section 7.2(a) to determine the impulse response of the system described by #1 of the list.

3. Repeat Problem 7.2 for the following transfer functions from the list:
(a) #2 (c) #14
(b) #3 (d) #15

4. Make a careful sketch of the impulse response of selections from the list, using graphical methods to determine the magnitude of each component.

5. For selections from the list, use graphical estimates to determine the largest and the smallest components of the impulse response.

6. Use the graphical method to determine the partial fraction expansion of functions selected from the list.

7. Assuming that the functions of the list are $G(\mathbf{D})$ in

$$1 + KG(\mathbf{D}) = 0,$$

Prepare the root locus diagram for the selected entries.

8. Presuming that the denominator of the functions of the list, when set equal to zero, is the characteristic equation of a system, prepare the root locus for the case where the following coefficient is to vary instead of being a constant:
(a) The coefficient of the term in **D**.
(b) The coefficient of the zero-power term (originally, the constant).
(c) The coefficient of the highest power term in **D**.

9. For the root locii prepared in Problem 7 specify the conditions under which the system is stable, if any.

10. For the transfer functions selected from the list, write the separate expressions giving magnitude ratio and phase shift for sinusoidal excitation, in terms of the exciting frequency ω.

11. For the transfer functions selected from the list, use the graphical technique to prepare sinusoidal response diagrams, for both magnitude and phase.

12. Supposing that the transfer functions of the list are the "severed" or "open loop" transfer functions for closed loop systems when multiplied by an arbitrary gain constant K, predict the value of K which gives marginal stability of the system. Use graphical sinusoidal response procedures.

13. Use the final value theorem of Laplace transform theory to determine the final response of selected transfer functions with the following forcing function (Be sure to examine the applicability of the theorem):
(a) Impulse (c) ramp $y = t$
(b) Step $u(t)$ (d) sinusoid $y = \sin \omega t$.

14. Suppose the numerators of functions selected from the list become the entries in the following expressions representing coupled systems:

$$A_{11}(\mathbf{D})X_1 + A_{12}(\mathbf{D})X_2 = B_1(Y_1)$$
$$A_{21}(\mathbf{D})X_1 + A_{22}(\mathbf{D})X_2 = B_2(Y_2).$$

[The instructor will specify six functions to provide the functions $A_{11}(\mathbf{D})$, etc.]

(a) Determine the characteristic equation for the system.

(b) Find at least one root of the characteristic equation, in the vicinity of zero on the real axis, by using a graphical procedure.

(c) State whether or not the system has any natural modes.

(d) Evaluate the amplitude ratios (X_1/Y) and (X_2/Y) for $Y_1 = Y_2 = Y \sin \omega t$ at $\omega = 1.0$, 10 and 100 Hz. Does the nature of the response change markedly in this range?

The List

1. $\dfrac{1}{(1 + \mathbf{D})(1 + 3\mathbf{D})}$

2. $\dfrac{1}{\mathbf{D}^2 + 2\mathbf{D} + 10}$

3. $\dfrac{1}{(1 - \mathbf{D})(1 + 3\mathbf{D})}$

4. $\dfrac{1}{(1 + \mathbf{D})(1 + 3\mathbf{D})(1 + 7\mathbf{D})}$

5. $\dfrac{1}{\mathbf{D}(1 + \mathbf{D})(1 + 4\mathbf{D})}$

6. $\dfrac{1}{\mathbf{D}(1 + \mathbf{D})(1 + 100\mathbf{D})}$

7. $\dfrac{1}{\mathbf{D}(1 - \mathbf{D})(1 + 2\mathbf{D})}$

8. $\dfrac{1}{(1 + \mathbf{D})(\mathbf{D}^2 + 4\mathbf{D} + 29)}$

9. $\dfrac{1}{(1 + \mathbf{D})(\mathbf{D}^2 - 4\mathbf{D} + 29)}$

10. $\dfrac{1}{\mathbf{D}(\mathbf{D}^2 + 2\mathbf{D} + 10)}$

11. $\dfrac{1}{(1 + \mathbf{D})(1 + 2\mathbf{D})(\mathbf{D}^2 + 2\mathbf{D} + 10)}$

12. $\dfrac{1}{\mathbf{D}^2(\mathbf{D}^2 + 16)}$

13. $\dfrac{1}{(\mathbf{D}^2 + 4\mathbf{D} + 29)(\mathbf{D}^2 + 2\mathbf{D} + 10)}$

14. $\dfrac{1 + 2\mathbf{D}}{(1 + \mathbf{D})(1 + 3\mathbf{D})}$

15. $\dfrac{1 + 2\mathbf{D}}{\mathbf{D}^2 + 2\mathbf{D} + 10}$

16. $\dfrac{1 + 2\mathbf{D}}{(1 - \mathrm{D})(1 + 3\mathbf{D})}$

17. $\dfrac{1 + 2\mathbf{D}}{(1 + \mathbf{D})(1 + 3\mathbf{D})(1 + 7\mathbf{D})}$

18. $\dfrac{1 + 2\mathbf{D}}{\mathbf{D}(1 + \mathbf{D})(1 + 4\mathbf{D})}$

19. $\dfrac{1 + 2\mathbf{D}}{\mathbf{D}(1 + \mathbf{D})(1 + 100\mathbf{D})}$

20. $\dfrac{(1 + \mathbf{D})(1 + 5\mathbf{D})}{\mathbf{D}(1 - \mathbf{D})(1 + 2\mathbf{D})}$

21. $\dfrac{(1 + 2\mathbf{D})(1 + 5\mathbf{D})}{(1 + \mathbf{D})(\mathbf{D}^2 + 4\mathbf{D} + 29)}$

22. $\dfrac{(1 + 2\mathbf{D})(1 + 5\mathbf{D})}{(1 + \mathbf{D})(\mathbf{D}^2 - 4\mathbf{D} + 29)}$

23. $\dfrac{\mathbf{D}^2 + 2\mathbf{D} + 5}{\mathbf{D}(\mathbf{D}^2 + 2\mathbf{D} + 10)}$

24. $\dfrac{\mathbf{D}^2 + 2\mathbf{D} + 5}{(1 + \mathbf{D})(1 + 2\mathbf{D})(\mathbf{D}^2 + 2\mathbf{D} + 10)}$

25. $\dfrac{(1 + \mathbf{D})(1 + 5\mathbf{D})}{\mathbf{D}^2(\mathbf{D}^2 + 16)}$

26. $\dfrac{(1 + \mathbf{D})(1 + 5\mathbf{D})}{(\mathbf{D}^2 + 4\mathbf{D} + 29)(\mathbf{D}^2 + 2\mathbf{D} + 10)}$

27. $\dfrac{\mathbf{D}^2 + 2\mathbf{D} + 5}{(1 + \mathbf{D})(1 + 3\mathbf{D})}$

28. $\dfrac{\mathbf{D}(1 + 2\mathbf{D})}{\mathbf{D}^2 + 2\mathbf{D} + 10}$

29. $\dfrac{(1 + \mathbf{D})(1 - 3\mathbf{D})}{(1 - \mathbf{D})(1 + 3\mathbf{D})}$

30. $\dfrac{(1 + 2\mathbf{D})(1 + 4\mathbf{D})(1 + 5\mathbf{D})}{(1 + \mathbf{D})(1 + 3\mathbf{D})(1 + 7\mathbf{D})}$

31. $\dfrac{(1 + 2\mathbf{D})^3}{\mathbf{D}(1 + \mathbf{D})(1 + 4\mathbf{D})}$

32. $\dfrac{(1 + 2\mathbf{D})(1 + 3\mathbf{D})}{\mathbf{D}(1 + \mathbf{D})(1 + 100\mathbf{D})}$

33. $\dfrac{(1 - 2\mathbf{D})(1 + 3\mathbf{D})}{\mathbf{D}(1 - \mathbf{D})(1 + 2\mathbf{D})}$

34. $\dfrac{\mathbf{D}^2 + 2\mathbf{D} + 10}{(1 + \mathbf{D})(\mathbf{D}^2 + 4\mathbf{D} + 29)}$

35. $\dfrac{(1 + 10\mathbf{D})^2}{(1 + \mathbf{D})(\mathbf{D}^2 - 4\mathbf{D} + 29)}$

36. $\dfrac{(1 + \mathbf{D})(1 + 2\mathbf{D})(1 + 3\mathbf{D})}{\mathbf{D}(\mathbf{D}^2 + 2\mathbf{D} + 10)}$

37. $\dfrac{\mathbf{D}(1 + 2\mathbf{D})(D^2 + 2\mathbf{D} + 5)}{(1 + \mathbf{D})(1 + 2\mathbf{D})(\mathbf{D}^2 + 2\mathbf{D} + 10)}$

38. $\dfrac{(1 + 2\mathbf{D})^2}{\mathbf{D}^2(\mathbf{D}^2 + 16)}$

39. $\dfrac{(\mathbf{D}^2 + 2\mathbf{D} + 5)(\mathbf{D}^2 + 2\mathbf{D} + 18)}{(\mathbf{D}^2 + 4\mathbf{D} + 29)(\mathbf{D}^2 + 2\mathbf{D} + 10)}$

8 Extraordinary aspects of component characterization

The dynamic analysis methods of Chapters 5, 6, and 7 require that device characteristics be approximated with linear expressions. For many cases this is well justified and very effective. There are, however, a number of important situations in which a linearized analysis does not yield the information required for design. For example, some systems are caused to be unstable by effects for which no linear approximation is possible; backlash in gears is a notorious case. Also, the determination of start-up transients, where components operate over wide ranges, is usually impossible with linear methods; this problem is one of the basic subjects of Chapter 9.

In this chapter, the emphasis is on the aspects of component performance which transcend those considered thus far or which defy consideration by linear methods. We begin by summarizing some of the common kinds of non-linear phenomena which occur in components and make trouble for the analyst. Among these are saturation, dead zone, and hysteresis. We then consider a group of cases, purposely avoided earlier, in which terminal power cannot be evaluated as a simple $\Gamma\Delta$ product. These include components operating on compressible fluids, ac electrical machinery, and certain mechanical arrangements. The characterization of such components is discussed and an indication of network analysis given. The chapter is concluded with a discussion of some of the thermal aspects of component performance. There are cases where these effects dominate system behavior; several examples are shown.

8.1 Harmonic Generation

One of the gentler, but still important, effects of non-linearity in systems is the generation of harmonics. This is particularly important in high-fidelity audio systems where component non-linearity can bring about objectionable distortion of the sound output. It is also important in some feedback systems having relatively high-frequency resonances.

The basic origin of harmonics in systems can be seen by considering the approximation of a characteristic by means of polynomials. For example, suppose that two system variables are related, because of component and network characteristics, such that the following approximation is reasonable:

$$y = a_0 + a_1 x + a_2 x^2 + a_3 x^3.$$

If x is sinusoidal, such as $X \sin \omega t$, the following identities from trigonometry may be used to determine y:

$$\begin{aligned} \sin^2 \omega t &= \tfrac{1}{2}(1 - \cos 2\omega t) \\ \sin^3 \omega t &= \tfrac{1}{4}(3 \sin \omega t - \sin 3\omega t). \end{aligned}$$

Therefore,

$$y = (a_0 + \tfrac{1}{2}a_2 X^2) + (a_1 X + \tfrac{3}{4}a_3 X^3) \sin \omega t - \tfrac{1}{2}a_2 X^2 \cos 2\omega t - \tfrac{1}{4}a_3 X^3 \sin 3\omega t.$$

Clearly, the term in x^2 leads to a second harmonic (as well as a constant) while the term in x^3 leads to the third harmonic as well as some of the fundamental. Generally, each term in a polynomial expansion of a non-linear characteristic leads to a harmonic at n times the fundamental frequency, where n is the degree of the term. For most real devices, the first three to five terms of the polynomial approximation give a good fit to the characteristics; the higher-degree terms have very small coefficients (the a's above) and the higher harmonics have exceedingly small amplitudes. This is somewhat helpful in analysis, as we shall see in the next section.

As a simple example of the origin of harmonics, consider the simple transistor amplifier shown in Figure 8.1(a). Because transistor current amplification changes with collector current, the incremental gain of the amplifier will be different for various values of the average output voltage. This results in a voltage transfer relationship such as that shown at (b). Since the curve has quite a bit of curvature in the operating region, the second-degree term in its approximation will be relatively strong. The change in curvature from one end to the other will lead to a relatively strong third-degree term as well. Therefore, this amplifier will have fairly large second and third harmonics in its output if it is operated over the range suggested by the sinusoid at the bottom of the figure. Incidentally, it is a simple matter to observe these phenomena with an analog computer if one can obtain an audio amplifier, a speaker, and one or more computer multipliers capable of operating at audio frequencies. Arrange the computer to give a three- or four-term polynomial, drive it with a simple computer-constructed oscillator, and listen to the output by means of the amplifier and speaker. By adjusting the coefficients of the various terms the nature of the sound may be strongly influenced. Because of certain general properties of many systems, it is possible to learn something of their non-

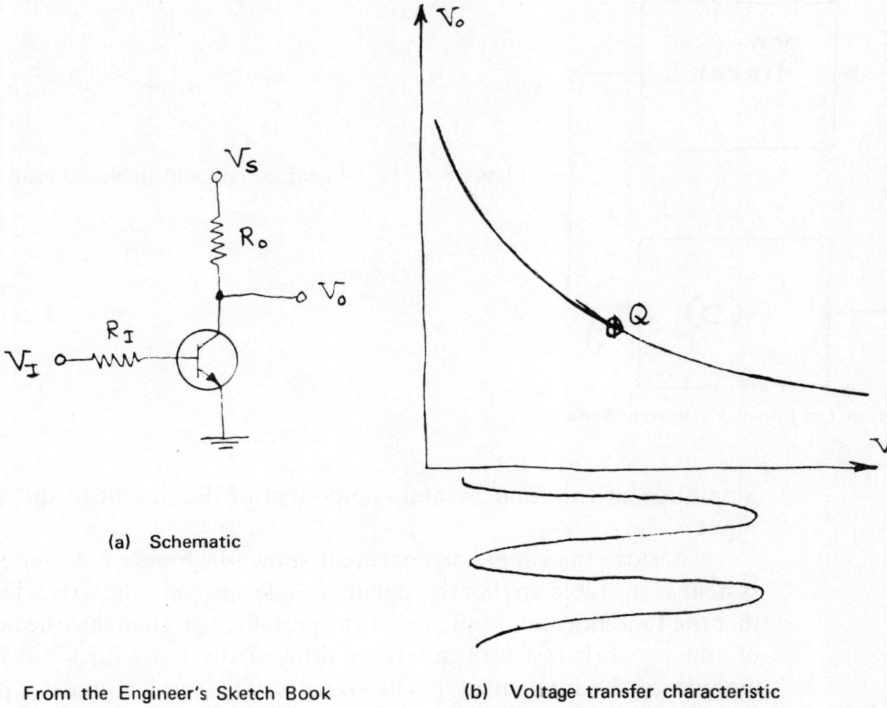

Figure 8.1 Audio amplifier.

linear behavior by considering the case of sinusoidal excitation and harmonic generation. This is the subject of the following section.

8.2 Describing Functions

A technique which may occasionally be used to determine the stability of closed-loop systems containing a non-linear element is based on harmonic analysis and is known as the describing function method. Here we shall not develop it in great detail; the objective is simply to acquaint the reader with its existence.

In most systems, especially those of the closed loop variety, the gain of most of the transfer functions falls off very rapidly beyond a certain frequency. The result of this is that, if harmonics are generated by non-linear components in the system, they tend to be filtered out by the other components. In this case, the non-linearity is of no consequence and we might as well use a linear approximation for the non-linear element. This linear approximation is the ratio of the

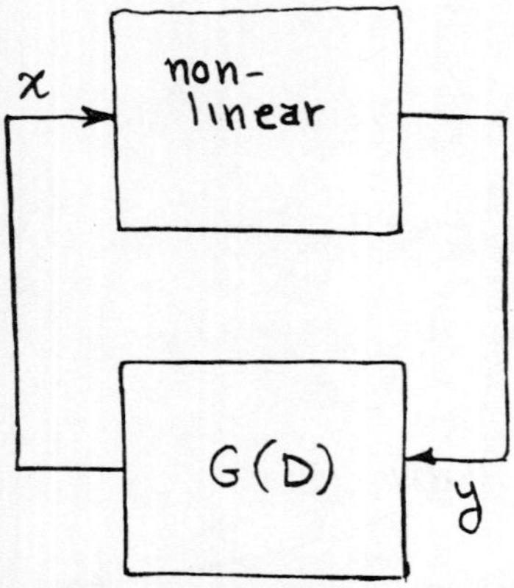

From the Engineer's Sketch Book

Figure 8.2 Non-linear component in closed loop system.

amplitude of the fundamental component of the output to the amplitude of input.

Consider the simple arrangement shown in Figure 8.2. Suppose that the system is unstable so that the signals x and y are periodic. Also, let us imagine that the fundamental frequency of the periodicity is such that the linear portion of the network (G) transmits very little of the harmonics but a significant amount of the fundamental. The signal x will then be almost a perfect sinusoid while the signal y will be the complex combination of the fundamental and its harmonics which are generated by the non-linear element. Although the harmonics are inputs to the linear portion of the network, they have no influence on the behavior. The transfer function G operates on the fundamental signal as if the harmonics were not there. (This is generally true of linear transfer functions; the principle of superposition pertains.)

Supposing that the non-linear component may be described by a polynomial, it is easy to show that the fundamental components of signals x and y must be exactly in or exactly out of phase. In this case, then, if the system is unstable the phase shift of the linear component must be zero or 180°; since the latter is more likely, let us assume it. From the frequency response diagram of the linear element, we can determine both the frequency and gain which exist at the 180° phase shift point. In order for the network to be oscillatory, the ratio of the fundamental amplitudes for the output and input of the non-linear element must be exactly the negative reciprocal of the linear element gain. The amplitude at which this occurs may be determined, if it exists, from an inspection of the expression for the ratio of the fundamental amplitudes. Thus, the frequency and amplitude of oscillation in the system may be determined.

It is important to note that the amplitude ratio changes with input ampli-

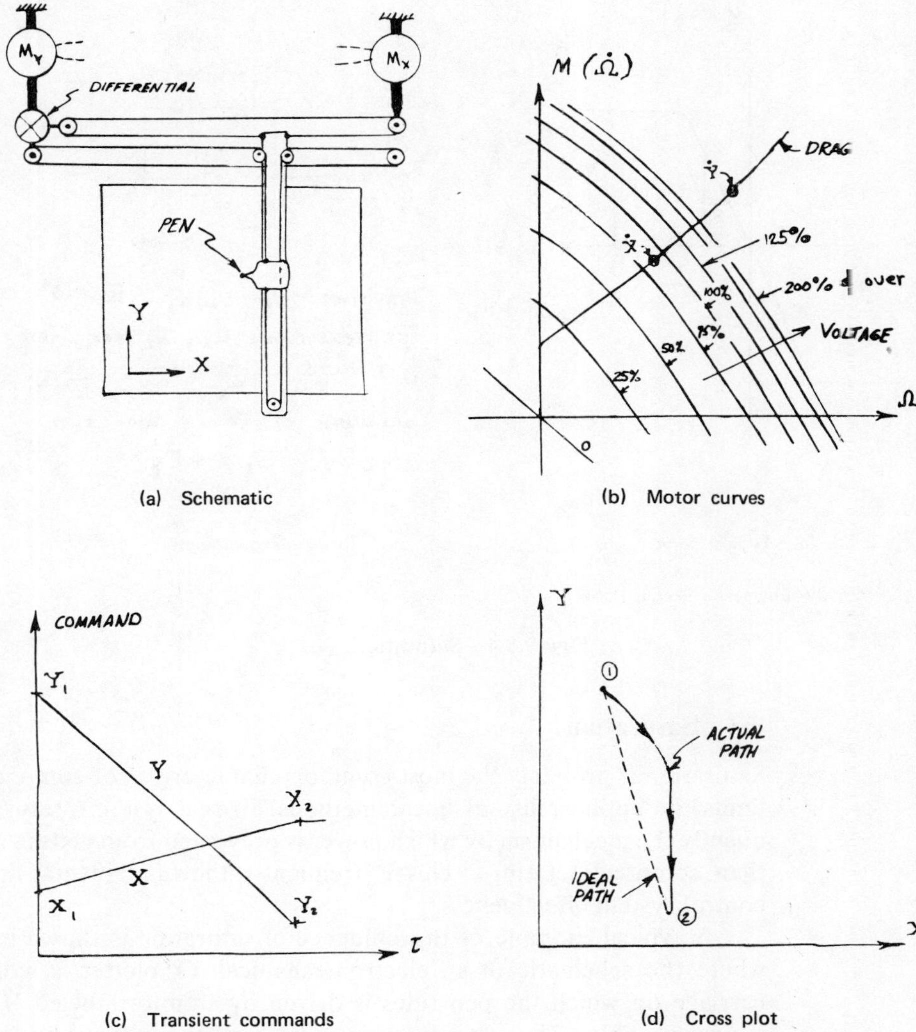

From the Engineer's Sketch Book

Figure 8.3 Electromechanical drafting machine.

tude for any non-linear component (See Figure 8.1). Also, for components with hysteresis there is a phase shift between the input and output. Finally, for components which include energy storage, the amplitude ratio and the phase shift are both functions of the frequency and input amplitude.

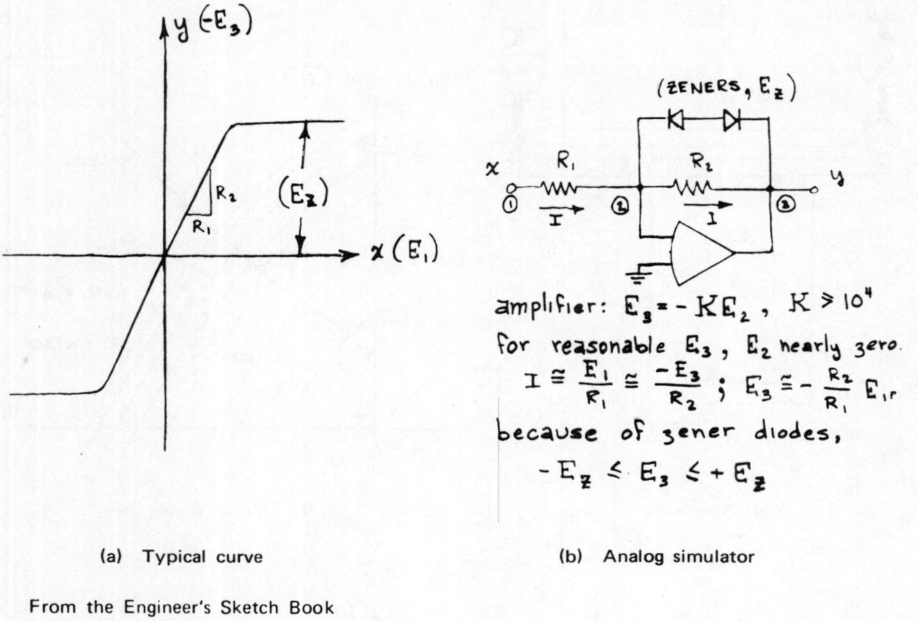

Figure 8.4 Saturation.

8.3 Saturation

Saturation is probably the most common characteristic of components which limits the applicability of linear methods. In real systems, saturation is frequently the mechanism by which power is prevented from getting much larger than component ratings. This is frequently the fundamental limitation on control system effectiveness.

A typical example of the influence of saturation is shown in Figure 8.3 where the schematic of an electromechanical *XY* plotter is indicated. The carriage on which the pen rides is driven by a motor labeled M_X through a system of cables. The pen is moved vertically on the carriage by another cable arrangement and the motor M_Y. A differential serves to isolate the *X* and *Y* motions. A typical set of mechanical characteristics for the motors is shown at (b) with the system drag curve superposed. In motors of this nature, the iron in the magnetic circuit saturates; this results in the motor curves being bunched up in the region where the voltages (and, therefore, the currents) are relatively high. If the recorder is supposed to draw a straight line as commanded by the signals shown at (c), and one of these signals requires a greater velocity than the machine is capable of, the resultant path will be similar to that shown in (d).

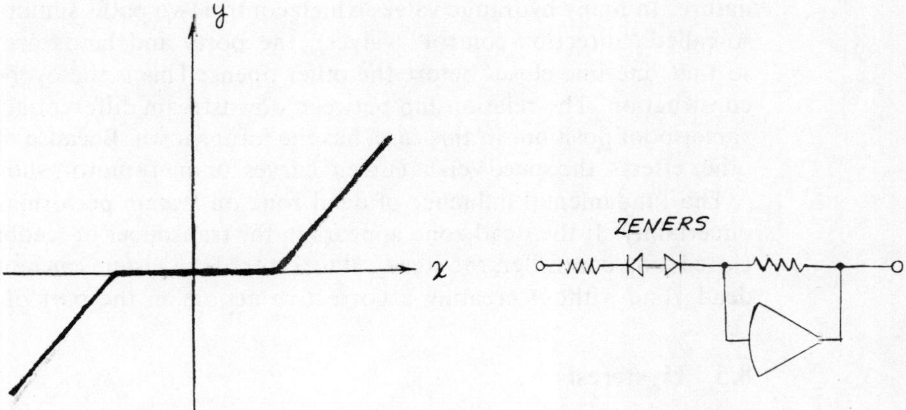

(a) Typical curve

(b) Analog simulator

From the Engineer's Sketch Book

Figure 8.5 Dead zone.

It is difficult, when designing *XY* recorders, to compensate for or avoid this problem.

The electronic amplifier network shown in Figure 8.4 may be used to simulate saturation for computing work. By using the artificial saturation provided by the Zener diodes, the amplifier itself is not required to saturate. This avoids problems with stabilized amplifiers and with those integrated circuit devices which tend to get lodged in the saturated condition.

8.4 Dead Zone

Many kinds of components exhibit a phenomenon known as dead zone. Fundamentally, the nature of the effect is that there is some region over which one terminal variable, normally proportional to another, remains at a fixed level. This is indicated in Figure 8.5 along with an electronic amplifier arrangement which may be used to simulate it. Common examples of this effect include mechanical arrangements in which stiction and other kinds of friction both operate; in this case x represents a force and y a velocity. In electrical systems, the voltage transfer characteristic for certain push-pull amplifiers has this nature. In many hydraulic valves which control two paths simultaneously (the so-called "direction-control" valves), the ports and lands are constructed so that one line closes before the other opens. This is the over-lapped valve construction. The relationship between downstream differential pressure and valve spool position, in this case, has the form shown. Because of stiction and other effects, the speed versus current curves for many motors show this nature.

The fundamental influence of dead zone on system performance is one of uncertainty. If the dead zone appears in the transducer or feedback loop of a closed loop controller, the steady-state error of the system can be as large as the dead zone without creating a corrective action on the part of the control.

8.5 Hysteresis

The well known hysteresis loop for magnetic materials typifies an effect which has many other forms. For example, when most materials are stressed, the stress-strain diagram forms a loop instead of being just a back-and-forth path. Some dielectric materials exhibit a similar kind of performance when the potential applied to them is varied cyclicly; in this case, the polarization versus field curves show a loop. In each case, the area inside the loop is proportional to energy which is lost in the cyclic process. Because of this, hysteresis in systems tends to play a role which is somewhat similar to that of damping.

An artificial means for developing hysteresis is shown in Figure 8.6. In this, a block slides with coulomb friction and is restrained by a spring. The force versus displacement curves for the arrangement are shown at (b). When the velocity of the block is negative, operation is on the upper curve; when it is positive, the lower curve applies. A transition from one to the other can be made at any point. Obviously, if this system is set to oscillating, the friction will soon remove all the energy and bring the oscillations to a halt.

Hysteresis may be simulated with a computing-type amplifier by using the arrangement shown at Figure 8.6(c). Here the amplifier must be capable of withstanding saturated operation for indefinite periods (many units are capable of this).

If the arrowheads on the graph of Figure 8.6(b) are reversed, a mythical device which generates, instead of dissipates, power is described. This is equivalent to negative damping. It may be simulated on a computer; the sketch at (d) shows how it may be used to build a square and triangular wave generator.

8.6 Power Dependent on Three Terminal Variables

An aspect of component behavior which requires further attention is that which occurs when the power at a terminal cannot be determined from a knowledge

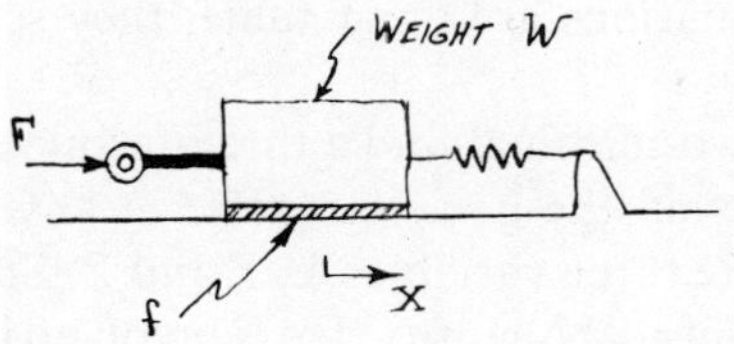

(a) Mechanical construction

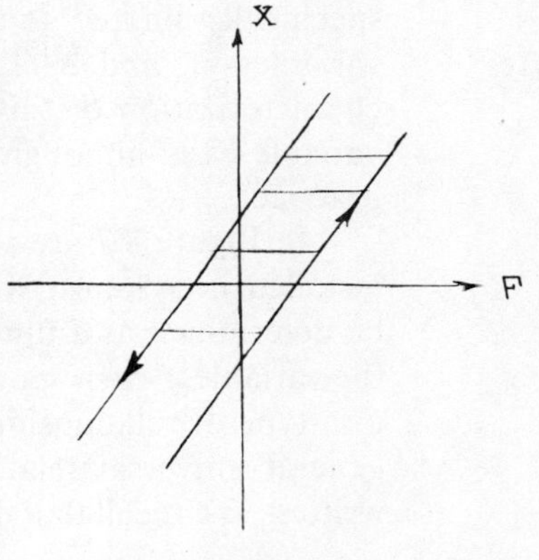

(b) Curves for (a)

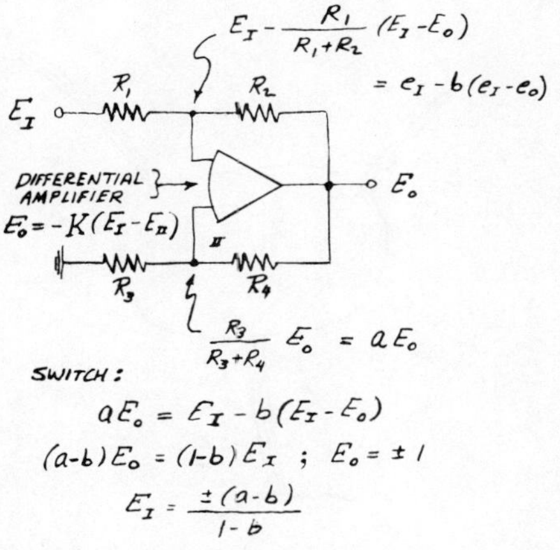

(c) Electronic simulator

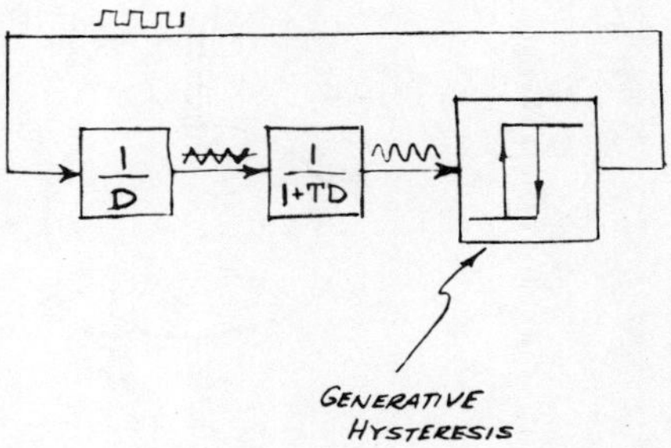

(d) Vibrator with negative hysteresis

From the Engineer's Sketch Book

Figure 8.6 Hysteresis.

of only two terminal variables. Examples of such components are those operating on compressible fluids, those in which chemical energy is transformed (such as gasoline engines), those in which mechanical power is transmitted in such a way that forces and velocities (or torques and rotational velocities) are not acting along the same lines, and all those in which power is transmitted in a periodic fashion, sinusoidally excited electrical machinery being a notable example. In the study of characterization of Chapter 3, the discussion was specifically limited to those cases where the power depended on only two variables—Γ and Δ. Let us begin this discussion by considering the static characterization of this new case. Here we shall go only as far as the three-variable case; although worse situations are imaginable, they seldom arise in practice.

In Figure 8.7 are shown a two-terminal and a three-terminal component installed in systems; we assume that the power transfer at any terminal can be determined as a function of the three variables X, Y and Z. Supposing that the variable X corresponds to a physical quantity that is involved in a conservation type of relationship and that variable Z is defined in a relative way, the general universal relationships shown at the bottom of the figure may be written. We recall that these would apply to both the component and the system

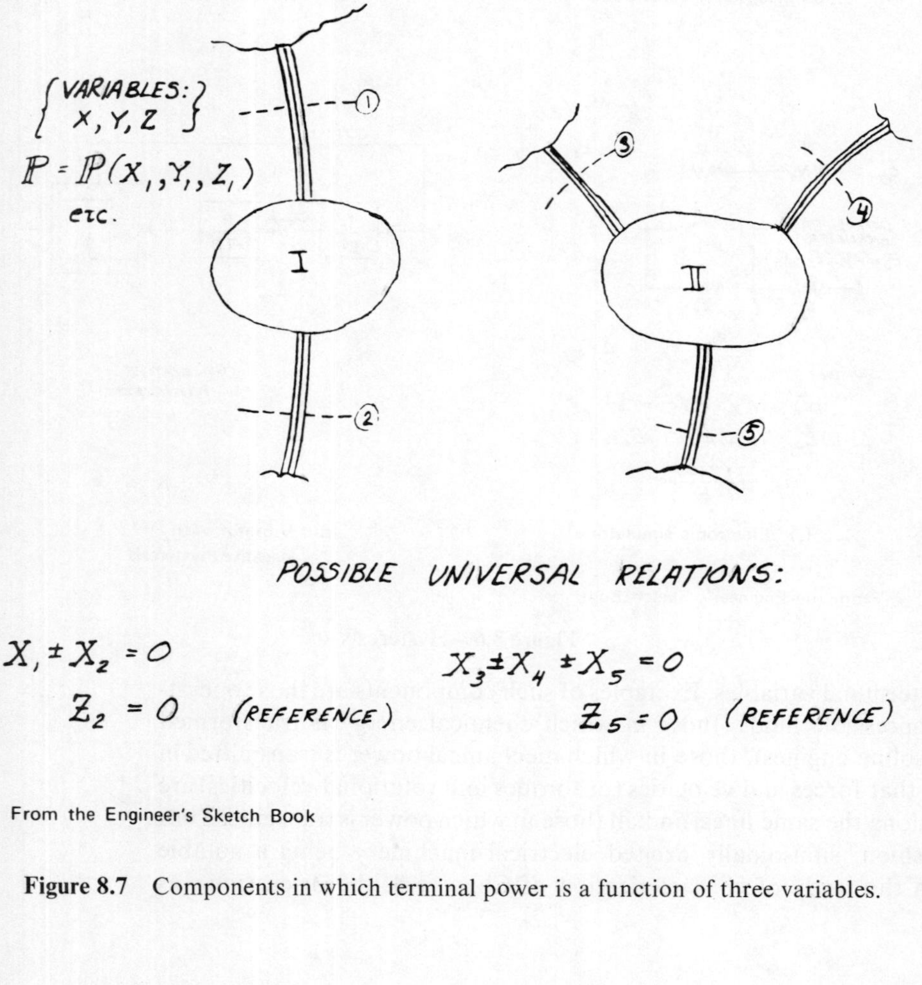

From the Engineer's Sketch Book

Figure 8.7 Components in which terminal power is a function of three variables.

to which it is connected. For the two-terminal case, there is a total of six variables; since there are two universal relationships, the characterizations must yield four independent relationships. We would guess that two would arise from the component and two from the system to which it is connected. For the three-terminal case, there are nine variables and two universal relationships; the characterizations would have to yield seven indpendent relationships. Under the foregoing assumptions, then, it is impossible for the component characterization to have the same number of dependent and independent variables. Thus, the characterization of the component and the system to which it is connected must be different in form. This is a somewhat distressing conclusion; let us contemplate it briefly. It might be possible that our assumptions were erroneous; perhaps the variable Y must also be a relative variable. However, when we check this possibility, it resolves the three-terminal situation but messes up the two-terminal case. Similarly, any assumed universal relationship, if applied to both cases, gives an unbalanced result for one. Therefore, we can only conclude that there is something inherently different about some kinds of three-variable components having the same number of terminals. For the case considered here, some three-terminal components would be characterized with four dependent and three independent terminal variables, while others would have three dependent and four independent variables. We are thus led to the thought that two components having the same characterization form could not operate together in a static equilibrium. These apparent quandaries are most readily resolved by the consideration of specfic cases, which we proceed to do in the following paragraphs. It turns out that dependent, independent variable selection, in this case, is related to the way power is transferred at the terminals. Roughly speaking, at a terminal where power enters a component the Y variable is independent; where power leaves, the Y variable is dependent. With two interconnected three-terminal components, the power transfer directions are complementary; hence, the characterizations are also. The reversibility of a component becomes significant; whether or not it will operate with other units depends on this.

8.7 Irregular Mechanical Power Transmission

A relatively simple, but important, case of three-variable power transfer in mechanical systems is shown in Figure 8.8. The figure indicates an automobile which we shall imagine to be spinning out with all the wheels locked. The velocity of the center of gravity of the vehicle is V_C while the spinning speed is Ω_C. A simple kinematic analysis allows the magnitude and direction of the sliding velocity at any particular tire to be determined. This is indicated at (b) in the figure. Because of the shape and tread of the tire, the friction force does

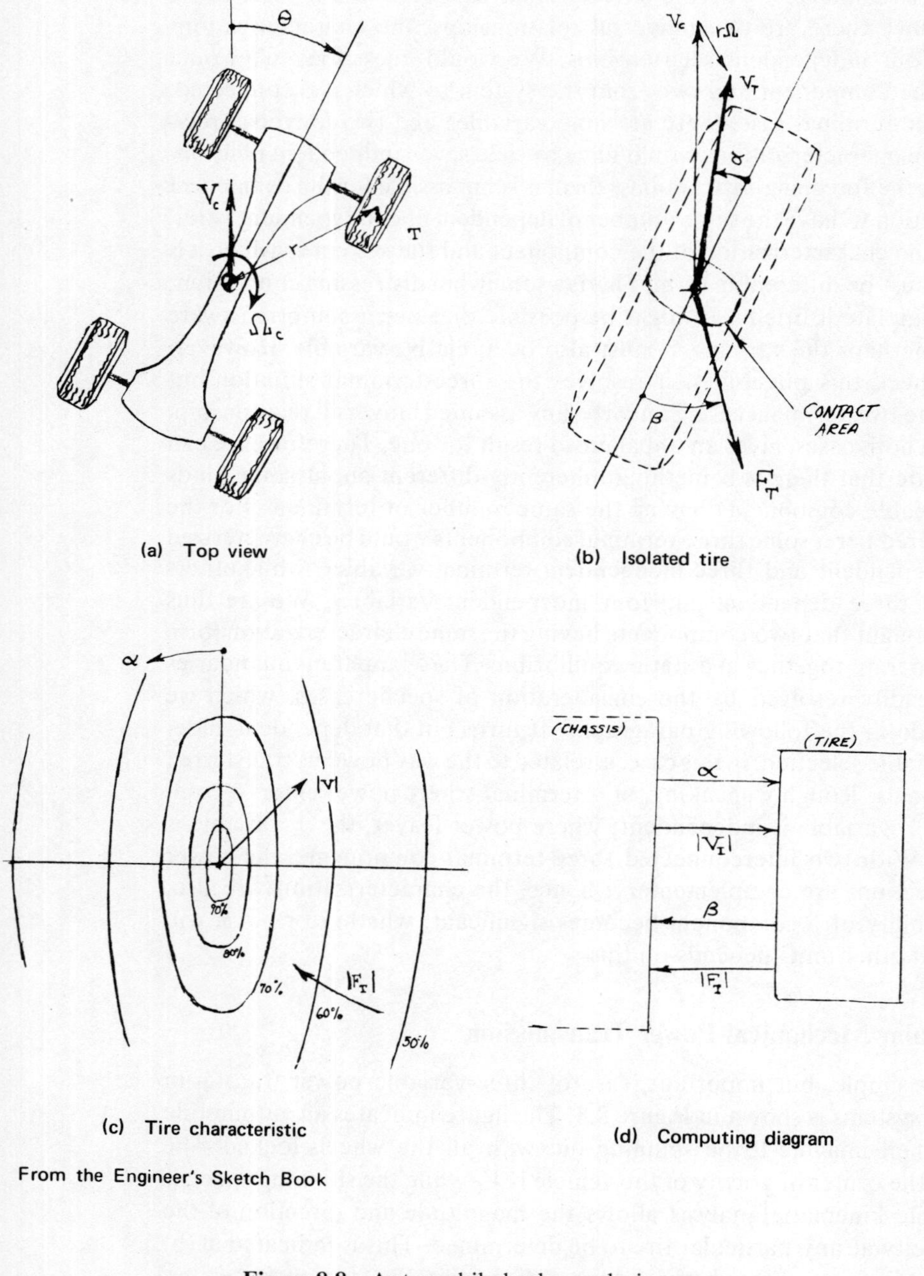

(a) Top view

(b) Isolated tire

(c) Tire characteristic

(d) Computing diagram

From the Engineer's Sketch Book

Figure 8.8 Automobile brake analysis.

not, in most likelihood, line up with the velocity. In particular, we would expect that the component of friction perpendicular to the tread would be larger than that parallel to it. This is indicated at (b) where the friction force F_T makes more of an angle with the tread line than does the sliding velocity V_T. In order to determine these effects for the purposes of a computer simulation, tests would have to be run with different values of the angle α and the sliding velocity V_T. These would result in a pair of functions, one for the magnitude of the friction force F_T and the other for its angle β. A possible kind of result from such a test, drawn with polar coordinates and normalized values for the friction force magnitude, is indicated at (c) of the figure. The angle β could also be shown on this graph. At (d) is shown a portion of a computing diagram for simulation of the vehicle. The power transfer between the tire and the road is given by $|V_T| \cdot |F_T| \cos(\alpha - \beta)$.

A somewhat tricky aspect of this example, which appears to make it look like a four-variable instead of a three-variable case, is that the component in question is mechanically variable, in the angle θ, through the twist that can be applied to the tire via a bending moment in the axle. A simpler form of two-terminal, three-variable translational mechanical dissipator would have simple ball-type connections above and below a pair of sliding plates confined by some frictionless mechanism to stay in an inertially fixed plane. Given the relative velocity between the balls, the force and its angle with the velocity would be determined. Bodies moving through gases and liquids represent this case quite well; there are a variety of towing situations in which this kind of power transfer is important.

8.8 Compressible Fluid Components

Compressible fluid—i.e., pneumatic—components have considerable applicability in system engineering and also exhibit the situation where power depends on more than two variables. Specifically, in order to determine the power transfer at a particular terminal of a gas-operated component, we must know both the state of the gas and the flow rate. Since two measurements must be made in order to determine the state, three are necessary to determine power. For most systems work, pressure and temperature are the most convenient variables with which to work; we shall use them here. The reader should keep in mind, however, that this combination is not the only possibility (density could be used in place of temperature, for example). In fact, in systems operating on saturated steam, this choice would be unworkable since pressure and temperature are not independent under the vapor dome. And, believe it or not, steam is still a very important medium for power transfer.

One of the biggest advantages of air-operated systems is that once power

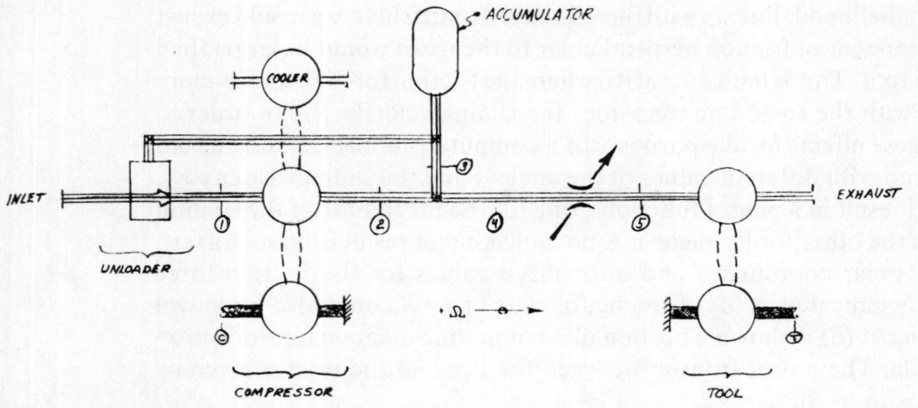

(a) Schematic diagram

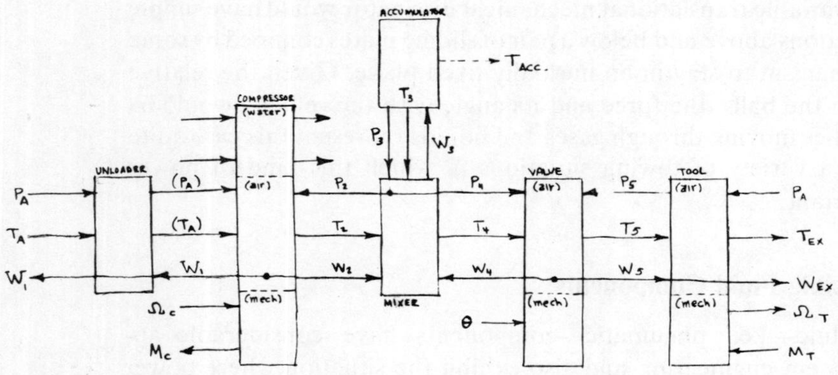

(b) Computing diagram (accumulator discharging)

From the Engineer's Sketch Book

Figure 8.9 Air compressor system.

has been delivered by air, it can be exhausted to the atmosphere, thereby avoiding the need for return lines. Also, leaks cause no particular danger or mess. These can be significant factors in industrial applications and often offset some of the disadvantages associated with pneumatics. The problems of limited response speed due to gas compressibility and the explosion hazard are the major compromises involved.

In order to examine the problems of pneumatic system analysis, let us consider the typical system shown at Figure 8.9(a). Here a compressor charges an accumulator which drives a tool of some kind through a variable orifice control. We assume that the compressor runs continuously; in order to avoid over-pressures and conserve power, an unloading valve is placed in the compressor inlet. When the accumulator pressure is at the preset value, the valve closes, the compressor can do no work and, therefore, take no power. The various terminals at which conditions must be known in order to carry out an analysis are indicated by circled numbers and letters. The air sides of the compressor, valve, and tool all illustrate the quandary which was encountered in the generalized analysis of section 8.6. For example, at the air side of the compressor, there are six variables: the two pressures, the two temperatures, and the two flow rates. Because of the conservation of matter, the two flow rates are not independent. We know that the weight flow rate into the machine is equal to the outrate, leaving five variables to be involved in characterizations. For an incompressible fluid machine, we can use relative pressures to compute power; there is no storage of energy within the fluid itself and, therefore, absolute pressures are of no concern. Here, however, this is not the case. power required to compress gas from 15 to 30 psia is different from that required to compress the same flow rate from 1015 to 1030 psia. Similarly, it is obvious that absolute temperatures, instead of relative values, must be used. Again, the two-terminal dissipator test makes this clear. In order to determine the power dissipated by a component through which gas flows, we must use absolute values of both temperature and pressure, since:

(1) At constant temperature, the gas must do mechanical work in the component in order to create dissipation. This work depends on the natural logarithm of the pressure ratio for ideal gases, and cannot be independent of the pressure reference. For non-ideal gases, the likelihood of this condition being satisfied is even less.

(2) At constant pressure, the amount of dissipation is proportional to the enthalpy change. For an ideal gas, this is given by $wc_p(T_2 - T_1)$ which is independent of the temperature reference;
w = weight flowrate,
c_p = specific heat at constant pressure,
T = temperature.

However, for any real gas, the specific heat is a function of temperature and the dissipation depends on absolute values.

In characterizing the compressor, then, there will be five variables associated with the air section. Obviously, the characterization will have an asymmetric form. Intuitively, it is obvious that the inlet temperature is physically independent of the other terminal variables; there is no way in which the compressor can influence the temperature of the air being delivered to it. This does *not* mean that we are forced to use inlet temperature as an independent variable in the characterization, but it does show that it is an admissable independent variable. Continuing the intuitive analysis, we may note that, given both the inlet and outlet pressures, we can determine the flow rate (supposedly we know the compressor displacement); this would also lead to a particular outlet temperature. Therefore, from physical reasoning, it appears that the proper characterization of the air section of the compressor has three independent and two dependent variables. For computational purposes, any selection would be satisfactory since there are no peculiarities associated with the machine performance in normal operating conditions. Note that the air section of the compressor looks like a power generator from the point of view of the other portions of the air circuit in the system; it has an excess of independent variables in its characterization.

A similar analysis of the control valve and tool gives reasonable selections for independent and dependent variables, the most straightforward approach being to take the temperature of an inflow as independent and that of an outflow as dependent. The pressure and flow-rate variables may be assigned as if the fluid were incompressible. Note that the tee between sections 2, 3, and 4 is a significant system component, unlike its corresponding devices in incompressible fluid systems. Here, when the accumulator is discharging, the gases at 2 and 3—which probably have different states—are mixed before flowing out at section 4. Therefore, the states are different at each of the terminals in this mode of operation. A reasonable computing diagram for the system is shown at (b) in Figure 8.9. (This is for the case where the accumulator is discharging and the unloader is open; other configurations would be required for other cases.)

Assuming we can now derive, from thermodynamic principles, the characteristics of the various components or that we have access to empirically determined characterizations, the process of analysis may begin. The principles of static analysis apply; the only difference between this and other problems we have considered is the matter of complexity. In a real system of this nature, the static state is probably never attained; the accumulator is either charging or discharging. In a properly designed system the compressor should be able to deliver enough air to supply the tool at rated pressure for any setting of the valve. However, in an overloaded system, there would be some equilibrium value for the accumulator pressure and for the compressor flowrate (equal to the tool flow rate).

For the dynamic analysis of pneumatic systems, we may apply small excursion analysis so long as the density variations of the gas are relatively small; the only significant difference between this kind of system and systems using incompressible fluids is the influence of temperature. In fluidic systems (those using relatively low-pressure and for signal, not power, transmission) the density variation is so small that the temperature effects are negligible. They may be analyzed as if they were incompressible fluid systems, in most cases.

8.9 ac Electrical Machinery

As is well known, the computation of electrical power in ac systems requires knowledge of voltage magnitude (rms), current magnitude (also rms) and phase angle. This has implications for component characterization and system analysis which we shall now investigate. First, it must be noted that our interest here is in the aspects of component and system performance in which the time constants and periods of oscillation are large compared to the period of the ac excitation. That is, we are concerned with the gross performance of the component or system; we do not investigate, for example, the details of the excitation waveforms.

The general nature of the situation may be seen by inspection of the characterizations of Figure 8.10. At (a) and (b) are the characteristics of a typical induction motor for constant frequency excitation. At (c) and (d) are the characteristics of an ac generator. In these, M denotes torque, Ω denotes rotational speed, E denotes voltage, I denotes current and P.F. is power factor. The motor, which is a four-terminal device, is characterized with two independent and three dependent variables. *Note*: the power factor is shown insensitive to voltage magnitude; this is not exactly true, but it is a good enough approximation for most engineering purposes. If the motor were a dc machine, its characterization would be the usual two-by-two arrangement; it is the significance of the power factor here that brings about the difference. The generator characteristics are shown for a fixed value of the shaft speed. In actuality, then, the generator characterization is in the form of two dependent and three independent variables.

Let us assume that the generator is driven by a constant-speed engine (so that its output has the frequency at which the motor has been characterized) and that the machines are used to transmit power from the constant-speed source to some mechanical load attached to the motor. In determining the static operating point of the system, we may begin by recognizing that the voltage, current, *and* power factor for the two electrical machines must be identical in static equilibrium. The static analysis is shown in Figure 8.11. If we superpose the characteristics of the mechanical load on the motor mechanical characteristics as at (a) we can determine the locus of $|E|$ versus

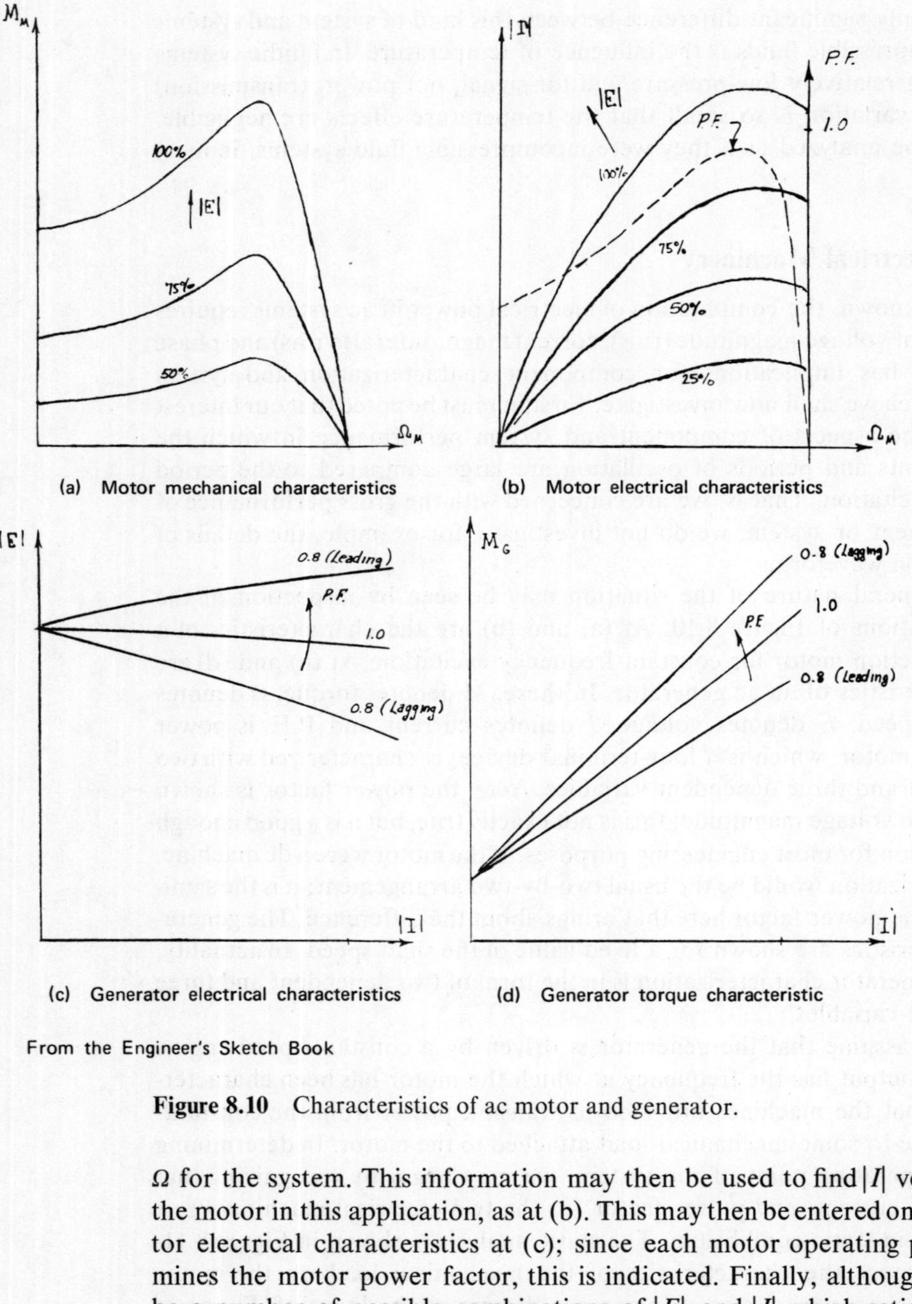

Figure 8.10 Characteristics of ac motor and generator.

Ω for the system. This information may then be used to find $|I|$ versus $|E|$ for the motor in this application, as at (b). This may then be entered on the generator electrical characteristics at (c); since each motor operating point determines the motor power factor, this is indicated. Finally, although there will be a number of possible combinations of $|E|$ and $|I|$ which satisfy both the

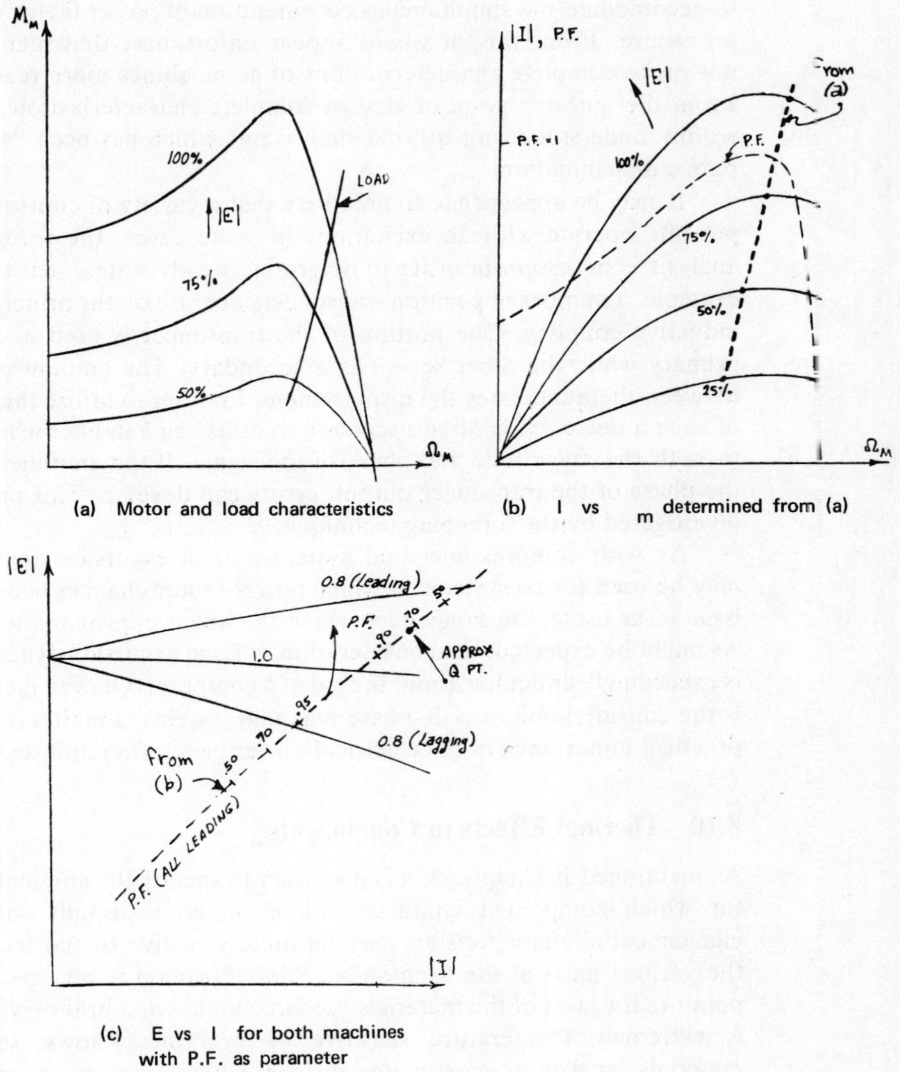

From the Engineer's Sketch Book

Figure 8.11 Static analysis of ac motor driven by a generator.

generator and motor, there will be only one at which the power factors are equal. This, of course, will be the static operating point.

Again, the ideas of static analysis do not change; it is simply necessary to accomodate the simultaneous consideration of power factor in the analysis procedure. From this, it would appear unfortunate that manufacturers do not make complete characterizations of ac machines more readily available. From the author's point of view, a complete characterization is much more readily understood and utilized than is one which has been "simplified" for public dissemination.

It may be appropriate to note here that a variety of control system components operate with ac excitation; in some cases, the foregoing kind of analysis is necessary in order to determine steady states, sensitivity, etc. For example, a number of position transducers operate on the principle of variable inductive coupling. One portion of the transducer is used as a transformer primary while the other serves as a secondary. The amount of ac coupling between them indicates the displacement. In order to utilize the output signal of such a device, it is often necessary to utilize an amplifier which is sensitive to both the magnitude and phase of the signal. If the amplifier input affects the phase of the transducer output, errors can develop. This problem can be investigated by the foregoing technique.

As with compressible fluid systems, small excursion analysis methods may be used for ac systems in which power-factor changes occur; the critical issue is, as usual, the range over which the linear approximations are useful. As might be expected, the consideration of large excursion dynamic operation is exceedingly difficult without the aid of a computer. Of even more complexity is the consideration of poly-phase machine systems, a matter of considerable practical importance in the electrical power generation industry.

8.10 Thermal Effects in Components

As mentioned in Chapter 3, it is necessary to specify the ambient temperature for which component characterizations apply. Especially with dissipative elements, the characteristics may be quite sensitive to the temperatures of the various parts of the component. Sliding friction is very sensitive to temperature for most of the materials used in clutches and brakes; viscous friction is extremely temperature sensitive as everyone knows; semiconductor materials, as well as most materials from which resistors are made, are also temperature sensitive. These effects can have some very important results, one of the most spectacular being thermal instability which we shall consider in the following section. Another important effect is the change in static operating point of a system; this is particularly troublesome in high-gain electronic amplifiers for dc applications. Temperature changes can also cause the sensitivity of certain transducers to vary, thereby destroying the calibration of instruments and feedback systems.

The exact description of how component characteristics change with changes in temperature is a difficult, if not impossible, task. In the static-state operation of a device, thermal equilibrium is usually attained very slowly and is a complex kind of thing, involving continuous variation of temperature throughout every body in the component. This distribution is sensitive to the manner in which heat is transferred away from the component and may change significantly "as the wind blows." Since humidity affects the heat transfer, the device may also serve as somewhat of a hygrometer. In critical applications, as with some transistors and some kinds of clutches or brakes, the thermal test conditions must be very carefully specified and must be similar enough to potential applications that the results will be useful in design.

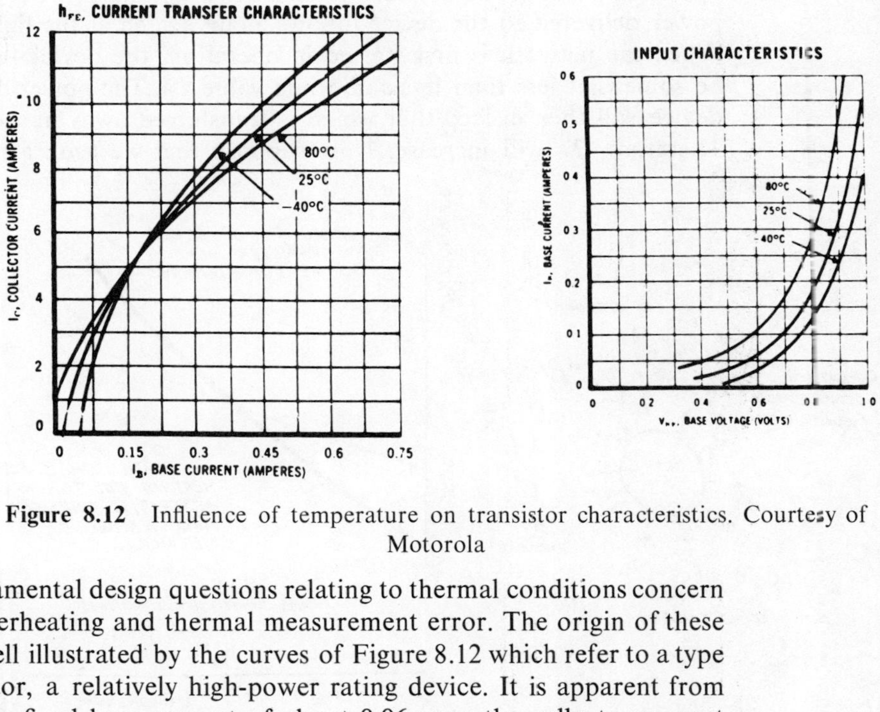

Figure 8.12 Influence of temperature on transistor characteristics. Courtesy of Motorola

The fundamental design questions relating to thermal conditions concern destructive overheating and thermal measurement error. The origin of these problems is well illustrated by the curves of Figure 8.12 which refer to a type 2N441 transistor, a relatively high-power rating device. It is apparent from these that, at a fixed base current of about 0.06 amp, the collector current can double as the temperature changes from $-40°$ to $+80°$C. Also, for a fixed base-to-emitter voltage of about 0.8, the base current can double for the same temperature change. This shows why it is very difficult to build a stable dc amplifier; it also explains the self-destruction of some transistor circuits, as we shall see next.

8.11 Thermal Instability

Thermal instability can occur in any power transfer medium. Basically, it is a situation where a system operates such that as one component heats up it is forced to accept even more power because of its changes in characteristics and a lack of compensation in the network. The situation is shown in general terms in Figure 8.13. Here some device "*D*" is shown in thermal equilibrium with its environs. The temperature at some particular point in the device (normally the hottest point) is used as an indication of the thermal state of the device. Usually, the heat transfer from the device to the surrounds is relatively well determined for any combination of T_D and T_A. This might have the nature of the heavier curve shown in part (b) of the figure. Since the device characteristics change with T_D, the static state of the network, and therefore the static operating point of the device, will change as T_D changes. Let us suppose that power delivered to the device follows a curve such as the lighter one at (b). When the network is first started in operation, the device temperature will be somewhat less than its equilibrium value T_Q. The power delivered to the device will then exceed that which is transferred away in the form of heat. Therefore, T_D will increase. Now suppose that we took a blow-torch and

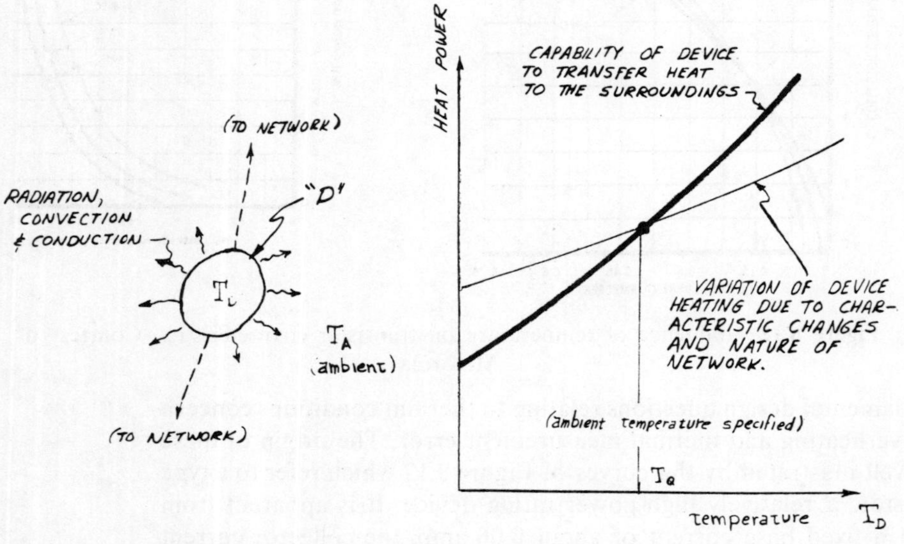

(a) Thermally sensitive component in network

(b) Power transfer curves

From the Engineer's Sketch Book

Figure 8.13 Thermal stability analysis.

heated D to above T_Q. When we take the torch away, the heat transfer will exceed the power delivered to the component and T_D will decrease. Therefore, the arrangement is thermally stable. *If the delivered power curve, however, is steeper than the curve giving heat transfer to the surrounds, the system is thermally unstable.*

Obviously, in order to determine thermal stability, we must make an analysis of the heat transfer system in which a component operates as well as an analysis of the static operating point variation as a function of component temperature.

As an example of this kind of analysis, let us look into the case of a heavy truck descending a long, steep grade. The author has some personal interest in this situation since he and several associates own and operate an old diesel truck with a low-bed trailer. The rig is used for moving antique railroad equipment, an obsession with the group. Some time ago we were working in the Los Angeles area, and made many trips over the infamous grapevine there. Coming down the hill, we were constantly reminded of potential disaster by the "bail-out" roads which have been provided every couple of miles. It is reported that some trucks, out of control, enter these roads at speeds of about 100 miles per hour. The run-outs are covered with a deep layer of sand and are arranged to take the truck up a hillside. Most drivers who have to use them die.

Some aspects of the situation are shown in Figure 8.14. At (b) is shown a typical friction coefficient versus temperature curve for brake lining material. It is clear that 550° F is the temperature at which disaster suddenly and without warning begins. At (c) is shown an indication of the analysis of the system; the numbers are strictly hypothetical. Since the driver is an integral part of the system in this case, we have to make an assumption about his strategy in descending the grade. Let us assume the simple approach where he decides ahead of time on the speed with which he will make the descent; he establishes this speed as he comes over the crest of the hill and then manipulates the brakes so as to maintain it. Since the weight of the truck is constant and we assume the hill to be of constant steepness, the amount of power which must be dissipated by the brakes is directly proportional to the speed of descent. This is clear from the horizontal portions of the brake power curves shown in part (c) of the figure. Obviously, at faster speeds, the brake temperature will have to be higher since more heat will have to be dissipated. Some reasonable values have been selected and the curve extrapolated into the region beyond 550°F; this is labeled the "cooling curve" for obvious reasons. With the numbers shown, there is a stable equilibrium at any speed up to about 35 mph.

This set of curves show why a driver who descends a hill at a somewhat

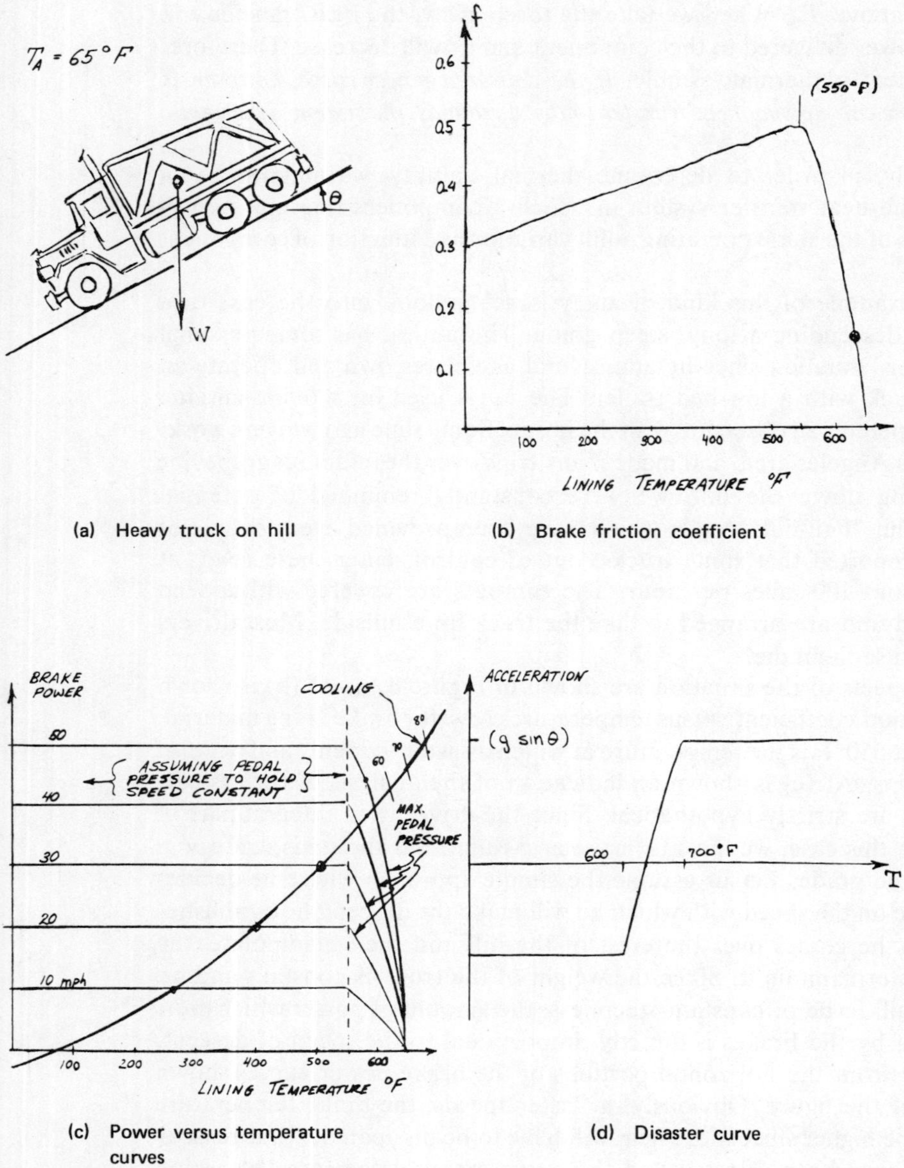

From the Engineer's Sketch Book

Figure 8.14 Truck brake analysis.

greater speed than others is being quite foolhardy; he may be leaving very little margin for safety. Let us consider the case of an eager runner who decides to bring her down at 50. For quite a while, there is no problem and no indication of trouble (although drivers behind him may see smoke coming off the brake drums). Then the lining temperature reaches 550° F, the driver feels a loss of braking power, and he clamps down with full pressure on the brake pedal. Can he do any good at this point? We can see that, at a lining temperature of 600° F and a speed of 50 mph, the brake power is still higher than the cooling curve; therefore, the brakes are still heating and the coefficient of friction is still decreasing. At the possible thermal equilibrium near 620° F, the coefficient is so much less than that required to hold the load that the truck must be accelerating; this is indicated roughly at (d). By this time it is too late; the ball game is over. In this case, although thermal stability seems to exist over the whole operating range, the system itself is unstable above a certain temperature.

The exact case indicated by Figure 8.12 may be observed if a power transistor such as the 2N441 is used to control a tungsten filament lamp (such as a 12-volt brake light bulb). These bulbs have very low resistance when cold and not an awful lot more when hot. Referring to the transistor temperature curves of Figure 8.12, let us imagine that the base circuit is open so that absolutely no current flows in it. Since no transistor is a perfect open circuit, even when the base current is precisely zero, there will be a small amount of collector current. This current will flow through a voltage drop equal to almost the full supply voltage as it goes from the collector to the emitter of the transistor (the transistor is supposedly off); this leads to some heating of the transistor. Now notice that an increase in the temperature leads to higher leakage current (if the base current is less than 0.15 amp). Therefore, the heating will increase and the temperature will rise higher. Unless very effective heat transfer facilities are utilized, this kind of circuit is almost certainly unstable thermally. (By this kind of circuit I mean one with a very low value of resistance in the collector to supply connection.) Various methods for avoiding the difficulty are given in basic electronics books; we will not consider the problem further here.

8.12 Abnormal Operation of Components

There are three tricks of component application which can be exceedingly useful in some marginal design situations. We mention these briefly here.

In some instances, a component may be operated well beyond its continuous power rating if the operation is of short duration. The use of limited duty cycles for electrical motors is one example; the maximum allowable winding

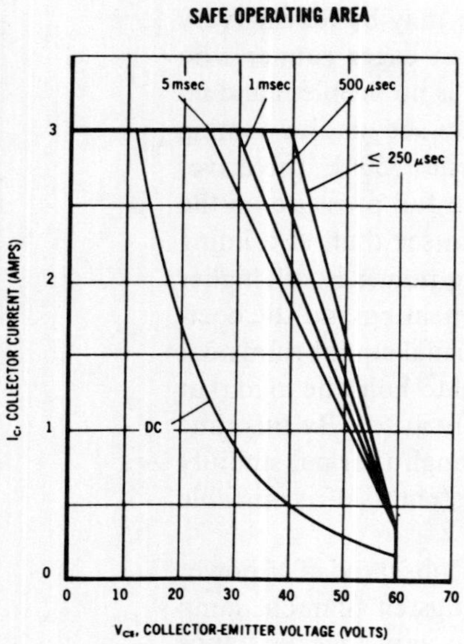

Figure 8.15 Pulse power rating of a transistor. Courtesy of Motorola.

temperature is the limiting factor in this. With transistors and other controllable dissipators, a similar possibility exists. For example, Figure 8.15 shows the dc operation limit for a typical transistor. For pulses of short duration, this limit may be exceeded to a significant extent as shown. Surprisingly, in this case the duty cycle is not significant. The failure mode of the transistor for this kind of operation is not gross heating, but a localized semiconductor effect known as secondary breakdown. Use of components at overrated conditions, if done knowledgeably, can have significant benefits (beyond cost reduction). For example, small transistors are faster than large ones, generally speaking. By using relatively small units beyond their dc ratings, some computer circuits can be made faster than possible with conventionally selected transistors. The same general situation holds true for, as one example, electromagnetic clutches where both the inertia and inductance are less in smaller sizes of similar units.

Another abnormal application of components is that which uses reverse power transfer conditions. For example, the dynamic braking of locomotives uses the traction motors as generators. The power which is generated is dissipated by huge resistors in the roof of the unit or, on electric railways, may be pumped back into the trolley wire to help drive another train. In this regard, it seems truly unfortunate that manufacturers seldom provide information on the reverse operation of their components; this would probably have the effect of originating new and unique applications if it were done. As one example, we might mention the internal combustion engine. Although many drivers utilize compression to hold their cars on hills or to assist in stopping in non-emergency situations, we never see the torque-speed curves for this mode of engine operation. It would seem that as more and more automatic control systems are developed, and more designers need this kind of information, we may see an improvement in the thoroughness of component characterizations.

A last trick which is sometimes useful is the exploitation of a secondary device characteristic. An example is the capacitance associated with a diode. This varies as the voltage applied to the diode is changed in the region of, roughly, 0.5 to 0.7. Some designers have used this as a means for changing the frequency of an oscillatory circuit; the action is very fast since it is all electronic. Another example is the resistance change of a tungsten filament as temperature changes; this may be used to limit the current in a circuit.

Problems

1. Suppose a device is characterized by the relationship

$$y = \tfrac{1}{3}(x + x^2 + x^3)$$

over the range $0 \leq x \leq 1$. What is the magnitude of the harmonics in the signal y if $x = \tfrac{1}{2}(1 + \sin \omega t)$?

2. Sketch the output waveform in the above problem, showing the component signals individually and using graphical addition to obtain the total signal.

3. Suppose that a signal $\sin \omega t + \sin 2\omega t$ is applied to a component with transfer function $1/(1 + TD)$, where $T = 1.5(1/\omega)$. What is the ratio of amplitudes of the 2ω and 1ω frequency components in the output?

4. Perform an analysis of the "dead-zone" simulator of Figure 8.5(b), using Figure 8.4(b) as a guide.

5. Derive an expression, in terms of the spring constant k, friction force fW, and half-amplitude of oscillation X_{max}, for the energy loss per cycle in the system of Figure 8.6(a). Also, write an expression for the power loss if the oscillation is at n cycles per second.

6. Suppose that the system of Figure 8.6(a) is given a good jolt and that a decaying oscillation of roughly five full cycles results, prior to rest. Show a reasonable history for the oscillation on the curves of Figure 8.6(b). Write the equations which govern the motion during any movement between two of the points where the motion reverses. Are these equations soluble by any of the methods presented in this text?

7. Carry out the argument associated with Figure 8.7, changing the assumptions as follows:

(1) Both the X and the Y variables are subject to conservation laws; Z is measured relatively.

(2) Only X is subject to a conservation law and none of the variables can be measured relatively.

8. Enter some reasonable estimates for the curves defining β in Figure 8.8(c). Sketch the velocities, forces, and angles for the left front wheel on a diagram such as that at (b). If this case were to be simply that of a normal two-terminal dissipator, what would have to be true of the angles α and β?

9. Redraw the computing diagram of Figure 8.9 for the case where the accumulator is charging. Retain the arbitrary condition that the temperature of the inflow to any device is to be taken as an independent variable.

10. Would it be possible, in analyzing electrical networks made up of resistors, capacitors, and inductors, to use voltage, current, and phase diagrams such as in Figure 8.11? For linear systems, the use of complex quantities for voltage, current, and impedance avoids the necessity for handling phase separately. Would it be necessary, in linearizing the system of Figure 8.11 about the quiescent point, to introduce complex notation or something equivalent?

11. Suppose that a system is thermally unstable but that the theoretical equilibrium temperature is high with respect to ambient. When the system is started, the temperature-sensitive device (as in Figure 8.13) will deliver more heat to the surrounds than is delivered to it by the network if the heat transfer curves do, in fact, have the relationship required for thermal instability. This implies that the device will get indefinitely cold. There is something wrong here; explain the situation.

12. In light of the above argument, is the theoretical equilibrium temperature of any self-destructing (by overheating) system ever greater than ambient?

13. Suppose a resistor is sensitive to temperature in such a way that the resistance decreases with increases in temperature. Is the unit a likely candidate for thermal self-destruction if:

(1) Operated at constant current?

(2) Operated at constant voltage?

14. Magnetically actuated clutches (such as in Figure 3.1) are subject to temperature effects roughly as in Figure 8.14(b); i.e., as temperature increases, the torque transmitted increases (to a limit) for given slip speed and actuating current. Are these units likely candidates for thermal self-destruction if operated:

(1) At constant torque?

(2) At constant slip speed?

9 Large excursion dynamic analysis

One of the richest aspects of system engineering—with respect to potential for system improvement, variety of techniques, and significance of analytic results—is that of large excursion analysis. Generally, it is a difficult field. The techniques that are available can be applied to only a limited range of problems, unless a computer is used to carry out the analysis. Usually, the methods yield only approximate results, whether or not a computer is used. In most cases, the pencil-and-paper conduct of a large excursion analysis tends to be tedious. Even with these compromises, however, there are major system engineering questions which may be satisfactorily considered with these methods. In this chapter, one such method is developed in detail.

The kinds of questions which must be answered by large excursion analysis are: Given a system that has a desired static operating point, will it start by itself? How long will it take to get started? Will it settle into the desired equilibrium, or go to some "accidental" equilibrium? Will it "hunt" instead of settling into any static equilibrium? Is it possible for some sets of initial conditions to make the system inoperable in its desired function? If the system is supposed to have cyclic operation, such as in one or another kind of vibrator, will the cycle *always* appear when the system is supposed to operate? For cyclic operation, how consistent will the period and amplitude of the cycle be?

The direct answers to these questions, as well as the considerable extra insight into system function which is given by the analysis, make the techniques for large excursion analysis of great interest to system engineers.

In selecting the method to be presented here, the author was influenced by two desires. One was to make the method consistent with the approaches taken in preceding chapters, in which considerable emphasis has been placed on the development, interpretation and use of the graphical characterizations of system components. This form of characterization appears to be universal in system engineering, while analytical representations are generally uncommon. Therefore, the analysis method developed here is basically graphical. The second influence was to make the method consistent with that used in computer studies since computer applications have become so important in engineering practice. This led to the choice of the state-variable formulation of the problems and to the emphasis on the nature of the various functions which appear and on the details of the solution process. It is hoped that the reader will notice, in each discussion and example, how a computer would go about the conduct of a parallel analysis. It is also hoped, however, that the reader will not discount the practical usefullness of the pencil-and-paper methods shown here.

Before commencing the discussion proper, some alternative methods for large excursion analysis are considered briefly.

9.1 Methods for Large Excursion Analysis

In many mechanical vibration problems, the influences which cause linear—or small excursion—analysis to be insufficiently accurate are due to geometrical complications. Although these may be difficult to work out analytically, it is always possible to develop an accurate mathematical representation of them. This allows any mechanical system of this nature to be described by a differential equation and, therefore, encourages an effort to determine the exact solution to the equation. These systems, and the resulting non-linear differential equations, have been studied for many years. Some researchers have spent their entire career in this general area. Some fine results have been developed and are summarized in many books and papers in the field of non-linear oscillations.

Since it is always possible to represent the characteristics of system components with some kind of mathematical function, perhaps developed from numerical methods, it is also always possible to determine a set of differential equations which describe a system. We may then attempt an exact solution of the set, but, judging from the experience of others, the likelihood of success is small. For one thing, the non-linear aspects of the equations are often more complicated than those which arise in mechanical vibration problems. For another, the order of the equations is often relatively high, while that for most mechanical problems which have yielded to the direct method of solution is only two. Although the practical usefulness of an exact solution is unquestioned, most engineers discount the approach as impractical because of difficulty and uncertain success. Since we would probably present the results of an exact solution in some kind of graphical form, most people are satisfied with any method which gives the same information and are willing to forego the elegance of a proper mathematical solution.

At the other extreme of elegance are some very gross approximation methods which have been developed and which are useful in certain cases. Among these is the "describing function method," discussed briefly in Chapter

8. In this, it is assumed that a system is sinusoidally excited and that frequencies higher than the excitation frequency are strongly attenuated by the action of energy storage and dissipation in the system. The various nonlinear components in the system are then replaced by approximately equivalent components which give the same gain and phase shift of the excitation frequency sinusoid as does the non-linear component. The transfer function of the equivalent component is amplitude dependent, unlike that for a linear component, and therefore contains a simple mechanism for representing the non-linear influence. System analysis is then carried out for different excitation amplitudes and frequency response diagrams may be prepared in the usual way. The method does allow system stability to be predicted in many cases; it should be familiar to the specialist in system engineering and may be found, described in detail, in many references (see Truxal*). We will not consider it further here since it is not generally applicable to systems and does not further the achievement of our purposes.

The method which is to be presented here is known as the "phase-plane" or "state-space" approach. The former usually implies a graphical method while the later suggests computer techniques. The concepts, however, are identical. The major feature of the method is that the mathematical description of system performance is written in such a way that only first-order differential equations appear; this is an automatic result of the use of state variables as defined in Chapter 5. It will be recalled that these variables provide a measure of the energy stored in a system. If a system is simulated on an analog computer using integrators exclusively as the dynamic elements, the state variables are the outputs of the integrators. The formulation of problems in this way is discussed in the following section.

9.2 State Space Analysis

When the dynamic analysis problem for a system has been properly formulated in the state-variable notation, the functional form of the relationships is parallel to that for the three-variable case shown here:

$$\dot{X} = \mathscr{P}(X, Y, Z)$$
$$\dot{Y} = \mathscr{Q}(X, Y, Z)$$
$$\dot{Z} = \mathscr{R}(X, Y, Z) \qquad \left\{\begin{matrix}\text{neglecting any}\\ \text{external forcing}\end{matrix}\right\}.$$

In terms of ordinary differential equations, the set represents a third-order case, and is often referred to in that way (the order may be readily confirmed

Control System Synthesis, John A. Truxal, McGraw-Hill Book Co.

by successive differentiation with respect to time and appropriate variable elimination). The basic statement made by the set of equations is that, given the state (the values of the state variables X, Y, and Z) at any instant, the way in which the state is changing is exactly specified. If we think of the set of state variables as representing a vector (here in three dimensions), the state velocity vector is completely determined as a function of each state vector. An important quality of this formulation is that the functions $\mathscr{P}$, $\mathscr{Q}$, and $\mathscr{R}$ are *static* functions of the state variables X, Y, and Z. These functions are derived from a *static* analysis of the system. Although they may be complicated functions, the concepts of their derivation are very basic. (This is one of the reasons we took so much care with the matters of static analysis in the earlier portions of the book.) That these are static functions is also apparent from a consideration of an analog computer simulation of the system. If we remove all the integrators from the simulation and replace their outputs with externally generated signals, the signals which previously were the integrator inputs must be static functions of the externally generated signals, since there are no dynamic components left in the simulation. Similarly, we may remove all the energy-storing elements from a physical system and replace them with the appropriate generator ($\boldsymbol{\Gamma}$ generator if the device has no $\boldsymbol{\Delta}$ in the static case; $\boldsymbol{\Delta}$ generator if there is no $\mathscr{R}$ in the static case). The result is a system which has no dynamic components; i.e., in which all the functional relationships are static. Incidentally, either of these techniques may be used, in practice, to physically generate the $\mathscr{P}$, $\mathscr{Q}$, and $\mathscr{R}$ functions if their analytical determination would be unusually troublesome.

In the way of anticipation, it might be noticed that the state-variable formulation of the system equations is very convenient if a numerical solution is to be carried out. Given the initial state of a system, the state after a small increment of time may be estimated by an evaluation of $\mathscr{P}$, $\mathscr{Q}$, and $\mathscr{R}$, estimation of the state change by multiplication by a small time increment, and estimation of the new state by addition of the change to the initial state. The process may be repeated indefinitely if a crude result is satisfactory.

In this chapter, we concentrate on second-order systems since the analysis can be carried out graphically without undue hardship.

A problem which has considerable instructional merit, with regard to state-space analysis, is the complete dynamic analysis of the transistor multi-vibrator shown in Figure 9.1. This network will be used in several of the following sections in order to illustrate various points. Here let us consider the formulation of its state-space analysis.

Assuming that the rate of vibration in the network of Figure 9.1 is relatively low, the small dynamic effects associated with the transistors may be neglected. The only energy storage in the system is then in the two capacitors.

The state variables must then be V_{AC} and V_{BD}. The state-variable formulation of the problem must be

$$\dot{V}_{AC} = \mathscr{P}\,(V_{AC}, V_{BD})$$
$$\dot{V}_{BD} = \mathscr{Q}\,(V_{AC}, V_{BD}).$$

A computing diagram which illustrates this formulation is shown at Figure 9.1(b).* Because the transistor characterizations, indicated as $\mathscr{Q}_1$ and $\mathscr{Q}_2$ (no direct relation to the state function $\mathscr{Q}$), are strongly non-linear over the range of operation of the circuit, the functional relationship between the integrator inputs, I_1 and I_2, and the state variables, V_{AC} and V_{BD}, are both quite non-linear and difficult to develop analytically. As we proceed, we shall attempt to develop some approaches to the problem which do not require a full evaluation of the functions $\mathscr{P}$ and $\mathscr{Q}$ for all possible values of the state variables. This can best be done by starting with some simpler cases and returning to this problem after some practice.

9.3 Single Energy-Storage Systems

Some of the basic concepts and techniques of state-space analysis are exhibited by problems which involve only a single energy storage component. Such systems are represented by the formulation.

$$\dot{X} = \mathscr{P}\,(X).$$

*Computing diagrams are used at several places in this chapter in order to emphasize the relationship between state variables and computing processes. The diagrams pertain directly to analog computation but are also helpful in the programming of a digital computer for system simulation. The blocks in these diagrams represent devices wherein the "output" is some specified mathematical function of the "input," both of which are some physical variable such as voltage, pressure, or rotational speed. Such components require an external source of power and must be designed so that the output is insensitive to the power demanded by the rest of the computing network. Thus, the computing diagram is a much different kind of thing than a schematic wherein the flow of power through a network is completely designated. We must be careful not to confuse the two kinds of diagram or to call things on a schematic diagram by the names "input" and "output." Standard symbols have been used on the diagrams here insofar as possible [see the integrator symbol at Figure 9.2(a)]. The circle with a cross represents an adder while rectangular blocks stand for general functions which are identified within the symbol. In every case, the computing diagram is developed by assuming that the system state variables are to be generated by an integrator. The appropriate input for each of the integrators is then determined by the usual techniques of static analysis and the appropriate computing arrangement drawn. In this, we must be careful to observe the rules of characterization governing the choice of independent and dependent variables.

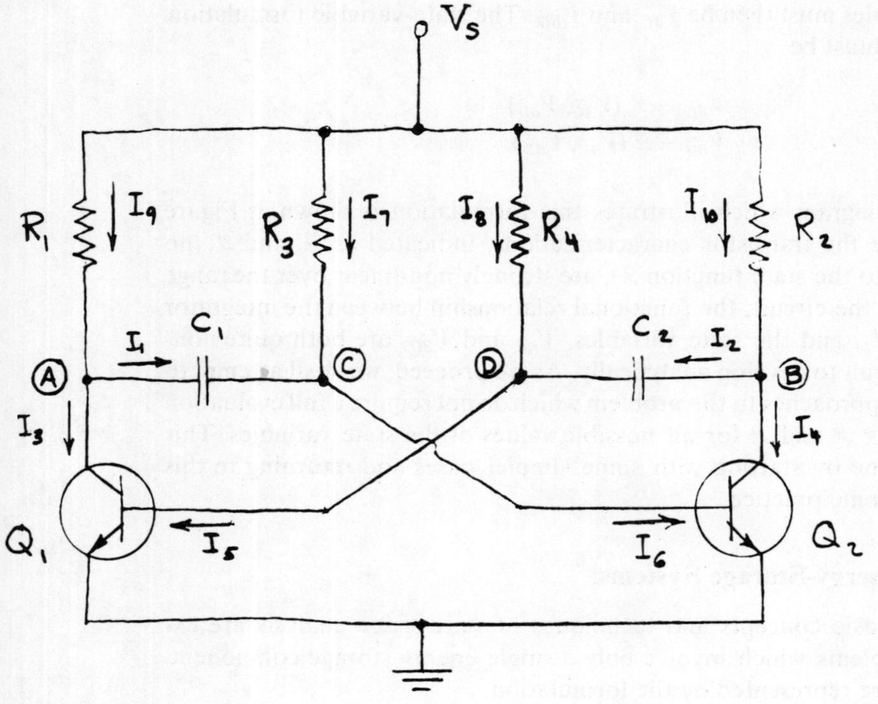

(a) Schematic

From the Engineer's Sketch Book

Figure 9.1 (*continued*)

The solution of this case may be carried out quite readily as indicated in Figure 9.2. At (a) is shown a computing diagram, which is extremely straightforward. At (b) is shown a representative function $\mathscr{P}(X)$ and a graphical solution approach. The problem to be solved here is the determination of the transient behavior of X. The final result appears on a graph of X versus time (t). Since the value of $\dot{X}$ is known for each value of X, from the function $\mathscr{P}(X)$, the slope of the solution at any point on the X-versus-t graph may be indicated. Here, normalized values for $\dot{X}$ have been used and a set of lines at different convenient slopes entered on the drawing. The values of X for which $\dot{X}$ has 80 percent of its maximum value have been determined. For these values of X, dashes having the proper slope have been indicated on the X-versus-t graph. By filling in the graph with a fairly dense pattern of such

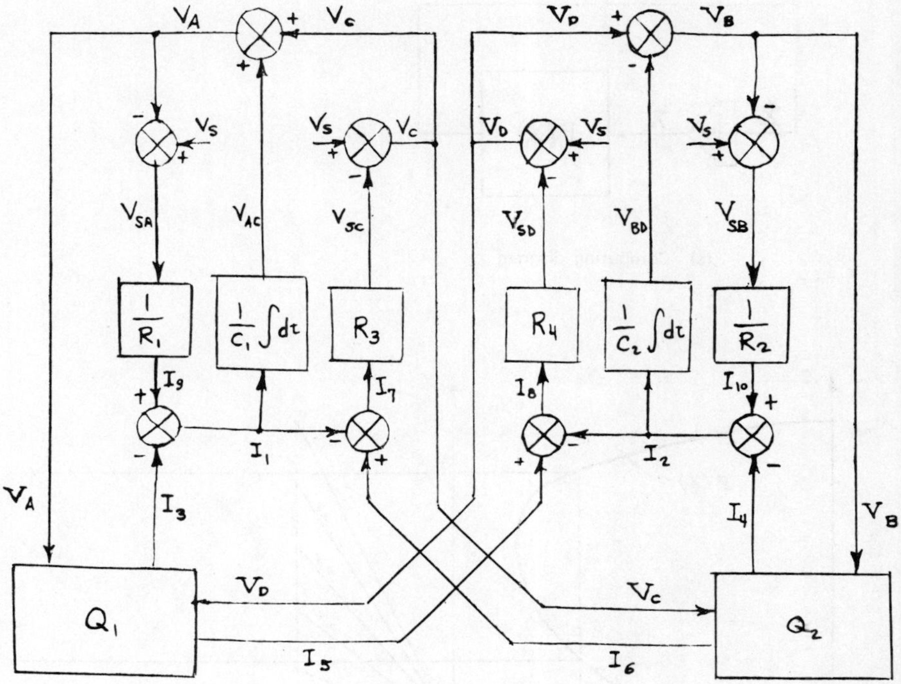

(b) Computing diagram

From the Engineer's Sketch Book

Figure 9.1 Multi-vibrator analysis by state variables.

dashes, the transient curve is made quite apparent and may be sketched easily.

An important aspect of the $\mathscr{P}(X)$ or $\dot{X}$-versus-X graph in this kind of problem is the way in which the $\dot{X} = 0$ line is crossed. A variety of cases are shown in Figure 9.3. If the crossing has a positive slope, it cannot represent a stable equilibrium. A small displacement from such a crossing, in the plus X direction, produces a positive velocity. This causes the displacement to increase, which results in even greater velocity. Similarly, a small negative displacement causes a negative rate of change and increased negative displacement. Conversely, a crossing that has negative slope represents a stable point. If the crossing has zero or infinite slope, the point has marginal stability; either $\dot{X}$ or X will be well defined, but the other variable will be

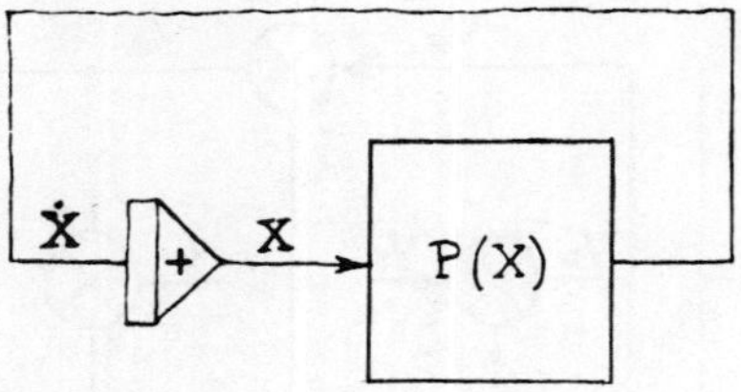

(a) Computing diagram

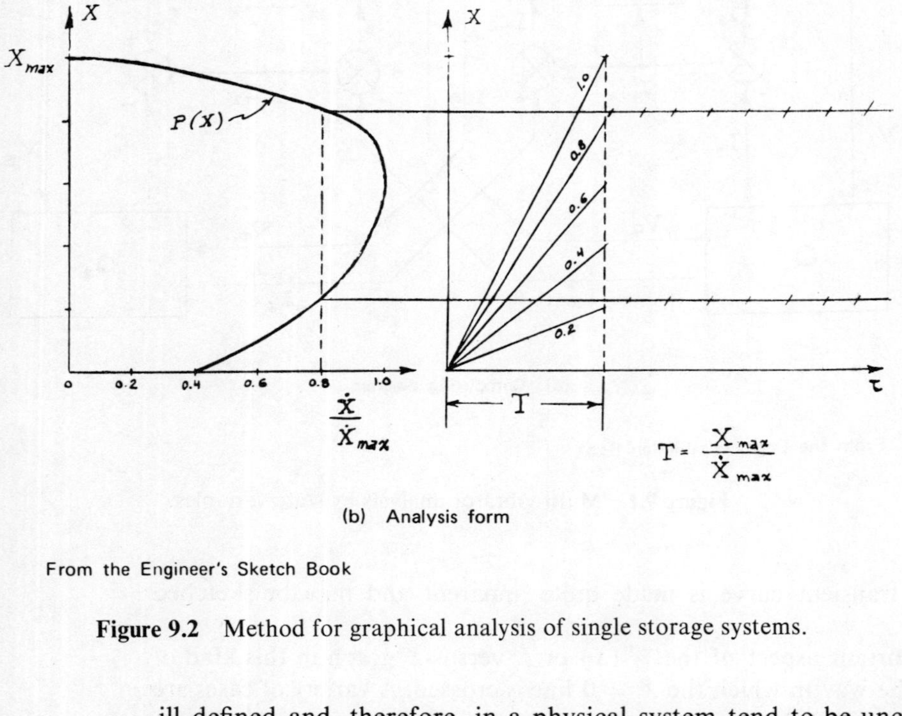

(b) Analysis form

From the Engineer's Sketch Book

Figure 9.2 Method for graphical analysis of single storage systems.

ill defined and, therefore, in a physical system tend to be uncertain. For a crossing with marginal stability, we must investigate how the curve bends as it leaves the $\dot{X} = 0$ axis in order to determine stability. Some cases are shown in the lower portion of (a).

If the function $\mathscr{P}(X)$ crosses the axis three times, as shown at (b), the system is bistable. It is obvious from this that any bistable system of the first order must have at least one unstable equilibrium point. It is helpful to know this when trying to invent such a system.

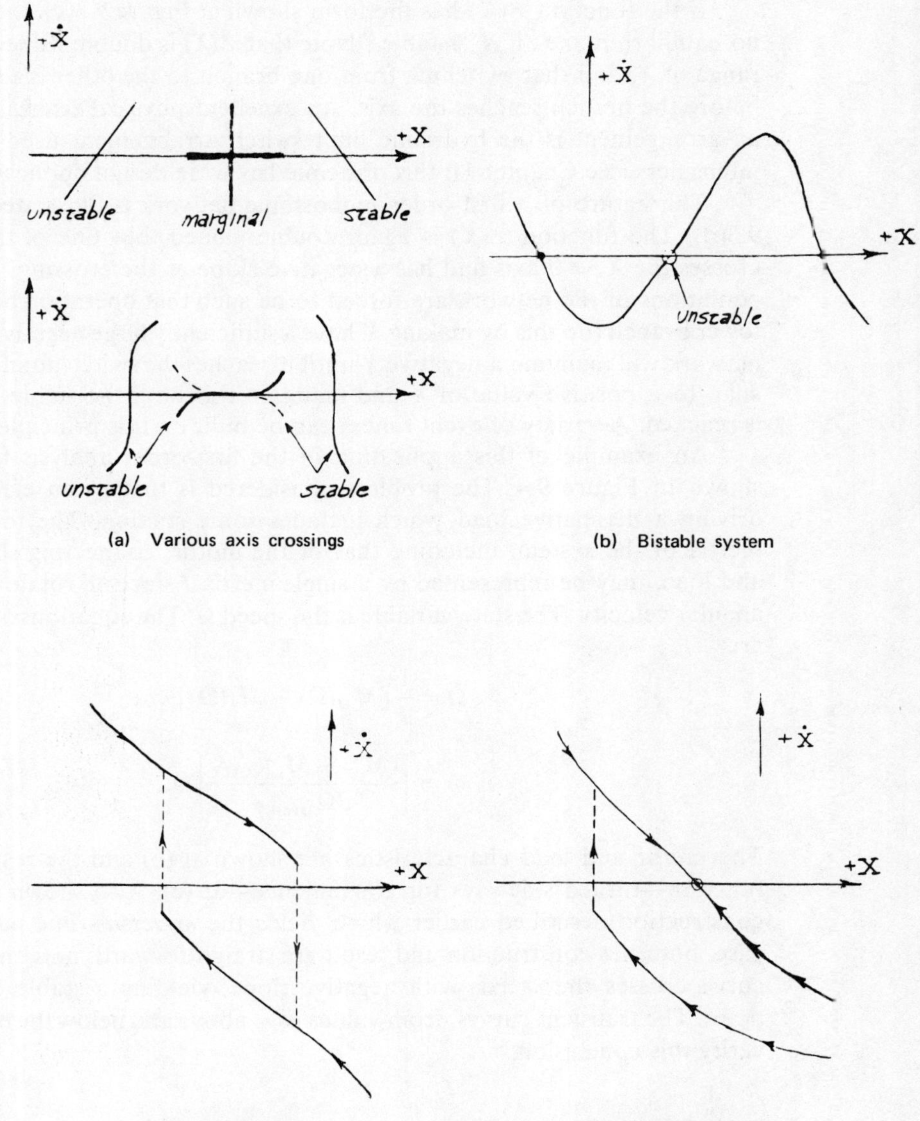

(a) Various axis crossings

(b) Bistable system

(c) Astable system

(d) Monostable system

From the Engineer's Sketch Book

Figure 9.3 Some special cases of single storage systems.

If the function $\mathscr{P}(X)$ has the form shown at Figure 9.3(c), the system has no equilibrium; i.e., it is "astable." Note that $\mathscr{P}(X)$ is double valued over some range of X, and that switching from one branch to the other is accomplished before the branch reaches the axis. An excellent physical realization of such an arrangement is the hydraulic limit switch arrangement used in pressure intensifiers (see Chapter 1); this principle has wide design applicability.

The nature of a first-order monostable network is illustrated by Figure 9.3(d). The function $\mathscr{P}(X)$ is again double valued, but one of the branches crosses the $\dot{X} = 0$ axis and has a negative slope at the crossing. If the initial conditions of the network are forced to be such that operation begins on the lower branch (do this by making $\dot{X}$ have a sufficiently large negative value), the network will maintain a negative $\dot{X}$ until it reaches the switch point. It will then snap to a positive value of $\dot{X}$ and maintain this until the stable equilibrium is reached. A variety of event timers can be built on this principle.

An example of this application of the first-order analysis technique is shown in Figure 9.4. The problem considered is that of an electric motor driving a dissipative load which includes some stiction. The total rotating inertia of the system, including that of the motor, connecting shaft, clutch, and load, may be represented by a single inertia J since all rotate at the same angular velocity. The state variable is the speed Ω. The equations of the system are:

$$\dot{\Omega} = \frac{1}{J}\left[M_M(\Omega) - M_L(\Omega)\right]$$

$$\dot{\omega} = \left[\frac{(M_M - M_L)_{\text{STALL}}}{J\Omega_{M(\text{n.l.})}}\right] m.$$

The motor and load characteristics are shown at (b) and the resulting $\mathscr{P}(\omega)$ function—turned sideways for convenience—at (c). Also shown at (c) is the construction described earlier which yields the ω-versus-time curve. In this case, both the construction and result are straightforward; note that the $\mathscr{P}(\omega)$ curve crosses the ω axis with negative slope, yielding a stable equilibrium point. The transient curves, from values of ω above and below the equilibrium, verify this conclusion.

9.4 Double Energy-Storage Systems

If the energy stored in a system may be written as a function of two state variables, the system equations become

$$\dot{X} = \mathscr{P}(X, Y)$$
$$\dot{Y} = \mathscr{Q}(X, Y).$$

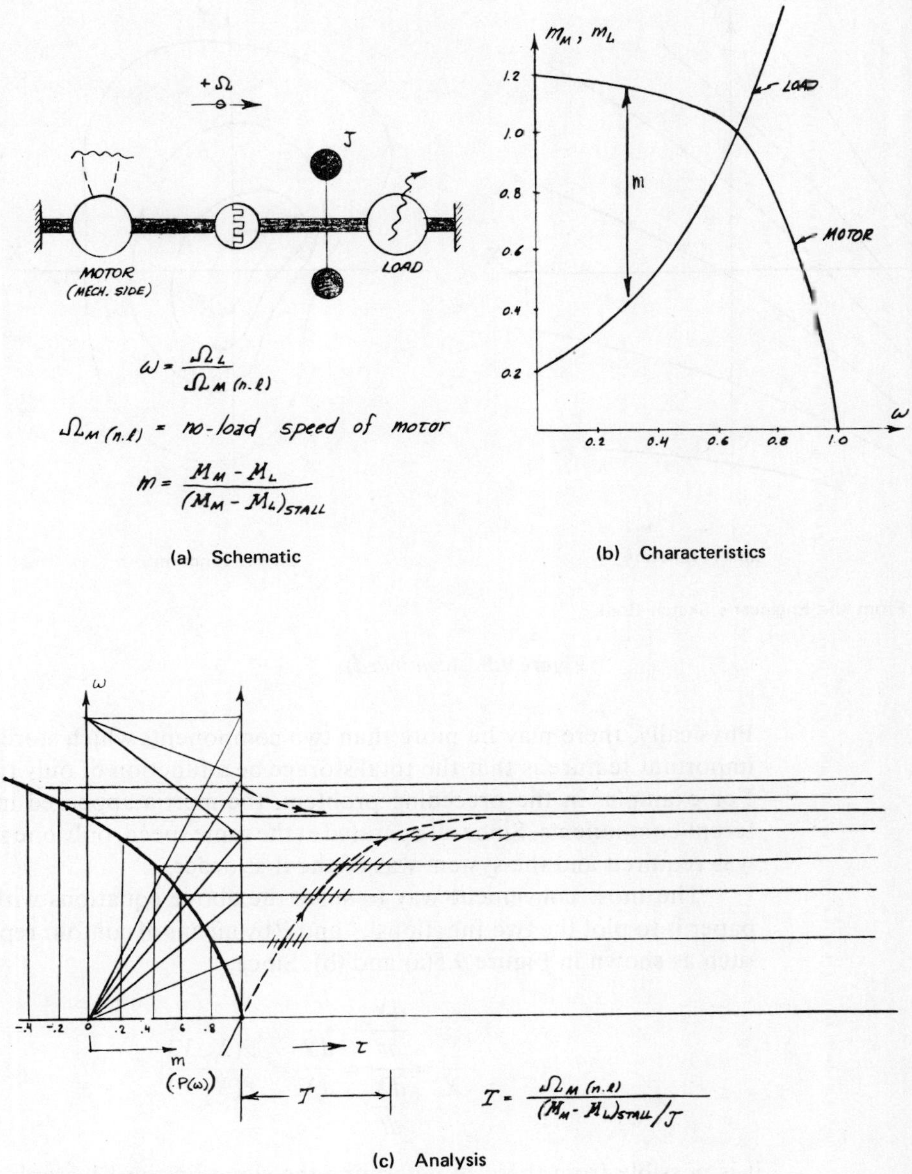

From the Engineer's Sketch Book

Figure 9.4 Motor start-up transient analysis.

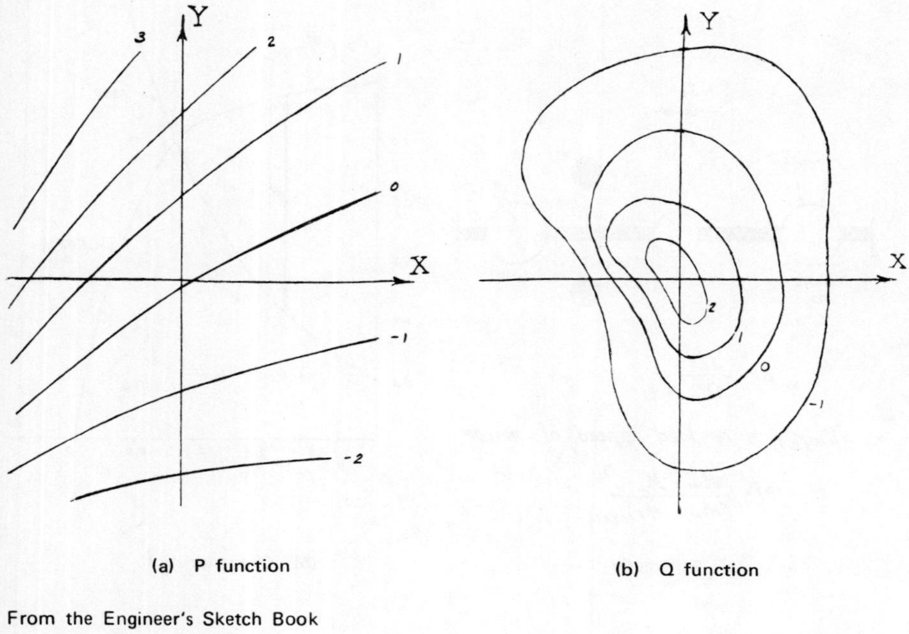

Figure 9.5 *(continued)*

Physically, there may be more than two components which store energy. The important feature is that the total storage be a function of only two variables. For example, in the preceding problem, the inertia appeared in several different components. Since all operated at the same speed, only one state variable was required and the system was of the first order.

The most convenient way to solve the above equations with pencil and paper is to plot the two functions $\mathscr{P}$ and $\mathscr{Q}$ by means of contour representations such as shown in Figure 9.5(a) and (b). Since

$$\frac{\dot{Y}}{\dot{X}} = \frac{\dfrac{dY}{dt}}{\dfrac{dX}{dt}} = \frac{dY}{dX} = \frac{\mathscr{Q}(X, Y)}{\mathscr{P}(X, Y)},$$

it is possible from these to determine the slope, in the XY graph, of any state change. For example, the two plots have been superposed at (c) of the figure. Since both $\dot{X}$ and $\dot{Y}$ are zero at points I and II, these are possible static equilibrium points for the system. At point A, both $\dot{X}$ and $\dot{Y}$ are minus one, therefore, a plot of X versus Y would have to go through this point with a slope of

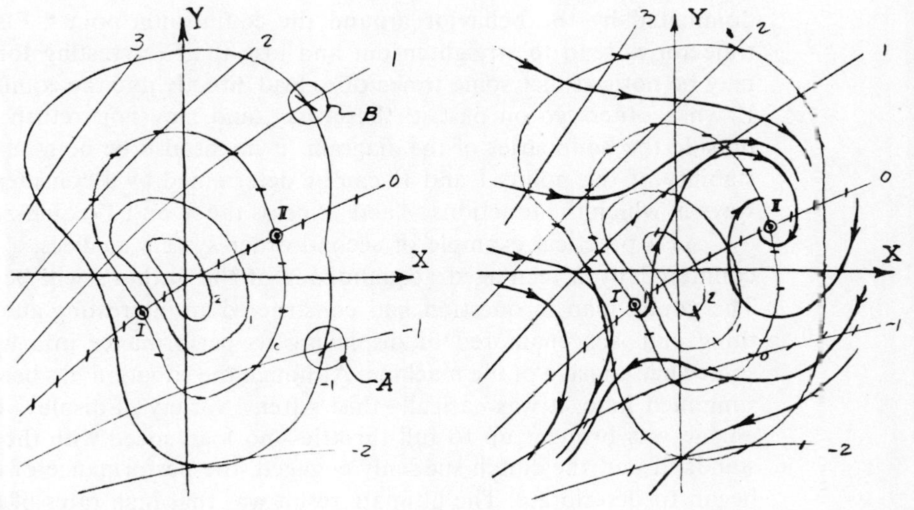

(c) Superposed functions with hachures

(d) Portraits

From the Engineer's Sketch Book

Figure 9.5 General form for graphical analysis of two-storage systems.

plus one. Similarly, the slope at point B would have to be minus one. Everywhere along the $\dot{Y} = 0$ contour, the slope must be zero while everywhere along the $\dot{X} = 0$ contour the slope must be infinite. At each intersection of the $\dot{X}$ and $\dot{Y}$ contours, the slope may be calculated and indicated with the aid of a small dash.* This procedure has been carried out at (d) in the figure and several XY "trajectories" drawn. From these, it may be seen that point I is an unstable equilibrium while point II represents a stable static state. The tendency is for a state near I to change in a direction away from it while a state near II falls into it.

A variety of general features of non-linear dynamic performance appear on this hypothetical trajectory diagram. These will be discussed in detail later, but the reader might notice that, for example, the appearance of the diagram is

*We may use transparent paper to facilitate the work. By using an underlay with parallel ruled lines and a guide for the various slopes to be used, the transparent sheet may be aligned in such a way that the ruled lines all indicate a certain slope. The $\dot{X}$ and $\dot{Y}$ contours may then be inspected for intersections which have this slope and appropriate dash entered at each. Often, however, carefully made estimates of the dash direction are adequate, especially if we start with the zero, plus and minus one, and infinite slope markers. We may then ascertain, as we continue, that our estimates bear a reasonable relationship to these.

dominated by the behavior around the equilibrium points. Elsewhere, the trajectories tend to straighten out and lose their interesting forms. Also, it may be noticed that some trajectories lead directly into the equilibrium point II while others go on past it; these may, and may not, return via a circuit outside the boundaries of the diagram. It might also be pointed out that the stability of the points I and II can be determined by a consideration of the ways in which the functions $\mathscr{P}$ and $\mathscr{Q}$ cross the $\dot{X}$ and $\dot{Y}$-equal-zero planes.

As a practical example of second-order system analysis, a problem encountered by a teen-aged acquaintance of the author's will be considered. The young man in question had constructed an interesting automobile-like thing and was enamored of displaying its performance in a way that was somewhat abusive of the machine. Although the situation has been somewhat simplified here, it was basically this: After a variety of displays in which the engine was brought up to full throttle—no load speed with the vehicle at a standstill and the clutch suddenly engaged, the performance of the machine began to deteriorate. The ultimate result was that high rates of acceleration to the original maximum speed finally became impossible to achieve. Although the engine speed and power were sufficient, an attempt to get high acceleration always ended in a terminal state at which the vehicle speed was about half that expected and in which the clutch was slipping and heating in a very extreme way. For the first example here, we will consider the case just prior to the total breakdown of the acceleration capability.

A simplified schematic of the system is shown in Figure 9.6(a). The inertia of those parts which rotate at speeds kinematically related to the engine speed is shown as J_1, while that for the parts rotating with the drive shaft (parts of the differential, the rear axles, the wheels, the reflected inertia of the vehicle body) is J_2. The two halves of the system are joined by the clutch, which we shall admit might be slipping. The dissipative load which is indicated is all the friction of the drive-shaft side of the system, and would include a fair amount of stiction. Dissipation on the engine side of the clutch will be accounted for in the engine characterization.

The state variables for the system are the engine and load speeds, Ω_E and Ω_L. A computing diagram for the system is shown at (b); the subscripts E, C, and L have obvious significance. Since it was impractical to measure the component characteristics directly, the estimates shown by (c) and (d) were used to consider the case. The engine characteristic is of a typical full-throttle form, and shown a power peak of about 200 horsepower. The load characteristic was chosen to give an equilibrium speed, assuming the clutch was not slipping, of about 43 mph in low gear (that used for the forementioned tests). Here the speed is accounted for by the revolutions per second of the driveshaft instead of the vehicle miles per hour; this is more convenient. The clutch characteristics, at (d), show the typical drop in the coefficient of friction as slipping speed is increased. This effect usually disappears above a certain speed, taken to be 60 RPS here. Note that the clutch is capable of transmitting the equilibrium point torque at either no slip or some relatively small value of slip.

The system equations have the form

$$\dot{\Omega}_E = \mathscr{P}(\Omega_E, \Omega_L)$$
$$\dot{\Omega}_L = \mathscr{Q}(\Omega_E, \Omega_L)$$

and, in terms of the component characteristics, are:

$$\dot{\Omega} = \frac{1}{J_I}[M_E(\Omega_E) - MC(\Omega E - \Omega L)]$$

$$\dot{\Omega}_L = \frac{1}{J_{II}}[M_C(\Omega_E - \Omega_L) - M_L(\Omega_L)].$$

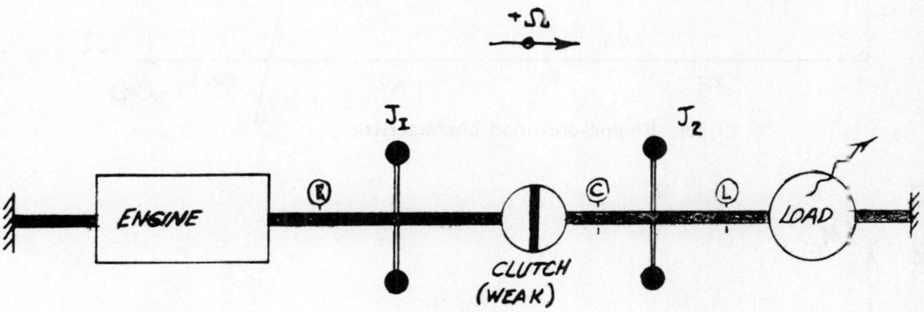

(a) Schematic and notation

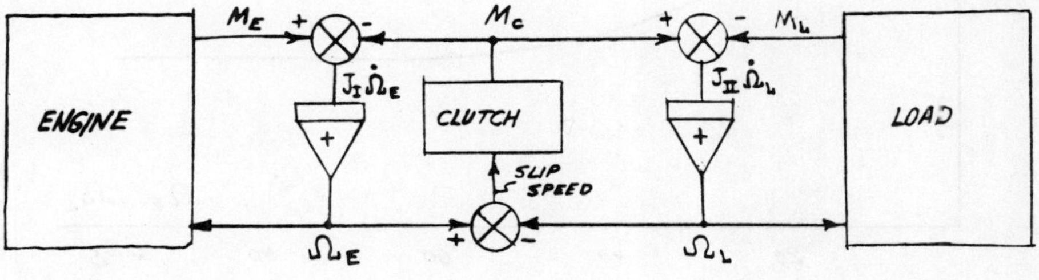

(b) Computing diagram

From the Engineer's Sketch Book

Figure 9.6 *(continued overleaf)*

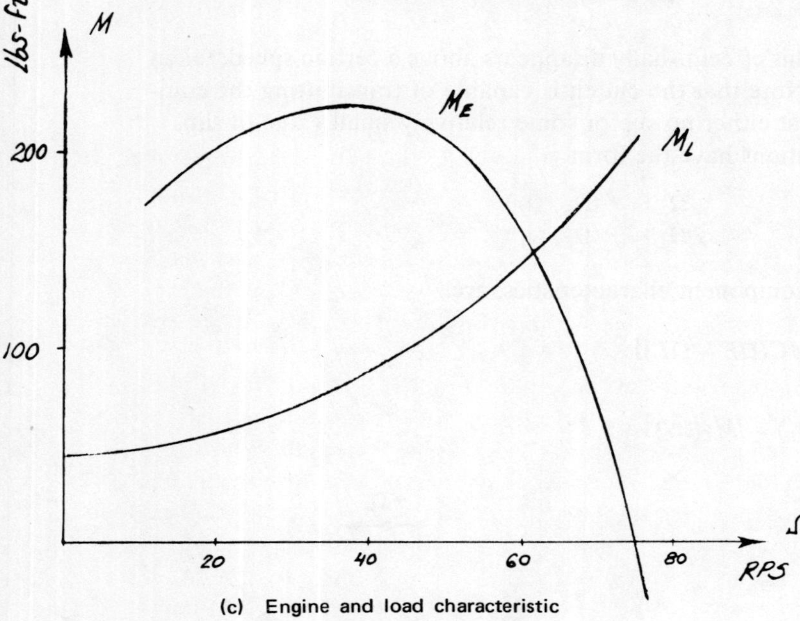

(c) Engine and load characteristic

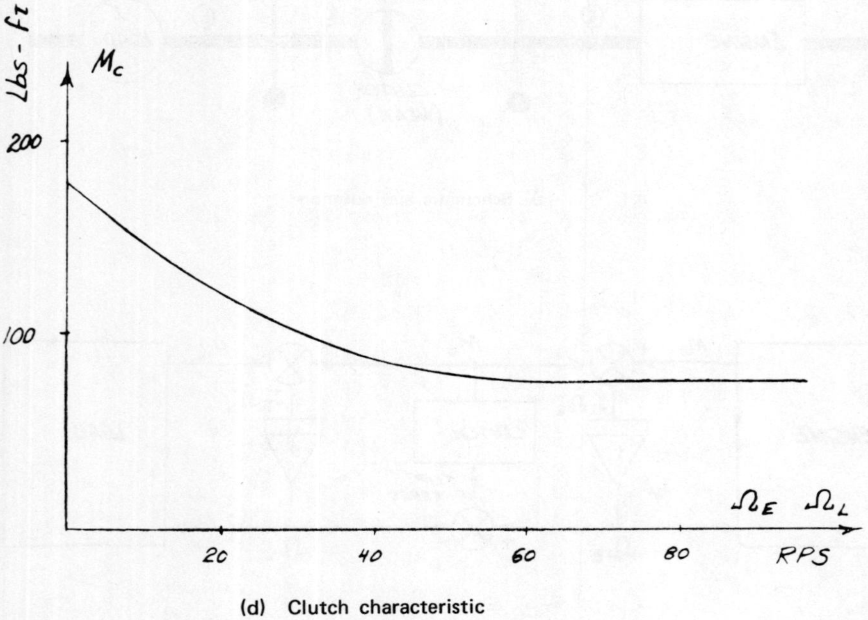

(d) Clutch characteristic

From the Engineer's Sketch Book

Figure 9.6 Two-inertia slipping clutch problem.

The $\mathcal{P}$ and $\mathcal{Q}$ functions may be evaluated by superposition of the component characteristics as indicated by the above equations. At Figure 9.7(a) is shown the derivation of the $\mathcal{P}$ function; the origin of the clutch characteristic is offset from that of the engine characteristic by an amount of Ω_L; for any vertical line on the graph the engine speed may be read directly and the torque difference obtained by inspection. In getting the contours for $\mathcal{P}$, a particular Ω_L can be set and the engine speeds for each level of torque difference read off. The combinations may then be entered on the contour plot. The $\mathcal{Q}$ function is generated similarly, except that the clutch characteristic must be turned end for end in order to make the quantities directly apparent.

The resulting contour plots are shown in Figures 9.7(c) and (d). The contours are superposed in Figure 9.8, and the slopes as well as some operating trajectories are also indicated there. It is clear that, in this condition of the components, the system operation is quite reasonable. Starting from an engine speed of 75 RPS and zero load speed, the engine speed holds fairly constant as the load speed comes up. Then, as the clutch slip speed starts to get small (approaching the no-slip line), the clutch torque increases, the engine slows down a bit as the clutch stops slipping (about 63 RPS), and then the final state is adopted by the system. Starting from lower engine speeds allows the clutch to stop slipping before the equilibrium is reached and also uses a portion of the Ω_L-versus-Ω_E plane in which Ω_L is much greater than that used by the driver. This would have been a far superior strategy, both from the performance and equipment maintainance point of view.

Some other interesting aspects of this system are considered in a later example and some of the chapter problems; in particular, the case of the more severely worn clutch and the unfortunate terminal state mentioned earlier will not be forgotten.

9.5 Trajectories in the Vicinity of Static Equilibrium

As mentioned in the preceding section, the trajectory diagram tends to be dominated by the behavior around the equilibrium points, stable or unstable. It so happens that we may ascertain this behavior, in a general sense, by linearizing the $\mathcal{P}$ and $\mathcal{Q}$ functions in the vicinity of the equilibrium points and applying the usual linear methods to determine the performance. Of course, this must be done judiciously. The $\mathcal{P}$ and $\mathcal{Q}$ functions in the preceding example were most irregular in this region, the derivatives being undefined at the equilibrium point, and linearization could be misleading if not thoroughly understood.

The basis of the linearization approach is the following analysis.

The defining relationships for the two-variable case are

$$\dot{X} = \mathcal{P}(X, Y)$$
$$\dot{Y} = \mathcal{Q}(X, Y).$$

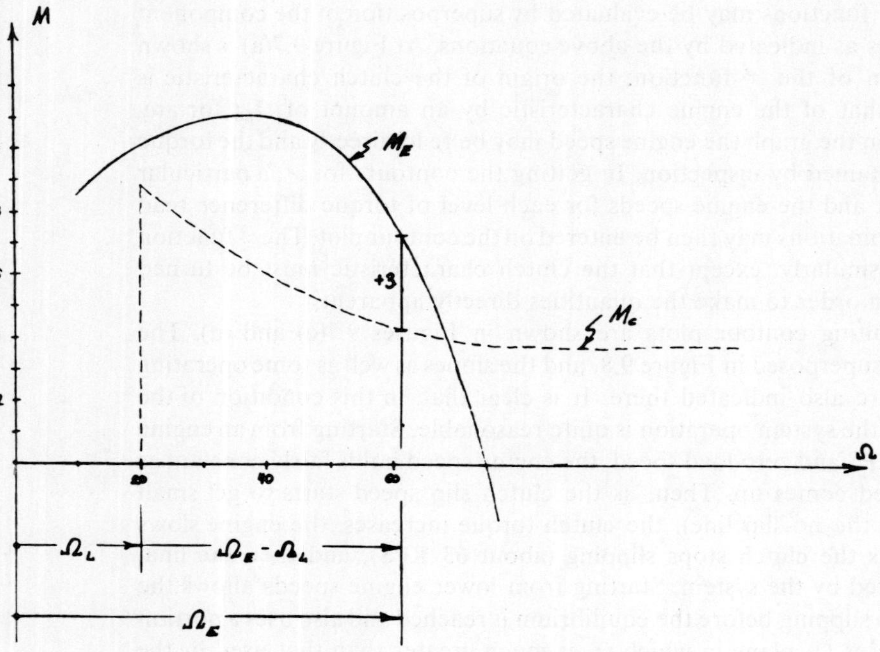

(a) Method for generating P functions

Figure 9.7 *(continued)*

The $\mathscr{P}$ and $\mathscr{Q}$ functions may be approximated, if they are sufficiently smooth, by the usual linearization:

$$\mathscr{P}(X, Y) \cong \mathscr{P}(X_0, Y_0) + \left.\frac{\partial \mathscr{P}}{\partial X}\right|_0 (X - X_0) + \left.\frac{\partial \mathscr{P}}{\partial Y}\right|_0 (Y - Y_0)$$

$$\mathscr{Q}(X, Y) \cong \mathscr{Q}(X_0, Y_0) + \left.\frac{\partial \mathscr{Q}}{\partial X}\right|_0 (X - X_0) + \left.\frac{\partial \mathscr{Q}}{\partial Y}\right|_0 (Y - Y_0).$$

In the static equilibrium condition,

$$\mathscr{P}(X_0, Y_0) = \mathscr{Q}(X_0, Y_0) = 0$$

Far from equilibrium points, the actual values of $\mathscr{P}$ and $\mathscr{Q}$ are large and therefore dominant; near equilibrium points the rate of change of the functions

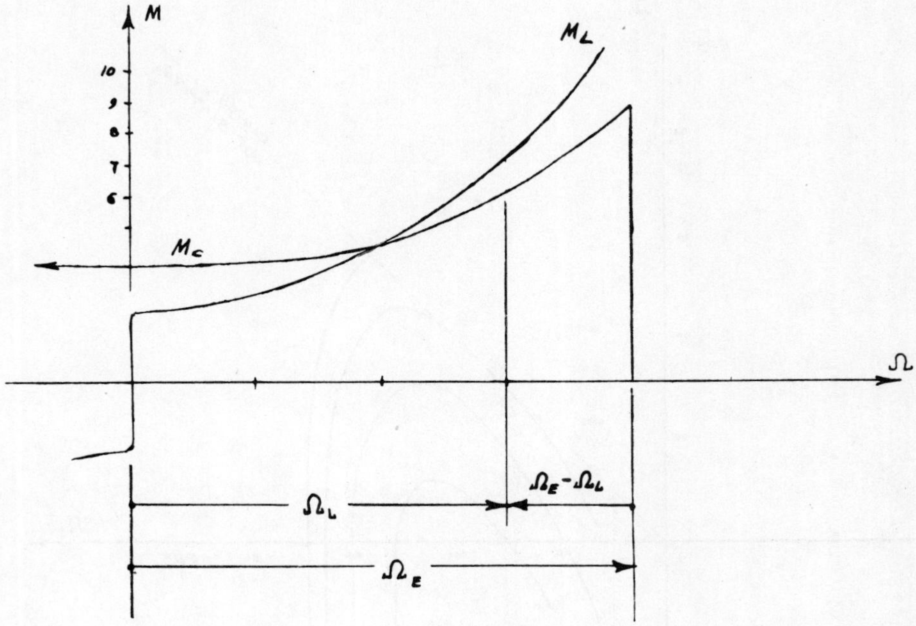

(b) Method for generating Q functions

From the Engineer's Sketch Book

Figure 9.7 *(continued overleaf)*

is more important so that the linearized representation gives considerable information. Using excursion variable notation, defined as usual,

$$x = X - X_0, \qquad y = Y - Y_0$$

the linearized relationships may be written more compactly as

$$\dot{x} = \frac{\partial \mathscr{P}}{\partial X}x + \frac{\partial \mathscr{P}}{\partial Y}y$$

$$\dot{y} = \frac{\partial \mathscr{Q}}{\partial X}x + \frac{\partial \mathscr{Q}}{\partial Y}y$$

and, by obvious identification, even more conveniently by

$$\begin{aligned} \dot{x} &= p_1 x + p_2 y \\ \dot{y} &= q_1 x + q_2 y. \end{aligned}$$

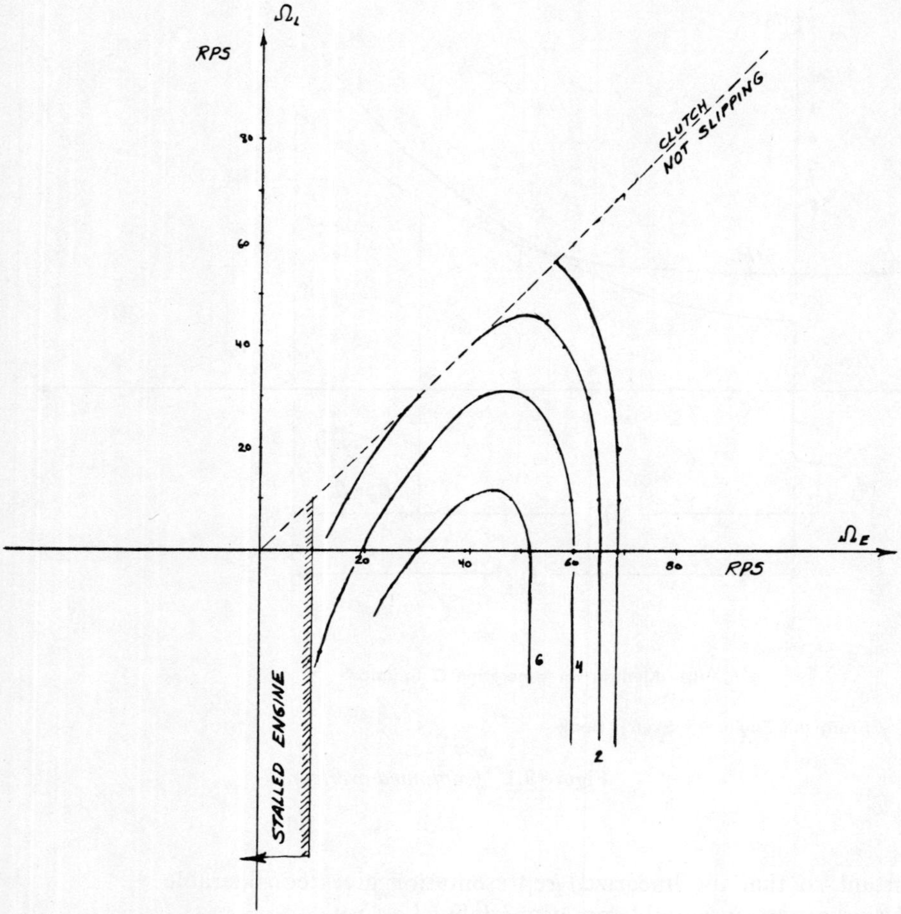

(c) P function contours

From the Engineer's Sketch Book

Figure 9.7 (*continued*)

This system of equations may be rewritten

$$\begin{aligned} (D-p_1)x - p_2 y &= 0 \\ - q_1 x + (D-q_2)y &= 0 \end{aligned}$$

so that the characteristic equation, the determinant of the coefficients above, becomes

$$D^2 - (p_1 + q_2)D + (p_1 q_2 - p_2 q_1) = 0.$$

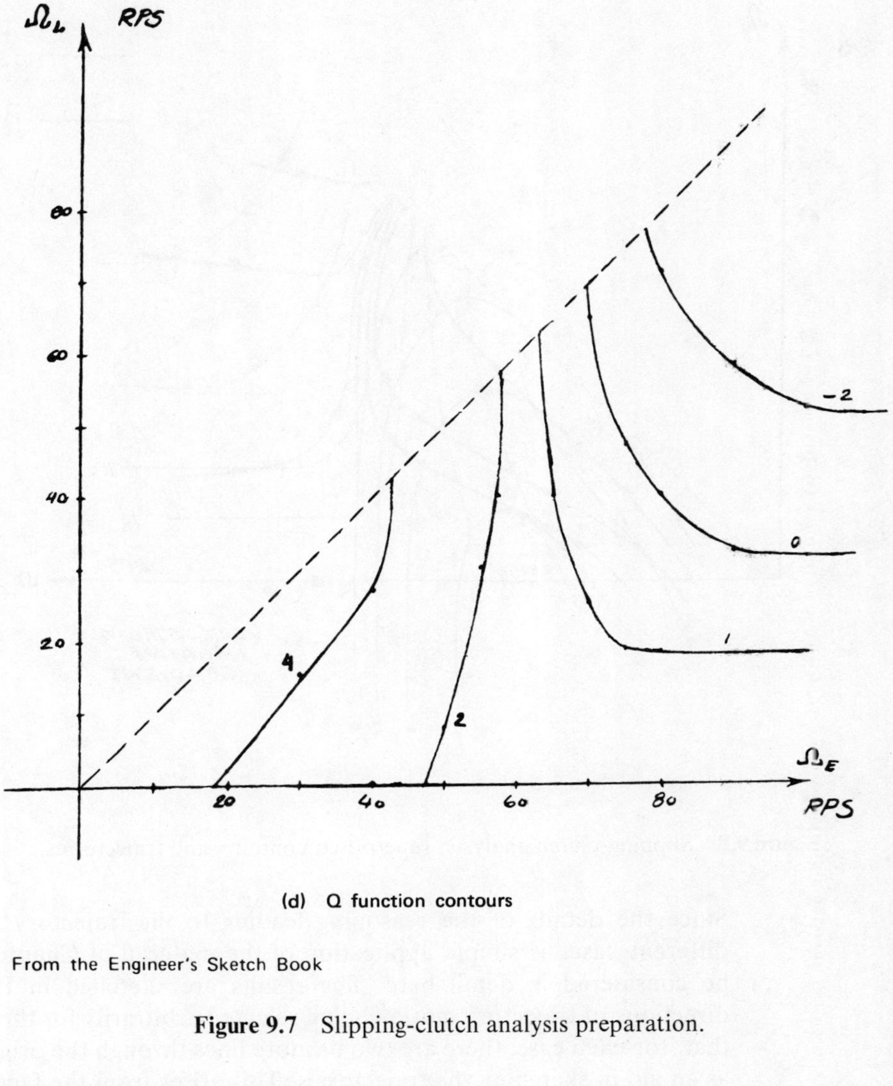

(d) Q function contours

From the Engineer's Sketch Book

Figure 9.7 Slipping-clutch analysis preparation.

The roots of the characteristic equation are then given by

$$r = \frac{p_1 + q_2}{2} \pm \sqrt{\left(\frac{p_1 + q_2}{2}\right)^2 - (p_1 q_2 - q_1 p_2)}.$$

The various cases which may arise depend on the sign of $(p_1 + q_2)$ and of

$$[(p_1 + q_2)/2]^2 - (p_1 q_2 - q_1 p_2).$$

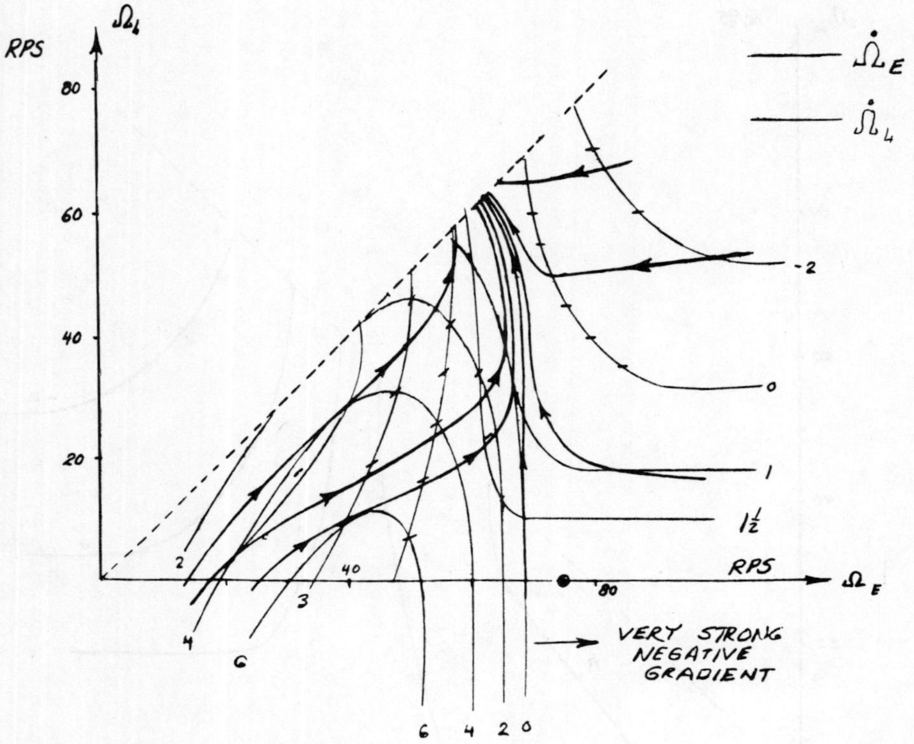

From the Engineer's Sketch Book

Figure 9.8 Slipping-clutch analysis; superposed contours and trajectories.

Since the details of the reasoning leading to the trajectory shapes for the different cases is simple application of the material of Chapter 6, it will not be considered in detail here. The results are sketched in Figure 9.9, the directions of trajectory motion being selected arbitrarily for the figure. Notice that, for each case, there are two or more lines through the origin which serve as an aid in sketching the trajectories. This arises from the fact that

$$\frac{\dot{y}}{\dot{x}} = \frac{q_1 + q_2\,(y/x)}{p_1 + p_2\,(y/x)}.$$

Hence, the slope on any line through the origin, when the system is linearized about the origin, is constant. It is a straightforward matter to determine the slope of the lines to which the trajectories are parallel or perpendicular. For the trajectories to be on a line through the origin,

$$m_t = m_1; \quad \frac{\dot{y}}{\dot{x}} = \frac{y}{x}; \quad m_t = \frac{q_1 + q_2(m_1)}{p_1 + p_2(m_1)} \qquad \begin{array}{l} m_t = \text{trajectory slope}, \\ m_l = \text{line slope}. \end{array}$$

This relationship is satisfied if

$$m_1 = -\frac{p_1 - q_2}{2p_2} \pm \sqrt{\left(\frac{p_1 - q_2}{2p_2}\right)^2 + \frac{q_1}{p_2}}.$$

For trajectories to be perpendicular to a line through the origin,

$$m_\tau = -\frac{1}{m}; \quad \frac{\dot{y}}{\dot{x}} = -\frac{x}{y}; \quad -\frac{1}{m_\tau} = \frac{q_1 + q_2 m_2}{p_1 + p_2 m_2}.$$

This is the case when

$$m_2 = -\frac{p_2 + q_1}{2q_2} \pm \sqrt{\left(\frac{p_2 + q_1}{2q_2}\right)^2 - \frac{p_1}{q_2}}.$$

For trajectories to have the slope of the line to which they are somewhere perpendicular,

$$m_3 = \frac{p_1 m_2 - q_1}{p_2 - q_2 m_2}.$$

The names associated with each of the trajectory patterns are commonly accepted and derive from similar patterns which arise in the study of the theory of equations. The physical significance of the various cases is as follows (the letters in parentheses referring to the portions of Figure 9.9):

(a) *Stable Node.* These trajectories represent the case where both roots of the characteristic equation are real and negative. The trajectories never cross the line having a slope of m_1; this line represents the set of initial conditions in which only one of the roots of the equation governs the motion. Of the two possible solutions for each of the slopes m_1 and m_2, there is only one real solution for each.

One special case that arises in some modes of system operation is that the x and y motions are uncoupled. The coefficients p_1 and q_2 are both zero, there are two real solutions for m_1 (zero and infinity), and all the trajectories remain in the quadrant of the diagram in which they start. Should the coefficients p_1 and q_2 be equal, there is an infinitude of solutions for m_1; all the trajectories are directed radially into the origin. This is obvious from the state variable representation for the case:

$$\dot{x} = -x$$
$$\dot{y} = -y$$

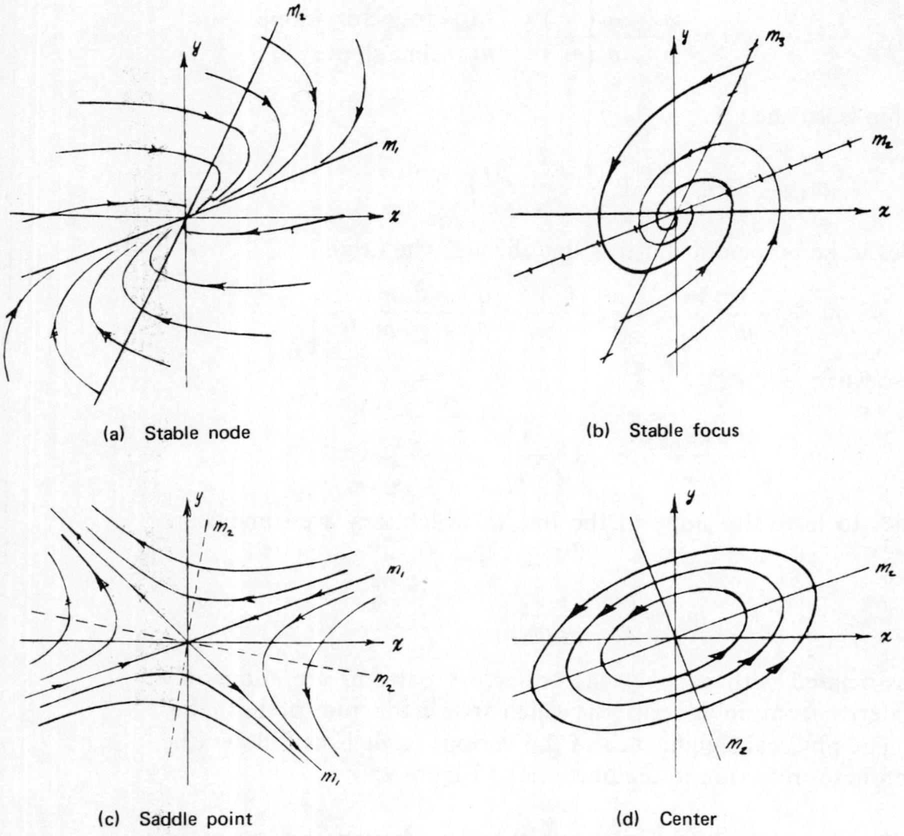

Figure 9.9 (*continued*)

(b) *Stable Focus*. Here the roots of the characteristic equation are conjugate complex and have a negative real part. There is a single real solution for m_2 and none for m_1. Of some help in sketching the trajectories is the line at a slope m_3; the slope of the trajectories along this line is the same as the slope of the line to which the trajectories are perpendicular. The response of both x and y to an initial condition is a decaying sinusoid. The special cases which arise (zero real part to the roots, imaginary parts zero) are contained in other cases discussed here.

(c) *Saddle Point*. This case arises when one root is positive and real while the other is negative and real. There are two solutions for both m_1 and m_2, as indicated by the sketch. If the system is given an initial condition lying on one of the radially inward trajectories, the motion will be exponential

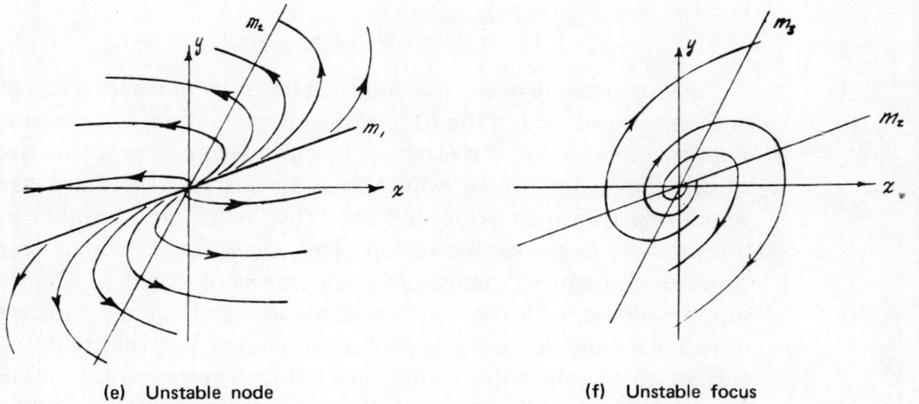

Figure 9.9 Trajectories in the vicinity of static equilibrium.

decay to the origin. In a physical system, the origin would not be a stable point since any disturbance—except one directly along the incoming trajectory—would start a motion which would leave the vicinity of the origin.

(d) *Center.* When the roots of the characteristic equation are complex conjugate with no real part, the motion of both x and y is pure sinusoidal. There are no solutions for the lines at a slope of m_1, but there are two for those at m_2; the trajectories are ellipses with the two lines as the major and minor axes. In the special case that $p_1 = q_2 = 0$ and $p_2 = -q_1$, the trajectories are circles.

(e) *Unstable Node.* This set of trajectories differs from that of the stable node only in that all the motion is away from the origin.

(f) *Unstable Focus.* This set of trajectories occurs when the roots of the characteristic equation are complex conjugate with a positive real part; the only difference between the pattern and that for the stable focus is the direction of motion.

Generally, the following rules apply:

For $(p_1 + q_2)$ (i) negative, the system is stable;
(ii) positive, the system is unstable
(in the vicinity of the equilibrium).

For $(p_1 + q_2)^2 < 4(p_1q_2 - p_2q_1)$
the trajectories form a focus.

If the quantities are equal, the trajectories are spirals.

For $(p_1 + q_2)^2 > 4(p_1q_2 - p_2q_1)$
the trajectories form a node.

As an example of the use of this material, let us return to the multi-vibrator problem introduced at the start of the chapter. We can learn something of the state-space behavior of this device by considering its performance around any equilibrium points which it may have. In this case there are three: the two in which one transistor is on and the other is off, and another in which each transistor is in the active region. This *might** occur, for example, with the collector-to-emitter voltages of each transistor equal to about one-half the supply voltage, with the two capacitors charged equally, and with the supply resistors R_3 and R_4 (see Figure 9.1) providing just the right amount of base current to maintain the equilibrium. Since the transistor characteristics are very non-linear in the region of the two extreme equilibria, it would probably

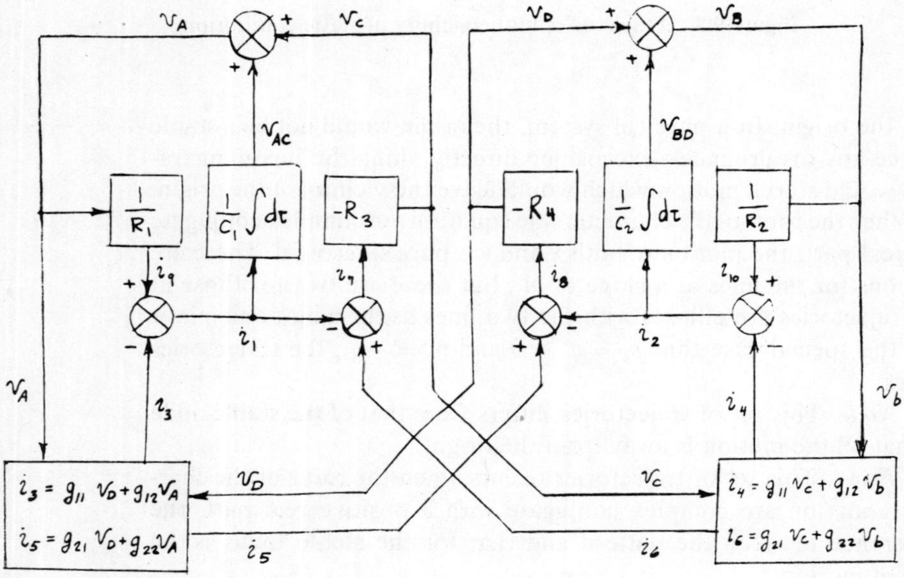

(a) Excursion variable computing diagram

From the Engineer's Sketch Book

Figure 9.10 *(continued)*

*Actually, however, the circuit design must be such that the equilibrium is with both transistors saturated "on."

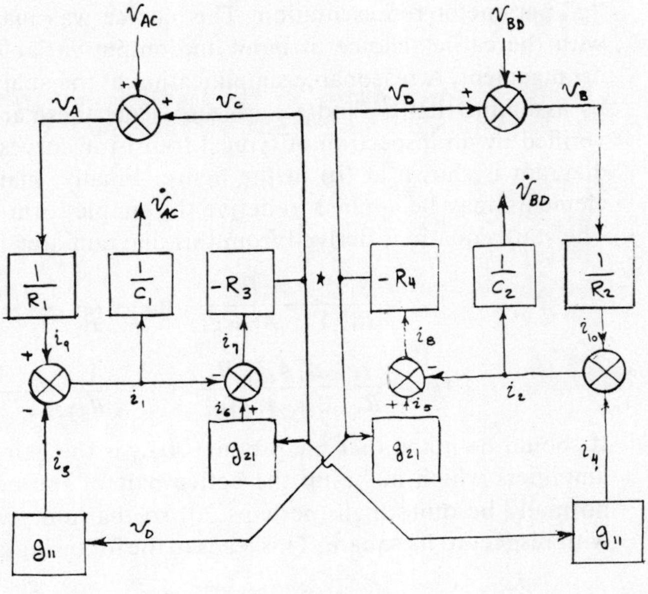

(b) Simplified computing diagram

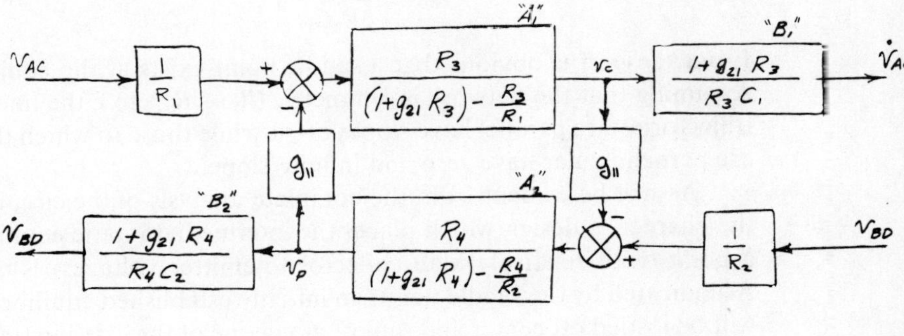

(c) Reduced computing diagram

From the Engineer's Sketch Book

Figure 9.10 Multi-vibrator analysis in the vicinity of the symmetric state.

be dangerous to apply a linearized analysis at those points. At the given point, however, the network can quite reasonably be linearized. Therefore, we shall proceed with a consideration of that case.

A small excursion computing diagram for the network is shown at Figure 9.10(a). Here the transistor characteristics have been approximated by the "g" parameter representation. This choice was made so as to be consistent with the earlier choice of input and output variables for the full computing arrangement. A reasonable simplification of the analysis may be accomplished by assuming that g_{12} and g_{22} are zero. That these are extremely small may be verified by an inspection of typical transistor curves. The resulting simplified diagram is shown at (b) in the figure. Finally, manipulations of computing elements may be applied to derive the simple form of diagram shown at (c). The state equations derived from this diagram are:

$$\dot{v}_{AC} = \frac{1}{R_1}\,\frac{A_1B_1}{1 + A_1A_2g_{11}^2}\,v_{AC} - \frac{1}{R_2}\,\frac{A_1A_2g_{11}B_1}{1 - A_1A_2g_{11}^2}\,v_{BD}$$

$$\dot{v}_{BD} = \frac{1}{R_1}\,\frac{A_1A_2g_{11}B_2}{1 + A_1A_2g_{11}^2}\,v_{AC} - \frac{1}{R_2}\cdot\frac{A_1B_2}{1 - A_1A_2g_{11}^2}\,v_{BD}.$$

It should be noted that the product AG_{11} is the gain of each of the transistor amplifiers which make up the active part of the network. This gain would normally be quite high (perhaps 30) so that unity would be quite negligible with respect to its square. This leads to the following simplified state equations:

$$\dot{v}_{AC} = \frac{1}{\beta R_1 C_2}\,v_{BD} \qquad \beta \cong \frac{g_{11}}{g_{21}}$$

$$\dot{v}_{BD} = \frac{1}{\beta R_2 C_1}\,v_{AC};$$

From these, it is obvious that a saddle point exists at the equilibrium point. Assuming that the network is symmetric ($R_1 = R_2$, etc.), the lines to which the trajectories are parallel have slopes of ± 1 while those to which the trajectories are perpendicular have zero and infinite slope.

As will be evident from the complete analysis of the circuit at the end of the chapter, a design which places the intermediate static equilibrium at such a point (with relatively high collector-to-emitter voltages) is unsatisfactory. As indicated by the saddle point, an initially established equilibrium condition will be jostled off center and run off along one of the existing trajectories. The circuit will then sit stubbornly in the corner of the region where this analysis prevails.

9.6 State-Variable Representations for Cascaded Integrations

When it is desired to use the state-variable formulation for problems in which a high-order, ordinary differential equation in one variable represents the physical system, the following general structure arises (considering a third-order case):

$$\dot{X} = \mathscr{P}(X, Y, Z)$$
$$\dot{Y} = X$$
$$\dot{Z} = Y.$$

This may be seen by considering the equation

$$\dddot{u} + a_2\ddot{u} + a_1\dot{u} + a_0 u = 0.$$

If the following identifications are made:

$$x = \ddot{u}$$
$$y = u$$
$$z = u$$

the state variable formulation of the problem becomes

$$\dot{x} = -a_2 x - a_1 y - a_0 z$$
$$\dot{y} = x$$
$$\dot{z} = y.$$

As might be expected, the absence of complexity in the functions other than the first one leads to simplifications in the conduct of a state space analysis.

The general influence of this kind of situation in a second order system is exemplified by the system shown in Figure 9.11(a). This is an arrangement which is quite useful in analog computation since it provides an almost perfect sinusoidal output of invariant amplitude. The system equation is

$$\ddot{z} - 2\zeta\dot{z} + \omega_n^2 z = 0; \qquad z \leqq 1.$$

If we make the identifications

$$x = \dot{z}$$
$$y = z,$$

the state variable formulation becomes

$$\dot{x} = 2\zeta x - \omega_n^2 y$$
$$\dot{y} = x \text{ for } y < 1, \dot{y} = 0 \text{ for } y = 1 \text{ and } x > 0.$$

If it were not for the limitation imposed by the lower selector ($z \leq 1$), the system would be completely linear and the linearization coefficients would be

$$p_1 = 2\zeta \qquad p_2 = -\omega_n^2$$
$$q_1 = 1 \qquad q_2 = 0.$$

For small (positive) values of ζ, an unstable system with trajectories in the form of a focus is indicated. It may be seen that the slope of the trajectories must be

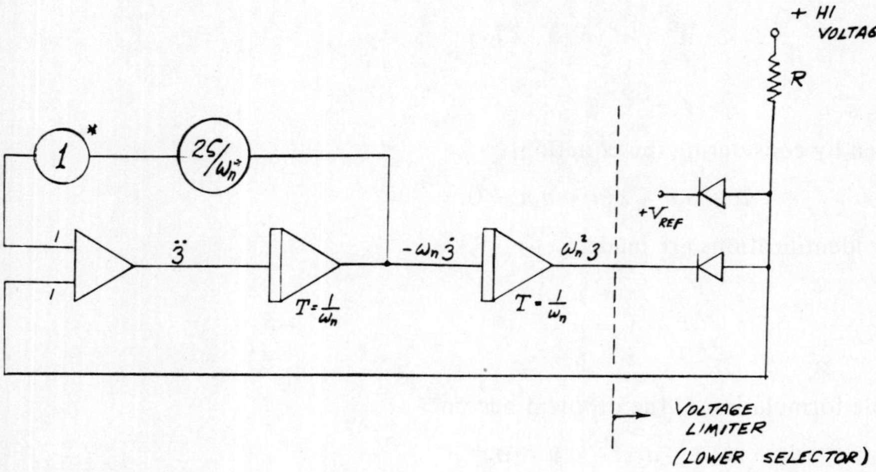

(a) Schematic

From the Engineer's Sketch Book

Figure 9.11 (*continued*)

infinite at the crossing of the x axis and that y must be increasing for all $x > 0$.

These results are true for any case of this form. One loop of the trajectory, started at $x = 0$ and $y = +1$ for convenience, is shown in part (b) of Figure 9.11. Due to the instability of the system, the trajectory tends to increase its distance from the origin with each loop. However, when it comes around into the first quadrant and reaches a value of $y = 1$, it can no longer change; so the value of x drops to zero. The loop is then negotiated again, since all the trajectories below the y axis cause motion in the same general direction. If the waveforms were perfect sinusoids, the loops would be ellipses or circles. The distortion of the loop is the price that is paid for the stabilization of the amplitude. This may be made exceedingly small, however, especially in high-quality computers. The distance along which the trajectory is straight is proportional to the negative damping coefficient ζ. This may be adjusted to such a small value that the trajectory just kisses the limiting line at $y = 1$. As shown in the computing diagram, two potentiometers might be used in series, the first set to a very small value and the second used to allow a very sensitive trimming of the adjustment. As usual a price is paid for this advantage also; the very low amount of instability causes the oscillator to be very slow in coming from the $x = 0, y = 0$ state up to the desired operating point. This may be overcome by enforcing the proper initial conditions on the integrators.

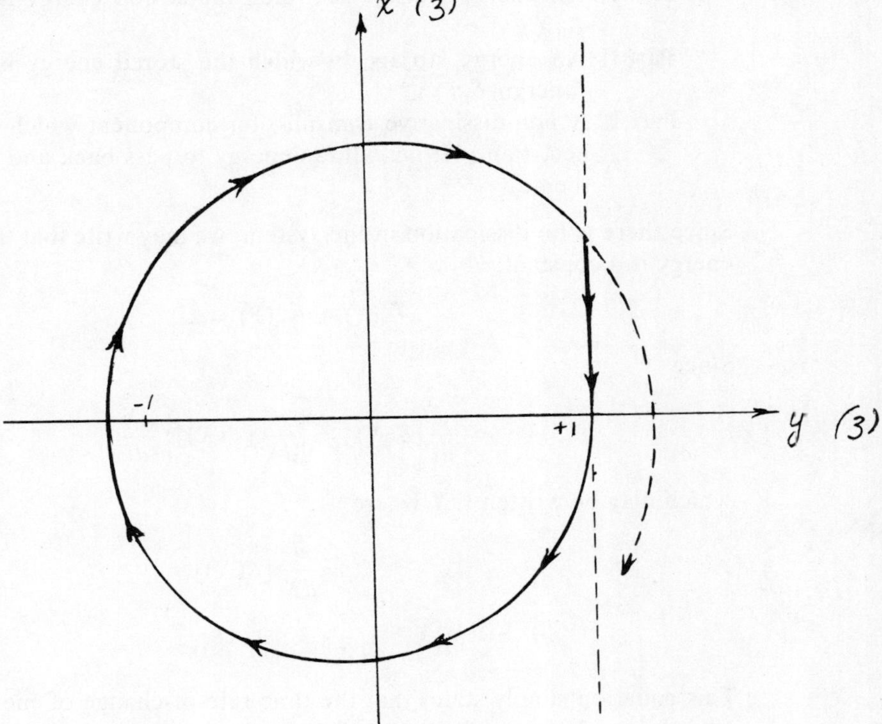

(b) Trajectories

From the Engineer's Sketch Book

Figure 9.11 Stabilized Sinusoidal Oscillator.

9.7 Non-Dissipative Systems

A type of system which has received considerable attention in the field of mechanics, particularly by those persons interested in non-linear vibration theory, is that which is known as "conservative"; i.e., is without dissipation. This type of system is of some interest in the general area of system engineering because some systems are "almost" conservative and because the analysis of conservative systems is a special case of the kind of analysis we are considering here. As usual, we limit our discussion to the second-order system.

A general, second-order, conservative system consists of three parts. Let us identify these parts as follows:

Part I An energy storage in which the stored energy is given by the $\mathscr{E}_{\mathrm{I}}(X)$.

Part II An energy storage in which the stored energy is given by the function $\mathscr{E}_{\mathrm{II}}(Y)$.

Part T A non-dissipative transmission component which serves to connect I and II; i.e., allows energy to pass back and forth between them.

Since there is no dissipation in the system, we may write that the total stored energy is a constant:

$$\mathscr{E}_{\mathrm{I}}(X) + \mathscr{E}_{\mathrm{II}}(Y) = C.$$

Since

$$\frac{d}{dt}[\mathscr{E}(X)] = \left\{\frac{d}{dX}[\mathscr{E}(X)]\right\}\frac{dX}{dt}$$

which may be written $\mathrm{E}'\dot{X}$ where

$$\mathscr{E}' = \frac{d}{dX}[\mathscr{E}(X)]$$

$$\mathscr{E}_{\mathrm{I}}'\dot{X} + \mathscr{E}_{\mathrm{II}}'\dot{Y} = 0.$$

This equation simply states that the time rate-of-change of energy stored in I plus that of II is equal to zero. Put another way, the power into I plus the power into II must be zero. From this, we see that

$$\frac{\dot{Y}}{\dot{X}} = -\frac{\mathscr{E}_{\mathrm{I}}'}{\mathscr{E}_{\mathrm{II}}'}.$$

The relationship may be made to appear more familiar by using

$$\mathscr{Q}(X) = \mathscr{E}_{\mathrm{I}}'(X) \quad \text{and} \quad \mathscr{P}(Y) = \mathscr{E}_{\mathrm{II}}'(Y).$$

It is instructive to consider the linearized representation of the such a system. The procedure is as usual:

$$\mathscr{Q}(X)\dot{X} + \mathscr{P}(Y)\dot{Y} = 0$$

$$[\mathscr{Q}(X_0) + q_1 x](\dot{X}_0 + \dot{x}) + [\mathscr{P}(Y_0) + p_2 y](\dot{Y}_0 + \dot{y}) = 0$$

if X_0, Y_0 is an equilibrium state, $\dot{X}_0 = \dot{Y}_0 = \mathscr{P}(X_0) = \mathscr{Q}(Y_0) = 0$ and

$$q_1 x\dot{x} + p_2 y\dot{y} = 0$$

since

$$\frac{d}{d\tau}\left(\frac{x^2}{2}\right) = x\dot{x}$$

$$q_1x^2 + p_2y^2 = c.$$

Therefore, if q_1 and p_2 have the same sign, the trajectories in the vicinity of an equilibrium are ellipses. If the signs are different, the curves are hyperbolas. The same result would be determined by applying the general results of section 9.5. Here, the fact that $p_1 + q_2$ is zero causes the possible trajectory forms to be only centers or saddles, which we have found. In this case, the orientation of the figures is symmetric with the x and y axes.

Recalling the multi-vibrator analysis temporarily, it may be noted that the trajectories around the unstable equilibrium point were in the form of a saddle. The approximate relationship used to determine this had $p_1 + q_2 = 0$ which represents a conservative system. Although the transistor arrangement is assuredly *not* conservative, the *incremental change* in dissipation, as the state variables vary in the region of the point, is negligible with respect to the incremental change in the power exchange to and from the two capacitors. To a large extent, this is due to the amplification effects in the circuit. Hence, the system behaves essentially as if it were conservative (*in this region*).

The actual analysis of conservative systems is quite straightforward since it is not necessary to work out the contours of the $\mathscr{P}$ and $\mathscr{Q}$ functions. The basic form of such an analysis is indicated by Figure 9.12. The computing diagram and notation is shown at (a), the energy functions and their derivatives at (b) and (c), and the initial steps in the construction of the trajectory diagram at (d). Note that there is only a single stable equilibrium point (marked S); this occurs at a point where the two energy functions are simultaneously at local minimums. This, of course, agrees with a very basic physical principle. The other equilibrium points are either centers or saddles; the alternative is indicated for several. This is determined from the slopes of the energy functions; i.e., the $\mathscr{P}$ and $\mathscr{Q}$ functions. The slope of one trajectory point is indicated (point a); the necessary sign inversion should also be noted.

An example of conservative system analysis is shown in Figure 9.13 where the case of a simple pendulum is worked out. Here the kinetic energy-storage element is perfectly linear; this causes the analysis to be symmetric about the θ axis. It may be seen that the possible equilibrium points are with the pendulum either straight up or straight down. The up position can be obtained by careful balancing, but is unstable. The down position represents a center on the trajectory diagram; the ellipses surrounding it show the approximately sinusoidal motion which occurs for small excursions.

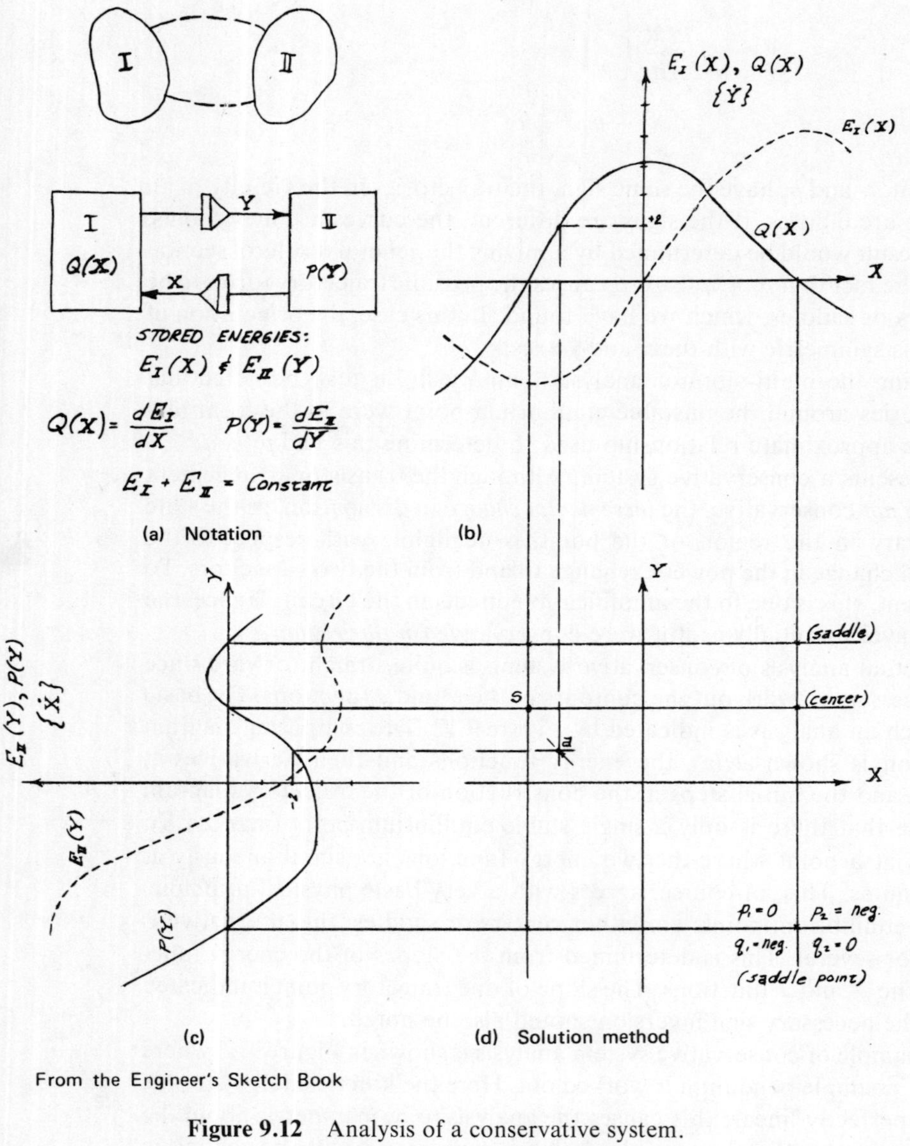

From the Engineer's Sketch Book

Figure 9.12 Analysis of a conservative system.

9.8 Systems with Distinct Linear Modes

In some systems, the state variable description of the dynamic performance consists of two or more sets of linear equations. Each set pertains to a particular region of the state space; in second-order systems, the trajectory diagram is

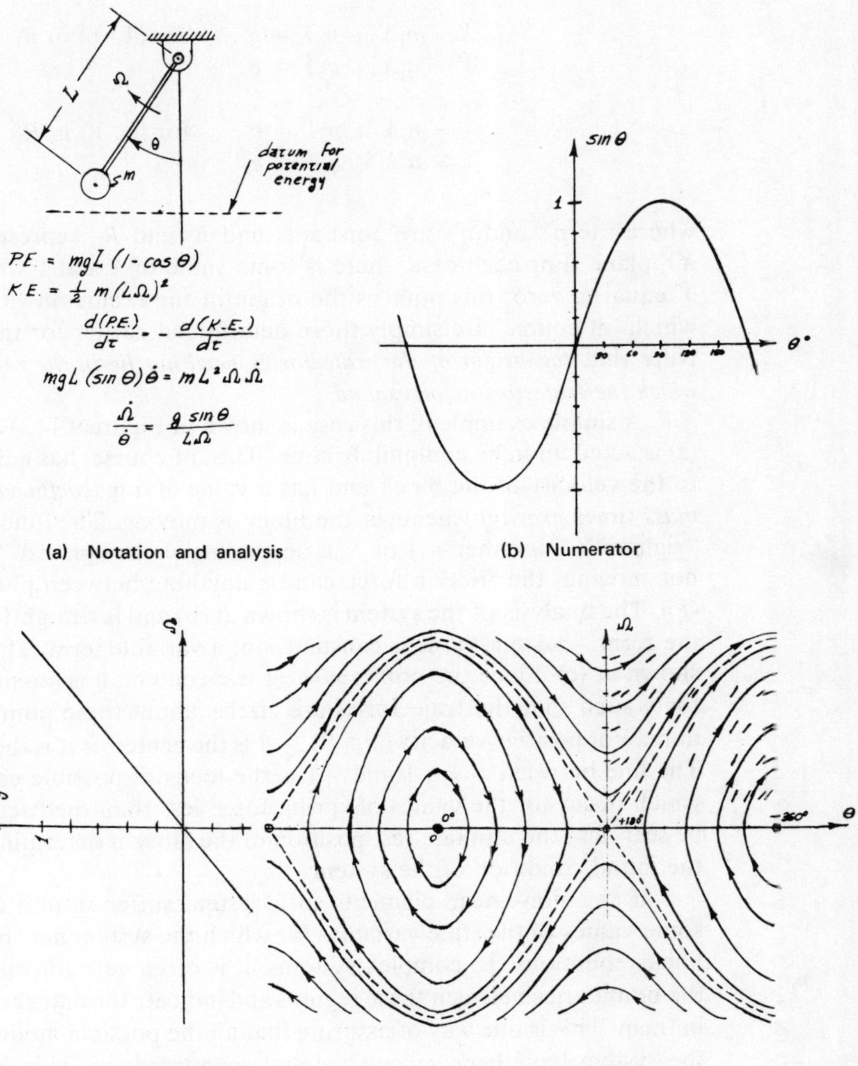

(a) Notation and analysis (b) Numerator

(c) Denominator (d) Trajectories

From the Engineer's Sketch Book

Figure 9.13 Analysis of pendulum.

divided into two or more areas representing the states for which the various sets of equations apply. The general formulation of such a second-order, two-mode system is as follows:

$$\begin{aligned} \dot{X} &= p_1 X + p_2 Y + c_1 \qquad \text{for } (X,\ Y) \text{ in } R_{\text{I}} \\ \dot{Y} &= q_1 X + q_2 Y + c_2 \end{aligned}$$

$$\begin{aligned} \dot{X} &= p_3 X + p_4 Y + c_3 \qquad \text{for } (X,\ Y) \text{ in } R_{\text{II}} \\ \dot{Y} &= q_3 X + q_4 Y + c_4 \end{aligned}$$

where the p's and q's are constants and R_{I} and R_{II} represent regions of the XY plane. For each case, there is some value of X and Y which makes $\dot{X}$ and $\dot{Y}$ equal to zero; this point is the origin of the excursion variable trajectories which, of course, are simply those determined earlier for the linearized case. Note that *the origin of the trajectories need not lie in the region* (R_{I} *or* R_{II}) *to which the trajectories correspond.*

A simple example of this case is shown in Figure 9.14. The block shown at (a) is acted upon by coulomb friction. This, of course, has a direction opposite to the velocity of the block and has a value of fmg (*coefficient of friction* times *mass* times *gravity*) whenever the block is moving. The function sgn (X), read "signum X," is either + 1 or − 1, depending on the sign of $\dot{X}$. When the block is not moving, the friction force can be anything between plus and minus fmg (F). The analysis of the system is shown at (b) and is straightforward; note that the term $-\omega_n^2$ sgn (y) is a constant, not a variable term. The trajectories are shown at (c). Since the points $x = \pm 1$ are centers, it is possible to find a scale for y such that the trajectories are circles about these points. This has been done. For positive velocity ($y > 0$), − 1 is the center; + 1 is the center for $y < 0$. The line between $x = +1$ and -1 is the locus of possible equilibrium points which occur for the values of spring force less than the friction force. It may be seen that the ultimate rest position of the block is determined completely by the initial condition of the system.

If the non-linear elements in a system suffer saturation, there will be some values of the state variables for which the system may be represented by linear equations. In complex systems, it is often very informative to identify the boundaries between these regions and indicate the nature of the trajectories in them. This is one way of assuring that all the possible modes of operation of the system have been recognized and accounted for. This is clearly brought out by the following example.

The schematic diagram of our old acquaintance, the multivibrator, is shown again in Figure 9.15(a). This circuit behaves linearly whenever both transistors are saturated (either full on or full off) since there are no other non-linear elements in the circuit. If we can identify the various regions of the V_{AC}, V_{BD} plane in which saturation occurs, the development of the complete trajectory diagram will be greatly facilitated. This can be done as indicated in the remainder of the figure. In the lower half of (a) is shown the equivalent circuit

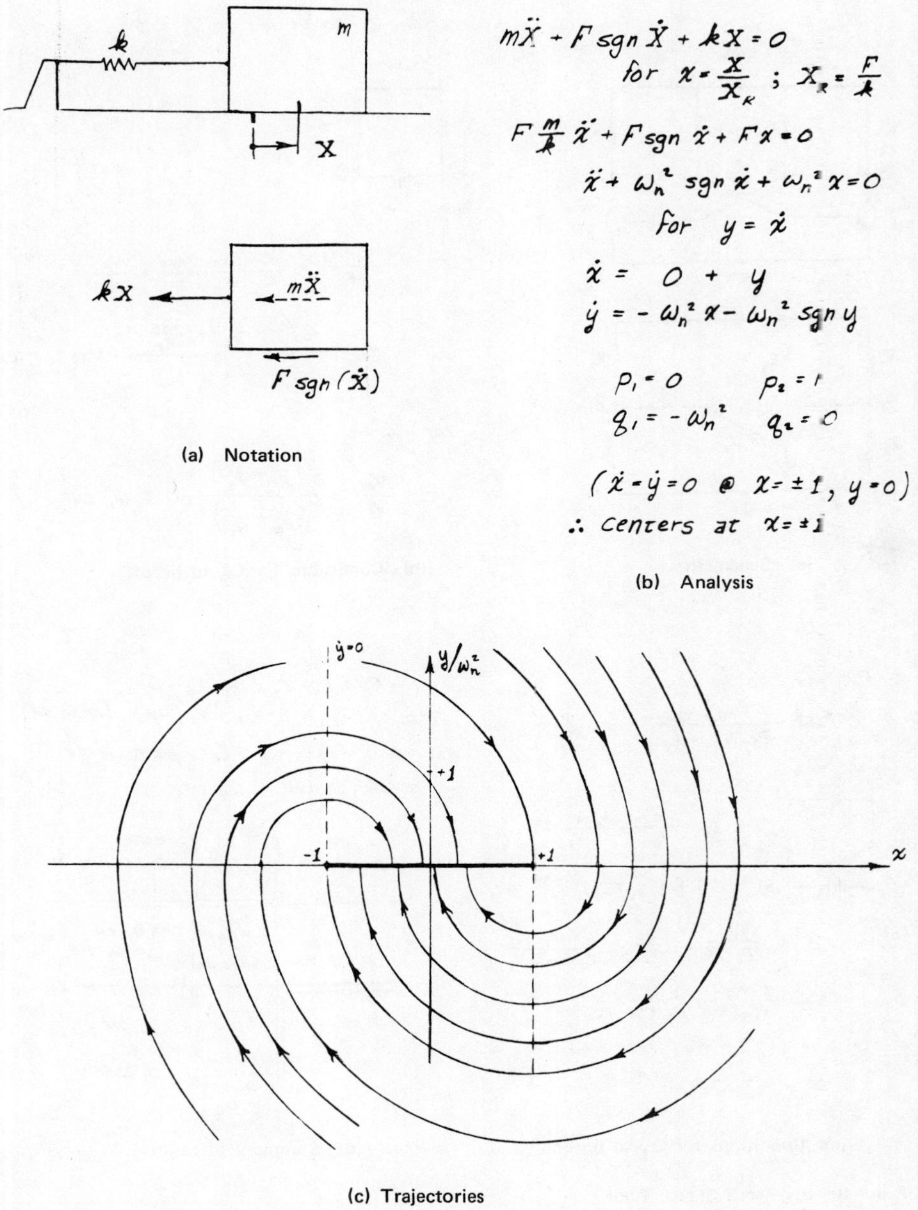

From the Engineer's Sketch Book

Figure 9.14 Analysis of spring-mass system with Coulomb friction.

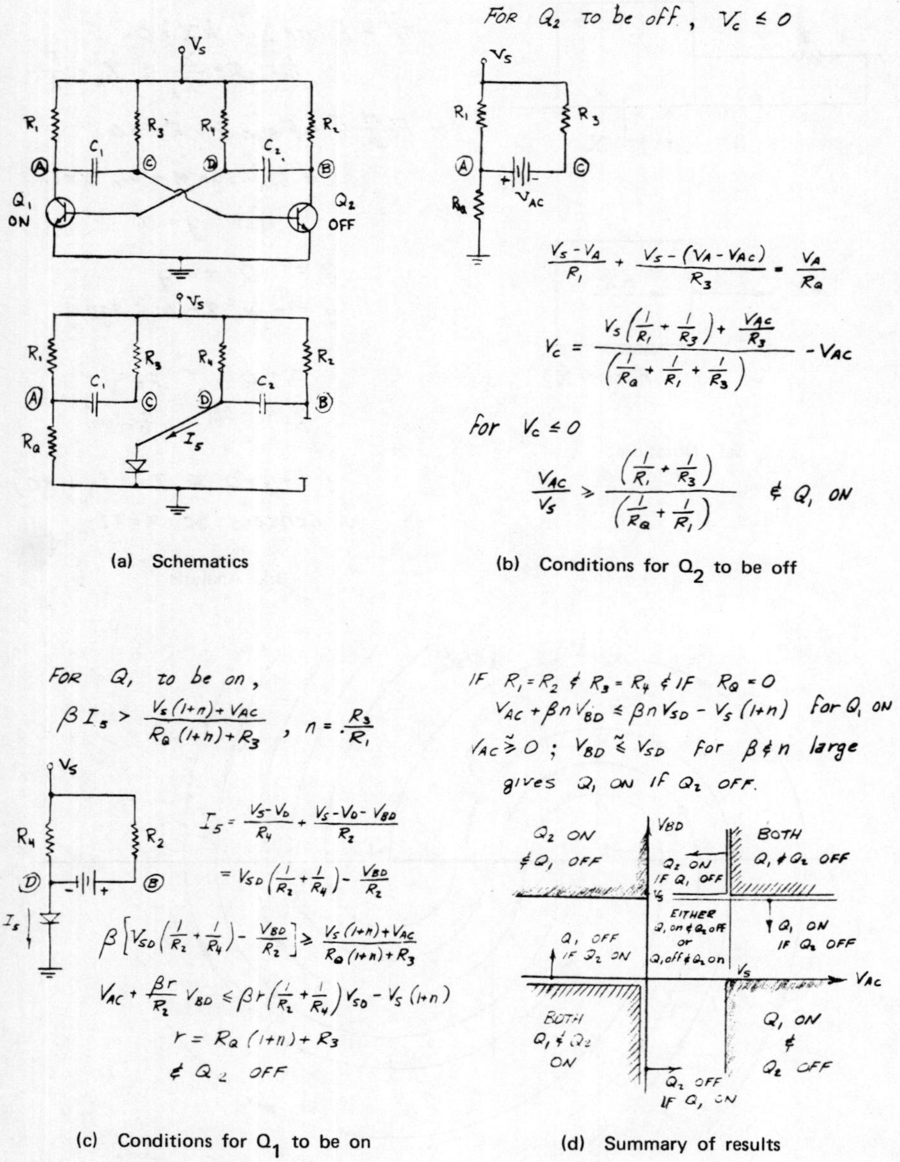

(a) Schematics

(b) Conditions for Q_2 to be off

(c) Conditions for Q_1 to be on

(d) Summary of results

From the Engineer's Sketch Book

Figure 9.15 Determination of operating mode regions for multi-vibrator.

for the case that Q_1 is fully conductive and Q_2 is fully non-conductive. A small resistance, R_Q, has been shown in place of Q_1 since a transistor is never a perfect short circuit. A very large resistance might be shown in place of Q_2 because of the leakage through a transistor when it is supposed to be off. However, this is so small that the effect is totally negligible in this analysis. At (b) is shown the analysis for the condition which must apply if Q_2 is to be off; this is simply that V_C be less than or equal to zero. When Q_1 is on, this will occur whenever the given inequality is satisfied. The condition for Q_1 to be on, which is that the base current I_5 must be large enough to saturate the transistor for the largest collector current which might flow, is shown at (c). The conditions in which both are on or both are off are determined by similar analyses. The result of this is shown at (d) and will be used as a framework for the final trajectory diagram.

9.9 Odds and Ends

In the development of a full trajectory diagram for a system, practically all the techniques of the preceeding sections are utilized. We can illustrate this by finishing the multi-vibrator analysis and looking a bit further into the problem of the slipping clutch.

The sets of linear, first-order differential equations which describe the multi-vibrator behavior in the various saturated transistor conditions of the circuit are derived in Figure 9.16. For the case that both transistors are saturated on, which is considered at (a), the final state is with V_{AC} equal to V_{BD} and the level being very small with respect to V_S. A representative value is -0.5 volt. The nature of the trajectories for this case is that of a stable node; if the corresponding components on opposite sides of the circuit have equal values, the trajectories are radial straight lines into the final state. For both transistors off, as shown at (b), the results are similar although the time constants are different. For one on and one off, as shown at (c), the final state is with one voltage large and negative and the other large and positive. The trajectories form a stable node; since the time constants cannot be equal (for a workable design of the circuit), the trajectories are not straight. The case for Q_1 and Q_2 off is shown at (d) in the figure; here only the portion of the trajectories below and to the right of the final state are shown. The other quadrants are, of course, symmetric about this point.

A superposition of all of our knowledge about the circuit is indicated in Figure 9.17. First, the V_{AC}, V_{BD} plane has been subdivided into the regions of different saturation conditions for the transistors. In the clearly defined regions such as where both are on, both are off, and one is on with the other off, the trajectories are readily indicated. In the first quadrant region where

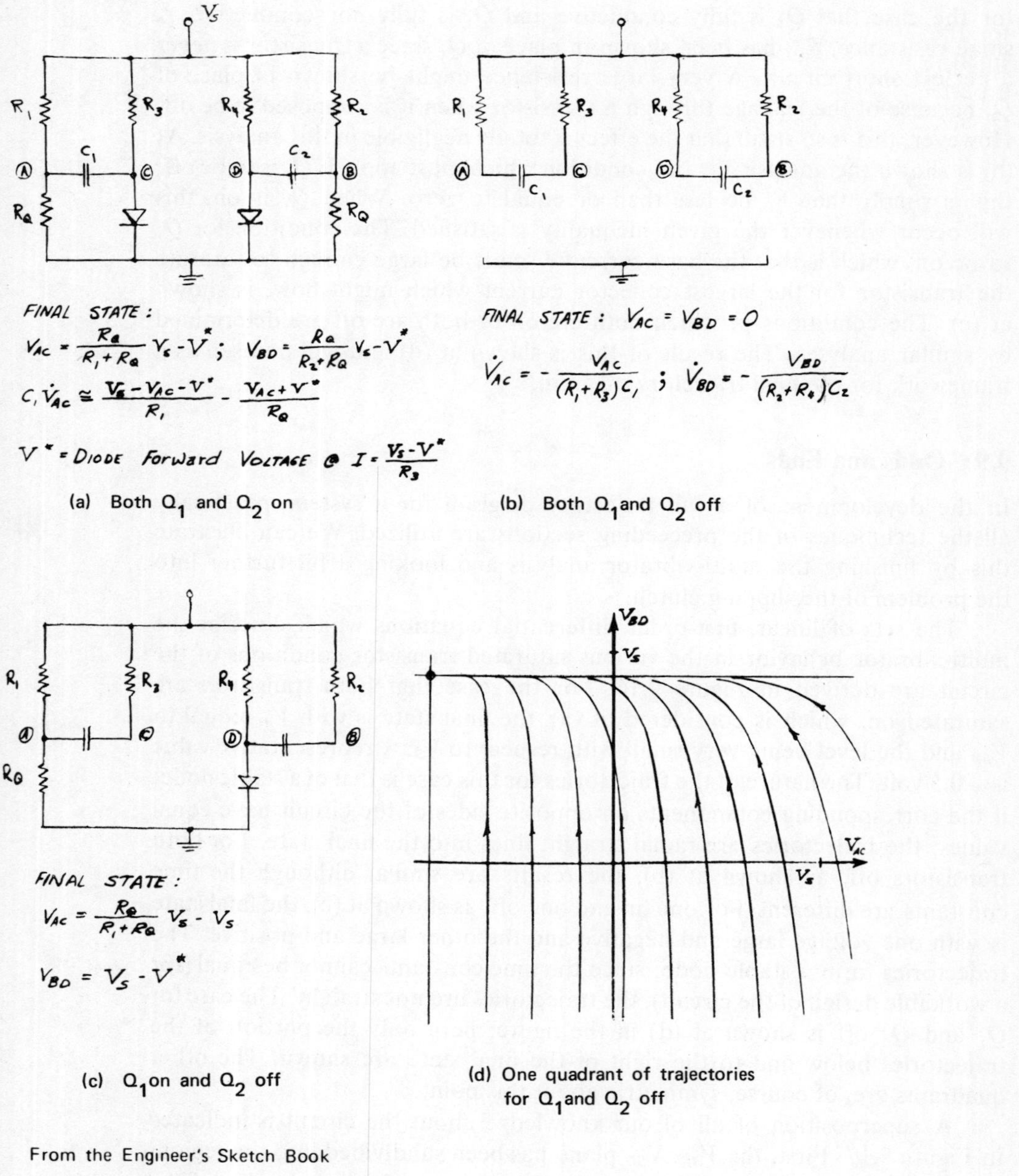

(a) Both Q_1 and Q_2 on

(b) Both Q_1 and Q_2 off

(c) Q_1 on and Q_2 off

(d) One quadrant of trajectories for Q_1 and Q_2 off

From the Engineer's Sketch Book

Figure 9.16 Portions of multi-vibrator analysis.

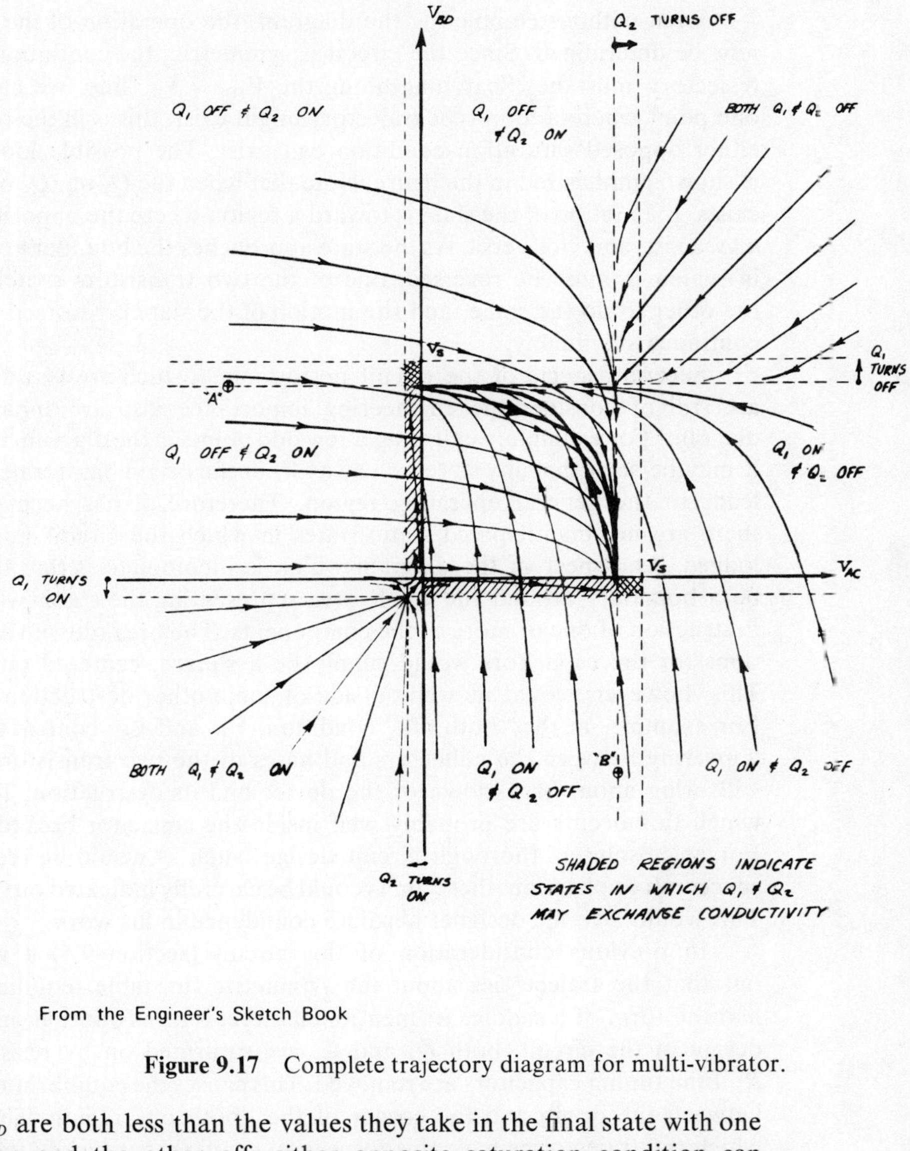

From the Engineer's Sketch Book

Figure 9.17 Complete trajectory diagram for multi-vibrator.

V_{AC} and V_{BD} are both less than the values they take in the final state with one transistor on and the other off, either opposite saturation condition can exist. For example, with $V_{AC} = V_{BD} = \frac{1}{2}V_S$, if Q_1 is on, then the base voltage of Q_2 must be negative and it must be off. Similarly, if Q_2 is on, the base voltage of Q_1 must be negative and *it* must be off. Therefore, either condition can prevail at the state $\frac{1}{2}V_S$, $\frac{1}{2}V_S$ and we must indicate both sets of trajectories over the whole region in which this can exist.

Even without completing the diagram, the operation of the circuit may now be determined. Since the circuit is symmetric, the continuous operating trajectory must be. Searching along the $V_{AC} = V_{BD}$ line, we can find only one point where a loop-type trajectory might exist; this is in the region where either opposed saturation condition can exist. The possible loop trajectory is shown emphasized in the figure. Note that when the Q_1 on, Q_2 off condition exists, the motion of the state is toward a region where the opposite condition *must* exist, and vice versa. As the state approaches the boundary of the region in which it must be reversed, one of the two transistors switches, forcing the other to do the same, and the motion of the state is reversed. This action continues indefinitely.

Several aspects of the circuit performance which are less dramatic but are still of considerable engineering importance also are apparent on the diagram. For example, neglecting a few odd points in the diagram temporarily, it may be seen that any state well away from the origin has a trajectory which leads to the general operating region. Therefore, it has been proved that there are no unanticipated static states in which the circuit might become lodged. An aspect of the diagram which is incomplete is that there should be a boundary around the whole area representing the states which lead to destruction of one or more of the components. The breakdown voltage conditions for the capacitors would simply be a square, centered on the origin. This, however, would be well outside of some other destruction conditions. For example, in the "both off" condition, V_{AC} and V_{BD} control the voltages appearing between the collectors and bases of the two transistors; too much will bring about breakdown of the device and its destruction. The levels at which this occurs are probably well inside the capacitor breakdown levels. For an absolutely thorough circuit design, such as would be required for a spacecraft application, these limits could be carefully indicated on the diagram. This would give the designer absolute confidence in his work.

In previous consideration of this circuit (section 9.5) it was pointed out that the trajectories about the symmetric (unstable) equilibrium point had the form of a saddle. As mentioned there, it turns out that, in a practical design of the circuit, both Q_1 and Q_2 are saturated on by resistors R_3 and R_4 if the timing capacitors are removed. This causes the equilibrium point to lie between the origin and the corner of the "both on" region. The area over which the trajectories have this form is so small that it hardly influences the diagram.

In the shaded regions above and to the right of the corner of the "both on" region, one transistor is active while the other is saturated, either on or off. This is an extremely complicated region when all its details are considered. Here we simply show the general nature of the behavior. It should be noted,

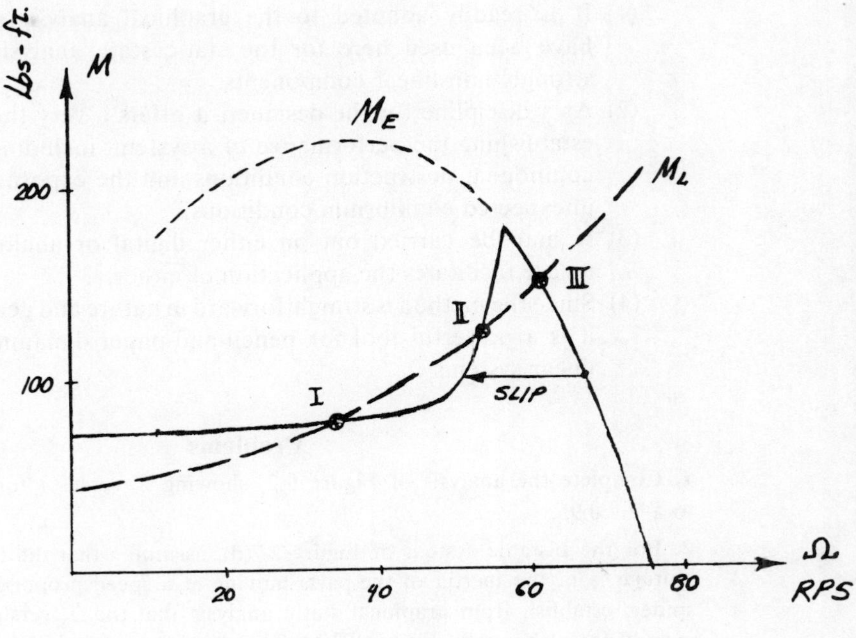

From the Engineer's Sketch Book

Figure 9.18 Characteristics of slipping-clutch which give poor performance.

however, that the saddle point is the reason that circuit does not get lodged at some point just below and to the left of the origin while the action in the one-active, one-saturated region is the explanation for the certain self-starting of the circuit.

Returning briefly to the slipping-clutch problem, from which several of the chapter excercise problems are derived, the conditions under which two stable equilibrium states arise is considered. In Figure 9.18 are shown some of the component characteristics for this case. The engine and load remain the same as in the earlier discussion, but the clutch condition has been modified so that there are three intersections in the engine-with-clutch characteristic and that of the load. The center equilibrium (II) is unstable and leads to jerkiness in the system. The right-hand point (III) is the normal operating condition in which the clutch does not slip; this can be obtained in the subject vehicle only by judicious management. Point (I) is the disastrous condition in which most of the engine power is being used to heat and destroy the clutch; this equilibrium speed is approximately one-half that of the normal conditon.

In summary of the state-space method, it may be said that the following advantages recommend its use:

(1) It is readily adapted to the graphical analysis procedures which have been used here for the static state analysis of systems with strongly non-linear components.
(2) As a discipline for the designer, it offers a very thorough means for establishing the performance of a system, including consideration of component destruction conditions and the exposure of unwanted or unexpected equilibrium conditions.
(3) It may be carried out on either digital or analog computers and its use facilitates the application of either.
(4) Since the method is straightforward in nature and generally applicable, it is a powerful tool for pencil-and-paper dynamic analysis of non-linear systems.

Problems

1. Complete the analysis of Figure 9.2, showing X versus t for the range $X^* = 0$ to $X^* = 0.95$.

2. For the bistable system of Figure 2.7(d), assuming that the fundamental energy storage is in the inertia of the parts moving at a speed proportional to that of the spider, establish from graphical static analysis that the $\dot{\Omega}_S$ versus Ω_s curve has the general form shown by Figure 9.3(b). The three $\dot{\Omega}_s = 0$ points and two $d\dot{\Omega}_s/d\Omega_s = 0$ points should be identified.

3. Repeat Problem 2 but for the hydraulic system of Figure 2.7(c), assuming the spool inertia to be the main energy storage. (V, the spool velocity, takes the place of Ω_s above.)

4. Show that the hydraulic timer of Figure 2.10(b), assuming that the accumulator is the only energy storage of significance, operates on the principle shown in Figure 9.3(d).

5. Establish that the hydraulic intensifier, Figure 1.26(c) and (d), does indeed operate on the principle of Figure 9.3(c).

6. Determine the time required for a fiywheel of 10 lb ft sec^2 inertia to be accelerated from rest up to 5400 RPM by the combination of the clutch shown in Figure 3.1 (actuated by 0.40 amp) and a synchronous motor running at 5400 RPM.

7. Expand the regions around points I and II in Figure 9.5(c), add contours of $\mathscr{P}$ and $\mathscr{Q}$ at plus and minus 0.1 (by graphical estimation) and sketch in several trajectories for each. Is it apparent that point I represents an unstable and point II a stable equilibrium?

8. Prepare a summary, as in Figure 9.3(a), showing how the stability of an equilibrium in a two-energy-storage problem may be determined from the $\mathscr{P}$ and $\mathscr{Q}$ functions in the vicinity of the similtaneous zero value lines. Use sketches of the functions which are point views of the zero valued contours.

9. For the example considered at Figure 9.7, assume the clutch characteristics are symmetric about the origin. Determine the trajectories which apply when the clutch and engine, without the brakes, are used to slow the car.

10. If you have access to an analog computer, or digital computer with graphical output, program the equations

$$\dot{x} = p_1 x + p_2 y$$
$$\dot{y} = q_1 x + q_2 y,$$

allowing for the four parameters to independently vary between plus and minus unity. Observe the various cases indicated in Figure 9.9.

11. Judging from the simplified equations derived from Figure 9.10(c), do you see any way in which a multi-vibrator circuit might be transformed into a sinusoidal oscillator? See Figure 9.9(d) and the conditions which bring it about.

12. Prepare state-variable formulations of the following differential equations:

$$(1)\quad \dddot{x} + 2\ddot{x} + 3\dot{x} + 4x = f(t)$$
$$(2)\quad \ddot{x} + 2\dot{x} + 30x + \dot{y} + 7y = f_1(t)$$
$$2\ddot{x} + 0.1\dot{x} + 7x + \ddot{y} + 14y = f_2(t)$$

13. Enter the types of equilibrium which exist at the remaining points of Figure 9.12, sketch the trajectories in the vicinity of all the equilibria, enter such trajectory slope information as can be found without undue effort, and sketch the overall trajectory pattern for the system.

14. Draw the trajectory diagrams for a frictionless spring-mass system in which the spring has a normalized characteristic $F^* = (X^*)^3$.

15. Sketch the trajectories of the system of Figure 9.14 for the case where a dash-pot works in series with the spring. Assume that the dashpot constant is small so that the roots of the characteristic equation are complex conjugates with relatively large imaginary parts; i.e., the damping ratio is small.

16. Suppose that a system is described by the equations

$$\ddot{x} + 0.2\dot{x} + x = F$$
$$F = -\text{sgn}\,(x + k\dot{x}).$$

(This represents a typical "contactor control" system.) Prepare trajectory diagrams for the following cases:

(1) $k = 0$.
(2) $k = +0.2$.
(3) $k = -0.2$.
(4) $k = +10$.
(5) $k = -10$.

Note: The phase plane will be split into two regions by an inclined line through the origin; the inclination depends on k.

17. In Figure 9.17.

(1) Are the points between the origin and $V_{BE} = V_{AC} = 0.8V_S$ saddle points? If not, why do they look like they do?

(2) Trace the history of a state which begins at $V_{BD} = -V_S$ and $V_{AC} = 0.05\ V_S$.

(3) Assuming the collector-to-emitter breakdown voltage of the transistors is $2V_S$, that the maximum allowable base-to-emitter voltage is $1.2V_S$, and that the capacitor breakdown voltage is $3V_S$, trace the boundary of the allowable operating region. Reference to Figures 9.16 (b) and (c) will be useful.

18. Prove that, in Figure 9.18, points I and III represent stable while II represents unstable equilibria.

Index